International Series in Operations Research & Management Science

Volume 374

The book series **International Series in Operations Research and Management Science** encompasses the various areas of operations research and management science. Both theoretical and applied books are included. It describes current advances anywhere in the world that are at the cutting edge of the field. The series is aimed especially at researchers, advanced graduate students, and sophisticated practitioners.

The series features three types of books:

• Advanced expository books that extend and unify our understanding of particular areas.

• Research monographs that make substantial contributions to knowledge.

• Handbooks that define the new state of the art in particular areas. Each handbook will be edited by a leading authority in the area who will organize a team of experts on various aspects of the topic to write individual chapters. A handbook may emphasize expository surveys or completely new advances (either research or applications) or a combination of both.

The series emphasizes the following four areas:

Mathematical Programming: Including linear programming, integer programming, nonlinear programming, interior point methods, game theory, network optimization models, combinatorics, equilibrium programming, complementarity theory, multiobjective optimization, dynamic programming, stochastic programming, complexity theory, etc.

Applied Probability: Including queuing theory, simulation, renewal theory, Brownian motion and diffusion processes, decision analysis, Markov decision processes, reliability theory, forecasting, other stochastic processes motivated by applications, etc.

Production and Operations Management: Including inventory theory, production scheduling, capacity planning, facility location, supply chain management, distribution systems, materials requirements planning, just-in-time systems, flexible manufacturing systems, design of production lines, logistical planning, strategic issues, etc.

Applications of Operations Research and Management Science: Including telecommunications, health care, capital budgeting and finance, economics, marketing, public policy, military operations research, humanitarian relief and disaster mitigation, service operations, transportation systems, etc.

This book series is indexed in Scopus.

Konstantin M. Zuev

Fundamentals of Statistical Inference

Foundations of Data Analysis

Springer

Konstantin M. Zuev
Department of Computing
and Mathematical Sciences
California Institute of Technology
Pasadena, CA, USA

ISSN 0884-8289 ISSN 2214-7934 (electronic)
International Series in Operations Research & Management Science
ISBN 978-3-032-03847-0 ISBN 978-3-032-03848-7 (eBook)
https://doi.org/10.1007/978-3-032-03848-7

This Springer imprint is published by the registered company Springer Nature Switzerland AG
The registered company address is: Gewerbestrasse 11, 6330 Cham, Switzerland

This book is dedicated to my mother, Galina, without whom it would not have been written, to my wife, Liza, who made the writing pleasant and worry-free, to my daughter, Sophia, who loves cats, and to you, my dear readers, without whom writing the book would not make any sense. I hope you will all enjoy the book. Each in your own way.

Fig. 0.1 Rainbow Cat by Sophia Zueva (5 years old). Including this drawing ensures that the first three persons mentioned in the dedication will like the book.

Foreword

Many years after my graduation from Caltech with a physics degree, I decided to audit Caltech's statistics class because as the co-founder and CEO of an AI software startup, I wanted to be well-versed in one of the foundational pieces of AI. As a physics student and in my previous career, I had taken some statistics courses, but I felt that I needed to go deeper given my new occupation, hence the desire to audit this course. It is then that I met Professor Zuev, who was teaching the course.

After attending a few lectures, and doing some of the problem sets, it became clear that Prof. Zuev taught statistics in a very different way from all of the professors that I had previously come into contact with. In fact, his teaching style and the way he approached the material was different from all the statistics textbooks that I had previously read.

His approach to statistics was unique in that he emphasized understanding the material by looking at topics from various different angles, and in the process weaving in the history of the people that founded statistics, and then reinforcing this learning by coming up with well-thought-out problems that required deep thinking. His explanations were always very clear, and you could also tell that he really cared about the subject and cared about students truly learning the material. I was not surprised to find out shortly thereafter that Professor Zuev had won several teaching awards at Caltech. I was also not surprised to find out that his course had received very high ratings from the students.

This book gives the reader the chance to learn statistics from a great teacher who really understands the subject, approaches it in a unique way, and truly cares about students learning the material.

Michael Amori
Co-Founder and CEO of Virtualitics Inc.
Pasadena, California, USA

Preface

I am looking for someone to share in an adventure that I am arranging, and it's very difficult to find anyone.
– J.R.R. Tolkien, *The Hobbit, or There and Back Again*

What is Statistical Inference? Opinions vary. In fact, there is a wide spectrum of attitudes toward statistical inference ranging from pure theoreticians, proving asymptotic efficiency and searching for most powerful tests, to wild practitioners, blindly reporting p-values [1] and claiming statistical significance for scientifically insignificant results. Even among the most prominent statisticians, there is no consensus: some discuss the relative importance of the main goals of statistical inference [2], others comment on the differences between "mathematical" and "algorithmic" cultures of statistical modeling [3], yet others argue that mathematicians should not even teach statistical inference [4]. The absence of a unified view on the subject led to different approaches and philosophies: there are both frequentist and Bayesian statistical inference, parametric and nonparametric, mathematical, computational, applied, etc. To complicate the matter, machine learning and other fields are bringing more and more new tools and ideas for data analysis and inference [5].

In this book, *statistical inference* is viewed as a branch of *mathematical engineering* that studies ways of extracting useful information about a phenomenon of interest from the data generated by that phenomenon for learning, prediction, and decision-making in the presence of uncertainty. Statistical inference is not mathematics per se because it is intimately related to real data. Mathematics is abstract, elegant, and can often be useful in applications; statistical inference is concrete, messy, and always useful. It is especially useful in the recently emerged transdisciplinary field of *data science*.

What is Data Science? Due to its strong transdisciplinary nature, there is no single widely accepted definition of data science. Nevertheless, it is more or less universally agreed that data science is concerned with the collection, processing, and analysis of data with the objective of extracting and communicating actionable knowledge in the form of business decisions, public policies, scientific models, and predictions [6, 7]. Although data science integrates concepts and methods from many different areas, including applied mathematics (especially probability, linear algebra, and optimization), computer science (especially artificial intelligence, machine learning, and distributed systems), electrical engineering (especially signal processing), network science, information and communication theory, control and decision theory, it is most deeply rooted in statistical inference. Essentially, data science originated from statistical inference and then evolved to include knowledge from machine learning, artificial intelligence, and many other fields. As an example, consider the classical linear regression, one of the most important methods for studying dependence, and perhaps the most popular statistical technique. In recent years, many other methods have been developed: neural networks, support vector machines, tree-based methods, to name but a few. Although these new fancy methods sometimes (but not as often as one may expect!) outperform the good old regression, many of them are really just modifications of regression. So when it comes to data analysis, using advanced data science tools, without a solid background in statistical inference, the best one could hope for is to blindly use some of the built-in functions implemented in modern programming languages without truly understanding what they do, how and why they work or don't work[1]. This would increase the risk of an unfortunate event of turning into a "wild practitioner" mentioned above. Statistical inference is thus of fundamental importance for modern data science.

[1] "The one who wants things done easier and anyhow, just quicker, is a fool." – Janusz Korczak, *Playful Pedagogy*.

With numerous applications in business, engineering, medicine, finance, physical and social sciences, data science is one of the most popular and fast developing fields in both academia and industry. As a result, data scientist is consistently ranked as the best job [8]. The single most important difference between a modern data scientist and a classical general statistician is perhaps that the former is proficient in many programming languages and algorithms and has a good understanding of software engineering. However, although it is necessary for a data scientist to have strong programming and analytical skills, it is not sufficient. Understanding the most fundamental ideas, concepts, and methods of statistical inference is a must for any modern data scientist.

About the Book. This book aims at providing a concise, reader-friendly, yet rigorous and thought-provoking introduction to the field of statistical inference and its applications. As opposed to classical books on statistics, where there is a strong emphasis on proofs, this book focuses on developing statistical thinking, intuitive understanding of the subject, and specific applications of data analysis. As a corollary, although present in the book, proofs are rather sketches of proofs and are not of paramount importance. Their main role is to provide the intuition and rationale behind the corresponding methods. The main focus is on the methods of statistical inference and their scope and limitations for real-world applications. On the other hand, statistical inference is not simply a toolbox that contains ready-made answers to all data-related questions. Almost always, as in solving engineering problems — hence the connection to mathematical engineering —- statistical inference and analysis of new data sets requires adjustment of existing tools or even developing completely new methods. To enable readers to modify existing methods and develop new ones, the book not only explains how the standard inferential methods work, but also *why*, when, and under what assumptions.

I have tried to write this book in a logical yet entertaining manner, where all definitions, concepts, and notions are well-motivated (as opposed to coming out of the blue, as often happens), rigorously yet clearly described, and intuitively explained. I have strived to make sure that readers are always aware of the big picture of what is going on, even when reading about some very concrete technical details. To prepare readers for further exploration of statistical inference and its applications, I have included carefully selected references to useful books, articles, open-access papers, databases, and websites. I sincerely hope that this book will

- Provide future and junior statisticians, data scientists, and other researchers who encounter data in their work with a necessary background in statistical inference,
- Develop readers' statistical thinking and intuitive feel for the subject,
- Introduce them to the most fundamental ideas, concepts, and methods of statistical inference, and explain why these methods work and when they don't.

After *working* through the book, readers should be able to understand[2] most (admittedly not all) contemporary papers that use modern statistical inference and to perform well-founded statistical analyses themselves.

[2] "Understanding is a creative process." – Sir David MacKay.

This book has grown out of a course on statistical inference that I have been teaching at the California Institute of Technology (Caltech) since 2016. In 2018 it became a core required course for the *Information and Data Sciences* option at Caltech. I hope the book will be adopted as a text for semester-long courses on statistical inference and data science applications at various universities and colleges. I expect the corresponding courses to be taught in departments of Statistics, Data Science, Applied Mathematics, and Computer Science. Courses based on this book could also be taught in other science and engineering departments if instructors prepare additional problems and data sets specifically tailored to their area (e.g. biology, astronomy, social science, civil engineering, etc.).

Readership and Prerequisites. The book is written primarily for:

- Junior statisticians, data scientists, and other researchers who encounter data in their work and want to learn how to make inferences from the data,
- Upper-level undergraduate students in mathematical sciences (pure and applied mathematics, statistics, information and data science, and computer science),
- Graduate students in non-mathematical sciences, engineering, medicine, and other fields who need to analyze data in their research.

Since this is an introductory text on statistical inference, no prior knowledge of statistics is assumed. However, readers are assumed to have a solid understanding of *probability*, which is — in the context of statistical inference — a language for describing and modeling variability in the data and uncertainty associated with the phenomenon of interest that generated the data. Here is a list of probabilistic concepts (typically covered in any introductory probability course) readers are assumed to be familiar with: random variable and its realizations, cumulative distribution function, probability mass function, probability density function, specific distributions (such as uniform, Bernoulli, binomial, and normal), expectation, variance, covariance, standard deviation, joint and conditional distributions, conditional expectation and variance, independence, Markov's and Chebyshov's[3] inequality, the law of total probability, the law of total expectation, the law of large numbers, the central limit theorem. Although brief probability refreshers are provided at the end of some chapters when necessary, they should not be regarded as sufficient substitutes for a proper introductory text on probability. Their main purpose is to provide a quick reminder on particular notions and results from probability theory for those who have already seen them but may have forgotten some details. The last chapter also assumes an introductory knowledge of *linear algebra*.

[3] For "Chebyshev" vs "Chebyshov" see Remark 3.5 in Probability Refresher 3.5 at the end of Chapter 3.

How to Read this Book. I would suggest reading each chapter twice. First time: glancing through, ignoring sidenotes and remarks, examining figures, and trying to grasp the main idea and understand the big picture of what is going on. Second time: with a pencil and paper, working through all the details, constructing examples and counterexamples, trying to find errors and typos, and blaming me for explaining easy things in an unnecessarily complicated way[4].

[4] "Don't just read it; fight it!" – Paul Halmos.

The book has two types of examples:

- "Toy examples," which are theoretical examples whose goal is to illustrate the corresponding concepts and methods in the simplest possible settings, and
- "Case studies," which show how the corresponding methods work in realistic situations within specific application domains.

The case studies are based on research and consulting projects, and drawn from various fields where statistical data analysis is frequently applied, including social sciences, sports analytics, population genetics, medicine, climatology, network science, economics, finance, and cosmology. When reading about these case studies, try to imagine how the corresponding ideas and methods would apply in other application domains.

I strongly believe that the most effective way of learning is self-study. As Ray Bradbury put it, "teachers are to inspire; librarians are to fulfill." This is especially true in mathematical and statistical sciences: the only way to learn is through active engagement with the material and mastery of problems. I wrote this book to inspire my readers, to show them why statistical inference is interesting, beautiful, and useful. To reinforce concepts and methods introduced in the text and to provide open-ended ways for further exploration, I have prepared and selected from other sources 54 carefully crafted theoretical and practical problems[5]. Not too many to avoid overwhelming readers, but enough for obtaining a good understanding of statistical inference and its applications. Solving and attempting to solve these problems is perhaps the most important part of working with this book. I encourage readers to first work on the problems by themselves, vigorously and creatively, and then check the solutions provided at the end of the book.

[5] "Judge a book by its exercises, and you cannot go far wrong." – George Cobb [9].

Finally, I would greatly appreciate it if readers could inform me about found errors and typos by emailing kostia@caltech.edu. All other book-related comments and remarks can be sent to the same email address. I look forward to your feedback!

References

1. Aschwanden, C.: Statisticians found one thing they can agree on: it's time to stop misusing p-values. (2016).
   http://fivethirtyeight.com/features/statisticians-found-one-thing-they-can-agree-on-its-time-to-stop-misusing-p-values/
2. Shmueli, G.: To explain or to predict? Statistical Science 25(3), 289–310 (2010).
   http://doi.org/10.1214/10-STS330
3. Breiman, L.: Statistical modeling: the two cultures. Statistical Science 16(3), 199–231 (2001).
   http://doi.org/10.1214/ss/1009213726
4. Moore, D.S.: Should mathematicians teach statistics? The College Mathematics Journal 19(1), 3–7 (1988).
   http://doi.org/10.2307/2686686
5. O'Connor, B.T.: Statistics vs. machine learning, fight! Blog on Artificial Intelligence and Social Science. (2008).
   http://http://brenocon.com/blog/2008/12/statistics-vs-machine-learning-fight/
6. Castrounis, A.: What is data science, and what does a data scientist do? (2017).
   https://www.kdnuggets.com/2017/03/data-science-data-scientist-do.html
7. Donoho, D.: 50 years of data science. Journal of Computational and Graphical Statistics 26, 745–766 (2017).
   https://doi.org/10.1080/10618600.2017.1384734
8. Piatetsky, G.: Data scientist — best job in America, 3 years in a row. (2018).
   https://www.kdnuggets.com/2018/01/glassdoor-data-scientist-best-job-america-3years.html
9. Cobb, G.W.: Introductory textbooks: a framework for evaluation. Journal of the American Statistical Association 82, 321-339 (1987).
   https://doi.org/10.1080/01621459.1987.10478436

Konstantin M. Zuev
Pasadena, California
August 2025

Acknowledgments

It takes more than an author to make a book.
– William B. Irvine, *A Guide to the Good Life*

Several people have contributed, in one way or another, to the creation of this book. I would like to express my sincere and warm gratitude to them.

I thank Professor Mathieu Desbrun for granting me a teaching-free fall term in 2015, which allowed me to write lecture notes that served as a skeleton for this book. I thank Professors Adam Wierman and Chris Umans, executive officers (aka department heads) of the Department of Computing and Mathematical Sciences at the California Institute of Technology, for their strong and continued support of teaching faculty in general and of myself in particular.

I thank the Caltech undergraduate and graduate students who took my course, *IDS 157 Statistical Inference*, over the last nine years for their keen interest, hard work, and good questions. Special thanks to Dave Luo, Nadia Volovich, Dongil Lee, Rona Yu, Ameera Abdelaziz, Po-Chih Chen, Elijah Cole, and Tatiana Brailovskaya for their useful comments and for finding typos and errors in the lecture notes.

A very special thanks to Pavlos Stavrinides, Yingying Gong, Ritvik Teegavarapu, and Jack Tao for reading portions of the first draft of the book and providing me with their valuable feedback.

I would also like to sincerely thank the following colleagues for helping to obtain or granting me permission to reproduce the photos of famous statisticians that are sprinkled throughout the book: Cindy Kirby (Stanford University), Elyse Gustafson (Institute of Mathematical Statistics), John Aldrich (University of Southampton), Lee Hayes and Horace Vella (University of Adelaide), Silke Okon (Oberwolfach Research Institute for Mathematics), and Sergey Polikarpov (Steklov Mathematical Institute).

I am also very grateful to a superb Springer team: Camille Price, Jialin Yan, Faith Su, and Shobha Karuppiah for giving me multiple extensions on the book's due date, and for their patience and assistance in preparing the book for publication.

Last but not least, I thank my mother, Galina Zueva, my wife, Liza Bradulina, and my daughter, Sophia Zueva, for their love, encouragement, and support.

Contents

Acronyms and Notation

BLUE	best linear unbiased estimate
CDF	cumulative distribution function
CI	confidence interval
CLT	central limit theorem
DKW	Dvoretzky-Kiefer-Wolfowitz (inequality)
EDA	exploratory data analysis
eCDF	empirical cumulative distribution function
GLRT	generalized likelihood ratio test
GRM	general regression model
IID	independent and identically distributed
IQR	interquartile range
KL	Kullback-Leibler (distance or divergence)
KS	Kolmogorov-Smirnov (statistic, test)
LLN	law of large numbers
LRT	likelihood ratio test
LTE	law of total expectation
LTV	law of total variance
MLE	maximum likelihood estimate
MLR	multiple linear regression
MMLR	multivariate multiple linear regression
MOE	margin of error
MOM	method of moments
MSE	mean squared error
N-SLR	normal simple linear regression
OLS	ordinary least squares
PCC	Pearson's correlation coefficient
PD	positive definite (matrix)
PDF	probability density function
PMF	probability mass function
PSD	positive semi-definite (matrix)
RSS	residual sum of squares

SCC	sample correlation coefficient
SLR	simple linear regression
SRS	simple random sampling
StrRS	stratified random sampling
SVD	singular value decomposition
se	standard error
sd	standard deviation
span	span of a set of vectors
s.t.	subject to (in a constrained optimization problem)
sup	supremum, the least upper bound of a set
SVD	singular value decomposition
SW	Shapiro-Wilk (test)
Q-Q	quantile-quantile (plot)
UMP	uniformly most powerful (test)
WLS	weight least squares
wRSS	weighted residual sum of squares

$!$	factorial function, $n! = 1 \times 2 \times \ldots \times n$
$\in$	"belongs," e.g. $a \in A$, element a belongs to set A
$\notin$	"does not belong," e.g. $a \notin A$, element a does not belongs to set A
$\varnothing$	empty set
$\mathbb{1}$	vector of ones, $\mathbb{1} = (1, \ldots, 1)^T$
$\mathbb{1}_A$	indicator function of set A, $\mathbb{1}_A(x) = 1$ if $x \in A$ and $\mathbb{1}_A(x) = 0$ if $x \notin A$
$\forall$	"for all," e.g. $\forall x \in A$, for all elements x in set A
$\exists$	"exists," e.g. $\exists x \in A$, there exists element x in set A
$\uparrow$	"increases" or "expands," e.g. $x \uparrow$, x increases; or $\mathscr{R} \uparrow$, set $\mathscr{R}$ expands
$\downarrow$	"decreases" or "shrinks," e.g. $x \downarrow$, x decreases; or $\mathscr{R} \downarrow$, set $\mathscr{R}$ shrinks
$\xrightarrow{\mathbb{P}}$	"converges in probability," see Probability Refresher 2.12
$\xrightarrow{\mathbb{D}}$	"converges in distribution," see Probability Refresher 2.12
$\Rightarrow$	"implies," e.g. $x > 1 \Rightarrow x > 0$
$\Leftrightarrow$	"if and only if" or "equivalent," e.g. $x > 1 \Leftrightarrow x - 1 > 0$
$\sim$	"distributed according to," see Remark 1.9
$\dot{\sim}$	"distributed approximately according to," see Remark 1.10
$\overset{\text{iid}}{\sim}$	"independent and identically distributed," e.g. $Z_1, \ldots, Z_n \overset{\text{iid}}{\sim} \mathscr{N}(0,1)$
$\gg$	"much greater than," e.g. $N \gg n$ means that N is much greater than n
$\ll$	"much less than," e.g. $n \ll N$ means that n is much less than N
$\lVert \cdot \rVert$	Euclidean norm, if $v \in \mathbb{R}^n$, then $\lVert v \rVert = \sqrt{\sum_{i=1}^n v_i^2}$
$\nabla_x f$	gradient of function $f : \mathbb{R}^n \to \mathbb{R}$, i.e. $\nabla_x f = (\frac{\partial f}{\partial x_1}, \ldots, \frac{\partial f}{\partial x_n})^T$
∞	positive infinity, negative infinity is denoted by "$-\infty$"
$\propto$	"proportional," e.g. $y \propto x$ means that $y = cx$ for some constant c
$[\,]$	denotes the integer part of a number, e.g. $[3.14] = 3$ and $[2.71] = 2$
$\{\}$	denotes a set, e.g. $\{x_1, \ldots, x_N\}$ is the set consisting of $x_1, \ldots, x_N$
$\mid$	"such that," e.g. $\{i \mid x_i = 0\}$ is the set of indices i such that $x_i = 0$
$\setminus$	set difference, e.g. $A \setminus B$ is the set of elements in A but not in B, i.e. $A \setminus B = \{x \mid x \in A \text{ and } x \notin B\}$

α	size of a hypothesis test
$\alpha_{\min}$	alternative notation for the p-value $\mathfrak{p}$
$100(1-\alpha)$	confidence level of a confidence interval I_α
β	vector of regression coefficients
$\beta(\theta)$	power function of a hypothesis test
$\beta^\dagger(\theta)$	power function of a UMP test
$\Gamma(z)$	Gamma function
δ	coefficient of variation
δ_{ij}	Kronecker delta, $\delta_{ij}=1$ if $i=j$, and $\delta_{ij}=0$ if $i \neq j$
$\delta(x)$	Dirac delta function
Θ	parameter space
θ	parameter or a vector of parameters of a distribution or a population
$\theta_\star$	true value of parameter θ
θ_0	hypothesized value of parameter θ
$\hat{\theta}$	estimate of parameter θ computed from the data
μ	mean of a distribution or population mean
μ_X	mean of a random variable or random vector X
Σ_X	covariance matrix of random vector X
σ^2	variance of a distribution or population variance
$\hat{\sigma}_u^2$	unbiased estimate of the variance
$\varkappa$	skewness of a distribution
κ	kurtosis of a distribution
$\Lambda(X)$	likelihood ratio statistic computed from the data $X=(X_1,\ldots,X_n)$
ρ_{XY}	Pearson's correlation coefficient between random variables X and Y
$\Phi(z)$	standard normal CDF, $\Phi(z)=\Phi(z\|0,1)$
$\Phi(z\|\mu,\sigma^2)$	CDF of a normal distribution with mean μ and variance σ^2
$\phi(z)$	standard normal PDF, $\phi(z)=\phi(z\|0,1)$
$\phi(z\|\mu,\sigma^2)$	PDF of a normal distribution with mean μ and variance σ^2
χ_q^2	chi-squared distribution with q degrees of freedom

A_{ij}	(i,j) element of matrix A
A^T	transposition of matrix A, $A_{ij}^T=A_{ji}$
A^+	pseudoinverse (the Moore-Penrose inverse) of matrix A
B	coefficient matrix in multivariate multiple linear regression
$\mathscr{B}(p)$	Bernoulli distribution with parameter p
$\mathscr{B}(n,p)$	Binomial distribution with parameters n and p
$\mathscr{B}(\alpha,\beta)$	Beta distribution with parameters α and β
$\mathbb{B}[\hat{\theta}]$	bias of random estimate $\hat{\theta}$
$\mathrm{coIm}A$	coimage of matrix A, if A is $n\times m$, then $\mathrm{coIm}A=\{A^Tx \mid x\in\mathbb{R}^n\}$
$\mathrm{coKer}A$	cokernel of matrix A, if A is $n\times m$, then $\mathrm{coKer}A=\{x\in\mathbb{R}^n \mid A^Tx=0\}$
$\mathrm{Cov}(X,Y)$	covariance between random variables X and Y
$D(X)$	Kolmogorov-Smirnov statistic computed from the data $X=(X_1,\ldots,X_n)$
$\det A$	determinant of matrix A
E	prediction error, or a vector or a matrix of errors
$\mathscr{E}(\lambda)$	exponential distribution with parameter λ (mean $1/\lambda$)

$\mathbb{E}[X]$	expected value (mean) of random variable X
$\mathbb{E}[Y\|X]$	conditional expected value of Y given X
$\mathscr{F}$	stochastic model
$\mathscr{F}(d_1, d_2)$	F-distribution with parameters d_1 and d_2
$F(x)$	cumulative distribution function of a random variable
$\hat{F}_n(x)$	empirical cumulative distribution function of a sample of size n
$f(x)$	probability density function of a random variable
$\mathfrak{f}$	statistical functional or F-statistic (depending on the context)
$G(X)$	t-test statistic computed from the data $X = (X_1, \ldots, X_n)$
H_0	null hypothesis
H_1	alternative hypothesis
$H(x)$	Heaviside step function
$\mathbb{H}$	hat matrix
$\mathscr{I}(\theta)$	Fisher information
$\mathbb{I}_d$	identity matrix of size $d \times d$
I_α	confidence interval with confidence level $100(1-\alpha)\%$
$\mathrm{Im}A$	image of matrix A, if A is $n \times m$, then $\mathrm{Im}A = \{Ax \mid x \in \mathbb{R}^m\}$
$\mathscr{K}$	Kolmogorov distribution
K	number of bins in a histogram
$\mathrm{Ker}A$	kernel of matrix A, if A is $n \times m$, then $\mathrm{Ker}A = \{x \in \mathbb{R}^m \mid Ax = 0\}$
$\mathscr{L}_n(\theta)$	likelihood function
$\ell_n(\theta)$	log-likelihood function
log	natural logarithm
$\mathscr{M}(n, p)$	multinomial distribution with parameters n and p
m	median of a distribution or population median
m_q	q^{th} moment of a probability distribution
$\hat{m}_q$	q^{th} sample moment of a sample
$\mathscr{N}(\mu, \sigma^2)$	normal (Gaussian) distribution with mean μ and variance σ^2
$\mathscr{N}(0, 1)$	standard normal distribution, $\mathscr{N}(0,1) = \mathscr{N}(\mu = 0, \sigma^2 = 1)$
$\mathscr{N}(\mu, \Sigma)$	multivariate normal distribution with parameters μ and Σ
$\mathscr{N}(0, \mathbb{I}_d)$	d-dimensional standard normal distribution
$\mathbb{N}$	set of natural numbers $\{1, 2, 3, \ldots\}$
n	sample size
$\mathscr{P}$	target population
$\mathscr{P}_{boot}$	bootstrap population
$\mathscr{P}_n$	set of all subsets of $\mathscr{P}$ of size n
$\mathscr{P}(\lambda)$	Poisson distribution with parameter λ (mean λ)
$\mathbb{P}$	probability distribution
$\mathbb{P}(A)$	probability of event A
$\mathbb{P}(A\|B)$	conditional probability of A given B
$\mathbb{P}_\pi$	permutation distribution
$\mathfrak{p}$	p-value
$p(x)$	probability mass function of a random variable
Q_1	first (lower) quartile of the sample $x_1, \ldots, x_n$
Q_3	third (upper) quartile of the sample $x_1, \ldots, x_n$

$\mathscr{R}$	rejection region of a hypothesis test
$\mathbb{R}$	set of real numbers
$\mathbb{R}^d$	d-dimensional real space
$r(x)$	regression function
r_x	range of the sample $x_1,\dots,x_n$
r_{xy}	sample correlation coefficient between samples $x_1,\dots,x_n$ and $y_1,\dots,y_n$
$\mathrm{rk}A$	rank of matrix A
$\mathscr{S}$	sample from the target population $\mathscr{P}$
$\mathscr{S}^{(b)}$	bootstrap sample
$\mathfrak{S}_n$	set of all permutations of the set $\{1,\dots,n\}$
$s(X)$	test statistic, any function of the data $X=(X_1,\dots,X_n)$
s_x	sample standard deviation of the sample $x_1,\dots,x_n$
$\mathscr{T}(n)$	Student's t-distribution with n degrees of freedom
$T(X)$	Student's t-statistic computed from the data $X=(X_1,\dots,X_n)$
$t_{n,q}$	q^{th} quantile of the t-distribution with n degrees of freedom
$\mathrm{tr}A$	trace of matrix A (sum of its diagonal elements)
$\mathscr{U}[a,b]$	uniform distribution on the interval $[a,b]$
$\mathbb{V}[X]$	variance of random variable X
$\mathbb{V}[Y\|X]$	conditional variance of Y given X
$W(X)$	Wald statistic computed from the data $X=(X_1,\dots,X_n)$
w	width, e.g. bin width in a histogram or width of a boxplot
$\mathscr{X}$	sample space, i.e. the range of values of a random variable or vector
$\mathbb{X}$	design matrix or predictor matrix in multiple linear regression
X	random variable or random vector
$\bar{x}$	sample mean of the sample $x_1,\dots,x_n$
$\bar{x}_\alpha$	α-trimmed mean of the sample $x_1,\dots,x_n$
$\tilde{x}$	sample median of the sample $x_1,\dots,x_n$
$x_{\min}$	smallest value in the sample $x_1,\dots,x_n$
$x_{\max}$	largest value in the sample $x_1,\dots,x_n$
$x_{(1)},\dots,x_{(n)}$	order statistics of $x_1,\dots,x_n$
$\mathbb{Y}$	response matrix in multivariate multiple linear regression
Z	standard normal random variable, $Z\sim\mathscr{N}(0,1)$
z_q	q^{th} quantile of the standard normal distribution, $z_q=\Phi^{-1}(q)$
$\mathfrak{z}_j$	z-score associated with regression coefficient β_j

1
Summarizing Data

Whatever is worth saying, can be stated in fifty words or less.
– Stanisław Ulam

Data is[1] at the heart of statistical inference. The most basic element of data is a single observation, x, a number. Usually real data comes in the form of a (very long) list of numbers. Even if the original data is more complex — a text, a curve, or an image — we will assume that we can always convert it to a set of n numerical observations $x_1,\ldots,x_n$, called a *sample*.

[1] For "data is" vs "data are" see [1].

To get a better feel for the data in hand, it is often useful (especially if the sample size n is large) to summarize it numerically and graphically. This can bring some insights about the data and help us to generate some hypotheses about the data and the underlying phenomenon that produced it. In this chapter, we will discuss some of the most often used summary statistics[2].

[2] In statistics, any quantity $s = s(x_1,\ldots,x_n)$ that can be calculated from data is called a *statistic*.

1.1 Histogram

A *histogram* is an iconic and perhaps the best-known statistical plot. Examples of histograms are shown in Fig. 1.1. To construct a histogram from sample $x_1,\ldots,x_n$:

1. Divide the interval $[x_{\min}, x_{\max}]$ into K disjoint intervals $B_1,\ldots,B_K$, called *bins*. Here, $x_{\min}$ and $x_{\min}$ denote the smallest and the largest values in the sample.
2. Denote the number of observations in B_k by n_k, so that $\sum_{k=1}^{K} n_k = n$.
3. For each bin B_k, plot a column over B_k so that the area of the column is the proportion n_k/n of the data points that fall into that bin. The height h_k of the column over B_k is therefore $h_k = \frac{n_k/n}{|B_k|}$, where $|B_k|$ denotes the length of B_k.

Remark 1.1. The total area of a histogram constructed this way. i.e. the area of all columns of the histogram, is 1[3]. Such histograms are called *normalized*. Sometimes non-normalized histograms are used, where the area of the column over B_k is simply the number of observations n_k. In this case, the total area of the histogram is n.

[3] Indeed, $\sum_{k=1}^{K} \frac{n_k}{n} = 1$.

Supplementary Information The online version contains supplementary material available at https://doi.org/10.1007/978-3-032-03848-7_1.

K. M. Zuev, *Fundamentals of Statistical Inference*,
International Series in Operations Research & Management Science 374,
https://doi.org/10.1007/978-3-032-03848-7_1

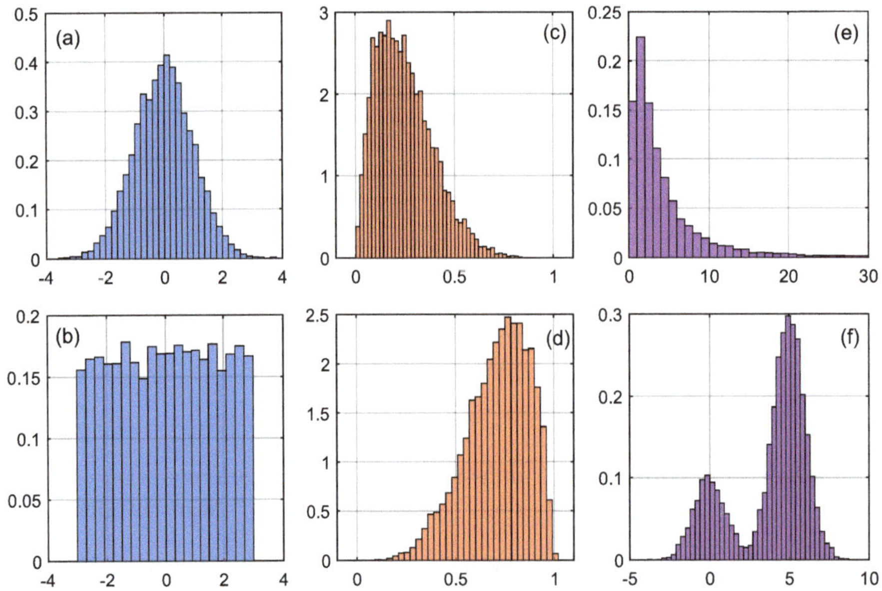

Fig. 1.1 Examples of histograms: (a) symmetric, "bell-shaped"; (b) symmetric, uniform; (c) right-skewed; (d) left-skewed; (e) unimodal; (f) multimodal.

Remark 1.2. If the data $x_1,\dots,x_n$ was generated by independent sampling from a probability density function (PDF) $f(x)$, then a normalized histogram constructed from the data can be viewed as an approximation of $f(x)$ [2].

As soon as bins are specified, constructing a histogram is straightforward. So the question is *how to choose bins*. There is no unique recipe for choosing bins: a good choice depends on the data. Let's try to understand what "good" means.

The main purpose of a histogram is to represent the *shape* of the sample: symmetry, skewness, modality, etc., see Fig. 1.1 for examples[4]. Let's assume for simplicity that all bins have equal width[5] $|B_k| = w$:

$$\begin{aligned} B_1 &= [x_{\min}, x_{\min} + w), \\ B_2 &= [x_{\min} + w, x_{\min} + 2w), \\ &\dots \\ B_K &= [x_{\max} - w, x_{\max}], \end{aligned} \tag{1.1}$$

where $x_{\min} = \min\{x_1,\dots,x_n\}$ and $x_{\max} = \max\{x_1,\dots,x_n\}$ are the smallest and the largest values in the sample. In this case, the total number of bins is

$$K = \frac{x_{\max} - x_{\min}}{w}. \tag{1.2}$$

The number of bins K in a histogram can drastically affect its visual appearance. If K is too small (the bin width w is too large), then the histogram will be too

[4] It may look strange that the histograms in panels (c) and (d) are called right- and left-skewed, respectively, and not the other way around. The reason for that is that "right" and "left" refer to the corresponding *long tails*, rather than to the positions of histogram peaks.

[5] Sometimes, it is useful to consider bins with different widths: narrower bins in the center of the data (where presumably there are more data points) and wider bins in the tails.

smooth[6]. On the other hand, if K is too large, then the histogram will be too rough[7]. This effect is illustrated in Fig. 1.2 with a normally distributed sample.

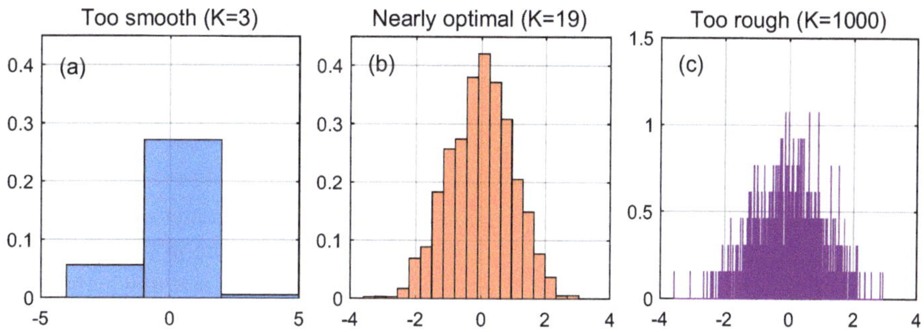

Fig. 1.2 Histograms of a normally distributed data $x_1,\ldots,x_n$, $n = 10^3$ with different number of bins: (a) $K = 3$ (b) $K = 19$, obtained via Scott's rule (1.3), and (c) $K = 1000$.

[6] If $K = 1$, then the histogram will consist of a single column over the interval $[x_{\min}, x_{\max}]$, and will simply represent the spread, not the shape of the data.

[7] In the extreme case, each bin will contain one or none of the data points, and the histogram will be a collection of narrow columns of the same height placed over the data points. This "high-resolution" histogram will represent the sample itself, rather than its shape.

Thus, histograms with either too few or too many bins do not represent the shape of the data well. Although there are no "best" values for the number of bins and the bin width, over the years, several simple heuristic rules for the approximately optimal values of K and w have been developed. For example, if the sample $x = \{x_1,\ldots,x_n\}$ appears to be approximately normally distributed[8], then we can use Scott's rule [3] for the bin width:

[8] That is we expect the histogram to be symmetric and bell-shaped as in Fig. 1.1(a) or Fig. 1.2(b).

$$w = 3.5\frac{s_x}{n^{1/3}}, \tag{1.3}$$

where s_x is the sample standard deviation defined in Section 1.2.2. Another useful rule of thumb for the bin width, which does not assume sample normality and is less sensitive to outliers, is the Freedman-Diaconis rule [2]:

$$w = 2\frac{\text{IQR}}{n^{1/3}}, \tag{1.4}$$

where IQR is the interquartile range of the sample introduced in Section 1.2.2.

Remark 1.3. A popular heuristic rule for the number of bins K, which is often recommended in statistics textbooks and implemented in statistical packages[9], is the Sturges formula [4]:

[9] For example, it is the default option in the `hist` function in R.

$$K \approx 1 + \log_2 n. \tag{1.5}$$

This rule assumes that the sample is approximately normally distributed and works reasonably well for small samples with $n < 200$. However, for large n it produces oversmoothed histograms. Moreover, while both (1.3) and (1.4) have solid statistical foundation, it has been shown that the argument leading to (1.5) is wrong [5].

In general, heuristics like (1.3) and (1.4) can be used as starting points for exploring the shape of the data with histograms. Varying number of bins and considering bins with different widths may reveal interesting insights about the data[10].

[10] "Beware of finding what you're looking for." – Richard Hamming.

1.2 Numerical Summaries

Numerical summaries provide quantitative information about the data. The two main numerical characteristics of a sample are its *center* and *spread*.

1.2.1 Measures of Center

A measure of center is a statistic that represents the "physical" center of the sample $x = \{x_1, \dots, x_n\}$, which can be often interpreted as a typical value in the sample or its most common observation[11]. One such measure is the *sample mean*, which is the average of the sample:

$$\bar{x} = \frac{1}{n} \sum_{i=1}^{n} x_i. \tag{1.6}$$

The sample mean is one of the most widely-used measures of center, but it has one serious drawback: it is sensitive to outliers. An *outlier* is an observation $x_i \in x$ that is far from the rest of the data[12]. There are two sources if outliers: natural variability in the data[13] and measurement errors. The sensitivity of the sample mean with respect to outliers follows directly from its definition (1.6): by changing the value of just one data point, say x_1, we can make $\bar{x}$ arbitrary small or large. Therefore, the sample mean of a sample that contains outliers could be a poor measure of its center.

An alternative measure of center, which is *robust*[14] to outliers, is the *median*. Intuitively, the median $\tilde{x}$ is the point that divides the data in half. To define the median more rigorously, we need to order the sample. The *order statistics* of $x_1, \dots, x_n$ are the same values but arranged in increasing order:

$$x_{(1)} \leq x_{(2)} \leq \dots \leq x_{(n)}. \tag{1.7}$$

Here, $x_{(1)} = x_{\min}$ is the smallest values in the sample, $x_{(2)}$ is the second smallest, etc., and $x_{(n)} = x_{\max}$ is the largest value. The median is then defined as follows[15]:

$$\tilde{x} = \begin{cases} x_{(\frac{n+1}{2})}, & \text{if } n \text{ is odd,} \\ \frac{1}{2}\left(x_{(\frac{n}{2})} + x_{(\frac{n}{2}+1)}\right), & \text{if } n \text{ is even.} \end{cases} \tag{1.8}$$

Remark 1.4. Some introductory statistics textbooks state that $\bar{x} > \tilde{x}$ for the right-skewed data and $\bar{x} < \tilde{x}$ for the left-skewed data. If fact, this is not always true and there are many counterexamples [6].

The main drawback of the median is the opposite of the drawback of the mean: $\tilde{x}$ is too insensitive to the changes in the sample values. Suppose for simplicity, that the sample size is odd, $n = 2k - 1$, then the median is the kth order statistic, $\tilde{x} = x_{(k)}$. Changing the values of $x_{(1)}, \dots, x_{(k-1)}$ in any way, as long as they do not exceed

[11] Often, but not always. Consider, for example, a sample, consisting of values uniformly spread on the intervals $[-2, -1]$ and $[1, 2]$. Then, zero is an approximate center of this sample, but it can't be viewed as a typical value.

[12] Outliers are often easy to spot in histograms: they are in the tails.

[13] For example, Bill Gates is an outlier in the data on people's wealth.

[14] That is, "insensitive."

[15] Convince yourself that $\tilde{x}$ defined this way indeed splits the data in half.

$x_{(k)}$, and changing the values of $x_{(k+1)}, \dots, x_{(n)}$ in any way, as long as they do not become smaller than $x_{(k)}$, does not affect the median $\tilde{x} = x_{(k)}$. A natural question is then whether we can find a measure of center which is a trade-off between two "extremes," the sensitive mean and the robust median.

A compromise between the mean and the median is given by a *trimmed mean*[16]. While the sample mean $\bar{x}$ is the average of all observations, we can think of the median $\tilde{x}$ as the average of the middle one or two observations as if the rest observations were discarded. The α-trimmed mean $\bar{x}_\alpha$ is defined as follows: discard $100\alpha\%$ of the observations on each side of the ordered sample $x_{(1)} \leq x_{(2)} \leq \dots \leq x_{(n)}$ and take the mean of the remaining middle $100(1-2\alpha)\%$ of the observations.

[16] It is also known as the *Olympic average*, since it is used in Olympic figure skating for computing a skater's final score.

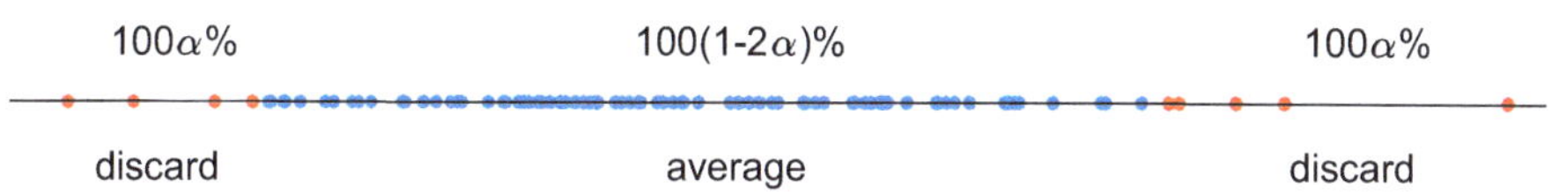

Fig. 1.3 Computing the α-trimmed mean $\bar{x}_\alpha$: discard $200\alpha\%$ of the data in the tails (orange points) and average the remaining middle data (blue points).

Mathematically,

$$\bar{x}_\alpha = \frac{x_{([n\alpha]+1)} + \dots + x_{(n-[n\alpha])}}{n - 2[n\alpha]}, \qquad \alpha \in (0, 0.5), \tag{1.9}$$

where $[s]$ denotes the integer part of s. The 0-trimmed mean is then the standard sample mean, $\bar{x}_0 = \bar{x}$, and the median can be viewed as the 0.5-trimmed mean. Thus, the trimmed mean $\bar{x}_\alpha$ interpolates between the mean $\bar{x}$ and the median $\tilde{x}$, as α goes from 0 to 0.5. In most applications, the values $\alpha \in [0.05, 0.25]$ are typically used[17].

[17] In Olympic figure skating, the single highest and lowest scores are dropped and the remaining scores get averaged. This helps to reduce the impact of biased or corrupt judges.

1.2.2 Measures of Spread

A measure of center is often accompanied by a measure of spread that quantifies the "scatteredness" of the sample, i.e. how far an individual sample value x_i can be from the center of the data. The simplest measure of spread is the *range*, which is the difference between the largest and smallest values,

$$r_x = x_{(n)} - x_{(1)}. \tag{1.10}$$

The range ignores most of the data and is very sensitive to outliers.

One of the most popular measures of spread is the *sample standard deviation*:

$$s_x = \sqrt{\frac{1}{n}\sum_{i=1}^{n}(x_i - \bar{x})^2}. \tag{1.11}$$

Remark 1.5. For the reasons discussed in Section 4.3, the sample standard deviation is sometimes defined as $s_x = \sqrt{\frac{1}{n-1}\sum_{i=1}^{n}(x_i - \bar{x})^2}$. But if n is large, then the difference between the two definitions is negligible.

Although the standard deviation s_x is also sensitive to outliers, it is more robust than the range r_x, and is often reported together with the mean $\bar{x}$.

A measure of spread that usually accompanies the median is the *interquartile range* (IQR), which is the difference between the upper and lower quartiles of the sample,

$$\mathrm{IQR} = Q_3 - Q_1, \tag{1.12}$$

where Q_1 is the first (lower) quartile that splits lowest 25% of the sample and Q_3 is the third (upper) quartile that splits highest 25% of the sample[18].

[18] The second quartile is the median, $Q_2 = \tilde{x}$.

1.2.3 Five-Number Summary

The *five-number summary* is a collection of numerical statistics that simultaneously provides information about the center and spread of a sample. The five numbers are:

1. the smallest value in the samle $x_{(1)} = x_{\min}$,
2. the first quartile Q_1,
3. the median $\tilde{x} = Q_2$,
4. the third quartile Q_3, and
5. the largest value in the sample $x_{(n)} = x_{\max}$.

1.3 Boxplot

A *boxplot* is a graphical version of the five-number summary. It provides information about the shape of the data, its center and spread, and shows potential outliers. To create a boxplot from the data $x_1, \ldots, x_n$:

1. Draw a box with the bottom edge placed at the first quartile Q_1 and the top edge placed at the third quartile Q_3. Half of the data lies in the box, since the height of the box is $Q_3 - Q_1 = \mathrm{IQR}$. The width of the box does not represent anything[19].
2. Draw a horizontal line through the box at the median $\tilde{x} = Q_2$.
3. Place two caps: at the largest data point that is less than or equal to $Q_3 + 1.5 \times \mathrm{IQR}$ and at the smallest data point that is greater than or equal to $Q_1 - 1.5 \times \mathrm{IQR}$.
4. Extend whiskers (dashed lines in Fig. 1.4(b)) from the ends of the box to the caps. Observations that lie between the caps are not considered as outliers.
5. The observations that fall outside the caps are considered as outliers[20]. Plot them individually with $\cdot$, $+$, $*$, $\times$, or your favorite symbol.

[19] If boxplots are used for comparing two or more samples, then the width of a boxplot w can be used for representing the sample size n. Conventionally $w \propto \sqrt{n}$ [7].

[20] Outliers are defined as points x_i that are either too large, $x_i > Q_3 + k \times \mathrm{IQR}$, or too small, $x_i < Q_1 - k \times \mathrm{IQR}$, with $k = 1.5$ as a usual choice.

An example of a boxplot is shown in Fig. 1.4(b).

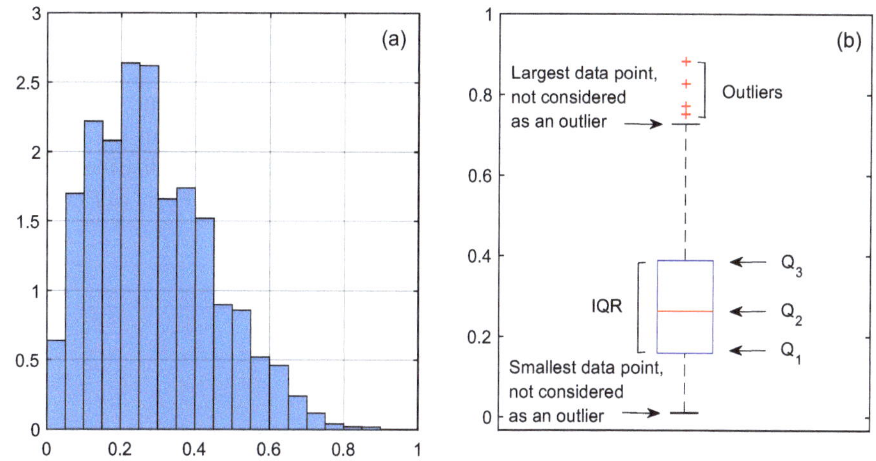

Fig. 1.4 A histogram (panel (a)) and a boxplot (panel (b)) of sample $x_1, \dots, x_n$, $n = 10^3$, generated from the beta distribution with parameters $\alpha = 2$ and $\beta = 5$.

A boxplot can be viewed as both graphical and numerical summary of the data. It contains slightly more information than the five-number summary: a boxplot shows all outliers, not just the largest and the smallest values in the sample. Histograms and boxplots nicely complement each other: while a histogram is visually more expressive in describing the shape of the sample[21], boxplots provide important quantitative information about the data, such as the median and IQR. Boxplots are especially useful when comparing related samples. For examples, salaries of data scientists at Google and Microsoft, lengths of flight delays of different airlines, ice cream consumption by boys and girls, etc.

[21] For example, a boxplot does not indicate whether there are gaps in the data or multiple modes (peaks).

1.4 Empirical CDF

Recall that the *cumulative distribution function* (CDF) $F(x)$ of a random variable X is defined as the probability that the value of X is less than or equal to x,

$$F(x) = \mathbb{P}(X \leq x). \tag{1.13}$$

An *empirical CDF* (eCDF) $\hat{F}_n(x)$ of sample $x_1, \dots, x_n$ is a "data analogue" of the "theoretical" CDF $F(x)$. It is defined as the fraction of data points x_i that are less than or equal to x:

$$\hat{F}_n(x) = \frac{\text{number of data points } \leq x}{n} = \frac{1}{n}\sum_{i=1}^{n} H(x - x_i), \tag{1.14}$$

where $H(x)$ is the Heaviside step function:

$$H(x) = \begin{cases} 0, & x < 0, \\ 1, & x \geq 0. \end{cases} \tag{1.15}$$

The eCDF $\hat{F}_n(x)$ is thus a step function (a piecewise constant function) that jumps up by $\frac{1}{n}$ at each of the data points x_i[22]. It is schematically shown in Fig. 1.5(a). Figure 1.5(b) shows the eCDF of a sample drawn from the uniform distribution on $[0,1]$. Notice how closely the eCDF resembles the uniform CDF, $\hat{F}_n(x) \approx F(x)$. As we will see in Section 4.2, this is not a coincidence.

[22] If the value of x_i appears k times in the sample, then the eCDF jumps up by $\frac{k}{n}$ at this value.

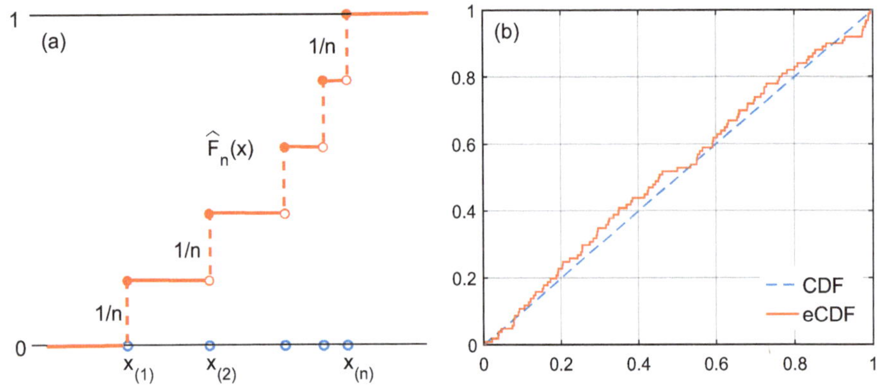

Fig. 1.5 (a) A schematic representation on an empirical CDF $\hat{F}_n(x)$ of sample $x_1, \ldots, x_n$; (b) The eCDF (solid line) of a sample of size $n = 100$ drawn form the uniform distribution on $[0,1]$ and the "theoretical" uniform CDF $F(x)$ (dashed line).

Compared with histograms and boxplots, eCDFs are rarely used as graphical summaries per se, since it is more difficult to read them and they convey less qualitative information about the shape of the data. However, the eCDF plays a fundamental role in nonparametric inference, in particular, in estimating the values of statistical functionals[23]. The eCDF is also used for constructing Q-Q plots, which are an indispensable graphical tool for exploring the distribution of the data.

[23] See Section 4.1

1.5 Q-Q Plots

Remarkably many samples are approximately normally distributed[24]. Visually, approximate normality of a sample means that its histogram is bell-shaped: it has a single peak (mode), it is not skewed to the right or left, and it is symmetric about its center. Examples are shown in Fig. 1.1(a) and Fig. 1.2(b).

[24] The normal distribution and the reasons for this empirical observation are briefly reviewed in Probability Refresher 1.8 at the end of this chapter.

Suppose now that, given the data $x_1, \ldots, x_n$, we want to check if it is approximately distributed according to the standard normal distribution. In other words, if the standard normal CDF $\Phi(z)$ is a good theoretical model for the data, and $x_1,\dots,x_n$ can be approximately viewed as realizations of the standard normal random variable $Z \sim \mathcal{N}(0,1)$. We can start by plotting the histogram of $x_1,\dots,x_n$ and see if it is bell-shaped. But the problem with this approach is that usually histograms do not allow to see clearly what happens in the *tails* of the data distribution, i.e. around $x_{(1)}$ and $x_{(n)}$: do the tails decay faster or slower than the normal tails? In the former case, the data tails are often referred to as *short tails*, whereas in the latter case they are called *long tails* or *heavy tails*.

Remark 1.6. Why is the tail behavior important?[25]
In applications concerned with the typical behavior of the underlying system or process, the tails of the data may not be important. However, if we are interested in the extreme system's behavior, then the data tails could be of paramount importance. Suppose, for example, that $x_1,\dots,x_n$ represent the responses of a system of interest to certain excitations[26]. Furthermore, assume that we don't like large values of x_i, and if X is a system response, then we call an even $X > x^*$ a system failure, where x^* is some critical threshold. The goal is to estimate the probability of system failure, i.e. to estimate $\mathbb{P}(X > x^*)$.

If the value of x^* corresponds to a "typical" response, that is if x^* is somewhere in the middle of the data, then we can accurately estimate the failure probability using the eCDF of the sample[27] $x_1,\dots,x_n$:

$$\mathbb{P}(X > x^*) = 1 - \mathbb{P}(X \le x^*) = 1 - F(x^*) \approx 1 - \hat{F}_n(x^*) = \frac{\text{number of } x_i > x^*}{n}. \quad (1.16)$$

Suppose instead that the critical threshold x^* is much greater than a typical system response, $x^* \gg \bar{x}$, and, in particular, $x^* > x_{(n)}$. This means that the system failure is a *rare* event. In this case, (1.16) gives a trivial estimate $\mathbb{P}(X > x^*) \approx 0$. If the system failure is not only rare, but also a high-impact event[28], then we may need to estimate its probability more accurately by a small but non-zero value.

If the histogram of $x_1,\dots,x_n$ appears to be bell-shaped, as in Fig. 1.6, then a naive tail-ignoring approach would be to fit a normal distribution to the data[29] and use the fitted distribution $\mathcal{N}(\hat{\mu},\hat{\sigma}^2)$ for estimating the failure probability:

$$\mathbb{P}(X > x^*) = 1 - \mathbb{P}(X \le x^*) = 1 - F(x^*) \approx 1 - \Phi(x^* \mid \hat{\mu},\hat{\sigma}^2). \quad (1.17)$$

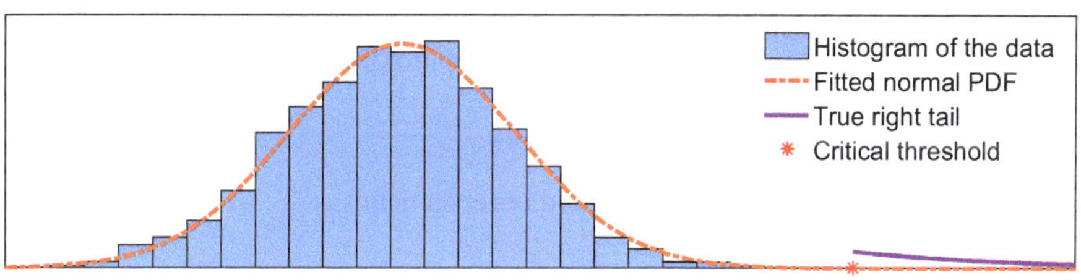

Fig. 1.6 The histogram of the data, the fitted normal distribution, and the true right tail of the data. If the data tail is longer than the fitted tail, then (1.17) leads to underestimation.

[25] "Don't ask questions – that was the first rule for a quiet life with the Dursleys." – J.K. Rowling, *Harry Potter and the Sorcerer's Stone.*

[26] For example, responses of a building to earthquake excitations, or levels of cholesterol as responses to unhealthy diets, or stock price changes due to market perturbations.

[27] As we will learn in Section 4.3, estimate (1.16) is called a *plug-in estimate*.

[28] For example, a building collapse, high cholesterol, or financial crisis.

[29] For example, using the method of moments discussed in Chapter 7 or the method of maximum likelihood described in Chapter 8.

However, if the true right tail is longer than the fitted normal tail, i.e. if it decays slower, as schematically shown in Fig. 1.6, then estimating the failure probability $\mathbb{P}(X > x^*)$ by the probability under the normal tail $1 - \Phi(x^* \mid \hat{\mu}, \hat{\sigma}^2)$ will lead to a substantial underestimation. In this case, the system failure would be perceived as a much more rare event than it actually is, which can lead to dramatic consequences. Understanding the tail behavior is thus important in many applications. A rigorous yet intuitive introduction to heavy-tailed distributions,accessible to anyone with a background in elementary probability, is provided in [8].

A *quantile-quantile (Q-Q) plot* is a graphical method that allows to explore the tail behavior better than histograms and helps to compare the data distribution with any theoretical distribution, in particular, with the standard normal distribution. As the name suggests, Q-Q plots are based on the notion of quantile.

The qth *quantile* of the standard normal distribution is a number z_q such that

$$\Phi(z_q) = q, \qquad \text{where } 0 < q < 1. \tag{1.18}$$

In other words, z_q is a number such that the probability mass (the area under the PDF curve) supported by the interval $(-\infty, z_q)$ is exactly q. Figure 1.7 clarifies this definition[30].

[30] Sometimes quantiles are also called "percentiles" in the literature.

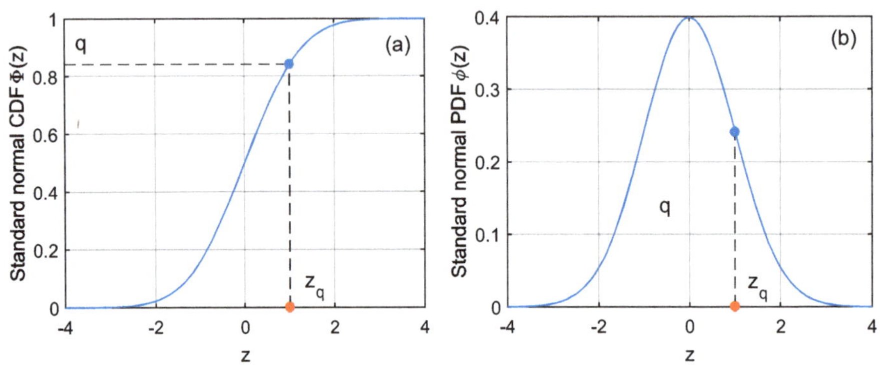

Fig. 1.7 (a) The standard normal quantile z_q in terms of the CDF: $\mathbb{P}(Z \leq z_q) = \Phi(z_q) = q$; (b) The standard normal quantile z_q in terms of the PDF: $\mathbb{P}(Z \leq z_q) = \int_{-\infty}^{z_q} \phi(z) = q$.

If sample $x_1, \ldots, x_n$ is distributed according to the standard normal distribution, then we expect[31] that its eCDF $\hat{F}_n(x)$ must be approximately equal to the standard normal CDF $\Phi(x)$,

[31] Thanks to Theorem 4.1 in Section 4.2.

$$\hat{F}_n(x) \approx \Phi(x). \tag{1.19}$$

Notice that by definition of the eCDF and order statistics $x_{(1)}, \ldots, x_{(n)}$,

$$\hat{F}_n\left(x_{(1)}\right) = \frac{1}{n}, \quad \hat{F}_n\left(x_{(2)}\right) = \frac{2}{n}, \quad \ldots, \quad \hat{F}_n\left(x_{(n)}\right) = \frac{n}{n} = 1. \tag{1.20}$$

On the other hand, as it follows from (1.18),

$$\Phi\left(z_{\frac{1}{n}}\right) = \frac{1}{n}, \quad \Phi\left(z_{\frac{2}{n}}\right) = \frac{2}{n}, \quad \ldots, \quad \Phi\left(z_{\frac{n}{n}}\right) = \frac{n}{n} = 1. \tag{1.21}$$

Therefore, the kth order statistics $x_{(k)}$ and the $\left(\frac{k}{n}\right)^{\text{th}}$ standard normal quantile $z_{\frac{k}{n}}$ must be approximately the same,

$$x_{(k)} \approx z_{\frac{k}{n}}, \qquad k = 1, \ldots, n. \tag{1.22}$$

This motivates the following definition. The *normal-quantile plot* for the sample $x_1, \ldots, x_n$ is a collection of n points on the xy-plane with coordinates

$$\left(z_{\frac{k}{n}}, x_{(k)}\right), \qquad k = 1, \ldots, n. \tag{1.23}$$

If the plotted points fall roughly on the line $y = x$, then this indicates that the sample is distributed approximately according to the standard normal distribution.

Remark 1.7. There is one little technical problem with definition (1.23): if $k = n$, then $z_1 = \Phi^{-1}(1) = +\infty$. There are may ways to get around this. One consists of using $z_{\frac{k}{n+1}}$ instead of $z_{\frac{k}{n}}$ in (1.23), to make sure that all standard normal quantiles are finite and can be plotted[32].

[32] Statistical packages use $z_{\frac{k-0.5}{n}}$, $z_{\frac{k-0.3}{n+0.4}}$, $z_{\frac{k-0.375}{n+0.25}}$, etc.

As an illustration, Fig 1.8(a) shows the normal-quantile plot for the data $x_1, \ldots, x_n$ sampled from the standard normal distribution. Departures from standard normality are indicated by systematic deviations from the line $y = x$. Examples of different departures are illustrated in Fig. 1.8(b), and (c).

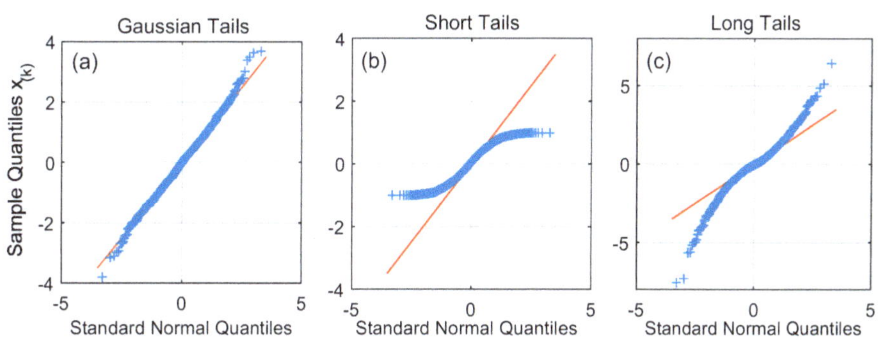

Fig. 1.8 Normal-quantile plots for the data $x_1, \ldots, x_n$ sampled from (a) standard normal distribution, (b) uniform distribution on $[-1, 1]$ (short tails), and (c) Laplace distribution with the PDF $f(x) = \frac{1}{2}e^{-|x|}$ (long tails). The sample size in all examples is $n = 10^3$.

The concept of a normal-quantile plot can be extended for comparing the distribution of the data $x_1, \ldots, x_n$ with any probability distribution $F(x)$, not necessarily standard normal. To do that, we simply need to plot the order statistics $x_{(k)}$, which can be interpreted as the $\left(\frac{k}{n}\right)^{\text{th}}$ quantiles of the eCDF[33], versus the $\left(\frac{k}{n}\right)^{\text{th}}$ quantiles

[33] Since $\hat{F}_n\left(x_{(k)}\right) = \frac{k}{n}$.

$\xi_{\frac{k}{n}}$ of distribution $F(x)$, defined by analogy with (1.18) as $F\left(\xi_{\frac{k}{n}}\right) = \frac{k}{n}$. This way we can construct, for example, a uniform-quantile plot, an exponential-quantile plot, etc. The more closely points $\left(\xi_{\frac{k}{n}}, x_{(k)}\right)$, $k = 1,\dots,n$, lie on the line $y = x$, the more accurately distribution $F(x)$ describes the distribution of the data[34].

[34] What can we say about the data distribution if the points fall on the line $y = ax + b$ instead of $y = x$? You will explore this question in Problem 1.3 at the end of this chapter.

Moreover, Q-Q plots can be used for comparing the distributions of two different samples. By analogy with the previous Q-Q plots, to compare the distributions of samples $x_1,\dots,x_n$ and $y_1,\dots,y_n$, we need to plot their sample quantiles $(x_{(1)}, y_{(1)}), \dots, (x_{(n)}, y_{(n)})$ on the xy-plane. Any systematic deviations from the line $y = x$ will indicate differences in the distributions of the two samples.

1.6 Case Study: CO_2 Emissions

Life is uncertain. Eat dessert first.

– Ernestine Ulmer

Since the end of the eighteenth century, after James Watt had invented the first practical steam engine that triggered the Industrial Revolution, carbon dioxide (CO_2) concentrations in the atmosphere and surface oceans have been constantly rising. The main reason for the increase in CO_2 concentration is the burning of fossil fuels, such as oil, natural gas and coal.

Carbon dioxide is a *greenhouse gas*. Greenhouse gases absorb heat (infrared radiation) emitted from the Earth to space and re-emit it in all directions, including back to the Earth. This warms the Earth and the resulting warming mechanism is called the *greenhouse effect*. The dominant contributor to the total greenhouse effect is water vapor (responsible for $\approx 50\%$ of the effect), followed by clouds ($\approx 25\%$) and then CO_2 ($\approx 20\%$) [9]. The greenhouse effect makes the Earth a comfortable planet to live on: without greenhouse gases, the average annual temperature on the Earth's surface would be about $-18°C$ ($0°F$) instead of much more pleasant $15°C$ ($59°F$) [10]. Burning fossil fuels, however, increases Earth's natural greenhouse effect by adding more carbon dioxide into the atmosphere. And strengthening the greenhouse effect can increase the average global temperature, which, in different places, can lead to pleasant, unpleasant, and completely unforeseen consequences.

For a realistic (and humorous) discussion of our energy options, strategies for getting off fossil fuels (which are a finite resource), and fundamental physical limits to sustainable energy, I strongly recommend the excellent book by Sir David MacKay [11][35]. This book is about facts, not emotions that are so often associated with global warming discussions; it is about numbers, not adjectives.

[35] The book is freely available online at www.withouthotair.com.

In this Case Study, we will summarize data [12] on total CO_2 emissions from the burning of fossil fuels by different countries during various years. The data is represented by an $n \times m$ matrix E, where $n = 192$ is the number of countries, $m = 264$ is the number of years (ranging from 1751 until 2014), and E_{ij} is the total CO_2 emission (in thousands of tones) by country i during year j. Figure 1.9 shows the total carbon dioxide emissions for the entire world ($\sum_{i=1}^{n} E_{ij}$ versus j) as well as individual contributions of top seven countries (E_{ij} versus j for fixed i) with the largest cumulative emissions for the period 1751-2014: the US, China, Russia, Germany, the UK, Japan, and India.

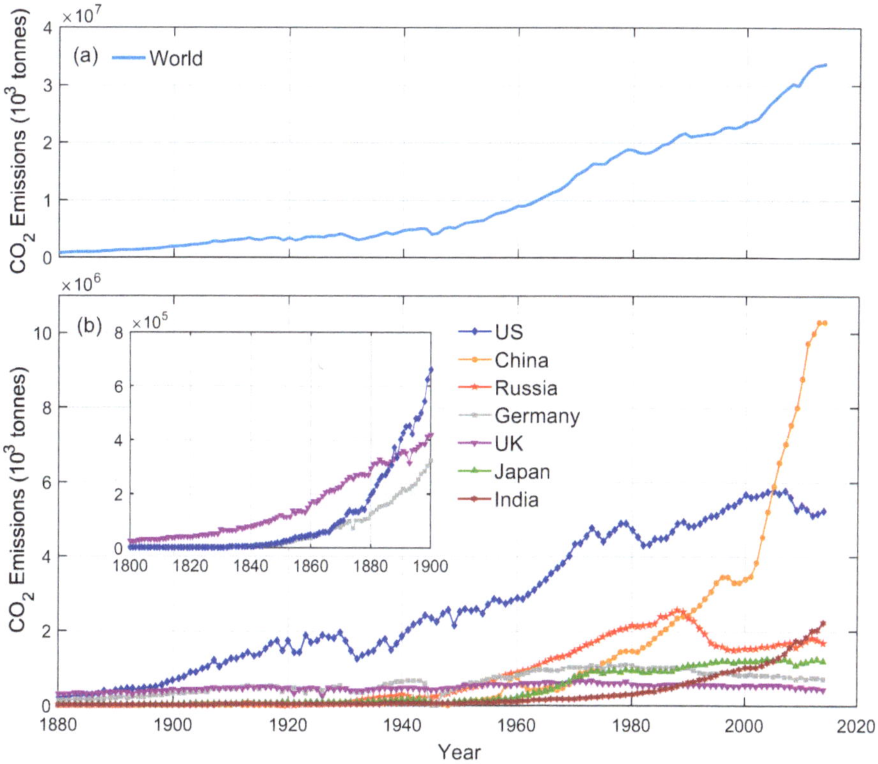

Fig. 1.9 Carbon dioxide emissions from the burning of fossil fuels for (a) the entire wold and (b) top emitters: the United States, China, Russia, Germany, the United Kingdom, Japan, and India.

Initially, the UK was the largest contributor to the CO_2 pollution, which is not surprising since the Industrial Revolution began in Great Britain. But as the revolution has spread to the rest of the world, the UK was outpaced by the US around 1890, which was subsequently outpaced by China around 2005. It is interesting to observe how the behavior of CO_2 emissions changed due to important historical and economic events: a sharp drop in the German emission after the end of World War II in 1945; decrease in the Russian emission after the Soviet Union collapse in 1991; fast increase of the Chinese emission after several economic reforms in the early 2000s and joining the World Trade Organization in 2001, that led to export growth, new energy demands, and, ultimately, skyrocketing CO_2 emissions.

To summarize this data, first let's look at the carbon dioxide emissions during 2014, the last year in the data set. Figure 1.9 (b) suggests that emissions vary a lot from country to country and cover several orders of magnitude. Indeed, while the smallest emission in 2014 was in Tuvalu[36], 11×10^3 tonnes of CO_2, the largest was in China, 10.3×10^9 tonnes of CO_2. It is thus more meaningful to study emissions on a logarithmic scale. So, let

$$x_i = \log(E_{im}), \qquad i = 1 \ldots, n, \tag{1.24}$$

be the logarithm of the total CO_2 emission by country i during 2014. The mean $\bar{x}$, range r_x, standard deviation s_x, IQR, and the five-number summary of $x_1, \ldots, x_n$ are shown in Table 1.1. Figure 1.10 summarizes the 2014 log-emissions graphically.

Table 1.1 Numerical summaries of the log-emissions of CO_2 during 2014.

$\bar{x}$	r_x	s_x	IQR	$x_{(1)}$	Q_1	$\tilde{x}$	Q_3	$x_{(n)}$
9.35	13.75	2.53	3.39	2.40	7.68	9.33	11.07	16.15

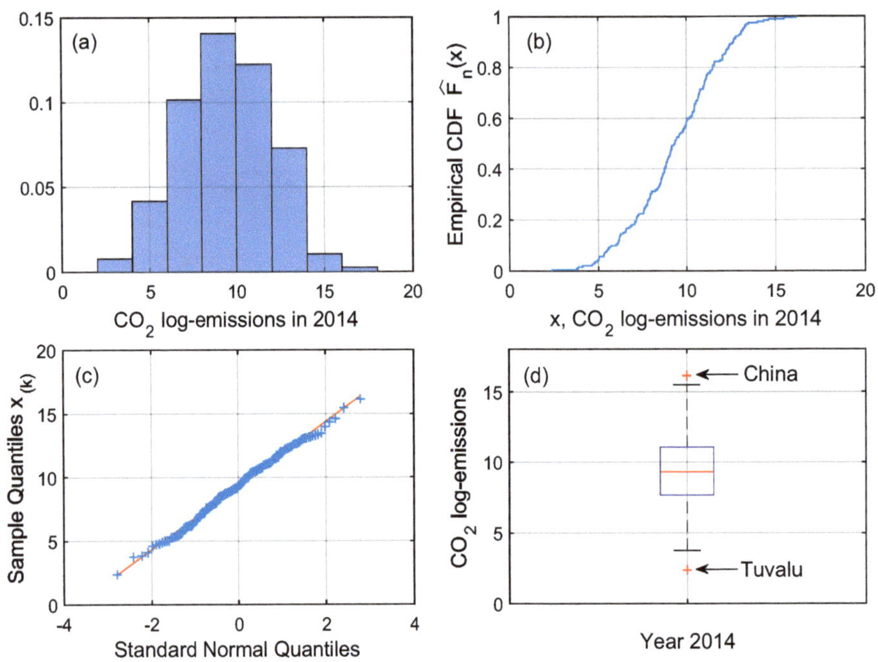

Fig. 1.10 Graphical summaries of the log-emissions of CO_2 in 2014: (a) histogram, (b) empirical CDF, (c) normal-quantile plot, and (d) boxplot.

The sample mean and median are approximately equal, $\bar{x} \approx \tilde{x}$, the histogram is bell-shaped, the boxplot is symmetric about the median, and the normal-quantile plot falls roughly on a straight line. These observations strongly suggest that the sample $x_1, \ldots, x_n$ is approximately normally distributed[37]. Indeed, both the *Kolmogorov-Smirnov test* (KS test) and the *Shapiro-Wilk test*[38] (SW test) confirm

[36] A small Polynesian island country located in the Pacific Ocean, between Hawaii and Australia.

[37] See Problem 1.3 at the end of this chapter.

[38] We will discuss these tests in Section 10.3 and Remark 10.9.

the hypothesis that the log-emissions are normally distributed. This means that the carbon dioxide emissions during 2014, $E_{1m}, \ldots, E_{nm}$, are approximately distributed according to a *log-normal distribution*[39].

Next, let's summarize *cumulative* carbon dioxide emissions by different countries during 1751-2014. In this case, a logarithmic scale is also more appropriate. So, let

$$t_i = \log\left(\sum_{j=1}^{m} E_{ij}\right), \qquad i = 1\ldots,n, \tag{1.25}$$

be the logarithm of the cumulative CO_2 emission by country i. Numerical and graphical summaries of the log-cumulative-emissions $t_1, \ldots, t_n$ are given in Table 1.2 and Figure 1.11, respectively.

Table 1.2 Numerical summaries of the log-cumulative-emissions of CO_2 during 1751-2014.

$\bar{t}$	r_t	s_t	IQR	$t_{(1)}$	Q_1	$\tilde{t}$	Q_3	$t_{(n)}$
12.78	14.35	2.78	4.08	5.39	10.79	12.75	14.87	19.74

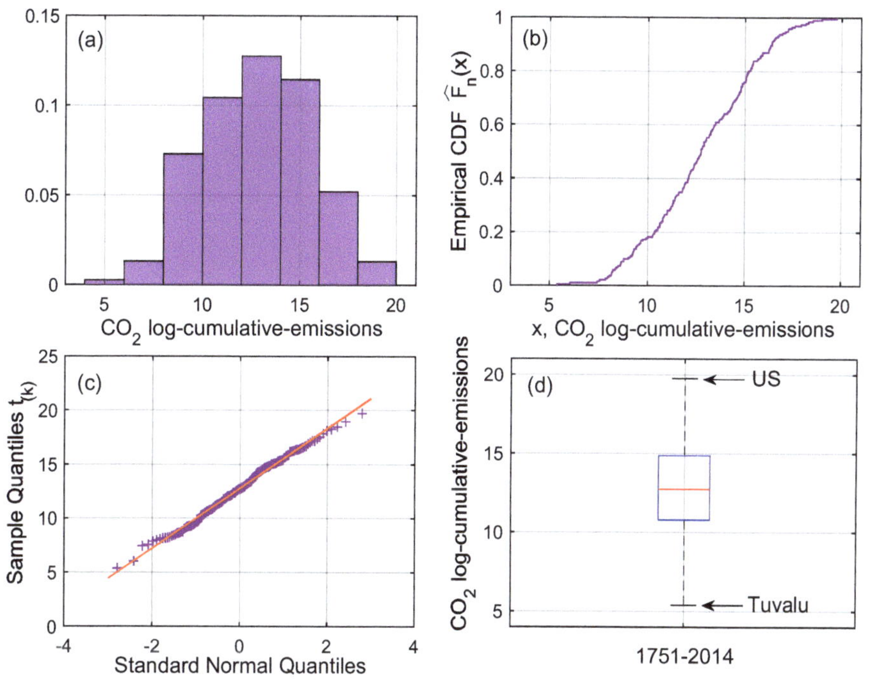

Fig. 1.11 Graphical summaries of the log-cumulative-emissions of CO_2 during 1751-2014: (a) histogram, (b) empirical CDF, (c) normal-quantile plot, and (d) boxplot.

As in the previous example, the histogram, eCDF, normal-quantile plot, and boxplot suggest — and both the KS and SW tests confirm — that the sample $t_1, \ldots, t_n$ is

[39] Recall that a random variable X is log-normally distributed if $Y = \log X$ is normally distributed.

approximately normally distributed. However, the mean and the spread of the cumulative CO_2 emissions are larger than that of the emissions in 2014. Another worth mentioning observation is that while the largest contribution to the carbon dioxide pollution in 2014 was made by China, see the outlier in Fig. 1.10 (d), the largest commulative contribution was still made by the US, see Fig. 1.11 (d).

To explore the CO_2 emissions by different countries in the past few decades, Fig. 1.12 shows the boxplots and empirical CDFs of log-emissions in 1974, 1984, 1994, 2004, and 2014.

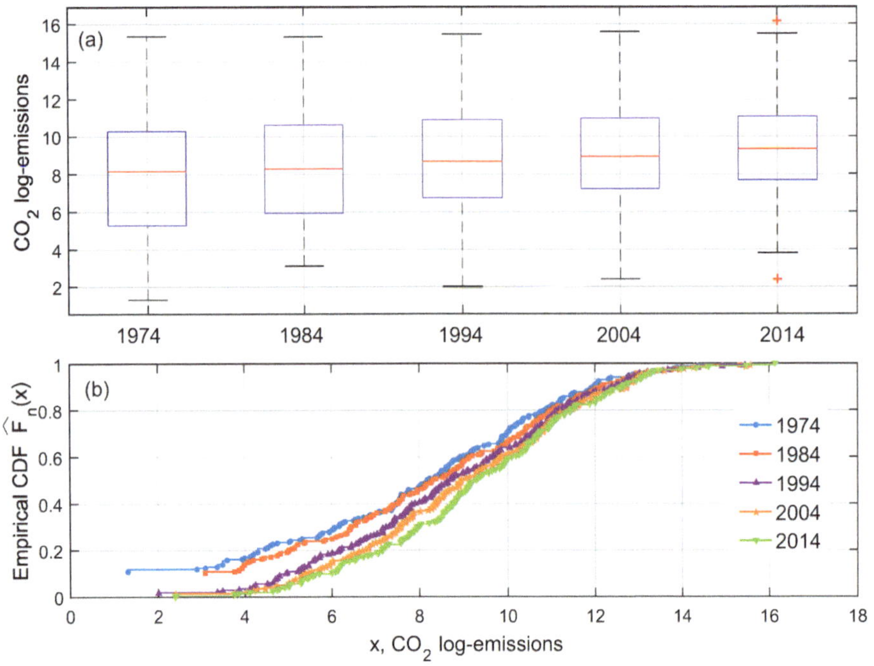

Fig. 1.12 Comparison of the CO_2 log-emissions in 1974, 1984, 1994, 2004, and 2014 via (a) boxplots and (b) empirical CDFs.

Remark 1.8. Some countries had zero carbon dioxide emissions in the past years[40]. For example, in 1974 there were 21 country with zero emissions, in 1984 there were 20 such countries, in 1994 there were only three, in 2004 only two (Liechtenstein and South Sudan[41]), finally, in 2014 all countries were burning fossil fuels and emitting CO_2. Consequently, the log-emissions for those countries in the corresponding years equal to $-\infty$. But this is not a problem[42], since boxplots and eCDFs are not sensitive to outliers. As long as $x_{\min}$ is the smallest value in a sample, its exact value does not affect the median $\tilde{x}$, the first and the thirds quartiles Q_1 and Q_3, and the IQR, which determine the corresponding boxplot. Similarly, since the eCDF is based on the order statistics (see Fig. 1.5 (a)), the smallest value $x_{\min} = x_{(1)}$ affects only the very end of the left tail, and not the rest of the eCDF. Ignoring countries with zero

[40] More precisely, their CO_2 emissions were rounded to zero since they were negligible.

[41] South Sudan gained its independence and was recognized as a separate country only in 2011.

[42] Computing the sample mean would be a problem.

CO_2 emissions, however, is a bad idea, since this would artificially increase the median and quartiles, and, as a result, would move the corresponding boxplot up. This effect would be especially pronounced for 1974 and 1984, the years with many countries with zero CO_2 emissions.

Both boxplots and eCDFs in Fig. 1.12 indicate that the carbon dioxide pollution grows with time, as expected. Another observation is that the IQR, which is the height of the box that contains half of the countries, shrinks towards the monotonically increasing median. This means that with time, countries pollute more on average and become more similar in terms of their amounts of CO_2 emissions.

Finally, let's compare the carbon dioxide emissions of six continents: Africa, Asia, Europe, North America, Oceania, and South America[43]. Figure 1.13 shows boxplots of the logarithms of the CO_2 emissions across different countries grouped by continent: for 2014 in panel (a), and during 1751-2014 in panel (b).

[43] The seventh continent, Antarctica, does not contribute to CO_2 emissions.

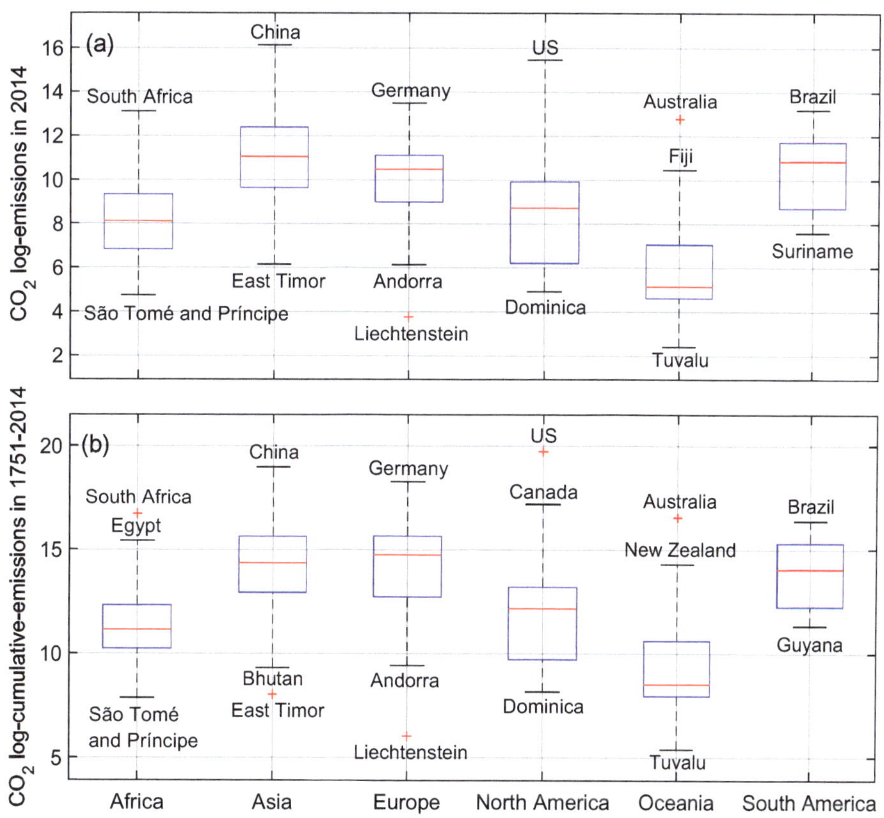

Fig. 1.13 Comparison of the carbon dioxide emissions by continents: (a) CO_2 log-emissions during 2014 and (b) CO_2 log-cumulative-emissions between 1751 and 2014.

To compliment the above boxplots that characterize the shape of the continental log-emissions, Figure 1.14 depicts in panel (a) the time evolution of the relative contributions of the continents to carbon dioxide pollution. Panels (b) and (c) show the total emissions in 2014 and cumulative emissions during 1751-2014, respectively.

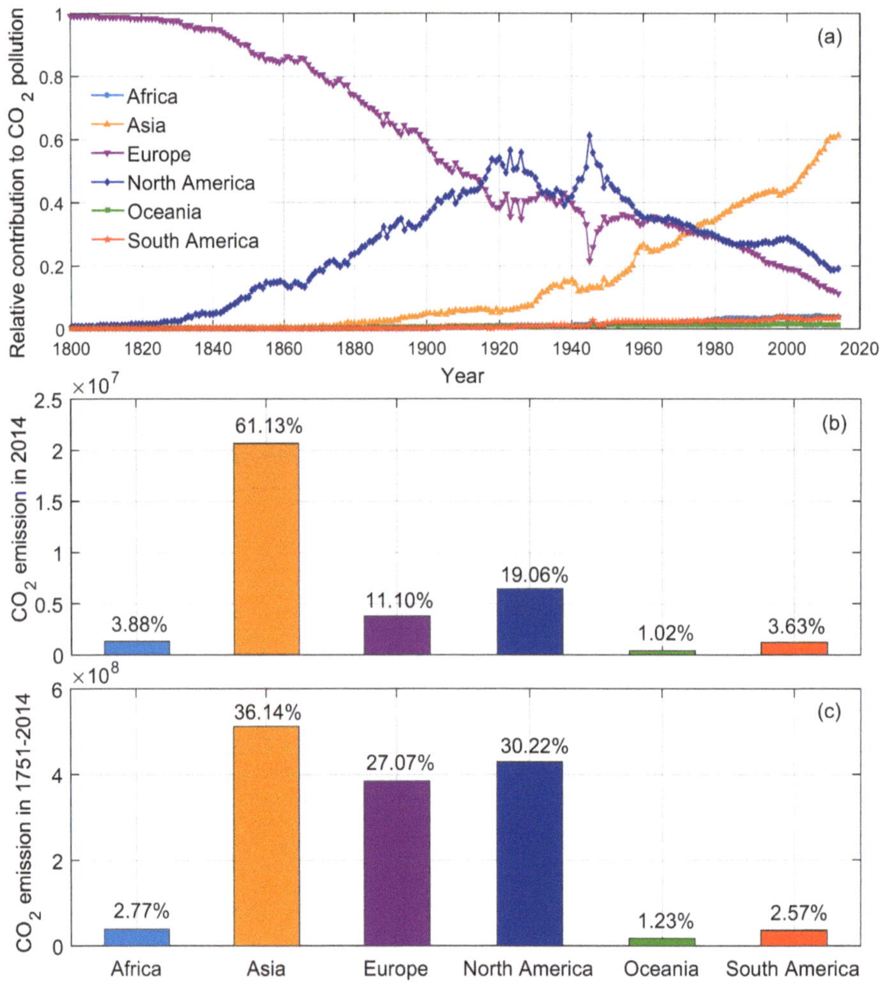

Fig. 1.14 (a) Proportions of the carbon dioxide emissions by different continents versus time; (b) Total CO_2 emissions in 2014; (c) Cumulative CO_2 emissions during the period 1751-2014.

Several observations are in order. Despite some local fluctuations, we can observe three clear trends in Fig. 1.14 (a): the European relative contribution to the CO_2 emissions decreases with time, the Asian contribution increases, and the North American contribution first increases, until 1920, then fluctuates for about 30 years, and then decreases after 1950. At the same time, relative contributions of Africa, Oceania, and South America remain negligible. Notice that both South American boxplots in Fig. 1.13 lie quite high. In particular, the median emission from South America is comparable to that of Asia and Europe and substantially higher than the median emission from North America. This does not contradict to the very small relative South American contribution to the carbon dioxide pollution observed in Fig. 1.14: just the US and China emit considerably more CO_2 than South America. For example, in 2014, the US alone contributed more than four times as much CO_2 as the entire South America, and China emitted approximately twice as much as the US. On the other hand, although Africa, Oceania, and South America have overall relatively small contributions, CO_2 pollution of their top emitters — South Africa, Australia, and Brazil — is comparable with the pollution produced by top Asian, European, and North American countries.

In this Case Study, we discussed several summaries of data [12] on carbon dioxide emissions from the burning of fossil fuels by different countries during various years. Problem 1.5 at the end of this chapter will ask you[44] to perform a similar analysis of CO_2 emissions *per person*. Yes, China emits a lot of CO_2 in total and Qatar a little[45], but will these conclusions remain unchanged if we take the populations of countries into account and look at the emissions per person? Explore!

[44] "...after all this is not my adventure." J.R.R. Tolkien, *The Hobbit, or There and Back Again.*

[45] In 2014, Qatar's CO_2 emission was about 1% of China's emission.

1.7 Further Reading

In this chapter, we have considered several standard numerical and graphical ways to summarize data. But they represent just the tip of the iceberg. There exists a rich variety of tools for summarizing and exploring the data, such as scatterplots[46], time series plots, odds ratio, parallel coordinates, and principal component analysis, to name but a few. Moreover, summarizing data can be (and should be) viewed as a part of *Exploratory Data Analysis* (EDA), an approach to data analysis that tries to uncover the underlying structure of the data, formulate some hypotheses about the phenomenon that generated the data, and come up with reasonable assumptions about the data that can be used later, at the inferential stage of data analysis. The EDA approach was introduced and promoted by mathematician, statistician, and one of the first data scientists John Tukey, who strongly believed that "data analysis is intrinsically an empirical science... defined by a ubiquitous problem rather than by a concrete subject" [13]. The EDA methodology is mostly based on graphical techniques and often used in data mining, big data analytics, and data science. Tukey's seminal work,

[46] We will discuss scatterplots in Chapter 11 in the context of regression analysis.

John Tukey (1915 – 2000), an inventor of a boxplot and one of the first data scientists. Reprinted with the permission of the Institute of Mathematical Statistics.

- Tukey, J.W.: *Exploratory Data Analysis*. Addison-Wesley (1977) `https://doi.org/10.1002/bimj.4710230408`,

although a bit outdated, still remains one of the best conceptual introductions to EDA. For a modern treatment of the subject,

- Cook, D., Swayne, D.F.: *Interactive and Dynamic Graphics for Data Analysis*. Springer Science & Business Media (2007) `https://doi.org/10.1007/978-0-387-71762-3`

is strongly recommended.

1.8 Probability Refresher: Normal Distribution

The *standard normal curve*, also known as the bell curve or the Gaussian curve[47], is shown in Fig. 1.7(b) and defined as

$$\phi(z) = \frac{1}{\sqrt{2\pi}} \exp\left(-\frac{z^2}{2}\right). \tag{1.26}$$

The coefficient $\frac{1}{\sqrt{2\pi}}$ does not have any sacral meaning, it is simply a normalizing constant that guarantees that the full area under the curve is exactly one,

$$\int_{-\infty}^{+\infty} \phi(z)dz = 1. \tag{1.27}$$

Recall that any non-negative function that integrates to one can be viewed as a *probability density function* (PDF) of a continuous random variable. Therefore, $\phi(z)$ can be interpreted as a PDF of a random variable, commonly denoted by Z, and called the *standard normal random variable*. The PDF $\phi(z)$ fully determines probabilistic properties of Z. Namely, for any set $A \subset \mathbb{R}$,

$$\mathbb{P}(Z \in A) = \int_A \phi(z)dz. \tag{1.28}$$

The standard normal random variable Z has zero mean and unit variance:

$$\begin{aligned} \mu &= \mathbb{E}[Z] = \int_{-\infty}^{+\infty} z\phi(z)dz = 0, \\ \sigma^2 &= \mathbb{V}[Z] = \mathbb{E}[(Z-\mu)^2] = \int_{-\infty}^{+\infty} z^2\phi(z)dz = 1. \end{aligned} \tag{1.29}$$

Because of the zero mean and unit variance, the following notation is often used to indicate that Z is distributed according to the standard normal distribution:

$$Z \sim \mathcal{N}(0,1). \tag{1.30}$$

Its cumulative distribution function (CDF), i.e. the *standard normal CDF*, is often denoted by $\Phi(z)$. It is shown in Fig. 1.7(a) and given by

$$\Phi(z) = \mathbb{P}(Z \leq z) = \int_{-\infty}^{z} \phi(z)dz. \tag{1.31}$$

Remark 1.9. We will use symbol "$\sim$" to denote that a random variable is "distributed according to" a certain distribution. For example, $X \sim F(x), X \sim f(x)$, and $X \sim p(x)$ mean, respectively, that random variable X has CDF $F(x)$, PDF $f(x)$, and probability mass function (PMF) $p(x)$. Statements $Z \sim \phi(z)$, $Z \sim \Phi(z)$, and $Z \sim \mathcal{N}(0,1)$ are all equivalent and mean that Z is the standard normal random variable.

[47] V.I. Arnold's principle states: "If a notion bears a personal name, then this name is not the name of the discoverer" [14]. Arnold's principle is applicable to itself, since it was stated by S.M. Stigler before Arnold and by R.K. Merton before Stigler. The Gaussian curve was first discovered in 1733 by Abraham de Moivre, 44 years before Carl Gauss was born.

Abraham de Moivre (1667 – 1754), a friend of Isaac Newton and a discoverer of the Gaussian curve.

At first glance it may not be clear why the normal distribution is called "normal." Indeed, intuitively, a uniform distribution seems to be "more normal" than strangely-looking distribution (1.26). It is due to the *central limit theorem* (CLT), the standard normal distribution $\mathcal{N}(0,1)$ is the single most important distribution in probability and statistics. Informally and very roughly[48], the CLT states that the properly shifted and rescaled sum $\sum_{i=1}^{n} X_i$ of random variables $X_1, \ldots, X_n$, that satisfy certain assumptions, is approximately standard normal:

[48] The CLT is reviewed more formally in Probability Refresher 2.12.

$$\frac{\sum_{i=1}^{n} X_i - a}{b} \dot{\sim} \mathcal{N}(0,1). \tag{1.32}$$

This result together with an empirical observation that many phenomena in Nature can be accurately modeled by sums of random variables explain why remarkably many histograms are approximately bell-shaped and follow the normal curve.

Remark 1.10. We will use symbol "$\dot{\sim}$" to denote that a random variable is "distributed *approximately* according to" a certain probability distribution. For example, $X \dot{\sim} \mathcal{N}(0,1)$ means that a random variable X is approximately normally distributed. Often $\dot{\sim}$ becomes exact $\sim$ in a certain limit (for instance, as the sample size $n \to \infty$).

It is straightforward to generalize the standard normal distribution $\mathcal{N}(0,1)$ to allow arbitrary mean and variance. A random variable X is called *normal* with mean μ and variance σ^2, denoted $X \sim \mathcal{N}(\mu, \sigma^2)$, if its PDF is

$$\phi(x|\mu,\sigma^2) = \frac{1}{\sqrt{2\pi}\sigma} \exp\left(-\frac{(x-\mu)^2}{2\sigma^2}\right). \tag{1.33}$$

Here are some useful properties of normally distributed random variables:

1. If $X \sim \mathcal{N}(\mu,\sigma^2)$, then $Z = \frac{X-\mu}{\sigma} \sim \mathcal{N}(0,1)$. This operation of shifting and rescaling of a random variable to make it zero mean and unit variance is often used in practice and called *standardization*.
2. If $X \sim \mathcal{N}(\mu,\sigma^2)$, then the probability that X belongs to an interval (a,b) can be computed as follows:

$$\begin{aligned}\mathbb{P}(a < X < b) &= \mathbb{P}\left(\frac{a-\mu}{\sigma} < \frac{X-\mu}{\sigma} < \frac{b-\mu}{\sigma}\right) \\ &= \mathbb{P}\left(\frac{a-\mu}{\sigma} < Z < \frac{b-\mu}{\sigma}\right) = \Phi\left(\frac{b-\mu}{\sigma}\right) - \Phi\left(\frac{a-\mu}{\sigma}\right).\end{aligned} \tag{1.34}$$

   Here we used standardization: $Z = \frac{X-\mu}{\sigma} \sim \mathcal{N}(0,1)$.
3. If $X_1 \sim \mathcal{N}(\mu_1,\sigma_1^2)$, $X_2 \sim \mathcal{N}(\mu_2,\sigma_2^2)$, and X_1 and X_2 are *independent*, then their linear combination $X = \alpha_1 X_1 + \alpha_2 X_2$, where α_1 and α_2 are constants, is also normally distributed with mean $\alpha_1\mu_1 + \alpha_2\mu_2$ and variance $\alpha_1^2\sigma_1^2 + \alpha_2^2\sigma_2^2$,

$$\alpha_1 X_1 + \alpha_2 X_2 \sim \mathcal{N}(\alpha_1\mu_1 + \alpha_2\mu_2, \alpha_1^2\sigma_1^2 + \alpha_2^2\sigma_2^2). \tag{1.35}$$

4. The normal distribution $\mathcal{N}(\mu, \sigma^2)$ follows the so-called *68-95-99.7 rule*: intervals $(\mu - \sigma, \mu + \sigma)$, $(\mu - 2\sigma, \mu + 2\sigma)$, and $(\mu - 3\sigma, \mu + 3\sigma)$ contain approximately 68%, 95%, and 99.7% of the probability mass. In other words, if $X \sim \mathcal{N}(\mu, \sigma^2)$, then

$$\begin{aligned}
\mathbb{P}(\mu - \sigma < X < \mu + \sigma) &= \int_{\mu-\sigma}^{\mu+\sigma} \phi(x|\mu, \sigma^2) dx \approx 0.68, \\
\mathbb{P}(\mu - 2\sigma < X < \mu + 2\sigma) &= \int_{\mu-2\sigma}^{\mu+2\sigma} \phi(x|\mu, \sigma^2) dx \approx 0.95, \\
\mathbb{P}(\mu - 3\sigma < X < \mu + 3\sigma) &= \int_{\mu-3\sigma}^{\mu+3\sigma} \phi(x|\mu, \sigma^2) dx \approx 0.997.
\end{aligned} \tag{1.36}$$

The 68-95-99.7 rule is illustrated in Fig. 1.15. Although, theoretically speaking, the support of the normal distribution is the entire real line $\mathbb{R}$ and for any interval $(a, b) \subset \mathbb{R}$ the theoretical probability $\mathbb{P}(X \in (a, b)) \neq 0$, practically, most of the probability is supported by the interval $(\mu - 3\sigma, \mu + 3\sigma)$. This means that most (99.7%) of the realizations of X will belong to that interval, and observing $X \notin (\mu - 3\sigma, \mu + 3\sigma)$ is a relatively rare event.

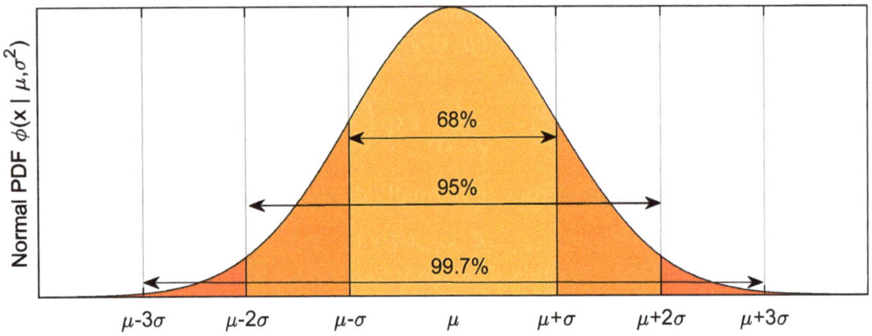

Fig. 1.15 The 68-95-99.7 rule. The tails of the normal distribution decay very fast.

Problems

1.1. Data Summaries and Linear Transformations

In this problem, you will explore how numerical data summaries change under a linear transformation of the data. Consider a sample $x_1, \ldots, x_n$. Let $\bar{x}, \tilde{x}, s_x$, and IQR_x denote, respectively, the sample mean, median, standard deviation, and the interquartile range of the sample. Let

$$y_i = \alpha + \beta x_i, \tag{1.37}$$

where α and β are constants. Express $\bar{y}, \tilde{y}, s_y$, and IQR_y in terms of $\bar{x}, \tilde{x}, s_x$, and IQR_x.

1.2. Optimization Interpretation of $\bar{x}$ and $\tilde{x}$

The sample mean $\bar{x}$ and median $\tilde{x}$ have interesting optimization interpretations. Show that

$$\bar{x} = \arg\min_{\alpha \in \mathbb{R}} \sum_{i=1}^{n} (x_i - \alpha)^2, \tag{1.38}$$

$$\tilde{x} = \arg\min_{\alpha \in \mathbb{R}} \sum_{i=1}^{n} |x_i - \alpha|. \tag{1.39}$$

Remark 1.11. Notation $\alpha^* = \arg\min_{\alpha \in A} f(\alpha)$ means that α^* is a value of α that minimizes the function $f(\alpha)$. That is, $f(\alpha^*) \leq f(\alpha)$ for any $a \in A$. Note that α^* is not necessarily unique.

1.3. Interpretation of Q-Q plots

Let $x_1, \ldots, x_n$ be a sample. We know that if the normal-quantile plot, i.e. a collection of points $\left(z_{\frac{k}{n}}, x_{(k)}\right)$, falls roughly on the line $y = x$, then the sample has approximately the standard normal distribution. What can you say about the distribution of the sample if points $\left(z_{\frac{k}{n}}, x_{(k)}\right)$, instead of $y = x$, fall on the line $y = ax + b$?

1.4. Readability of Q-Q plots

In this problem, you will investigate the "quality" of normal-quantile plots. Even if a sample $x_1, \ldots, x_n$ is drawn exactly from the standard normal distribution $\mathcal{N}(0,1)$, its normal-quantile plot will not fall *exactly* on the line $y = x$: there will be some deviations due to the randomness. If the sample size n is small, then these deviations can be substantial. On the other hand, the larger n is, the more stable and aligned with the straight line the normal-quantile plots are. It is therefore important to get some experience in how much deviation from the line $y = x$ for the normal-quantile plot of a $\mathcal{N}(0,1)$-distributed sample is acceptable. Let's investigate this issue.

(a) Draw a random sample of size $n = 20$ from the standard normal distribution $\mathcal{N}(0,1)$ and plot both the normal-quantile plot and the histogram. Do the points on the Q-Q plot appear to fall on the line $y = x$? Is the histogram symmetric, unimodal, and bell-shaped? Do this several times and summarize you observations.

(b) Repeat part (a) for samples of sizes $n = 50$, $n = 100$, and $n = 1000$.

(c) Based on the experiments with normal samples in parts (a) and (b), what is your personal estimate for the "critical" sample size n^*, such that for $\mathcal{N}(0,1)$-distributed samples of size $n \geq n^*$, the normal-quantile plots are stable enough to be easily interpreted (i.e. do not deviate substantially from $y = x$)?

1.5. CO_2 Emissions Per Person

In the Case Study in Section 1.6, we discussed and summarized the data on the total carbon dioxide emissions from the burning of fossil fuels by different countries. In this problem, you will work with CO_2 emissions per person. First, download the data from [15]. For each of 192 countries, the data shows a number of tonnes of CO_2 emitted per person in a given year during 1800-2014.

(a) Figure 1.9 (a) shows the carbon dioxide emissions for the US, China, Russia, Germany, the UK, Japan, and India — top seven countries with the largest cumulative emissions. Create a similar figure that shows the time evolution of per-person CO_2 emissions for these countries. In 2014, in terms of total emissions, the top three polluting countries were China, the US, and India. What are their ranks in terms of per-person emissions and which three countries did emit the largest amounts of CO_2 per person in 2014?

(b) Similar to the total CO_2 emissions, per-person emissions vary substantially from country to country and cover several orders of magnitude. While the largest emission is 45.4 tonnes of CO_2 per person, the smallest is only 0.044 tonnes of CO_2 per person. It is thus more convenient to work with the logarithms of per-person emissions, which we will refer to as the log-per-person emissions. Compute numerical summaries (mean, range, standard deviation, IQR, and the five-number summary) of the log-per-person emissions in 2014 and summarize them in a table similar to Table 1.1.

(c) Create a figure similar to Fig. 1.10 that summarizes the log-per-person emissions of CO_2 in 2014 graphically via a histogram, empirical CDF, normal-quantile plot, and boxplot. Together with the boxplot, show the individual contributions of the United States, China, Russia, Germany, the United Kingdom, Japan, and India. As we saw in Section 1.6, the log-emissions in 2014 are approximately normally distributed (see Fig. 1.10 and the discussion after that figure). Does this observation hold for the log-per-person emissions in 2014?

References

1. Grammarist: Data is or data are? The singular vs plural debate. https://grammarist.com/usage/data/.
2. Freedman, D., Diaconis, P.: On the histogram as a density estimator: L^2 theory. *Zeitschrift für Wahrscheinlichkeitstheorie und Verwandte Gebiete* 57(4), 453–476 (1981). https://doi.org/10.1007/BF01025868
3. Scott, D.W.: On optimal and data-based histograms. *Biometrika* 66(3), 605–610 (1979). https://doi.org/10.2307/2335182
4. Sturges, H.A.: The choice of a class interval. *Journal of the American Statistical Association* 21(153), 65–66 (1926). https://doi.org/10.1080/01621459.1926.10502161
5. Hyndman, R.J.: The problem with Sturges' rule for constructing histograms. Unpublished note (1995). https://robjhyndman.com/papers/sturges.pdf
6. von Hippel, P.T.: Mean, median, and skew: correcting a textbook rule. *Journal of Statistics Education* 13(2), (2005). https://doi.org/10.1080/10691898.2005.11910556
7. McGill, R., Tukey, J.W., Larsen, W.A.: Variations of box plots. *The American Statistician* 32(1), 12–16 (1978). https://doi.org/10.2307/2683468

8. Nair, J., Wierman, A., Zwart, B.: *The Fundamentals of Heavy Tails: Properties, Emergence, and Estimation.* Cambridge University Press (2022).
https://adamwierman.com/book/
9. Schmidt, G.A., Ruedy, R., Miller, R.L., Lacis, A.A.: The attribution of the present-day total greenhouse effect. *Journal of Geophysical Research* 115(D20), D20106 (2010).
https://doi.org/10.1029/2010JD014287
10. NASA GISS Science Briefs: Greenhouse gases: refining the role of carbon dioxide, (1998).
https://www.giss.nasa.gov/research/briefs/ma_01/
11. MacKay, D.J.C.: *Sustainable Energy – Without the Hot Air.* UIT Cambridge (2008).
www.withouthotair.com
12. Gapminder. Yearly CO_2 emissions (1000 tones).
https://www.gapminder.org/data/
Source link: https://cdiac.ess-dive.lbl.gov/
The data set is also available at https://link.springer.com/chapter/10.1007/978-3-032-03848-7_1
13. Tukey, J.W.: The future of data analysis. *The Annals of Mathematical Statistics* 33(1), 1–67 (1967).
https://doi.org/10.1214/aoms/1177704711
14. Arnold, V.I.: On teaching mathematics. *Russian Mathematical Surveys* 53(1), 229–236 (1998).
https://dsweb.siam.org/The-Magazine/All-Issues/vi-arnold-on-teaching-mathematics
15. Gapminder. CO_2 emissions (tones per person).
https://www.gapminder.org/data/
Source link: https://cdiac.ess-dive.lbl.gov/
The data set is also available at https://link.springer.com/chapter/10.1007/978-3-032-03848-7_1

2
Populations, Samples, and Inference

By a small sample we may judge of the whole piece.
– Miguel de Cervantes Saavedra, *Don Quixote*

The main goal of statistical inference is to learn about a phenomenon of interest by extracting useful information from the available data generated by that phenomenon. A particular case, often occurring in applications, is when we want to learn a property of a large *population*, but the data is available only for its small subset, called *sample*. Samples are obtained from populations by performing surveys, interviews, polls, or experiments. They are used for obtaining the desired information about populations. The point of sampling a large population is to reduce the cost and the amount of work it would take to examine the entire population. Although we may not realize it, we perform sampling on a regular basis: when tasting food while cooking[1], when making a blood test[2], when watching a movie trailer, etc. Other examples include political polls, health surveys, and audits of financial records, to name but a few. In this chapter, we will discuss how to make inferences *from a sample to a population*. We will also introduce several fundamental notions and methods of statistical inference that will be used in subsequent chapters.

[1] We don't need to eat an entire pot of soup to check whether it is salty enough.

[2] A medical technician does not need to drain us of blood to measure the red blood cell count.

2.1 A Bit of History

The first documented attempt to learn about a population using only limited information about its subset was made by the English haberdasher John Graunt in 1662. In his seminal book [1], he estimated the population of London by analyzing the statistics of deaths. Because of this contribution, John Graunt was elected to the Royal Society and is considered as one of the founders of modern demography. Another interesting example of early attempts to make inferences about a population from sampled data goes back to 1812. Pierre-Simon Laplace analyzed the accuracy of the estimated size of the French population [2]. An interesting historic review of how sampling became an accepted scientific method is described in [3].

John Graunt (1620-1674), an English haberdasher and a founder of demography.

Supplementary Information The online version contains supplementary material available at https://doi.org/10.1007/978-3-032-03848-7_2.

K. M. Zuev, *Fundamentals of Statistical Inference*,
International Series in Operations Research & Management Science 374,
https://doi.org/10.1007/978-3-032-03848-7_2

2.2 Terminology

Let's start by introducing some key terminology and notation.

- *Target population*: A population of interest, a group of objects, an ensemble of entities that we want to learn about. It is often called simply "population" for brevity. We will denote the target population by $\mathscr{P}$.
- *Population unit*: A member of the target population. For example, in human populations, population units are people.
- *Population size*: The total number of units in the target population, denoted by N. We assume that N is very large, $N \gg 1$.

Remark 2.1. In some applications, defining the target population and/or finding its size may not be trivial. For example, in a political poll, should the target population be all people eligible to vote or just all registered voters? If we are interested in studying a population of all HIV-infected people living in a given city, how to find the population size? Here, for simplicity, we assume that the target population $\mathscr{P}$ is well defined and its size N is known.

- *Unit characteristic*: Numerical information about population unit[3]. For unit i, we denote the value of its characteristic by x_i, $i = 1, \ldots, N$.
- *Population parameter*: A numerical summary of the unit characteristics across the entire population $\mathscr{P}$. Mathematically, a population parameter is a function of unit characteristics, $\theta = \theta(x_1, \ldots, x_N)$. A population parameter is the main quantity of interest, the property of the target population that we want to learn. There are many different population parameters one could be interested in, but here are four examples that are often used:
  1. Population mean, which measures the average unit characteristic:
  $$\mu = \frac{1}{N}\sum_{i=1}^{N} x_i. \tag{2.1}$$
  2. Population total:
  $$\tau = \sum_{i=1}^{N} x_i = N\mu. \tag{2.2}$$
  3. Population variance, which quantifies the variability of unit characteristics:
  $$\sigma^2 = \frac{1}{N}\sum_{i=1}^{N} (x_i - \mu)^2. \tag{2.3}$$
  4. Population standard deviation:
  $$\sigma = \sqrt{\frac{1}{N}\sum_{i=1}^{N} (x_i - \mu)^2}. \tag{2.4}$$

[3] For example, weight, height, age, sex, income, blood test result, etc.

In an "ideal survey,"[4] we would take the entire target population $\mathscr{P}$, measure the characteristics $x_1, \ldots, x_N$ for all units, and simply compute the parameter of interest $\theta(x_1, \ldots, x_N)$. This ideal, as almost all ideals, is rarely achieved in applications: either the population is too large, or measuring x_i is too expensive, or both. In practice, we select a *subset* of the target population, called sample, and *estimate* the population parameter using this subset. This is schematically shown in Fig. 2.1.

[4] Also called "census."

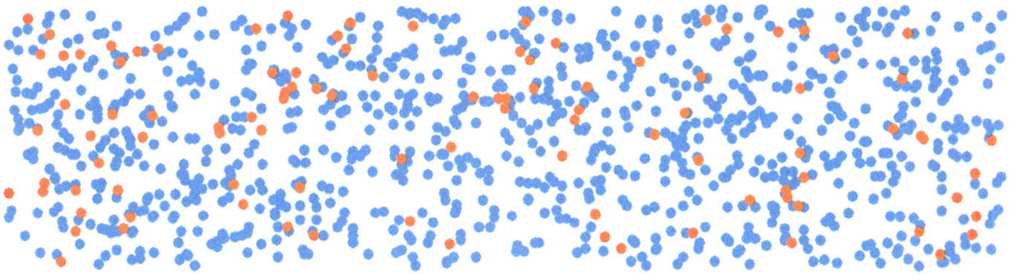

Fig. 2.1 The target population $\mathscr{P}$ (all points) and the selected sample $\mathscr{S} \subset \mathscr{P}$ (orange points).

- *Sample*: A subset $\mathscr{S}$ of the target population, $\mathscr{S} \subset \mathscr{P}$.
- *Sample unit*: A population unit selected for the sample, an element of $\mathscr{S}$.
- *Sample size*: The total number of units in the sample, denoted by n. Sample size is often much less than the population size, $n \ll N$. The fraction n/N is called the *sampling fraction*. The sampling fraction is typically very small, $n/N \ll 1$.

Let $\mathscr{P} = \{1, \ldots, N\}$ be the target population and $\mathscr{S} = \{s_1, \ldots, s_n\} \subset \mathscr{P}$ be a sample from $\mathscr{P}$[5]. Since we are not interested in the population units per se, but rather in their characteristics, we identify $\mathscr{P}$ and $\mathscr{S}$ with the corresponding populations of characteristic values, that is

[5] Each $s_i \in \{1, \ldots, N\}$ and all s_i are different, that is $s_i \neq s_j$ for $i \neq j$.

$$\mathscr{P} = \{x_1, \ldots, x_N\} \quad \text{and} \quad \mathscr{S} = \{x_{s_1}, \ldots, x_{s_n}\}. \tag{2.5}$$

To avoid heavy notation, we denote the ith sample unit x_{s_i} simply by X_i, and, thus,

$$\mathscr{S} = \{X_1, \ldots, X_n\} \subset \{x_1, \ldots, x_N\} = \mathscr{P}. \tag{2.6}$$

- *Sample statistic*: A numerical summary of the unit characteristics across sample $\mathscr{S}$. Mathematically, a sample statistic is any function of sampled characteristics, that is, any function of the data, $\hat{\theta} = \hat{\theta}(X_1, \ldots, X_n)$. The sample statistic $\hat{\theta}$ is constructed with the aim to estimate the corresponding population parameter θ. For example, a reasonable sample statistic for the population mean μ in (2.1) is the *sample mean*:
$$\overline{X} = \frac{1}{n} \sum_{i=1}^{n} X_i. \tag{2.7}$$
- *Selection rule*: A method for choosing a sample $\mathscr{S}$ from a target population $\mathscr{P}$.

Selection rules used in practice are often *probabilistic*, meaning that $X_1,\dots,X_n$ are selected at *random* according to some probability algorithm. Formally, a random sample $\mathscr{S}$ of size n obtained by a probabilistic selection rule from a population $\mathscr{P}$ can be written as follows:

$$\mathscr{S} \sim \mathbb{P}_{\mathscr{P}_n}, \tag{2.8}$$

where $\mathscr{P}_n$ is the set of all subsets of $\mathscr{P}$ of size n (all candidates for being selected as a sample) and $\mathbb{P}_{\mathscr{P}_n}$ is a probability distribution on $\mathscr{P}_n$ that describes the chances of subsets for being selected.

Why are probabilistic rules important?[6] As we will see, probabilistic selection rules allow not only to estimate a population parameter $\theta \approx \hat{\theta}$, but also to quantify the *accuracy of this estimation*. There are several popular methods for selecting a sample, such as stratified random sampling, cluster sampling, and systematic sampling. Here we consider the simplest probabilistic rule: *simple random sampling*.

[6] "You see, Ioganich..." Bykov began earnestly, making his expression as profound as possible, without the slightest idea of how he would finish the sentence. – Arkady and Boris Strugatsky, *The Land of Crimson Clouds*.

2.3 Simple Random Sampling

In simple random sampling (SRS), every subset of $\mathscr{P}$ consisting of n population units, i.e. every set in $\mathscr{P}_n$, has the same probability of being selected as a sample. In other words, $\mathbb{P}_{\mathscr{P}_n}$ is a *uniform* distribution on $\mathscr{P}_n$. Since there are

$$\binom{N}{n} = \frac{N!}{n!(N-n)!} \tag{2.9}$$

subsets in $\mathscr{P}_n$, the probability of being selected is $1/\binom{N}{n} = \frac{n!(N-n)!}{N!}$ for each subset. Intuitively, we first "mix up" the target population and then grab n units. Algorithmically, to draw a simple random sample $\mathscr{S}$ of size n from $\mathscr{P} = \{x_1,\dots,x_N\}$, we

1. Select X_1 from $\mathscr{P}$ uniformly at random.
2. Select X_2 from $\mathscr{P} \setminus \{X_1\}$ uniformly at random.
3. Select X_3 from $\mathscr{P} \setminus \{X_1,X_2\}$ uniformly at random.
4. Proceed like this until n units $\mathscr{S} = \{X_1,\dots,X_n\}$ are sampled.

In short, we draw n units one at a time *without replacement*[7].

[7] SRS with replacement is discussed in [4].

2.4 Estimating the Population Mean

So, let $\mathscr{S} = \{X_1,\dots,X_n\}$ be a SRS sample drawn from the population $\mathscr{P}$, and suppose that we want to estimate the population mean μ in (2.1). Let's consider the sample mean $\overline{X}$ in (2.7) as an estimate of μ.

Our first goal is to investigate the accuracy of estimation $\mu \approx \overline{X}$. Let's quickly look at their expressions again:

$$\mu = \frac{1}{N}\sum_{i=1}^{N} x_i \quad \text{and} \quad \overline{X} = \frac{1}{n}\sum_{i=1}^{n} X_i. \tag{2.10}$$

The expression for the sample mean $\overline{X}$ clearly mimics the expression for the population mean μ. But the key difference between μ and $\overline{X}$ is that while μ is a *deterministic* quantity, $\overline{X}$ is *random*, since $X_1,\dots,X_n$ are obtained through a random process. The accuracy of estimation $\mu \approx \overline{X}$ depends, therefore, on the properties of the random variable $\overline{X}$, whose properties are determined by the properties of sample units $X_1,\dots,X_n$. So, let's first study the random variables $X_1,\dots,X_n$.

The target population $\mathscr{P} = \{x_1,\dots,x_N\}$ may have repeated values, that is, some of x_i may have the same value. Let $y_1,\dots,y_M$, where $M \leq N$, denote all distinct values contained in $\mathscr{P}$. For example, if $\mathscr{P} = \{x_1 = 1, x_2 = 1, x_3 = 2, x_4 = 3, x_5 = 3\}$, then there are $M = 3$ distinct values: $y_1 = 1$, $y_2 = 2$, $y_3 = 3$. Let N_j be the number of population units with value y_j, $j = 1,\dots,M$. In the previous example, $N_1 = 2$, $N_2 = 1$, $N_3 = 2$. By definition, $\sum_{j=1}^{M} N_j = N$. The probability distribution of the random variable X_i is fully described by the following theorem.

Theorem 2.1. *Element X_i is a discrete random variable with values in $\{y_1,\dots,y_M\}$ and probability mass function*

$$\mathbb{P}(X_i = y_j) = \frac{N_j}{N}, \quad j = 1,\dots,M, \tag{2.11}$$

Its expectation and variance are

$$\mathbb{E}[X_i] = \mu \quad \textit{and} \quad \mathbb{V}[X_i] = \sigma^2, \tag{2.12}$$

where μ is the population mean and σ^2 is the population variance.

Proof. Since the sample $\mathscr{S} = \{X_1,\dots,X_n\}$ is obtained by choosing a set of n population units uniformly at random, each population unit x_k has the same probability of being selected as the ith sample unit X_i. That is, $\mathbb{P}(X_i = x_k) = \frac{1}{N}$ for any $k = 1,\dots,N$[8]. This implies (2.11):

[8] Problem 2.1 asks you to show this using the algorithmic definition of SRS.

$$\mathbb{P}(X_i = y_j) = \sum_{k:x_k=y_j} \mathbb{P}(X_i = x_k) = \sum_{k:x_k=y_j} \frac{1}{N} = \frac{N_j}{N}. \tag{2.13}$$

It is also instructive to see how (2.11) can be proved using the algorithmic definition of SRS. Since X_1 is drawn from $\mathscr{P}$ uniformly at random, for X_1 property (2.11) is obvious. The second sample unit X_2 is drawn from $\mathscr{P} \setminus \{X_1\}$ also uniformly at random. Using the *law of total probability*,

$$\begin{aligned}\mathbb{P}(X_2 = y_j) &= \mathbb{P}(X_2 = y_j \,|\, X_1 = y_j)\mathbb{P}(X_1 = y_j) + \mathbb{P}(X_2 = y_j \,|\, X_1 \neq y_j)\mathbb{P}(X_1 \neq y_j) \\ &= \frac{N_j - 1}{N-1}\frac{N_j}{N} + \frac{N_j}{N-1}\frac{N - N_j}{N} = \frac{N_j^2 - N_j + NN_j - N_j^2}{N(N-1)} = \frac{N_j}{N}.\end{aligned} \tag{2.14}$$

By induction on i it can be shown[9] that (2.11) holds for all $X_1,\dots,X_n$.

[9] It is a useful exercise!

Properties (2.12) are of course consequences of (2.11). For the expectation, we have:

$$\mathbb{E}[X_i] = \sum_{j=1}^{M} y_j \mathbb{P}(X_i = y_j) = \frac{\sum_{j=1}^{M} y_j N_j}{N} = \frac{\sum_{i=1}^{N} x_i}{N} = \mu. \tag{2.15}$$

Here we used the fact that

$$\sum_{j=1}^{M} y_j N_j = \overbrace{\underbrace{y_1 + \ldots + y_1}_{N_1 \text{ terms}} + \ldots + \underbrace{y_M + \ldots + y_M}_{N_M \text{ terms}}}^{N \text{ terms}} = \sum_{i=1}^{N} x_i. \tag{2.16}$$

The two sums, $\sum_{j=1}^{M} y_j N_j$ and $\sum_{i=1}^{N} x_i$, are simply two different ways to represent the population total (2.2), in terms of $y_1,\dots,y_M$ and in terms of $x_1,\dots,x_M$, respectively.

Finally, the variance

$$\begin{aligned}\mathbb{V}[X_i] &= \mathbb{E}\left[(X_i - \mathbb{E}[X_i])^2\right] = \mathbb{E}\left[(X_i-\mu)^2\right] = \sum_{j=1}^{M}(y_j-\mu)^2\mathbb{P}(X_i = y_j)\\ &= \frac{\sum_{j=1}^{M}(y_j-\mu)^2 N_j}{N} = \frac{\sum_{i=1}^{N}(x_i-\mu)^2}{N} = \sigma^2.\end{aligned} \tag{2.17}$$

Here, similar to (2.16), we used the fact that $\sum_{j=1}^{M}(y_j-\mu)^2 N_j = \sum_{i=1}^{N}(x_i-\mu)^2$. This completes the proof. □

As an immediate corollary of Theorem 2.1, we obtain the following result:

Corollary 2.1. *If* $\mathscr{S} = \{X_1,\dots,X_n\}$ *is a simple random sample, then*

$$\mathbb{E}[\overline{X}] = \mu. \tag{2.18}$$

Proof. Indeed, using (2.12),

$$\mathbb{E}[\overline{X}] = \mathbb{E}\left[\frac{1}{n}\sum_{i=1}^{n} X_i\right] = \frac{1}{n}\sum_{i=1}^{n}\mathbb{E}[X_i] = \frac{1}{n}\sum_{i=1}^{n}\mu = \mu. \tag{2.19}$$

□

Intuitively, this result tells us that, although $\overline{X}$ is random and its value changes from sample to sample, $\overline{X} = \mu$ on average[10]. In more detail, "on average" means that if we generate K simple random samples $\mathscr{S}_1,\dots,\mathscr{S}_K$ and compute the corresponding sample means $\overline{X}_1,\dots,\overline{X}_K$, then their average $\frac{1}{K}\sum_{k=1}^{K}\overline{X}_k \approx \mu$ and this approximation will become the exact equality in the limit $K \to \infty$.

[10] This is good news and justifies the description of $\overline{X}$ as a "reasonable sample statistic" for μ that we gave above.

The property of an estimate being equal to the estimated quantity on average is fundamental and deserves a special name and a proper definition.

Definition 2.1. Let θ be a population parameter and $\hat{\theta} = \hat{\theta}(X_1,\dots,X_n)$ be a random sample statistic that estimates θ. We say that $\hat{\theta}$ is *unbiased* if

$$\mathbb{E}[\hat{\theta}] = \theta. \tag{2.20}$$

Thus, $\overline{X}$ is an *unbiased* estimate of μ. Unbiasedness, although it is a desirable property, does not make an estimate $\hat{\theta}$ automatically a good estimate. Even if on average $\hat{\theta}$ equals to θ, its variability can be so large that its particular realizations can be far from θ, yielding inaccurate estimation of θ.

So, our next step is to investigate *how variable* $\overline{X}$ is. As a measure of variability of the sample mean, let's use the *standard error*[11] of $\overline{X}$,

$$\mathrm{se}[\overline{X}] = \sqrt{\mathbb{V}[\overline{X}]}. \tag{2.21}$$

The variance of the sample mean is

$$\begin{aligned}\mathbb{V}[\overline{X}] &= \mathbb{V}\left[\frac{1}{n}\sum_{i=1}^{n} X_i\right] = \frac{1}{n^2}\mathbb{V}\left[\sum_{i=1}^{n} X_i\right] = \frac{1}{n^2}\sum_{i=1}^{n}\sum_{j=1}^{n}\mathrm{Cov}(X_i,X_j) \\ &= \frac{1}{n^2}\sum_{i=1}^{n}\underbrace{\mathrm{Cov}(X_i,X_i)}_{\mathbb{V}[X_i]} + \frac{1}{n^2}\sum_{i\neq j}\mathrm{Cov}(X_i,X_j) = \frac{\sigma^2}{n} + \frac{2}{n^2}\sum_{i<j}\mathrm{Cov}(X_i,X_j).\end{aligned} \tag{2.22}$$

Remark 2.2. If sampling were done *with replacement* then $X_1,\dots,X_n$ would be *independent* random variables, and we would have:

$$\mathbb{V}[\overline{X}] = \frac{1}{n^2}\mathbb{V}\left[\sum_{i=1}^{n} X_i\right] = \frac{1}{n^2}\sum_{i=1}^{n}\mathbb{V}[X_i] = \frac{1}{n^2}\sum_{i=1}^{n}\sigma^2 = \frac{\sigma^2}{n}. \tag{2.23}$$

In SRS, however, sampling is done without replacement and this introduces dependence between the sample units. So, although $X_1,\dots,X_n$ are *identically distributed* (Theorem 2.1), they are *not independent*[12]. In particular, $\mathbb{V}\left[\sum_{i=1}^{n} X_i\right] \neq \sum_{i=1}^{n}\mathbb{V}[X_i]$.

To continue our computation in (2.22), we need to find the covariance between sample units X_i and X_j. But before finding the exact value of $\mathrm{Cov}(X_i,X_j)$, let's think: do we expect this covariance to be positive or negative? An intuitive interpretation of the covariance between random variables X and Y is the following: if $\mathrm{Cov}(X,Y) > 0$, then Y tends to increase as X increases; if $\mathrm{Cov}(X,Y) < 0$, then Y tends to decrease as X increases. Suppose X_i is large compared to other values in the target population. Then, since sampling is done without replacement, the large value of X_i is no longer available for X_j, and, therefore, X_j tends to be smaller than it would be if the value of X_i were available. Similarly, if X_i is small, then X_j tends to be large. Thus, we intuitively expect that $\mathrm{Cov}(X_i,X_j) < 0$[13]. The following theorem confirms our expectation.

Theorem 2.2. *If $i \neq j$, then the covariance between X_i and X_j is*

$$\mathrm{Cov}(X_i,X_j) = -\frac{\sigma^2}{N-1}. \tag{2.24}$$

[11] The standard error of an estimate is just another name for its standard deviation. "Standard deviation" is used more often in probability, where random variables are considered on their own, outside of estimation context; "standard error" is used more often in statistics, where random variables are used for estimating parameters.

[12] A collection $X_1,\dots,X_n$ obtained by simple random sampling is an example of *exchangeable* random variables, see Remark 10.15.

[13] Another somewhat indirect reason why we expect the covariance to be negative is because we expect SRS to be more efficient than sampling with replacement: sampling already sampled units would not provide new information about the population. That is, we expect the variance in (2.22) to be smaller than the variance in (2.23).

Proof. The covariance is

$$\begin{aligned}\text{Cov}(X_i,X_j) &= \mathbb{E}[X_iX_j]-\mathbb{E}[X_i]\mathbb{E}[X_j] = \sum_{s=1}^{M}\sum_{t=1}^{M} y_s y_t \mathbb{P}(X_i=y_s, X_j=y_t)-\mu^2 \\ &= \sum_{s=1}^{M}\sum_{t=1}^{M} y_s y_t \mathbb{P}(X_i=y_s)\mathbb{P}(X_j=y_t\,|\,X_i=y_s)-\mu^2.\end{aligned} \tag{2.25}$$

Both probabilities participating in the double sum are straightforward to find:

$$\begin{aligned}\mathbb{P}(X_i=y_s) &= \frac{N_s}{N}, \\ \mathbb{P}(X_j=y_t\,|\,X_i=y_s) &= \begin{cases}\frac{N_t-1}{N-1}, & \text{if } s=t,\\ \frac{N_t}{N-1}, & \text{if } s\neq t.\end{cases}\end{aligned} \tag{2.26}$$

Now we can substitute (2.26) into (2.25) and find the covariance. But this way is computationally quite tedious[14]. Here is a more elegant approach. An important observation, that follows from (2.26) and (2.25), is that $\text{Cov}(X_i,X_j)$ does not depend on i and j, that is $\text{Cov}(X_i,X_j)=\text{Cov}(X_1,X_2)$, for all $i\neq j$. Therefore,

[14] See [5], pages 207-208, for this computation.

$$\begin{aligned}\mathbb{V}[\overline{X}] &= \frac{\sigma^2}{n}+\frac{2}{n^2}\sum_{i<j}\text{Cov}(X_i,X_j) = \frac{\sigma^2}{n}+\frac{2}{n^2}\sum_{i<j}\text{Cov}(X_1,X_2) \\ &= \frac{\sigma^2}{n}+\frac{2}{n^2}\frac{n(n-1)}{2}\text{Cov}(X_1,X_2) = \frac{\sigma^2}{n}+\frac{n-1}{n}\text{Cov}(X_1,X_2).\end{aligned} \tag{2.27}$$

Notice that (2.27) holds *for any* $n=1\dots,N$. In particular, it holds for $n=N$, i.e. for the case where the sample $\mathscr{S}=\mathscr{P}$. But in this case, $\mathbb{V}[\overline{X}]=0$, since $\overline{X}=\mu$, and (2.27) reduces to

$$0=\frac{\sigma^2}{N}+\frac{N-1}{N}\text{Cov}(X_1,X_2). \tag{2.28}$$

From here,

$$\text{Cov}(X_i,X_j)=\text{Cov}(X_1,X_2)=-\frac{\sigma^2}{N-1}, \tag{2.29}$$

as promised in the statement of the theorem. □

Combining (2.27) with (2.29) leads to the following important corollary.

Corollary 2.2. *If* $\mathscr{S}=\{X_1,\dots,X_n\}$ *is a simple random sample, then*

$$\mathbb{V}[\overline{X}]=\frac{\sigma^2}{n}\left(1-\frac{n-1}{N-1}\right) \quad \textit{and} \quad \text{se}[\overline{X}]=\frac{\sigma}{\sqrt{n}}\sqrt{1-\frac{n-1}{N-1}}. \tag{2.30}$$

A few comments and observations are in order:

1. The factor $\left(1-\frac{n-1}{N-1}\right)$ is called the *finite population correction.*

2. The finite population correction is always less than one, since $n < N$. Therefore, the variance $\mathbb{V}[\overline{X}] < \sigma^2/n$. This means that SRS, as expected, is more accurate than sampling with replacement, for which the variance[15] is exactly σ^2/n.
3. If the sampling fraction n/N is small, then the finite population correction is approximately one, and

$$\mathbb{V}[\overline{X}] \approx \frac{\sigma^2}{n} \quad \text{and} \quad \text{se}[\overline{X}] \approx \frac{\sigma}{\sqrt{n}}. \tag{2.31}$$

This is intuitively expected: if the target population is large and the sample size n is very small compared to the population size N, then sampling with and without replacement are essentially identical, since the probability of resampling of already sampled units is negligible.
4. To double the accuracy of estimation $\mu \approx \overline{X}$, i.e. to reduce the standard error $\text{se}[\overline{X}]$ by half, the sample size n must be approximately quadrupled.
5. If the population standard deviation σ is small, i.e. the population units are not very dispersed, then even a relatively small sample $\mathscr{S} = \{X_1, \dots, X_n\}$ will yield a fairly accurate estimate $\mu \approx \overline{X}$. In general, the larger σ is, the larger n must be to achieve the same accuracy.

These observations are nice, but we have a problem: σ^2 is an *unknown* parameter. Therefore, we can't really compute the variance or the standard error of the sample mean using (2.30). So, our next goal is to estimate the population variance using simple random sampling.

[15] See Remark 2.2.

2.5 Estimating the Population Variance

Estimating the population variance σ^2 is important because of at least two reasons: a) it is an important population parameter by itself and b) it appears in the formula (2.30) for the standard error of the sample mean. So, if we want to compute $\text{se}[\overline{X}]$ or to determine the required sample size n to achieve a prescribed value of error, we must know the value of σ^2. In the next few pages, we will learn both how to estimate σ^2 and how to estimate $\text{se}[\overline{X}]$...[16] without estimating σ^2![17]

Recall that the population variance is

$$\sigma^2 = \frac{1}{N}\sum_{i=1}^{N}(x_i - \mu)^2. \tag{2.32}$$

It seems natural to use the following estimate:

$$\hat{\sigma}^2 = \frac{1}{n}\sum_{i=1}^{n}(X_i - \overline{X})^2, \tag{2.33}$$

which is simply a sample analog of (2.32). This estimate, however, is *biased*.

[16] Imagine a drum roll.

[17] "Can we buy all this in London?" Harry wondered aloud. "If yeh know where to go," said Hagrid. – J.K. Rowling, *Harry Potter and the Philosopher's Stone.*

Theorem 2.3. *The expected value of $\hat{\sigma}^2$ is given by*

$$\mathbb{E}[\hat{\sigma}^2] = \sigma^2 \frac{Nn-N}{Nn-n}. \tag{2.34}$$

Proof. To compute the expected value of $\hat{\sigma}^2$, it is convenient to expand the square and rewrite it as follows:

$$\hat{\sigma}^2 = \frac{1}{n}\sum_{i=1}^{n} X_i^2 - \frac{2}{n}\overline{X}\underbrace{\sum_{i=1}^{n} X_i}_{n\overline{X}} + \underbrace{\frac{1}{n}\sum_{i=1}^{n}\overline{X}^2}_{\overline{X}^2} = \frac{1}{n}\sum_{i=1}^{n} X_i^2 - \overline{X}^2. \tag{2.35}$$

Then, using (2.12), (2.18), (2.30), and the fact that for any random variable Y, its second moment $\mathbb{E}[Y^2] = \mathbb{V}[Y] + \mathbb{E}[Y]^2$, we have:

$$\begin{aligned}\mathbb{E}[\hat{\sigma}^2] &= \frac{1}{n}\sum_{i=1}^{n}\mathbb{E}[X_i^2] - \mathbb{E}[\overline{X}^2] = \frac{1}{n}\sum_{i=1}^{n}(\underbrace{\mathbb{V}[X_i]}_{\sigma^2} + \underbrace{\mathbb{E}[X_i]^2}_{\mu^2}) - (\mathbb{V}[\overline{X}] + \mathbb{E}[\overline{X}]^2) \\ &= \sigma^2 + \mu^2 - \frac{\sigma^2}{n}\left(1 - \frac{n-1}{N-1}\right) - \mu^2 = \sigma^2\frac{Nn-n-N+n}{Nn-n} = \sigma^2\frac{Nn-N}{Nn-n},\end{aligned} \tag{2.36}$$

□

Since the sample size $n < N$, the fraction $\frac{Nn-N}{Nn-n} < 1$, and, therefore,

$$\mathbb{E}[\hat{\sigma}^2] < \sigma^2. \tag{2.37}$$

In other words, $\hat{\sigma}^2$ tends to *underestimate* σ^2. On the other hand, Theorem 2.3 helps to construct an unbiased estimate for the population variance. Indeed, to cook up an unbiased estimate from $\hat{\sigma}^2$, all we need to do is to properly normalize it: $\hat{\sigma}^2 \to \alpha\hat{\sigma}^2$, where the constant $\alpha = (\frac{Nn-N}{Nn-n})^{-1}$ makes sure that $\mathbb{E}[\alpha\hat{\sigma}^2] = \sigma^2$.

Corollary 2.3. *If $\mathscr{S} = \{X_1, \ldots, X_n\}$ is a simple random sample, then*

$$\hat{\sigma}_u^2 = \hat{\sigma}^2\frac{Nn-n}{Nn-N} = \left(1 - \frac{1}{N}\right)\frac{1}{n-1}\sum_{i=1}^{n}(X_i - \overline{X})^2. \tag{2.38}$$

is an unbiased estimate of the population variance σ^2, $\mathbb{E}[\hat{\sigma}_u^2] = \sigma^2$.

Remark 2.3. Note that if both the population size N and the sample size n are very large, then the derived unbiased estimate approximately equals to the biased one:

$$n \gg 1,\ N \gg 1 \quad \Longrightarrow \quad 1 - \frac{1}{N} \approx 1,\ \frac{1}{n-1} \approx \frac{1}{n} \quad \Longrightarrow \quad \hat{\sigma}_u^2 \approx \hat{\sigma}^2. \tag{2.39}$$

Substituting (2.38) into (2.30) gives the estimates for the variance $\mathbb{V}[\overline{X}]$ and the standard error $\mathrm{se}[\overline{X}]$ of the sample mean:

$$\widehat{\mathbb{V}}[\overline{X}] = \frac{\hat{\sigma}_u^2}{n}\left(1 - \frac{n-1}{N-1}\right) \quad \text{and} \quad \widehat{\mathrm{se}}[\overline{X}] = \frac{\hat{\sigma}_u}{\sqrt{n}}\sqrt{1 - \frac{n-1}{N-1}}. \tag{2.40}$$

Remark 2.4. Note that while the estimate of the variance is unbiased, $\mathbb{E}[\widehat{\mathbb{V}}[\overline{X}]] = \mathbb{V}[\overline{X}]$, the estimate of the standard error is not, $\mathbb{E}[\widehat{\mathrm{se}}[\overline{X}]] \neq \mathrm{se}[\overline{X}]$, since $\hat{\sigma}_u$ (despite the subscript "$_u$") is a *biased* estimate of the population standard deviation, $\mathbb{E}[\hat{\sigma}_u] \neq \sigma$. This is because the square root of the expectation does not equal to the expectation of the square root, $\mathbb{E}[\hat{\sigma}_u] = \mathbb{E}[\sqrt{\hat{\sigma}_u^2}] \neq \sqrt{\mathbb{E}[\hat{\sigma}_u^2]} = \sigma$. Since this is often forgotten, it is important to reiterate: in general, and in this particular case, the operations of taking the expectation and computing the square root *do not commute*, $\mathbb{E}[\sqrt{Y}] \neq \sqrt{\mathbb{E}[Y]}$.

So, we know how to unbiasedly estimate the population mean $\mu \approx \overline{X}$ and how to estimate the standard error of this estimate, $\mathrm{se}[\overline{X}] \approx \widehat{\mathrm{se}}[\overline{X}]$, by unbiasedly estimating the population variance, $\sigma^2 \approx \hat{\sigma}_u^2$. But here is a strange at first glance question: can we estimate $\mathrm{se}[\overline{X}]$ *without* estimating σ^2?

Remark 2.5. Let's start with the following natural question: *why should we care?*[18] Indeed, we already know how to estimate σ^2 and, therefore, we can estimate $\mathrm{se}[\overline{X}]$! Why should we limit ourselves and try to estimate $\mathrm{se}[\overline{X}]$ without estimating σ^2? The problem is that for many parameters θ it is often relatively easy to construct an estimate $\hat{\theta}$, but difficult to derive an analytical expression (like (2.30)) for $\mathrm{se}[\hat{\theta}]$. For example, consider the unbiased estimate $\hat{\sigma}_u^2$ of σ^2. It is not clear how to find its standard error $\mathrm{se}[\hat{\sigma}_u^2]$. For such cases, where expressions for $\mathrm{se}[\hat{\theta}]$ are not available, we need an alternative, more *universal method*, for estimating standard errors.

[18] "I don't care" said Arthur coldly. – Douglas Adams, *The Hitchhiker's Guide to the Galaxy*.

The answer to the question asked before Remark 2.5 is: yes, we can! We can estimate $\mathrm{se}[\overline{X}]$ using the so-called *bootstrap method*.

2.6 The Bootstrap Method for Estimating the Standard Error

The bootstrap is a general *simulation-based* method, introduced by Bradley Efron in [6], for measuring variability of random estimates. It is a very popular method and often used in applications, since it requires no analytical calculations and is easy to implement. In later chapters, we will meet the bootstrap in different contexts. Here is our first encounter with the method: in the context of simple random sampling.

Bradley Efron, an American statistician and the father of the bootstrap method. Photo courtesy of the Department of Statistics at Stanford University.

To better understand the main idea behind the bootstrap, let's forget for the moment about SRS, and consider the following problem. Let Y be a random variable distributed according to some distribution $\mathbb{P}$. Suppose we can sample from $\mathbb{P}$, that is, we can generate *independent and identically distributed* (IID) random variables:

$$Y_1, \ldots, Y_B \overset{\text{iid}}{\sim} \mathbb{P}. \tag{2.41}$$

How to estimate the standard deviation of Y? We can do this using the *law of large numbers*[19] (LLN) as follows. By definition, the standard deviation

[19] The LLN, one of the two main achievements in probability theory, is briefly reviewed in Probability Refresher 2.12.

$$\mathrm{sd}[Y] = \sqrt{\mathbb{V}[Y]} = \sqrt{\mathbb{E}[(Y - \mathbb{E}[Y])^2]}, \tag{2.42}$$

and, by the law of large numbers,

$$\begin{aligned}
&\frac{1}{B}\sum_{i=1}^{B} Y_i \xrightarrow{\mathbb{P}} \mathbb{E}[Y] = \mu_Y, \quad \text{as } B \to \infty, \\
&\frac{1}{B}\sum_{j=1}^{B} (Y_i - \mu_Y)^2 \xrightarrow{\mathbb{P}} \mathbb{E}[(Y - \mu_Y)^2] = \mathbb{V}[Y], \quad \text{as } B \to \infty.
\end{aligned} \tag{2.43}$$

Therefore, if B is large, $B \gg 1$, then

$$\mathrm{sd}[Y] \approx \sqrt{\frac{1}{B}\sum_{j=1}^{B}\left(Y_j - \frac{1}{B}\sum_{i=1}^{B} Y_i\right)^2} \tag{2.44}$$

Now let's apply this idea to the random variable $Y = \overline{X}$, where $\overline{X}$ is the sample mean of $\mathscr{S} = \{X_1, \dots, X_n\}$ obtained by SRS. To do this, first, we need to generate B simple random samples from the target population $\mathscr{P}$:

$$\begin{aligned}
&\mathscr{S}^{(1)} = \{X_1^{(1)}, \dots, X_n^{(1)}\} \subset \mathscr{P}, \\
&\dots \\
&\mathscr{S}^{(B)} = \{X_1^{(B)}, \dots, X_n^{(B)}\} \subset \mathscr{P},
\end{aligned} \tag{2.45}$$

then compute the corresponding sample means:[20]

[20] They play the role of $Y_1, \dots, Y_B$ in (2.41).

$$\overline{X}^{(1)} = \frac{1}{n}\sum_{i=1}^{n} X_i^{(1)}, \quad \dots \quad \overline{X}^{(B)} = \frac{1}{n}\sum_{i=1}^{n} X_i^{(B)}, \tag{2.46}$$

and, finally, estimate $\mathrm{se}[\overline{X}]$ by analogy with (2.44):

$$\mathrm{se}[\overline{X}] \approx \sqrt{\frac{1}{B}\sum_{j=1}^{B}\left(\overline{X}^{(j)} - \frac{1}{B}\sum_{i=1}^{B} \overline{X}^{(i)}\right)^2}. \tag{2.47}$$

Looks good expect for one thing: the *total* sample size in (2.45) is nB, which is much larger than our original sample size n, $nB \gg n$. Therefore, this straightforward method for estimating $\mathrm{se}[\overline{X}]$ is not practically acceptable, since we assume that sampling n population units is the maximum we can afford because of the time or budget limitations[21]. Here is where the *bootstrap principle* comes into play.

[21] After all, if we could afford sampling nB units, we would use a more accurate estimate $\frac{1}{nB}\sum_{i=1}^{nB} X_i$ for estimating μ instead of less accurate $\frac{1}{n}\sum_{i=1}^{n} X_i$.

The intuition behind the bootstrap principle is the following. The main underlying assumption in making inferences from a sample $\mathscr{S}$ to the entire population $\mathscr{P}$ is that $\mathscr{S}$ *represents* $\mathscr{P}$ well. This means that $\mathscr{S}$, albeit being small, is similar to $\mathscr{P}$ with respect to the parameter of interest, the property of $\mathscr{P}$ that we want to learn. But given $\mathscr{S}$, we can create a new large population of size N by simply taking N/n

copies of each X_i[22]. It is called the *bootstrap population*:

$$\mathscr{P}_{boot} = \{\underbrace{X_1,\ldots,X_1}_{N/n},\ldots,\underbrace{X_n,\ldots,X_n}_{N/n}\}. \tag{2.48}$$

We can view $\mathscr{P}_{boot}$ as an approximation of the original target population $\mathscr{P}$. If n is extremely small, then of course this approximation is poor. For example, if $n = 1$, then $\mathscr{P}_{boot}$ consists simply of N copies of one sample unit X_1. But the larger the sample size n is, the more accurate the approximation $\mathscr{P}_{boot} \approx \mathscr{P}$ is. In the other limiting case, if $n = N$, then $\mathscr{P}_{boot} = \mathscr{P}$. The construction of the bootstrap population is schematically illustrated in Fig. 2.2.

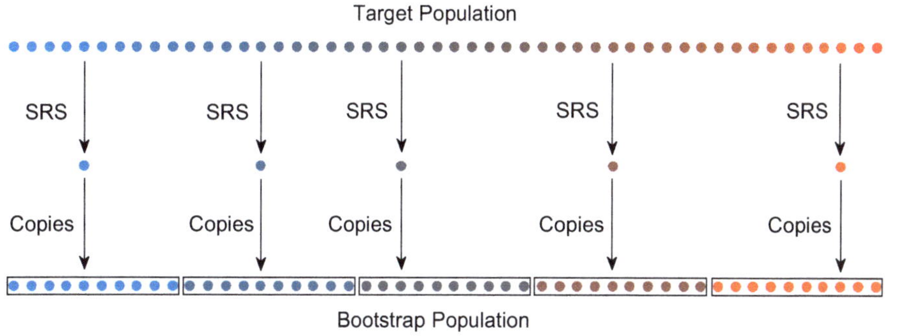

Fig. 2.2 The top row of points represents the target population $\mathscr{P}$; the middle row of points is a sample $\mathscr{S}$ obtained from $\mathscr{P}$ by simple random sampling; the bottom row of points is the bootstrap population $\mathscr{P}_{boot}$ obtained from $\mathscr{S}$ by copying its units.

The *bootstrap principle* states: use $\mathscr{P}_{boot}$ instead of $\mathscr{P}$ in (2.45). In other words, instead of sampling the target population in (2.45), the bootstrap[23] suggests to *reuse* our original sample $\mathscr{S} = \{X_1,\ldots,X_n\}$. This leads to the following algorithm for estimating the standard error se$[\overline{X}]$:

1. Draw B simple random samples from $\mathscr{P}_{boot}$:

$$\begin{aligned} \mathscr{S}^{(1)} &= \{X_1^{(1)},\ldots,X_n^{(1)}\} \subset \mathscr{P}_{boot}, \\ &\cdots \\ \mathscr{S}^{(B)} &= \{X_1^{(B)},\ldots,X_n^{(B)}\} \subset \mathscr{P}_{boot}, \end{aligned} \tag{2.49}$$

   Samples $\mathscr{S}^{(1)},\ldots,\mathscr{S}^{(B)}$ are called the *bootstrap samples*.

2. Compute the sample means of the bootstrap samples:

$$\overline{X}^{(1)} = \frac{1}{n}\sum_{i=1}^{n} X_i^{(1)}, \quad \ldots, \quad \overline{X}^{(B)} = \frac{1}{n}\sum_{i=1}^{n} X_i^{(B)}, \tag{2.50}$$

[22] For simplicity, we assume here that N/n is an integer. But if it is not, we can always round N/n to the nearest integer. A more neat way to get around the problem of N/n being non integer is discussed in Problem 2.7.

[23] This method derives its name from the old saying "to pull yourself up by your own bootstraps"[7].

The sample means $\overline{X}^{(1)},\dots,\overline{X}^{(B)}$ are called the *bootstrap replications* of $\overline{X}$.

3. Estimate the standard error of $\overline{X}$:

$$\mathrm{se}[\overline{X}] \approx \widehat{\mathrm{se}}_B[\overline{X}] = \sqrt{\frac{1}{B}\sum_{j=1}^{B}\left(\overline{X}^{(j)} - \frac{1}{B}\sum_{i=1}^{B}\overline{X}^{(i)}\right)^2}. \tag{2.51}$$

The estimate $\widehat{\mathrm{se}}_B[\overline{X}]$ is called the *bootstrap estimate* of the standard error.

2.6.1 Example: Gaussian Population

To illustrate the bootstrap method, let's consider a "Gaussian" population $\mathscr{P}$, where $x_1,\dots,x_N$ are independently drawn from the normal distribution $\mathcal{N}(\mu_0,\sigma_0^2)$ with mean $\mu_0 = 0$ and standard deviation $\sigma_0 = 10$, and the population size $N = 10^4$. In other words, we generate N realizations of $Z \sim \mathcal{N}(\mu_0,\sigma_0^2)$, "freeze" them, and denote the obtained values by $x_1,\dots,x_N$. The resulting population mean and population standard deviation are

$$\mu = \frac{1}{N}\sum_{i=1}^{N} x_i = 0.038 \quad \text{and} \quad \sigma = \sqrt{\frac{1}{N}\sum_{i=1}^{N}(x_i-\mu)^2} = 10.074. \tag{2.52}$$

As expected[24], the population parameters are close to the distribution parameters, $\mu \approx \mu_0$ and $\sigma \approx \sigma_0$.

[24] Thanks to the LLN.

Let $\mathscr{S} = \{X_1,\dots,X_n\}$ be a simple random sample from $\mathscr{P}$ of size $n = 10^2$. The obtained value of the sample mean is

$$\overline{X} = \frac{1}{n}\sum_{i=1}^{n} X_i = 0.204. \tag{2.53}$$

The estimate $\mu \approx \overline{X}$ is not very accurate due to the high heterogeneity ($\sigma = 10.074$) of the target population and small sample size (1% of the population size). The *exact* value of the standard error of $\overline{X}$ is given by (2.30)[25]:

[25] In this example, we can compute the exact value, since we know the population standard deviation σ.

$$\mathrm{se}[\overline{X}] = \frac{\sigma}{\sqrt{n}}\sqrt{1-\frac{n-1}{N-1}} = 1.002. \tag{2.54}$$

The *analytical* estimate (2.40) of the standard error is

$$\widehat{\mathrm{se}}[\overline{X}] = \frac{\hat{\sigma}_u}{\sqrt{n}}\sqrt{1-\frac{n-1}{N-1}} = \hat{\sigma}\sqrt{\frac{Nn-n}{Nn-N}}\frac{1}{\sqrt{n}}\sqrt{1-\frac{n-1}{N-1}}$$
$$= \sqrt{\frac{1}{n}\sum_{i=1}^{n}(X_i-\overline{X})^2}\sqrt{\frac{N-1}{Nn-N}}\sqrt{\frac{N-n}{N-1}} \tag{2.55}$$
$$= \sqrt{\frac{1}{n}\sum_{i=1}^{n}(X_i-\overline{X})^2}\sqrt{\frac{N-n}{Nn-N}} = 1.075.$$

The *bootstrap* estimate (2.51) of the standard error based on $B = 10^3$ bootstrap samples $\mathscr{S}^{(1)},\ldots,\mathscr{S}^{(B)}$ is

$$\widehat{\mathrm{se}}_B[\overline{X}] = \sqrt{\frac{1}{B}\sum_{j=1}^{B}\left(\overline{X}^{(j)} - \frac{1}{B}\sum_{i=1}^{B}\overline{X}^{(i)}\right)^2} = 1.044. \tag{2.56}$$

Interestingly, the obtained bootstrap estimate is even slightly more accurate than the analytical estimate. This is not always going to be the case. Given sample $\mathscr{S} = \{X_1,\ldots,X_n\}$, the analytical estimate $\widehat{\mathrm{se}}[\overline{X}]$ has a fixed deterministic value, but the bootstrap estimate $\widehat{\mathrm{se}}_B[\overline{X}]$ is random, since its value depends on the bootstrap samples $\mathscr{S}^{(1)},\ldots,\mathscr{S}^{(B)}$, which are chosen from the bootstrap population by a random process: simple random sampling.

To illustrate the variability of the bootstrap estimates, Fig. 2.3 shows the boxplots of the bootstrap estimates with $B = 10^2, 10^3$, and $B = 10^4$. The larger B is, the smaller the dispersion of the bootstrap estimates is.

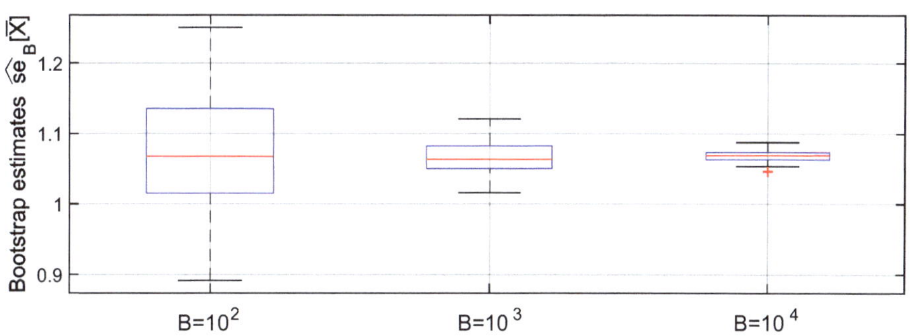

Fig. 2.3 Boxplots of bootstrap estimates $\widehat{\mathrm{se}}_B[\overline{X}]$. Each boxplot is constructed based on 100 bootstrap estimates. That is, (2.49), (2.50), and (2.51) were repeated 100 times for each value of B.

So, what do we know so far about the sample mean $\overline{X}$ of a SRS sample? We know its expected value[26] (Corollary 2.1), its variance and standard error (Corollary 2.2), and we also learned how to estimate the standard error both analytically and using the bootstrap method. Ideally, however, we would like to know the entire probability distribution[27] of $\overline{X}$, called *sampling distribution*, since it would tell us everything about the random variable $\overline{X}$ and, therefore, everything about the accu-

[26] Recall that the sample mean is an unbiased estimate of the population mean, $\mathbb{E}[\overline{X}] = \mu$.

[27] A random variable can't be fully described by its mean and variance.

racy of estimation $\mu \approx \overline{X}$. Next section shows how the sampling distribution of $\overline{X}$ can be approximated by a normal distribution.

2.7 Normal Approximation for the Sample Mean

First, let's recall the crowning achievement of probability theory: the *central limit theorem*[28] (CLT). Simply put, the classical CLT says that if $Y_1, \ldots, Y_n$ are *independent and identically distributed* (IID) random variables with the common mean $\mathbb{E}[Y_i] = \mu$ and variance $\mathbb{V}[Y_i] = \sigma^2$, and n is large, then their average $\overline{Y} = \frac{1}{n}\sum_{i=1}^{n} Y_i$ has a distribution which is approximately normal with mean $\mathbb{E}[\overline{Y}] = \mu$ and variance $\mathbb{V}[\overline{Y}] = \sigma^2/n$:

$$\overline{Y} \mathrel{\dot\sim} \mathcal{N}\left(\mathbb{E}[\overline{Y}], \mathbb{V}[\overline{Y}]\right) = \mathcal{N}\left(\mu, \frac{\sigma^2}{n}\right), \tag{2.57}$$

where symbol $\dot\sim$ means "approximately distributed according to."

So, the question is: can we apply the CLT to random variables $X_1, \ldots, X_n$ obtained from a finite population by simple random sampling? Formally speaking, "no." Indeed, first of all, although $X_1, \ldots, X_n$ are identically distributed (Theorem 2.1), they are not independent, since $\mathrm{Cov}(X_i, X_j) \neq 0$ (Theorem 2.2). Second, the CLT is a limiting result: the distribution of $\overline{Y}$ converges to normal as $n \to \infty$ and (2.57) holds for large n. In the SRS context, however, n is not necessarily large and, moreover, it makes no mathematical sense to tend n to infinity since $n < N$.

Nevertheless, it can be shown that if both $n \gg 1$ and $N \gg n$, that is, if n is large and N is much larger than n, then $\overline{X}$ is approximately normally distributed:

$$\overline{X} \mathrel{\dot\sim} \mathcal{N}(\mathbb{E}[\overline{X}], \mathbb{V}[\overline{X}]), \tag{2.58}$$

or, equivalently,

$$\frac{\overline{X} - \mu}{\mathrm{se}[\overline{X}]} \mathrel{\dot\sim} \mathcal{N}(0, 1). \tag{2.59}$$

The intuition behind this approximation is the following: if $N \gg n \gg 1$, then sample units $X_1, \ldots, X_n$ are nearly independent[29], the sample size n is large, and, therefore, the CLT holds approximately.

The CLT approximation (2.59) is a very strong result: it says that for *any* target population $\mathscr{P}$ and SRS sample $\mathscr{S} \subset \mathscr{P}$, as long as $N \gg n \gg 1$, the sample mean $\overline{X}$ is approximately normally distributed with mean μ, which is the parameter of interest, and standard error that we know how to estimate (analytically or using the bootstrap). This result has many useful applications. In particular, it can be used for quantifying the error of estimation $\mu \approx \overline{X}$ (Section 2.8) and for constructing *confidence intervals* for μ (Section 2.9). But before we discuss these applications, let's consider an example that illustrates the normal approximation (2.59) to the sampling distribution of $\overline{X}$.

[28] For a more detailed discussion of CLT see Probability Refresher 2.12.

[29] In particular, $\mathrm{Cov}(X_i, X_j) = -\frac{\sigma^2}{N-1} \approx 0$.

2.7.1 Example: Poissonian Population

To illustrate the CLT approximation (2.59) with a specific example, let's create a "Poissonian" population $\mathscr{P} = \{x_1, \dots, x_N\}$, in a similar way we have created the Gaussian population in Example 2.6.1. Namely, let $x_1, \dots, x_N$ be $N = 10^5$ independent realizations of a Poisson random variable $X \sim \mathscr{P}(\lambda_0)$ with parameter $\lambda_0 = 10$[30]. The obtained population mean and population standard deviation are

$$\mu = \frac{1}{N}\sum_{i=1}^{N} x_i = 10.015 \quad \text{and} \quad \sigma = \sqrt{\frac{1}{N}\sum_{i=1}^{N}(x_i - \mu)^2} = 3.158. \tag{2.60}$$

[30] The probability mass function of $X \sim \mathscr{P}(\lambda_0)$ is $\mathbb{P}(X = k) = e^{-\lambda_0}\frac{\lambda_0^k}{k!}$, where $k = 0, 1, 2, \dots$.

As in the Gaussian population example, thanks to the LLN, the population parameters approximately equal to the distribution parameters[31], $\mu \approx \lambda_0$ and $\sigma \approx \sqrt{\lambda_0}$.

[31] If $X \sim \mathscr{P}(\lambda_0)$ is a Poisson random variable with parameter λ_0, then its mean and variance are $\mathbb{E}[X] = \mathbb{V}[X] = \lambda_0$.

Let $\mathscr{S}^{(1)}, \dots, \mathscr{S}^{(K)}$ be $K = 10^3$ SRS samples of size $n = 10^3$ from $\mathscr{P}$ and $\overline{X}^{(1)}, \dots, \overline{X}^{(K)}$ be the corresponding sample means. If (2.59) holds, then the standardized sample means

$$Z^{(1)} = \frac{\overline{X}^{(1)} - \mu}{\mathrm{se}[\overline{X}^{(1)}]}, \ \dots \ , Z^{(K)} = \frac{\overline{X}^{(K)} - \mu}{\mathrm{se}[\overline{X}^{(K)}]} \tag{2.61}$$

should be approximately distributed according to the standard normal distribution. Random variables $\overline{X}^{(1)}, \dots, \overline{X}^{(K)}$ are IID and their common standard error is given by (2.30):

$$\mathrm{se}[\overline{X}^{(1)}] = \dots = \mathrm{se}[\overline{X}^{(K)}] = \frac{\sigma}{\sqrt{n}}\sqrt{1 - \frac{n-1}{N-1}} = 0.099. \tag{2.62}$$

Figure 2.4 shows the histogram and the normal-quantile plot for $Z^{(1)}, \dots, Z^{(K)}$. The histogram is approximately bell-shaped and symmetric about zero. The normal-quantile plot lies accurately on the line $y = x$. As discussed in Chapter 1, these observations strongly suggest that, in accordance with the CLT approximation (2.59), $Z^{(1)}, \dots, Z^{(K)}$ are distributed approximately according to the standard normal distribution. Moreover, both the Kolmogorov-Smirnov test and the Shapiro-Wilk test[32] accept the hypothesis that the standardized sample means $Z^{(1)}, \dots, Z^{(K)}$ are normally distributed.

[32] These tests are discussed in Section 10.3 and Remark 10.9.

2.8 Quantifying the Estimation Error

The normal approximation (2.59) to the sampling distribution of $\overline{X}$ can be used for quantifying the error of estimation $\mu \approx \overline{X}$. In real applications, the chance that the estimate is exact, $\overline{X} = \mu$, is negligible[33]. More formally, the probability

[33] This may happen in some special rare cases (i.e. if $n = N$), but in general, it is very unlikely.

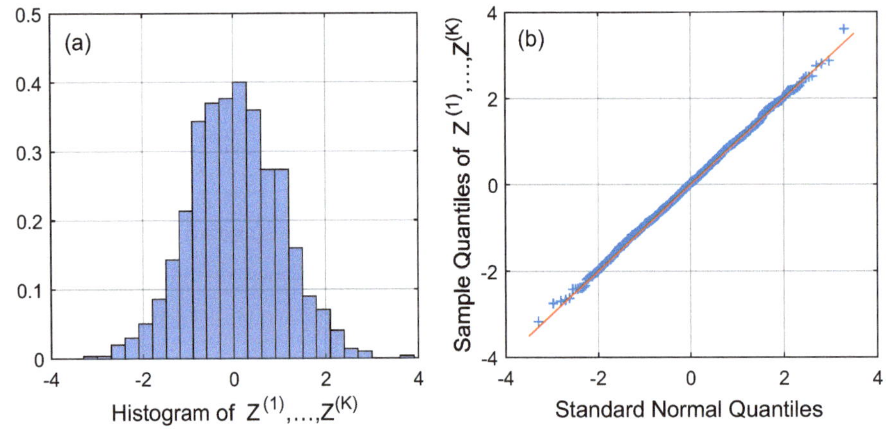

Fig. 2.4 The histogram (panel (a)) and the normal-quantile plot (panel (b)) for the standardized sample means $Z^{(1)},\ldots,Z^{(K)}$.

$$\mathbb{P}(\overline{X}=\mu)\approx 0. \tag{2.63}$$

An interesting question is then to estimate the probability that $\overline{X}$ belongs to a certain *interval* centered on μ. Let $\varepsilon>0$ be a *tolerance parameter*. Then the probability that $\overline{X}$ belongs to the interval $(\mu-\varepsilon,\mu+\varepsilon)$ can be estimated using (2.59) as follows:

$$\begin{aligned}\mathbb{P}\left(\mu-\varepsilon<\overline{X}<\mu+\varepsilon\right)&=\mathbb{P}\Bigg(-\frac{\varepsilon}{\mathrm{se}[\overline{X}]}<\underbrace{\frac{\overline{X}-\mu}{\mathrm{se}[\overline{X}]}}_{\dot\sim\mathcal{N}(0,1)}<\frac{\varepsilon}{\mathrm{se}[\overline{X}]}\Bigg)\\&\approx\Phi\left(\frac{\varepsilon}{\mathrm{se}[\overline{X}]}\right)-\Phi\left(-\frac{\varepsilon}{\mathrm{se}[\overline{X}]}\right)\\&=\Phi\left(\frac{\varepsilon}{\mathrm{se}[\overline{X}]}\right)-\left(1-\Phi\left(\frac{\varepsilon}{\mathrm{se}[\overline{X}]}\right)\right)\\&=2\Phi\left(\frac{\varepsilon}{\mathrm{se}[\overline{X}]}\right)-1.\end{aligned} \tag{2.64}$$

In the second and third equalities, we used, respectively, property (1.34) and the property of the standard normal CDF, $\Phi(-z)=1-\Phi(z)$ for any z, which stems from the symmetry of the standard normal distribution about zero, see Fig. 2.5.

The standard error on the right-hand side can be estimated analytically (2.40) or using the bootstrap. Note that if $\varepsilon\to 0$, then $\Phi\left(\frac{\varepsilon}{\mathrm{se}[\overline{X}]}\right)\to\Phi(0)=1/2$, and approximation (2.64) reduces to (2.63).

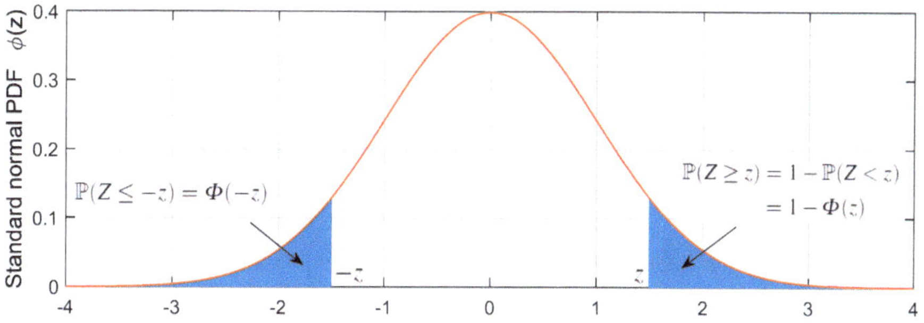

Fig. 2.5 The symmetry of the standard normal distribution about zero induces the property of the standard normal CDF: $\Phi(-z) = 1 - \Phi(z)$ for any z.

2.9 Confidence Intervals for the Population Mean

As we discussed in the previous section, the probability that the sample mean $\overline{X}$ equals exactly to the population mean μ is negligible, (2.63). We then fixed a deterministic interval centered on μ and estimated the probability that $\overline{X}$ belongs to that interval. Now let's consider the *inverse problem*: let's fix $\alpha \in (0,1)$ and find a *random* interval I_α that contains the population mean μ with probability $(1-\alpha)$,

$$\mathbb{P}(\mu \in I_\alpha) = 1 - \alpha. \tag{2.65}$$

Random interval I_α is called a *confidence interval* for the population mean and the value $100(1-\alpha)\%$ is called the *confidence level*[34].

[34] Usually, confidence levels of 90% ($\alpha = 0.1$) and 95% ($\alpha = 0.05$) are used.

Remark 2.6. The notion of a confidence interval is general and can be defined for any population or distribution parameter θ. If θ is a vector, then, instead of confidence intervals, we can define *confidence sets* that trap θ with a prescribed probability. We will discuss this in more detail in Chapter 3.

The normal approximation (2.59) to the sampling distribution of $\overline{X}$ allows to construct an approximate $100(1-\alpha)\%$ confidence interval I_α for μ. So, our goal is to find $I_\alpha = (A_\alpha, B_\alpha)$, where A_α and B_α are some random variables, the left and the right ends of I_α, such that

$$\mathbb{P}(A_\alpha < \mu < B_\alpha) = 1 - \alpha. \tag{2.66}$$

On the other hand, since $(\overline{X} - \mu)/\mathrm{se}[\overline{X}]$ is approximately standard normal,

$$\begin{aligned} \mathbb{P}\left(z_{\frac{\alpha}{2}} < \frac{\overline{X} - \mu}{\mathrm{se}[\overline{X}]} < -z_{\frac{\alpha}{2}}\right) &\approx \Phi\left(-z_{\frac{\alpha}{2}}\right) - \Phi\left(z_{\frac{\alpha}{2}}\right) \\ &= \left(1 - \Phi\left(z_{\frac{\alpha}{2}}\right)\right) - \Phi\left(z_{\frac{\alpha}{2}}\right) \\ &= 1 - 2\Phi\left(z_{\frac{\alpha}{2}}\right) = 1 - 2\frac{\alpha}{2} = 1 - \alpha, \end{aligned} \tag{2.67}$$

where z_q is the qth standard normal quantile, $\Phi(z_q) = q$, see Fig. 1.7.

Remark 2.7. The appearance of $z_{\frac{\alpha}{2}}$ on the left-hand side and $-z_{\frac{\alpha}{2}}$ on the right-hand side of the inequality may look like a mistake at the first glace, but it is not. Since $\alpha \in (0,1)$, $\alpha/2 \in (0,1/2)$, and $z_{\frac{\alpha}{2}} \in (-\infty,0)$. So, $z_{\frac{\alpha}{2}} < 0 < -z_{\frac{\alpha}{2}}$.

Rewriting (2.67) in the form of (2.66), we have:

$$\mathbb{P}\left(\overline{X} + z_{\frac{\alpha}{2}} \operatorname{se}[\overline{X}] < \mu < \overline{X} - z_{\frac{\alpha}{2}} \operatorname{se}[\overline{X}]\right) \approx 1 - \alpha. \tag{2.68}$$

Therefore,

$$I_\alpha = \left(\overline{X} + z_{\frac{\alpha}{2}} \operatorname{se}[\overline{X}],\ \overline{X} - z_{\frac{\alpha}{2}} \operatorname{se}[\overline{X}]\right) \tag{2.69}$$

is an approximate $100(1-\alpha)\%$ confidence interval for μ.

The confidence interval I_α is centered on the sample mean $\overline{X}$. Notice also that while the interval I_α is random (since its center $\overline{X}$ is random), its length is not. The length of I_α is equal to $-2z_{\frac{\alpha}{2}} \operatorname{se}[\overline{X}]$, and it increases as α decreases, since $z_{\frac{\alpha}{2}} = \Phi^{-1}(\alpha/2)$ is an increasing function of α (see Fig. (1.7)). This makes perfect intuitive sense: the more confidence we want to have that I_α traps μ (i.e. the smaller α is), the larger I_α should be.

It is instructive to consider the two limiting cases, where $\alpha \to 0$ and $\alpha \to 1$:

$$\begin{aligned} &\text{If } \alpha \to 0, \text{ then } z_{\frac{\alpha}{2}} = \Phi^{-1}\left(\frac{\alpha}{2}\right) \to -\infty \text{ and } I_\alpha \to (-\infty,\infty). \\ &\text{If } \alpha \to 1, \text{ then } z_{\frac{\alpha}{2}} = \Phi^{-1}\left(\frac{\alpha}{2}\right) \to 0 \text{ and } I_\alpha \to \overline{X}. \end{aligned} \tag{2.70}$$

In both cases, the defining property (2.65) of the confidence interval I_α reduces to intuitively expected results:

$$\begin{aligned} &\text{If } \alpha \to 0, \text{ then } \mathbb{P}(\mu \in I_\alpha) = 1-\alpha \text{ becomes } \mathbb{P}(-\infty < \mu < \infty) = 1. \\ &\text{If } \alpha \to 1, \text{ then } \mathbb{P}(\mu \in I_\alpha) = 1-\alpha \text{ becomes } \mathbb{P}(\overline{X} = \mu) = 0. \end{aligned} \tag{2.71}$$

Remark 2.8. Confidence intervals often have the following form:

$$I = \text{statistic} \pm \text{something}, \tag{2.72}$$

where "statistic" is some function of the data and "something" is called the *margin of error*. For the confidence interval (2.69), the statistic is $\overline{X}$ and the margin of error is $-z_{\frac{\alpha}{2}} \operatorname{se}[\overline{X}]$, which is half of the interval length.

Confidence intervals are often *misinterpreted*[35]. Suppose that we got a SRS sample $\mathscr{S} = \{X_1,\ldots,X_n\}$ from the target population $\mathscr{P}$, set the confidence level to, say 95% (i.e. $\alpha = 0.05$), plugged in all the numbers into (2.69) and obtained, for instance, the interval $(-1,1)$. Does it mean that μ belongs to $(-1,1)$ with probability 0.95? No, of course not: μ is a deterministic (not random) parameter, it either belongs to $(-1,1)$ or it does not. In other words, once the sample is drawn the deterministic *realization* $(-1,1)$ of the random confidence interval I_α is calculated[36],

[35] Unfortunately, even by professional scientists.

[36] In practice, the word "realization" is often omitted and intervals like $(-1,1)$ are described as "confidence intervals." It is important to keep in mind though that they are, in fact, deterministic realizations of random confidence intervals.

this realization either contains μ or it does not. It is no longer a matter of probability.

The correct interpretation of confidence intervals is the following. First, it is important to realize that (2.65) is a probability statement about the random confidence interval I_α, not about the deterministic population mean μ. Perhaps, it would be intuitively more clear to rewrite it as follows

$$\mathbb{P}(I_\alpha \ni \mu) = 1 - \alpha. \tag{2.73}$$

The frequency interpretation of this equation is that if we take many SRS samples $\mathscr{S}^{(1)}, \ldots, \mathscr{S}^{(K)}$ and compute the corresponding realizations $I_\alpha^{(1)}, \ldots, I_\alpha^{(K)}$ of the confidence interval I_α based on these samples, then about $100(1-\alpha)\%$ of these realizations will contain μ. The following example illustrates this.

2.9.1 Example: Bernoulli Population

Let $\mathscr{P} = \{x_1, \ldots, x_N\}$ be a Bernoulli population, where $x_1, \ldots, x_N$ are independent realizations of a Bernoulli random variable $X \sim \mathscr{B}(p_0)$ with parameter $p_0 = 1/2$, and $N = 10^5$. The population mean and standard deviation are

$$\mu = \frac{1}{N}\sum_{i=1}^{N} x_i = 0.501 \quad \text{and} \quad \sigma = \sqrt{\frac{1}{N}\sum_{i=1}^{N}(x_i - \mu)^2} = 0.500 \tag{2.74}$$

As in the Gaussian and Poissonian examples considered above, the LLN forces the population parameters to be approximately equal to the distribution parameters[37], $\mu \approx p_0$ and $\sigma \approx \sqrt{p_0(1-p_0)}$.

[37] If $X \sim \mathscr{B}(p_0)$, then its mean and variance are $\mathbb{E}[X] = p_0$ and $\mathbb{V}[X] = p_0(1-p_0)$.

Figure (2.6) shows $K = 100$ realizations of the 85% confidence interval $I_{\alpha=0.15}$ in panel (a) and $K = 100$ realizations of the 95% confidence interval $I_{\alpha=0.05}$ in panel (b), constructed based on (2.69) and 200 independent SRS samples of size $n = 10^3$.

The theory predicts that about 85% of intervals in panel (a) and 95% of intervals in panel (b) should contain the population mean μ, that is, approximately 85 and 95 intervals out of 100, respectively. The simulation results in Fig. (2.6) agree with theory well: exactly 85 intervals in panel (a) and 94 intervals in panel (b) contain μ.

The margins of errors, which determine the length of the intervals, in the two cases are:

$$\begin{aligned} \alpha = 0.15 \quad &\Longrightarrow \quad -z_{\frac{\alpha}{2}}\,\mathrm{se}[\overline{X}] = -\Phi^{-1}\left(\frac{\alpha}{2}\right)\frac{\sigma}{\sqrt{n}}\sqrt{1 - \frac{n-1}{N-1}} = 0.022, \\ \alpha = 0.05 \quad &\Longrightarrow \quad -z_{\frac{\alpha}{2}}\,\mathrm{se}[\overline{X}] = -\Phi^{-1}\left(\frac{\alpha}{2}\right)\frac{\sigma}{\sqrt{n}}\sqrt{1 - \frac{n-1}{N-1}} = 0.031. \end{aligned} \tag{2.75}$$

As expected, the length $-2z_{\frac{\alpha}{2}}\,\mathrm{se}[\overline{X}]$ of $I_{\alpha=0.05}$ (the interval with higher confidence level) is larger then the length of $I_{\alpha=0.15}$.

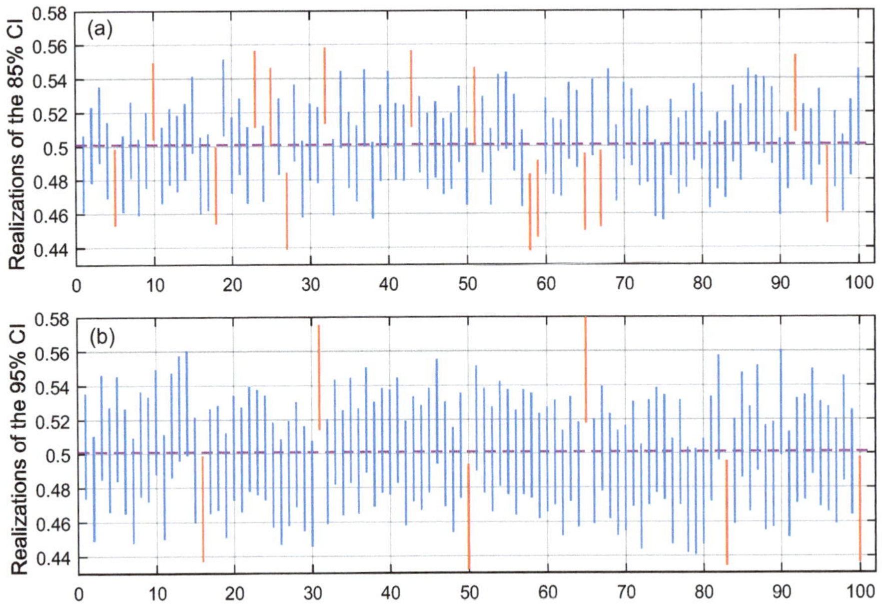

Fig. 2.6 Independent realization of the confidence interval (CI) I_α given by (2.69) for (a) $\alpha = 0.15$ and (b) $\alpha = 0.05$. The dashed line represents the population mean $\mu = 0.501$. The intervals that contain μ are blue and the interval that don't contain μ are orange.

In this example, for illustrative purposes we used the exact value of the standard error $\mathrm{se}[\overline{X}]$ given in Corollary 2.2. In applications, where the value of the population standard deviation σ is unknown, the standard error must be estimated either analytically (2.40) or using the bootstrap method discussed in Section 2.6.

2.10 Case Study: Body Mass Index of Olympics Athletes

Running is a healthy exercise and strengthens the body.
– Emanuel Kant, *On Education*

The *body mass index* (BMI) is a simple yet effective characteristic of human body which is often used to classify a person as underweight, normal weight, overweight, or obese. It is defined as a ratio of the body mass (in kilograms) over the squared body height (in meters),

$$\mathrm{BMI} = \frac{\mathrm{mass}}{\mathrm{height}^2}. \tag{2.76}$$

Although there are some minor variations in different countries, the commonly accepted BMI ranges and the corresponding medical conditions, proposed by the World Health Organization (WHO), are summarized in Table 2.10.

Table 2.1 BMI ranges and conditions as defined by the World Health Organization.

BMI (in kg/m^2)	Condition
< 18.5	Underweight
$[18.5, 25)$	Normal Weight
$[25, 30)$	Overweight
≥ 30	Obesity

According to the National Heart, Lung, and Blood Institute (NHLBI), overweight and obese people are at higher risk of developing serious health problems and diseases, including heart disease, high blood pressure, type 2 diabetes, gallstones, breathing problems, and certain types of cancer [8]. Moreover, the WHO reported that increasing degrees of overweight and obesity, measured by the BMI, lead to the increase of mortality rates, and estimated that at least 2.8 million people die each year worldwide as a result of being overweight or obese [9]. The WHO recommends individuals to maintain their BMI in the range 18.5 to 24.9 kg/m^2.

In this Case Study, we will explore the BMI of athletes participated in the 2016 Summer Olympic Games, that took place from 5 to 21 August in Rio de Janeiro, Brazil. More then 11,000 athletes from 207 countries representing 28 Olympic sports participated in the Games. The data set [10] contains the following information on the 11,538 Olympic athletes: name, nationality, sex, date of birth, height, weight, sport, and the number of gold/silver/bronze medals won by the athlete[38].

[38] This data was originally published at the official Rio 2016 website, www.rio2016.com. However, for some mysterious reason, as soon as the Games finished, the website was deleted (now it redirects to another website) and the data was removed. Fortunately, Matt Riggott web scraped the data while the website was still up and made it publicly available [11].

Remark 2.9. Data Cleaning.
Data cleaning is the process of correcting or removing incorrect or missing pieces of information from a data set. It is an important data pre-processing step which is often performed in real applications before making any inferences from the data. In the athlete data, about 6% of athletes have their mass or height (or both) measurements missing. All these athletes were removed from the data, since their BMI can't be calculated. What is worse, several athletes have obviously incorrect values of their mass or height measurements which results into extreme BMI values. For example, Yvette Lewis, a Panamanian athlete who competes in the 100 meter hurdles and the triple jump, can't weight 150 kg and have BMI of 49 kg/m^2 (according to Wikipedia, her true mass is 75 kg, and, therefore, her true BMI is 24.5 kg/m^2). To reduce the effect of incorrect mass and height measurements, several extreme BMI values (very large and very small) were manually checked and the corresponding masses and heights were corrected or removed (if the true values were not found).

Figure 2.7 graphically summarizes the BMI data for $N = 10,857$ Rio 2016 Olympic Athletes. Panels (a), (b), (c) show the histogram, the boxplot, and the normal-quantile plot, respectively, and panel (d) depicts BMI boxplots for each sport[39]. The first observation is that BMIs are not normally distributed. Instead, their distribution is right-skewed, the left tail is short, and the right tail is long. Gymnasts and weightlifters have the smallest and the largest median BMI, respectively. The

[39] The masses of all boxers are missing in the data set. That is why there are only 27 out of 28 Olympic sports in panel (d).

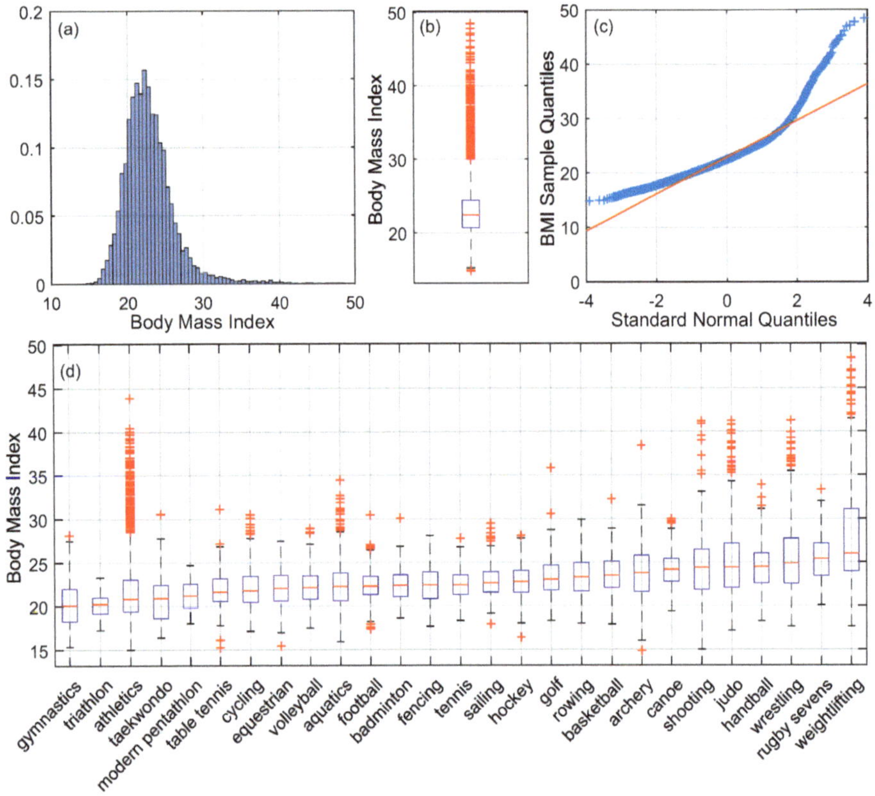

Fig. 2.7 The body mass index of the Rio 2016 Olympic athletes: (a) histogram, (b) boxplot, (c) normal-quantile plot, and (d) boxplots for each sport.

smallest BMI of 14.85 kg/m^2 belongs to Sultan Duzelbayev, a male archer from Kazakhstan, and the largest BMI of 48.43 kg/m^2 belongs ot Shaimaa Haridy, a female weightlifter from Egypt. The mean and the standard deviation of the population of athlete BMIs are

$$\mu_{\text{BMI}} = 22.86\ \text{kg}/\text{m}^2 \quad \text{and} \quad \sigma_{\text{BMI}} = 3.37\ \text{kg}/\text{m}^2. \tag{2.77}$$

Remark 2.10. Although the BMI is a useful and widely used indicator of obesity, as essentially all quantitative measures, it has a number of limitations and drawbacks. In particular, the BMI is based on the total body mass and does not distinguish between muscles and fat. As a result, the BMI tends to overestimate obesity for athletes, since their high muscle mass can classify them as overweight or obese. According to the definitions in Table 2.1, 5% of the Rio 2016 Olympic athletes are underweight, 76% have normal weight, 15% are overweight, and 3% are obese. These numbers suggest that the BMI thresholds must be adjusted for athletes and other very fit people with high muscle mass.

Suppose now that we don't have access to the entire "target" population $\mathscr{P}_{\text{BMI}}$ of athlete BMIs, and want to estimate the population mean μ_{BMI} using simple random sampling[40]. Let $\mathscr{S} = \{X_1, \ldots, X_n\} \subset \mathscr{P}_{\text{BMI}}$ be a SRS sample, and $\overline{X}$ be the corresponding sample mean. Figure 2.8(a) shows the variability of sample mean values (computed for 100 independent SRS samples) around the true value (2.77) for $n = 10^2$ and $n = 10^3$. As expected, sample means corresponding to the larger sample size $n = 10^3$ (orange dots) are more tightly focused around the true value (represented by the green dashed line), and, therefore, in general, are more accurate estimates of μ_{BMI}.

[40] For instance, this could be the case if the athlete data were not publicly available, and we would have to pay for the information about each athlete.

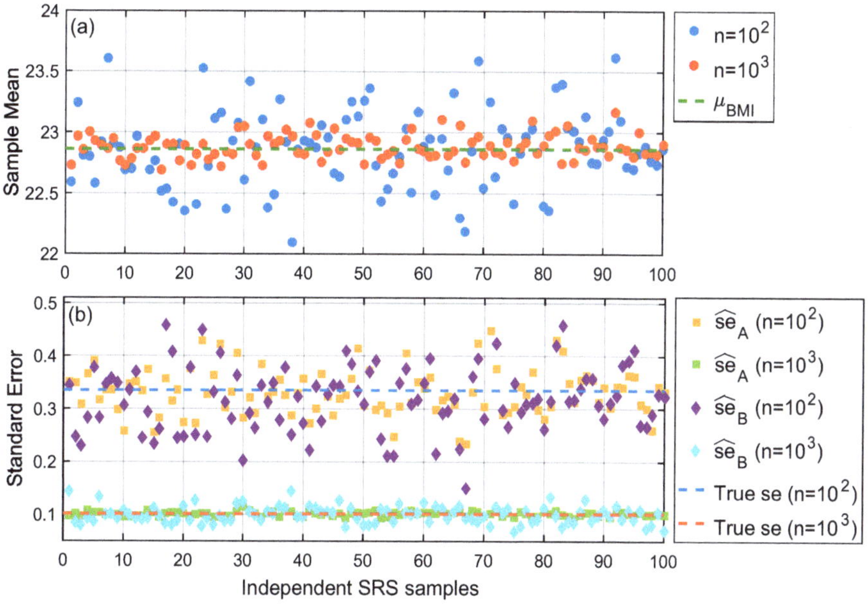

Fig. 2.8 Panel (a): Variability of the values of the sample means $\overline{X}$, computed for 100 SRS samples, around the true value $\mu_{\text{BMI}} = 22.86$ for $n = 10^2$ and $n = 10^3$. Panel (b): Variability of analytical ($\widehat{\text{se}}_A$) and bootstrap ($\widehat{\text{se}}_B$) estimates of the standard error of $\overline{X}$ around the true values $\text{se}[\overline{X}] = 0.34$ ($n = 10^2$) and $\text{se}[\overline{X}] = 0.10$ ($n = 10^3$).

Figure 2.8(b) shows the variability of both analytical (2.40) and bootstrap (2.51) estimates[41] of the standard error of $\overline{X}$ (computed based on the same 100 SRS samples) around the true value of the standard error, which is given by (2.29):

[41] Computed based on 10^3 bootstrap samples.

$$\text{se}[\overline{X}] = \frac{\sigma_{\text{BMI}}}{\sqrt{n}} \sqrt{1 - \frac{n-1}{N-1}} = \begin{cases} 0.34, & \text{if } n = 10^2, \\ 0.10, & \text{if } n = 10^3. \end{cases} \tag{2.78}$$

The standard error estimates for $n = 10^3$ (green squares and light blue rhombuses) vary much less around the true value $\text{se}[\overline{X}] = 0.10$ (represented by the orange dashed line) compared to the variability of the estimates for $n = 10^2$ (yellow squares and purple rhombuses) around the true value $\text{se}[\overline{X}] = 0.34$ (represented by the blue dashed line). This is because the larger the sample size n, the smaller the true value of the standard error $\text{se}[\overline{X}]$, and the more accurate both analytical and bootstrap estimates of the standard error[42].

[42] In the limiting case, where $n = N$, $\text{se}[\overline{X}] = 0$, and both analytical and bootstrap estimates of $\text{se}[\overline{X}]$ also equal zero.

Figure 2.9(a) shows how the probability (2.64) that the sample mean $\overline{X}$ belongs to the interval $(\mu_{\text{BMI}} - \varepsilon, \mu_{\text{BMI}} + \varepsilon)$,

$$\mathbb{P}\left(\mu_{\text{BMI}} - \varepsilon < \overline{X} < \mu_{\text{BMI}} + \varepsilon\right) \approx 2\Phi\left(\frac{\varepsilon}{\text{se}[\overline{X}]}\right) - 1, \tag{2.79}$$

versus the tolerance parameter $\varepsilon \in [0, 1]$. Each curve corresponds to one of the 100

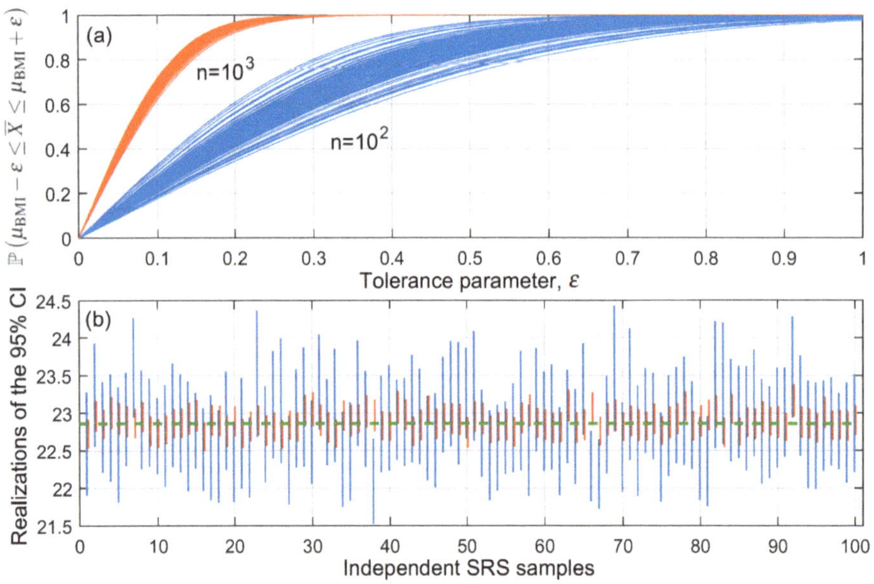

Fig. 2.9 Panel (a): The probability (2.79) versus the tolerance parameter ε for $n = 10^2$ (blue curves) and $n = 10^3$ (orange curves). Panel (b): 95% confidence intervals for μ_{BMI} (represented by the green dashed line) for $n = 10^2$ (blue intervals) and $n = 10^3$ (orange intervals).

analytical estimates of the standard error $\text{se}[\overline{X}]$, that appears in the right-hand-side of (2.79), obtained from the 100 independent SRS samples. As expected, for a given ε, the probabilities that correspond to the sample size $n = 10^3$ are larger than those that correspond to $n = 10^2$: the larger n is, the more accurately $\overline{X}$ estimates μ_{BMI}.

Finally, Figure 2.9(b) shows 100 realizations of the 95% confidence intervals for μ_{BMI} (green dashed line) for $n = 10^2$ (blue intervals) and $n = 10^3$ (orange intervals). As theory predicts, the orange intervals are shorter than the blue ones.

2.11 Further Reading

Simple random sampling considered in this chapter is the simplest sampling method for obtaining information about the target population. It provides a basis for more advanced sampling designs, such as *stratified random sampling*, *cluster sampling*, *systematic sampling*, etc.

For example, in stratified random sampling (StrRS), the target population is partitioned into sub-populations, called *strata*, which are then independently sampled using SRS. In many applications, stratification is natural. For instance, human populations, geographical areas, age groups, and professional occupations form natural strata. SrtRs is often used when, in addition to the information about the entire population, we are interested in learning about each individual sub-population. Moreover, estimates obtained from StrRS can be substantially more accurate than estimates from SRS if population units are are relatively homogeneous within each sub-population but heterogeneous across different sub-populations. If the total sample size we could afford is n and there are K strata, then we face an *optimal resource allocation* problem: how to choose the sample sizes n_k for each stratum, so that $\sum_{k=1}^{K} n_k = n$ and the standard error of the corresponding estimate is minimized? This problem leads to the so-called *Neyman allocation scheme*, but this is a different story[43] (which is told in both books recommended below).

A detailed discussion of sampling is given in a fundamental, yet accessible to readers with diverse statistical backgrounds, monograph

[43] "In the end, we'll all become stories." – Margaret Atwood, *Moral Disorder*.

- Lohr, S.L.: *Sampling: Design and Analysis*. CRC Press (2022). `https://www.sharonlohr.com/sampling-design-and-analysis-3e`

The following handbook will be especial useful to academic and industry researchers who are involved in practical survey research. It describes the whole survey sampling process, from design and data analysis to publication.

- Bethlehem, J.: *Applied Survey Methods: A Statistical Perspective*. Weley (2009). `https://doi.org/10.1002/9780470494998`

In addition, the website `http://www.applied-survey-methods.com/`, which accompanies the book, contains a number of computational tools and provides software packages, freely available for download; it is strongly recommended.

2.12 Probability Refresher: LLN and CLT

The *law of large numbers* (LLN) and the *central limit theorem* (CLT) are the two most important results in probability theory. Methods of modern statistical inference extensively use these results. Both LLN and CLT describe the limiting behavior of the sample mean $\overline{Y}$ of a sequence of random variables $Y_1, \ldots, Y_n$ as $n \to \infty$.

The law of large numbers provides a link between *randomness* and *determinism*. The LLN states that if $Y_1,\dots,Y_n,\dots$ is a sequence of independent and identically distributed (IID) random variables with the common mean $\mu = \mathbb{E}[Y_i]$, then the sample mean $\overline{Y}$ converges to μ in probability,

$$\overline{Y} = \frac{1}{n}\sum_{i=1}^{n} Y_i \xrightarrow{\mathbb{P}} \mu, \quad \text{as } n \to \infty. \tag{2.80}$$

The arrow $\xrightarrow{\mathbb{P}}$ means "converges in probability." Recall that in calculus, we say that a sequence of real numbers $x_1,\dots,x_n,\dots$ converges to a as $n \to \infty$, if for every $\varepsilon > 0$, we can find N such that $|x_n - a| < \varepsilon$ for all $n > N$. That is, all sequence members $x_{N+1}, x_{N+2},\dots$ lie in the ε-neighborhood of a. In probability, convergence of random variables is more subtle and the calculus definition does not make sense. Indeed, if $X_1,\dots,X_n,\dots$ are random variables, then $|X_n - a| < \varepsilon$ is an *event* that has some probability of occurrence, it may happen or may not. So, the deterministic notion of convergence must be updated for the case of random variables.

There are several reasonable ways to define the convergence of random variables. One of them is convergence in probability. We say that a sequence of random variables $X_1,\dots,X_n,\dots$ converges to a random variable X *in probability* as $n \to \infty$, denoted

$$X_n \xrightarrow{\mathbb{P}} X, \tag{2.81}$$

if for any $\varepsilon > 0$,

$$\lim_{n\to\infty} \mathbb{P}(|X_n - X| < \varepsilon) = 1. \tag{2.82}$$

Note that for every ε, the probability $\mathbb{P}(|X_n - X| < \varepsilon) = x_n$ is a real number, and the limit $\lim_{n\to\infty} x_n = 1$ in (2.82) is understood in the calculus sense. Intuitively, $X_n \xrightarrow{\mathbb{P}} X$ means that for large n, the value of X_n is very close to the value of X with large probability.

So, formally the LLN says that

$$\lim_{n\to\infty} \mathbb{P}(|\overline{Y} - \mu| < \varepsilon) = 1. \tag{2.83}$$

Intuitively this means that, although the values of individual IID random variables Y_i are uncertain and random (and could be highly dispersed), the value of their average $\overline{Y}$, although also uncertain and random, becomes more and more certain, more and more concentrated around μ, as n increases.

To illustrate the LLN, let's consider an IID sample $Y_1,\dots,Y_N$ of size $N = 10^4$ form the *Rademacher distribution*[44], i.e. $Y_i = \pm 1$ with probability $1/2$. Since $\mu = \mathbb{E}[Y_i] = 0$, the LLN predicts that the sample mean $\overline{Y} = \frac{1}{n}\sum_{i=1}^{n} Y_i$ will converge to zero as $n \to \infty$. Figure 2.10 shows five realizations of $\overline{Y}$ (computed based on five independent samples $Y_1,\dots,Y_N$) versus n for $n = 1,\dots,10^2$ (top panel), $n = 10^2,\dots,10^3$ (middle panel), and $n = 10^3,\dots,10^4$ (bottom panel). Each boxplot is computed based on 10^3 independent realizations of $\overline{Y}$ and shows the variability of the sample mean for a given value of n. As n increases, boxplots shrink and $\overline{Y}$ goes to zero, as expected.

The LLN was first proved by Jacob Bernoulli[45] [14] in 1713 for a special case,

44 It is named after Hans Rademacher (1892-1969), a German mathematician who immigrated to the US in 1934 because of the Nazis.

45 The Bernoulli family has produced the largest number (13!) of celebrated mathematicians in the history of mathematics.

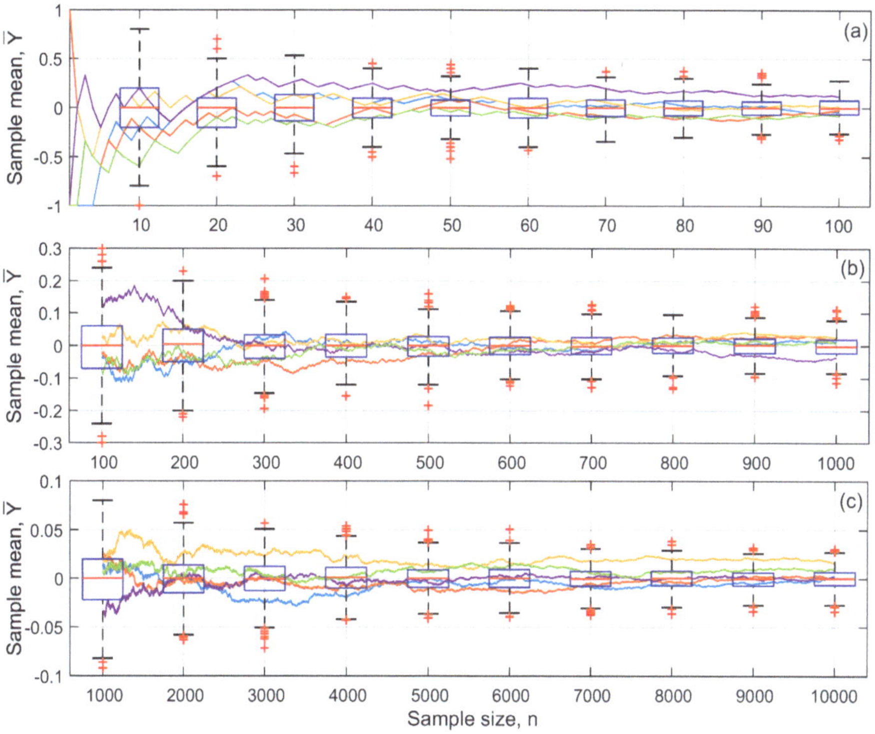

Fig. 2.10 The evolution of five independent realizations of the sample mean $\overline{Y}$ as a function of the sample size n. Each boxplot is based on 10^3 independent realizations of $\overline{Y}$ for a given value of n.

where all random variables are distributed according to the Bernoulli distribution, $Y_i \sim \mathscr{B}(p)$. The "empirical version" of this special case of LLN — if an experiment has two possible outcomes, "success" and "failure", and the probability of success is p, then the total number of successes in n independent trials of the experiment is approximately np — was, of course, known to many mathematicians and scientists well before Bernoulli[46]. But Bernoulli was the first who rigorously formalized and proved this statement. He fully realized its importance and called it his "Golden Theorem." The name "law of large numbers" was actually coined by Siméon Poisson in 1837. After Bernoulli, many mathematicians refined, extended, and generalized his LLN. The modern version of the LLN for arbitrary IID random variables formulated above was proved by Aleksandr Khinchin [15] in 1929.

Jacob Bernoulli (1654-1705), a Swiss mathematician who proved the first version of the law of large numbers.

Remark 2.11. There are, in fact, two versions of the LLN: *weak* and *strong*. The difference between the two versions is in the *type* of convergence of the sample mean $\overline{Y}$ to μ. In the weak law formulated above, the convergence is in probability. In the strong law, which is more subtle and that was proved by Andrey Kolmogorov [16] in 1930, $\overline{Y}$ converges to μ *almost surely*. Formally, almost sure convergence of the sample mean $\overline{Y}$ to μ means that

[46] For example, Gerolamo Cardano (1501-1576), an addicted gambler and one of the most influential mathematicians, scientists and philosophers of the Renaissance, used it in the games of chance.

$$\mathbb{P}\left(\lim_{n\to\infty}\overline{Y}=\mu\right)=1. \tag{2.84}$$

Almost sure convergence implies convergence in probability, hence the names of the laws. The principal difference between the weak LLN (2.83) and the strong LLN (2.84) is the following. In the weak LLN, no matter how large the sample size n is, there is still (a very small but non-zero) probability that $|\overline{Y}-\mu|>\varepsilon$. Whereas in the strong LLN, for any $\varepsilon>0$, there exists N such that for all $n>N$, the probability that $|\overline{Y}-\mu|>\varepsilon$ is zero.

If $Y_1,\ldots,Y_n$ are random variables, then so is their sample mean $\overline{Y}$. The LLN tells us that $\overline{Y}\approx\mu$ for large n, but it does not tell us anything about the *probability distribution* of $\overline{Y}$. The central limit theorem does.

The CLT states that if $Y_1,\ldots,Y_n,\ldots$ is a sequence of IID random variables with the common mean $\mu=\mathbb{E}[Y_i]$ and variance $\sigma^2=\mathbb{V}[Y_i]$, then the sample mean $\overline{Y}$ converges to $\mathcal{N}(\mu,\sigma^2/n)$ in distribution,

$$\overline{Y}=\frac{1}{n}\sum_{i=1}^{n}Y_i\xrightarrow{\mathbb{D}}\mathcal{N}\left(\mu,\frac{\sigma^2}{n}\right),\quad \text{as } n\to\infty. \tag{2.85}$$

The arrow $\xrightarrow{\mathbb{D}}$ means "converges in distribution," which is another type of convergence of random variables[47]. We say that a sequence of random variables $X_1,\ldots,X_n,\ldots$ converges to random variable X *in distribution* as $n\to\infty$, denoted

$$X_n\xrightarrow{\mathbb{D}}X, \tag{2.86}$$

if for any x,

$$\lim_{n\to\infty}F_n(x)=F(x), \tag{2.87}$$

where F_n and F are the CDFs of X_n and X, respectively.

So, the CLT says that for large n, the sample mean $\overline{Y}$ is approximately normally distributed with parameters μ and σ^2/n,

$$\overline{Y}\stackrel{\cdot}{\sim}\mathcal{N}\left(\mu,\frac{\sigma^2}{n}\right), \tag{2.88}$$

or, equivalently,

$$\frac{\overline{Y}-\mu}{\sigma/\sqrt{n}}\stackrel{\cdot}{\sim}\mathcal{N}(0,1). \tag{2.89}$$

Remark 2.12. The fact that $\mathbb{E}[\overline{Y}]=\mu$ and $\mathbb{V}[\overline{Y}]=\sigma^2/n$ for any n is trivial. Indeed,

$$\begin{aligned}\mathbb{E}[\overline{Y}]&=\mathbb{E}\left[\frac{1}{n}\sum_{i=1}^{n}Y_i\right]=\frac{1}{n}\sum_{i=1}^{n}\mathbb{E}[Y_i]=\frac{1}{n}\sum_{i=1}^{n}\mu=\mu,\\ \mathbb{V}[\overline{Y}]&=\mathbb{V}\left[\frac{1}{n}\sum_{i=1}^{n}Y_i\right]=\frac{1}{n^2}\sum_{i=1}^{n}\mathbb{V}[Y_i]=\frac{1}{n^2}\sum_{i=1}^{n}\sigma^2=\frac{\sigma^2}{n}.\end{aligned} \tag{2.90}$$

[47] Convergence in probability $X_n\xrightarrow{\mathbb{P}}X$ is a stronger property than convergence in distribution $X_n\xrightarrow{\mathbb{D}}X$. In other words, convergence in probability implies convergence in distribution.

The remarkable part of the CLT is that the limiting distribution of $\overline{Y}$ is normal *regardless* of the distribution of Y_i.

In terms of CDFs, the CLT looks as follows:

$$\lim_{n\to\infty} \mathbb{P}\left(\frac{\overline{Y}-\mu}{\sigma/\sqrt{n}} \leq z\right) = \Phi(z), \tag{2.91}$$

where $\Phi(z)$ is the standard normal CDF.

Remark 2.13. The CLT states the result at the level of CDFs: the CDF of $\overline{Y}$, properly normalized, converges to the standard normal CDF. A natural question is whether this result holds at the level of PDFs: if $Y_1,\dots,Y_n,\dots$ are continuous random variables, does the PDF of $\sqrt{n}(\overline{Y}-\mu)/\sigma$ converge to the standard normal PDF $\phi(z)$? Roughly speaking, the answer is "yes," but certain conditions of smoothness are necessary, since, generally, $F_n \to F$ does not imply $F_n' \to F'$.

To illustrate the CLT, as in the case of LLN, let's consider a collection $Y_1,\dots,Y_n$ of IID Rademacher random variables, i.e. $Y_i = \pm 1$ with probability $1/2$. Since $\mu = \mathbb{E}[Y_i] = 0$ and $\sigma^2 = \mathbb{V}[Y_i] = 1$, the CLT predicts that the CDF $F_n(z)$ of random variable

$$X_n = \frac{\overline{Y}-\mu}{\sigma/\sqrt{n}} = \sqrt{n}\,\overline{Y} = \frac{1}{\sqrt{n}}\sum_{i=1}^{n} Y_i \tag{2.92}$$

converges to the standard normal CDF, i.e. $F_n(z) \to \Phi(z)$ as $n \to \infty$, and, therefore, $F_n(z) \approx \Phi(z)$ for large n. For a given finite n, it is hard (although possible) to find $F_n(z)$ analytically, but it is easy to approximate $F_n(z)$ empirically by the eCDF $\widehat{F}_{n,N}(z)$ constructed based on N independent realizations of $X_n = \sqrt{n}\,\overline{Y}$. Thanks to Theorem 4.1 in Section 4.2, $\widehat{F}_{n,N}(z) \to F_n(z)$ as $N \to \infty$, and, therefore, $\widehat{F}_{n,N}(z) \approx F_n(z)$ for large N. Figure 2.11 shows how $\widehat{F}_{n,N}(z) \approx F_n(z)$ with $N = 10^3$ converges to $\Phi(z)$ as n increase from $n = 10$ (panel (a)) to $n = 10^4$ (panel (d)).

The first version of the CLT was formulated by Abraham de Moivre (the actual discoverer of the Gaussian curve, see Marginal Note 47 in Section 1.8) in 1733 for a very special case, where all random variables are distributed according to the Bernoulli distribution, $Y_i \sim \mathscr{B}(p)$, with $p = 1/2$. He presented his result privately to several close friends in a seven-page offprint [17]. The importance of his discovery, however, was not understood and it was essentially forgotten[48].

[48] Perhaps because de Moivre's result appeared to be too special and it was not clear at all that it, in fact, holds universally.

Remark 2.14. De Moivre's biography is quite tragic and some bits of it deserve to be mentioned at least briefly. He was born in France, but had to immigrate to England because of the religious persecution. In England, religious discrimination was replaced by the national one: despite being highly regarded for his work and being a friend of Isaac Newton himself, de Moivre could not get a faculty position because he was a foreigner. He discovered the Gaussian curve, proved the first version of the CLT, derived the Stirling formula for estimating a factorial, first defined and studied the Poisson distribution. Throughout his life de Moivre remained poor. His main income was from private tutoring and playing chess.

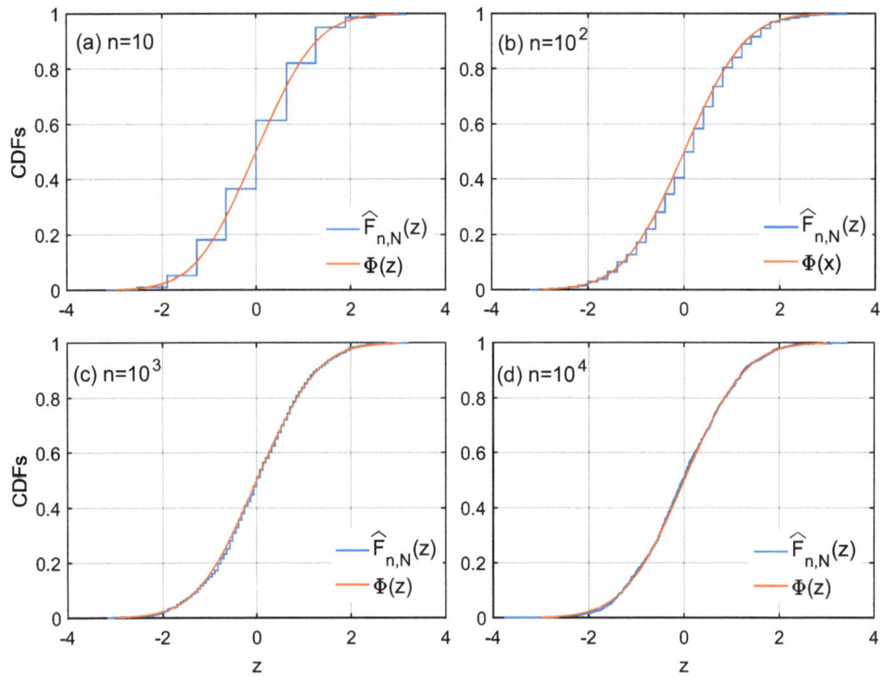

Fig. 2.11 The convergence of the eCDF $\widehat{F}_{n,N}(z) \approx F_n(z)$ to the standard normal CDF $\Phi(z)$ as n increases for (a) $n = 10$, (b) $n = 10^2$, (c) $n = 10^3$, and (d) $n = 10^4$. Each empirical CDF $\widehat{F}_{n,N}(z)$ is constructed based on $N = 10^3$ independent realizations of $X_n = \sqrt{n}\overline{Y}$.

The fundamental importance and universality of the CLT became apparent after the monumental work by Pierre-Simon de Laplace [2]. In his colossal *Théorie Analytique des Probabilités*, dedicated to Napoleon, Laplace not only extended de Moivre's version to the Bernoullian case with arbitrary p, but also discovered a more general version of the central limit theorem, developed the underlying mathematical machinery needed for proving modern versions of the CLT (characteristic functions, Laplace's method for approximating integrals), and described important statistical applications of the CLT.

Pierre-Simon de Laplace (1749-1827), a French mamathematician, physicist, astronomer, and philosopher, one of the founders of probability theory.

After Laplace, many prominent mathematicians worked on the CLT and its different versions [18]. The modern version formulated above, also know as the "classical" CLT, was rigorously proved by Aleksandr Lyapunov during 1900-1901 in a series of four papers [19, 20, 21, 22] on the "Theorem of Laplace". The name "central limit theorem" was coined by George Pólya only in 1920.

Remark 2.15. The classical CLT has a number of extensions and generalizations for cases where random variables $Y_1,\ldots,Y_n$ are not necessarily identically distributed or independent. One of the most often used in applications version is the *Lyapunov CLT* [22]. The theorem says that if $Y_1\ldots,Y_n,\ldots$ is s sequence of independent but not identically distributed random variables with $\mathbb{E}[Y_i] = \mu_i$ and $\mathbb{V}[Y_i] = \sigma_i^2$, then, if Lyapunov's condition is satisfied[49],

[49] Lyapunov's condition is a technical requirement that essentially says that the moments of $|Y_i|$ should not grow too fast.

$$\overline{Y} = \frac{1}{n}\sum_{i=1}^{n} Y_i \xrightarrow{\mathbb{D}} \mathcal{N}\left(\frac{1}{n}\sum_{i=1}^{n}\mu_i, \frac{1}{n^2}\sum_{i=1}^{n}\sigma_i^2\right), \quad \text{as } n \to \infty. \tag{2.93}$$

Note that if $Y_1 \ldots, Y_n, \ldots$ are actually IID with $\mathbb{E}[Y_i] = \mu$ and $\mathbb{V}[Y_i] = \sigma^2$, then

$$\frac{1}{n}\sum_{i=1}^{n}\mu_i = \mu \quad \text{and} \quad \frac{1}{n^2}\sum_{i=1}^{n}\sigma_i^2 = \frac{\sigma^2}{n}, \tag{2.94}$$

and Lyapunov's CLT (2.93) reduced to the classical version (2.85).

The CLT tells us that the sample mean $\overline{Y}$ converges to the normal distribution, but it does not tell how fast this happens. The rate of convergence is described by the *Berry-Esseen theorem* [23, 24]. Roughly speaking, the rate of convergence is $n^{-1/2}$. More precisely, if $Y_1 \ldots, Y_n, \ldots$ are IID with $\mathbb{E}[Y_i] = 0$, $\mathbb{E}[Y_i^2] = \mathbb{V}[Y_i] = \sigma^2$, and $\mathbb{E}[|Y_i^3|] = \rho$, then for all z and n

$$|F_n(z) - \Phi(z)| \leq C\frac{\rho}{\sigma^3\sqrt{n}}, \tag{2.95}$$

where $F_n(z)$ is the CDF of $\sqrt{n}\,\overline{Y}/\sigma$, that converges to $\Phi(z)$ by the CLT, and $C \approx 0.47$ is a universal constant that does not depend on z and n.

Problems

2.1. Simple Random Sampling
This problem will test your basic understanding of simple random sampling (SRS). Let $\mathscr{P} = \{x_1, \ldots, x_N\}$ be the target population, and $\mathscr{S} = \{X_1, \ldots, X_n\}$ a simple random sample from $\mathscr{P}$. Using the algorithmic definition of SRS in Section 2.3, compute the following probabilities:

(a) $\mathbb{P}(X_i = x_k)$,

(b) $\mathbb{P}(x_k \in \mathscr{S})$,

(c) $\mathbb{P}(X_1 = x_k,\ X_2 = x_s)$, where $k \neq s$,

(d) $\mathbb{P}(X_i = x_i,\ \text{for all } i = 1, \ldots, n)$.

2.2. Weighted Sample Mean
In Chapter 2, we focused on the sample mean $\overline{X} = \frac{1}{n}\sum_{i=1}^{n} X_i$ as an estimate of the population mean $\mu = \frac{1}{N}\sum_{i=1}^{N} x_i$. Let's consider a more general class of estimates:

$$\overline{X}_w = \sum_{i=1}^{n} w_i X_i, \tag{2.96}$$

where w_i are some fixed weights. Note that the standard sample mean $\overline{X}$ is a special case of (2.96) with $w_i = 1/n$.

(a) What condition should the weights $w_1,\ldots,w_n$ satisfy to guarantee that the estimate $\overline{X}_w$ is an *unbiased* estimate of the population mean μ?
(b) Among all unbiased estimates $\overline{X}_w$, find the most *efficient* estimate, that is the one with the smallest standard error. If you can't solve this problem rigorously, before checking the solution, think what answer you intuitively expect to get and why.

2.3. Estimating Proportions
In studying populations, the quantity of interest is often the proportion of population units that have a certain characteristic. For example, smoke or do not smoke, male or female, democrat or republican, like this book or not, etc. In this case, it makes sense for the unit characteristic x_i to be 1 or 0 depending on whether the characteristic of interest is present or not:

$$x_i = \begin{cases} 1, & \text{if the } i^{\text{th}} \text{ unit has the characteristic of interest,} \\ 0, & \text{if not.} \end{cases} \tag{2.97}$$

The population mean μ is then simply the proportion of ones, $\mu = \frac{1}{N}\sum_{i=1}^{N} x_i = \frac{\#1}{N}$.

(a) Show that the population variance is this case is

$$\sigma^2 = \mu(1-\mu). \tag{2.98}$$

(b) We know (Corollary 2.3) that $\hat{\sigma}_u^2$ is an unbiased estimate of σ^2. To compute this estimate, given by (2.38), we need to know *all* sample unites X_i. Use the result in part (a) to derive an unbiased estimate $\hat{\hat{\sigma}}_u^2$ of σ^2 that depends *only* on the sample mean $\overline{X}$, sample size n, and population size N.
(c) A Silicon Valley company has 300 employees. Out of 90 employees choosen by simple random sampling, 70 said that they like coffee and 20 that they don't. Using results in part (b), construct a 95% confidence interval for the proportion of company employees who like coffee.

2.4. Finding the Sample Size
A political poll in a large city is planned. The goal of the poll is to estimate the percentage of people supporting a certain bill. Find the minimum sample size n required to make a 95% confidence interval for the population percentage that is at most 4 percentage points wide[50]. The population variance is unknown: assume the worst-case scenario. The population size N is also unknown, but you can assume that it is large enough, so that the sampling fraction n/N is negligible and, therefore, the finite population correction can be ignored.

[50] This means that the length of the interval should be at most 0.04.

2.5. Variance of a Binary Population
Let $P = \{x_1,\ldots,x_N\}$ be the target population and suppose that population units x_i take only two values: α and β. Let π be the proportion of x_i's taking the value α.

(a) Find the population variance σ^2 in terms of α, β, and π.
(b) What value of π maximizes the population variance?

2.6. Matching the Accuracies of Surveys
A survey of physicians is planned with the goal to estimate the mean salary of doctors in the US. Doctors will be surveyed separately in each state. According to [12], there are $N_H = 9,464$ doctors in Hawaii and $N_C = 149,283$ doctors in California. A simple random sample of $n_H = 1000$ Hawaiian doctors is planned. Assume that the population variances σ_H^2 and σ_C^2 for the two populations of doctors are the same.

(a) What sample size n_C for the Californian survey should be used in order to achieve the same accuracy (measured by the standard error) as in the Hawaiian survey?

(b) What would be the answer in part (a) if $n_H = N_H$? Does this answer depend on the assumption $\sigma_H^2 = \sigma_C^2$?

2.7. The Bootstrap Method
In this problem, you will implement two versions of the bootstrap method for estimating the standard error of the sample mean: one is the standard bootstrap method described in Section 2.6 and the other is a modification of the standard method for the case where N/n is not integer. First, generate a Gaussian population $\mathscr{P} = \{x_1, \dots, x_N\}$ of size $N = 10^4$ in the same way as it was done in Example 2.6.1. Namely, generate N realizations $x_1, \dots, x_N$ of $Z \sim \mathcal{N}(\mu_0, \sigma_0^2)$ with $\mu_0 = 0$ and $\sigma_0 = 10$. The population $\mathscr{P}$ is the target population.

(a) Compute the population mean μ. Suppose now that you want to estimate μ using simple random sampling with the sample size $n = 700$. Draw a simple random sample $\mathscr{S} = \{X_1, \dots, X_n\}$ from the population $\mathscr{P}$ and compute the sample mean $\overline{X}$. Compute the exact standard error of $\overline{X}$. This exact value can be used as a reference value against which the estimates of $\mathrm{se}[\overline{X}]$ can be compared.

(b) Assume now that sample $\mathscr{S}$ is your data and you don't know the entire population $\mathscr{P}$, and, therefore, your can't compute the exact value of $\mathrm{se}[\overline{X}]$. Implement the bootstrap method, described in Section 2.6, to estimate $\mathrm{se}[\overline{X}]$. Since N/n is not integer, round it to the nearest integer to construct the bootstrap population $\mathscr{P}_{boot}$. Use $B = 10^3$ bootstrap samples. Compute the bootstrap estimate $\widehat{\mathrm{se}}_B[\overline{X}]$.

(c) Consider the following modification of the bootstrap method for the case where N/n is not integer, and, therefore, the size of $\mathscr{P}_{boot}$ is $[N/n] \times n \neq N$, where $[a]$ denotes the integer part of a. The new method differs from the original only in the way the bootstrap samples are generated, the rest is the same. In the new method, *two* bootstrap populations are created: the first takes k copies of each unit and the second takes $k+1$ copies of each unit,

$$\begin{aligned} \mathscr{P}_{boot}^{(1)} &= \{\underbrace{X_1, \dots, X_1}_{k}, \dots, \underbrace{X_n, \dots, X_n}_{k}\}, \\ \mathscr{P}_{boot}^{(2)} &= \{\underbrace{X_1, \dots, X_1}_{k+1}, \dots, \underbrace{X_n, \dots, X_n}_{k+1}\}, \end{aligned} \tag{2.99}$$

where $k = \lfloor N/n \rfloor$ is the greatest integer less than or equal to N/n. Then the population size N can be written as follows:

$$N = k \times n + r, \quad 0 < r < n, \tag{2.100}$$

where r is the remainder after division of N by n. To create a bootstrap sample, first, one of the two bootstrap populations is chosen. The first population $\mathscr{P}^{(1)}_{boot}$ is selected with probability

$$p = 1 - \frac{r}{n}, \tag{2.101}$$

and the second population $\mathscr{P}^{(2)}_{boot}$ is chosen with the remaining probability $1 - p$. Then a simple random sample of n units is taken from the chosen bootstrap population to form a bootstrap sample. In other words, to generate a bootstrap sample $\mathscr{S}^{(b)}$, $b = 1, \ldots, B$, we set $\mathscr{P}_{boot} = \mathscr{P}^{(1)}_{boot}$ with probability p and $\mathscr{P}_{boot} = \mathscr{P}^{(2)}_{boot}$ with probability $1 - p$, and then proceed as in the standard bootstrap method[51].

[51] If N/n is integer, then $r = 0$, $p = 1$, and the modified method reduces to the standard bootstrap.

Remark 2.16. The main idea behind the selection probability (2.101) is to ensure that the *expected* size of now *random* bootstrap population $\mathscr{P}_{boot}$ is N. Indeed,

$$\begin{aligned} \mathbb{E}[\text{size of } \mathscr{P}_{boot}] &= p\mathbb{E}[\text{size of } \mathscr{P}^{(1)}_{boot}] + (1-p)\mathbb{E}[\text{size of } \mathscr{P}^{(2)}_{boot}] \\ &= pnk + (1-p)n(k+1) = nk + n - pn \\ &= nk + r = N. \end{aligned} \tag{2.102}$$

This means that on average, bootstrap samples are draw from a bootstrap population of the same size N as the original target population $\mathscr{P}$.

Remark 2.17. If $N \gg n$, then this version of bootstrap is very similar to the modification of bootstrap proposed by Peter Bickel and David Freedman in [13]. In their method, the first bootstrap population is selected with probability

$$\tilde{p} = \left(1 - \frac{r}{n}\right)\left(1 - \frac{r}{N-1}\right). \tag{2.103}$$

If $N \gg n$, then $N \gg r$, the ratio $r/(N-1) \approx 0$, and $\tilde{p} \approx p$.

Implement the modified bootstrap method with $B = 10^3$ bootstrap samples and compute the bootstrap estimate $\widehat{\text{se}}^*_B[\overline{X}]$ of $\text{se}[\overline{X}]$.

(d) To explore the variability of the bootstrap estimates, run both algorithm in (b) and (c) 100 times, and plot two boxplots (one for each method) for the obtained estimates $\widehat{\text{se}}_B[\overline{X}]$ and $\widehat{\text{se}}^*_B[\overline{X}]$ together with the reference exact value of $\text{se}[\overline{X}]$ computed on part (a).

2.8. The Body Mass Index of Male and Female Olympic Athletes
In the Case Study in Section 2.10, we estimated the mean BMI (body mass index) of the Rio 2016 Olympic athletes using simple random sampling. In our study, we did not differentiate between male and female athletes. In this problem, you will estimate the mean BMI for male and female Olympic athletes separately. First, download the athletes data from [10].

(a) Clean the data by first removing all athletes whose height or mass is missing. Then compute all BMIs, sort them, and, if necessary, correct the heights and masses of athletes with the 5 smallest and 5 largest BMIs[52]. Use the Wikipedia articles dedicated to the corresponding athletes for checking their heights and masses. After cleaning the data, recompute all BMI values.

(b) Construct two boxplots for BMIs of male and female athletes and compute the true values of the mean male BMI, $\mu^{\rm m}_{\rm BMI}$, and the mean female BMI, $\mu^{\rm f}_{\rm BMI}$.

(c) Estimate the population parameters $\mu^{\rm f}_{\rm BMI}$ and $\mu^{\rm m}_{\rm BMI}$ using simple random samples $\mathscr{S}_X = \{X_1, \ldots, X_n\} \subset \mathscr{P}_{\rm f}$ and $\mathscr{S}_Y = \{Y_1, \ldots, Y_n\} \subset \mathscr{P}_{\rm m}$, where $\mathscr{P}_{\rm f}$ and $\mathscr{P}_{\rm m}$ are the populations of female and male BMIs and the sample size $n = 500$. Estimate the standard errors of the estimates both analytically and using the bootstrap method with 10^3 bootstrap samples, construct the 95% confidence intervals for $\mu^{\rm f}_{\rm BMI}$ and $\mu^{\rm m}_{\rm BMI}$ based on samples $\mathscr{S}_X$ and $\mathscr{S}_Y$, and plot them together with the true values of $\mu^{\rm f}_{\rm BMI}$ and $\mu^{\rm m}_{\rm BMI}$ computed in part (b).

(d) Find five athlete(s) of the same sex as yours with the BMI closest to your BMI.

[52] Of course, this procedure will not remove all errors in the data, but at least several most obvious ones will be corrected.

References

1. Graunt, J.: Natural and Political Observations Mentioned in a Following Index, and Made Upon the Bills of Mortality. London (1662). In: *Mathematical Demography. Biomathematics*, vol 6. Springer, Berlin, Heidelberg. `https://doi.org/10.1007/978-3-642-81046-6_2`
2. Laplace, P.S.: *Théorie Analytique des Probabilités*. Paris (1812). Available free online from `https://archive.org/details/thorieanalytiqu01laplgoog/page/n4`
3. Bethlehem, J.: The rise of survey sampling. CBS Discussion Paper: 1572-0314. *The Hague: Statistics Netherlands* (2009). `https://hdl.handle.net/11245/1.312955`
4. Lohr, S.L.: *Sampling: Design and Analysis*. CRC Press (2022). `https://www.sharonlohr.com/sampling-design-and-analysis`
5. Rice, J.A.: *Mathematical Statistics and Data Analysis*, 3rd edition. Brooks/Cole (2007). `https://www.stat.berkeley.edu/~rice/Book3ed/index.html`
6. Efron, B.: Bootstrap methods: another look at the jackknife. *The Annals of Statistics* 7(1), 1–26 (1979). `https://doi.org/10.1214/aos/1176344552`
7. Diaconis, P., Efron, B.: Computer-intensive methods in statistics. *Scientific American* 248(5), 116–129 (1983). `https://www.vanderbilt.edu/psychological_sciences/graduate/programs/quantitative-methods/quantitative-content/diaconis_efron_1983.pdf`
8. National Heart, Lung, and Blood Institute: Why Is a Healthy Weight Important? `https://www.nhlbi.nih.gov/health/educational/lose_wt/`
9. World Health Organization: Mean Body Mass Index (BMI). `https://www.who.int/data/gho/data/themes/topics/topic-details/GHO/body-mass-index`
10. Kaggle. 2016 Olympics in Rio de Janeiro. `https://www.kaggle.com/rio2016/olympic-games`. Source link: `https://github.com/flother/rio2016`. The data set is also available at `https://link.springer.com/chapter/10.1007/978-3-032-03848-7_1`

11. Riggott, M.: Data for the 2016 Olympic Games in Rio de Janeiro. https://flother.is/2017/olympic-games-data/
12. Young, A., Chaudhry, H.J., Pei, X., Arnhart, K., Dugan, M., Snyder, G.B.: A census of actively licensed physicians in the United States, 2016. *Journal of Medical Regulation* 103(2), 7–21 (2017). https://www.fsmb.org/siteassets/advocacy/publications/2016census.pdf
13. Bickel, P.J., Freedman, D.A.: Asymptotic normality and the bootstrap in stratified sampling. *The Annals of Statistics* 12(2), 470–482 (1984). https://doi.org/10.1214/aos/1176346500
14. Bernoulli, J.: *Ars Conjectandi* (1713). English translation: E.D. Sylla, *The Art of Conjecturing*, Johns Hopkins University Press (2006).
15. Khinchin, A.Y.: Sur la loi des grands nombres. *Comptes rendus de l'Académie des Sciences* 188, 477-479, (1929).
16. Kolmogorov, A.N.: Sur la loi forte des grands nombres. *Comptes rendus de l'Académie des Sciences* 191, 910-912 (1930).
17. de Moivre, A.: Approximatio ad summam terminorum binomii $(a+b)^n$ in seriem expansi. 7 pages offprint (1733).
18. Fischer, H.: *A History of the Central Limit Theorem: From Classical to Modern Probability Theory*. Springer, New York, NY (2011). https://doi.org/10.1007/978-0-387-87857-7
19. Lyapunov, A.M.: Sur une proposition de la théorie des probabilités. *Bulletin de l'Académie Impériale des Sciences de St.-Pétersbourg* (5)13, 359-386 (1900).
20. Lyapunov, A.M.: Sur un théorème du calcul des probabilités. *Comptes rendus hebdomadaires des séances de l'Académie des Sciences de Paris* 132, 126-128 (1901).
21. Lyapunov, A.M.: Nouvelle forme du théorème sur la limite de probabilité. *Mémoires de l'Académie Impériale des Sciences de St.-Pétersbourg VIIIe Série, Classe Physico-Mathématique* 12, 1-24 (1901).
22. Lyapunov, A.M.: Une proposition générale du calcul des probabilités. *Comptes rendus hebdomadaires des séances de l'Académie des Sciences de Paris* 132, 814 (1901).
23. Berry, A.C.: The accuracy of the Gaussian approximation to the sum of independent variates. *Transactions of the American Mathematical Society* 49(1), 122–136 (1941). https://doi.org/10.1090/S0002-9947-1941-0003498-3
24. Esseen, C.-G.: On the Liapunoff limit of error in the theory of probability. *Arkiv för Matematik, Astronomi och Fysik.* A28: 1–19 (1942).

3
Modeling and Inference: A Big Picture

Stochastic models and statistical reasoning are more relevant to the world... than exact models and logical reasoning.
– David Mumford, *The Dawning of the Age of Stochasticity* [1]

Suppose we are interested in studying a certain phenomenon that can be schematically represented as follows:

Furthermore, suppose we collected some data – a set of inputs and the corresponding responses – by observation or experiment. What can we learn, or *infer*, about the phenomenon from the data? There are two fundamental goals of data analysis:

1. *Understanding:* We want to understand the mechanism of how the responses are associated with the inputs. What is "inside" the Nature "black box"?[1]
2. *Prediction:* We want to learn how to predict the response to a future input.

These goals are related: we can potentially use our understanding of the phenomenon for constructing an accurate prediction algorithm, and we can use the prediction algorithm for testing and validating our understanding.

The central idea of statistical inference is to replace the Nature "black box", i.e. the unknown mechanism that Nature uses to associate the responses to the inputs, with a *stochastic model*:

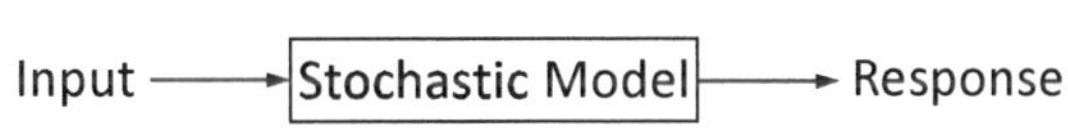

The key feature of a stochastic model[2] is that the observed *variability* in the data is modeled by *probability distributions*, which form the building-blocks of the model. In other words, the data is viewed as a collection of realizations of random variables. In this chapter, we will define two main types of stochastic models and

[1] "There are questions that you don't want to spoil with an answer." – Aleksandr Asmolov.

[2] Stochastic models are also called statistical models.

Supplementary Information The online version contains supplementary material available at https://doi.org/10.1007/978-3-032-03848-7_3.

K. M. Zuev, *Fundamentals of Statistical Inference*,
International Series in Operations Research & Management Science 374,
https://doi.org/10.1007/978-3-032-03848-7_3

provide a big picture of several fundamental concepts and problems of statistical inference.

3.1 Stochastic Models

Suppose that the data consists only of "responses" $X_1,\dots,X_n$. A *stochastic model* $\mathscr{F}$ is then simply a set of cumulative distribution functions (CDFs) F for $X_1,\dots,X_n$. By modeling the data with $\mathscr{F}$ we *assume*[3] that $X_1,\dots,X_n$ is an IID sample from one of the CDFs in $\mathscr{F}$:

[3] It is important to remember that this is an assumption and it could be wrong.

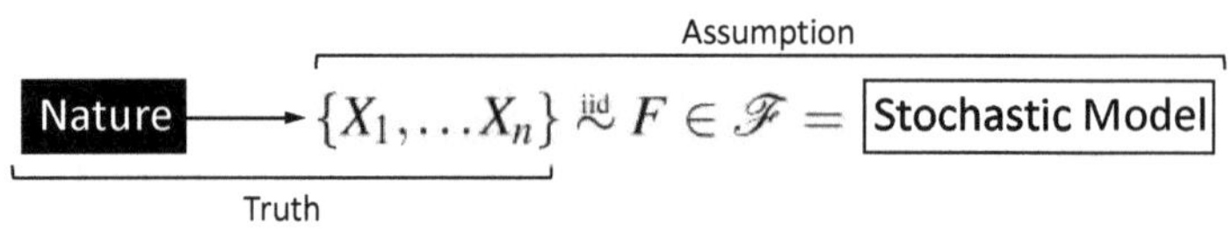

The goal is then to infer F, or some properties of F (such as the mean and the variance) from the data. If our assumption is correct and $\mathscr{F}$ is a suitable model for the data, then the inferred properties of F could be useful for learning about the underling phenomenon of interest and the mechanism that generated the data.

Remark 3.1. If it is reasonable to consider $X_1,\dots,X_n$ as realizations of continuous or discrete random variables, a stochastic model $\mathscr{F}$ can be defined as a set of probability density functions (PDFs) or probability mass functions (PMFs), respectively.

3.1.1 Parametric and Nonparametric Models

There are two big classes of stochastic models: *parametric* and *nonparametric*[4]. A stochastic model $\mathscr{F}$ is called parametric if it can be parameterized by a finite number of parameters. For example, if we decide to model the data by a normal distribution, then our stochastic model $\mathscr{F}$ will be a two-dimensional parametric model:

[4] Which leads to two subfields of statistics: parametric and nonparametric statistics.

$$\mathscr{F} = \left\{ \phi(x|\mu,\sigma^2) = \frac{1}{\sqrt{2\pi}\sigma} \exp\left(-\frac{(x-\mu)^2}{2\sigma^2}\right), \quad \mu,\sigma^2 \in \mathbb{R} \right\}. \tag{3.1}$$

Here, the mean μ and the variance σ^2 are the model parameters. In general, a parametric model has the following from

$$\mathscr{F} = \{F(x|\theta), \quad \theta \in \Theta\}, \tag{3.2}$$

where θ is an unknown parameter, or a vector of parameters, that belongs to the *parameter space* $\Theta \subset \mathbb{R}^d$, and $F(x|\theta)$ is a CDF parameterized by θ. In *parametric inference*, we want to infer θ from the data.

Quite naturally, a stochastic model $\mathscr{F}$ is called nonparametric if it is not parametric, that is if $\mathscr{F}$ cannot be parameterized by a finite number of parameters. For example, if we decide to model the data by a distribution with zero mean, then our model $\mathscr{F}$ will be nonparametric:

$$\mathscr{F} = \left\{F : \int x dF(x) = 0\right\}. \tag{3.3}$$

Remark 3.2. If you feel uncomfortable with the (Riemann-Stieltjes) integral in (3.3)[5], then let's further assume that the CDF F is sufficiently smooth[6], so that the probability distribution defined by F is continuous with the PDF f, and $F' = f$. Then $\int x dF(x) = \int x F'(x) dx = \int x f(x) dx$, and the model becomes

$$\mathscr{F} = \left\{f : \int x f(x) dx = 0\right\}. \tag{3.4}$$

[5] We'll discuss its meaning in Remark 4.1.

[6] More formally, continuously differentiable.

Taking this example to extreme and throwing away the zero mean assumption, we obtain the most general nonparametric stochastic model,

$$\mathscr{F} = \{\text{all CDFs}\}. \tag{3.5}$$

At first glance, this model may look silly, but it is not. In fact, this model is often used in practice when not much is known about the mechanism that generated the data. Essentially, the model in (3.5) says that all we assume is that $X_1, \ldots, X_n$ is an IID sample from some distribution. In the forthcoming chapters we will see that we can learn a lot from the data even under this seemingly weak assumption.

Remark 3.3. The IID assumption is, in fact, not that weak as it may initially appear. For example, it immediately allows us to use the law of large numbers (LLN) and the central limit theorem (CLT)[7], the two main workhorses of statistical inference. Intuitively, the IID assumption states that all X_i provide some pieces of information about the underlying phenomenon, and these pieces have the same quality and are independent of each other. Although it is a very useful and convenient modeling assumption that allows to construct powerful statistical methods, it is usually hard to fully justify it in practice. Over the years, many heuristic diagnostics and formal tests have been developed for checking the IID assumption, such as the entropy-based tests [2, 3] and generalized runs tests [4]. However, these techniques are typically more efficient in disproving the IID assumption rather than in validating it. Whether the IID assumption is reasonable depends not only on the data per se, but also on how the data was collected and how it will be used. Nevertheless, the IID assumption is often made[8] in statistical inference, and the following, somewhat pragmatic approach, is often adopted. Let's assume that the IID assumption is satisfied, and let's learn about the phenomenon of interest from the data and under this assumption. If the inferred knowledge would allow us to make accurate predictions and good, knowledge-driven decisions, this would indirectly validate, if not the IID assumption itself, then at least its use in the inferential process.

[7] See Probability Refresher 2.12, for a brief review of LLN and CLT.

[8] After all, we can't do inference without making assumptions [5].

Historically, parametric models were developed first and used more often, since most methods for the analysis of nonparametric models were not feasible in practice because of the limited computing power. This has changed after computing became cheaper and more widely available, and nowadays both parametric and nonparametric models are extensively used in applications. These two classes of stochastic models have their own advantages and disadvantages.

Advantages of parametric models:

1. If a parametric model $\mathscr{F}$ is correct[9], then parametric methods are more effective (we can infer more from the same amount of data) than their nonparametric counterparts.
2. Usually parametric models are easier to interpret then nonparametric models, since their parameters often have "physical" meanings.
3. Parametric models are generally easier to analyze analytically.

[9] This means that there exists $\theta_\star \in \Theta$, often called the "true value," such that the corresponding CDF $F(x|\theta_\star) \in \mathscr{F}$ indeed adequately describes the data.

Advantages of nonparametric models:

1. Nonparametric models have a lower risk of misspecification[10] compared to parametric models. Informally, nonparametric models are "larger" or "richer" than parametric models (recall that the latter can be finitely parameterized, while the former cannot be). Large models are more likely to contain a correct distribution that accurately describes the data.
2. In the cases where there is not much theory behind the data generating process, nonparametric models are usually easier to formulate.
3. Typically, nonparametric models and methods are less sensitive to outliers.

[10] Misspecification is a choice of a model $\mathscr{F}$ that does not contain a distribution that adequately describes the data.

Stochastic modeling is an art. It is based on a proper incorporation of the scientific knowledge about the underlying phenomenon into the model, and on finding a balance between the model complexity and realisticity on one hand, and the ability to analyze the model analytically or numerically on the other hand. The choice of a stochastic model also depends on the problem and the required answer so that different models may be appropriate for a single data set.

3.2 Types of Problems in Statistical Inference

There are three fundamental types of problems in statistical inference: *point estimation*, constructing *confidence intervals* and sets, and *hypothesis testing*. In this chapter, we will briefly consider all these problems, and illustrate the differences between them with an iconic coin flipping example. In the rest of the book, we will study these problems in detail in different contexts.

3.2.1 Point Estimation

In *point estimation* the goal is to find a single "best guess" (point estimate) of some unknown quantity of interest from the data. If θ is the quantity of interest, then its point estimate is usually denoted by $\hat{\theta}$. For example, if $\theta = \mu$ is the population mean, then the sample mean $\hat{\theta} = \overline{X}$ is a point estimate of μ. In general, the quantity of interest θ could be a population parameter, a model parameter, or a vector of parameters, or even a CDF, PDF, PMF, or some other function of interest. Whatever θ is, the goal of point estimation is to find $\hat{\theta}$ such that $\hat{\theta} \approx \theta$.

Let $X_1, \ldots, X_n$ be data which is modeled as an IID random sample from a distribution $F(x \mid \theta) \in \mathscr{F}$, where $\mathscr{F}$ is a parametric model. A point estimate $\hat{\theta}_n$ of parameter θ, by definition, is some function of the data:

$$\hat{\theta}_n = s(X_1, \ldots, X_n). \tag{3.6}$$

It is convenient to enhance the notation for a point estimate $\hat{\theta}$ with a subscript $_n$, since it reminds us that the estimate is obtained from the data of sample size n. This will be useful when we study the asymptotic properties of the estimate as $n \to \infty$.

It is important to highlight the conceptual difference between θ and $\hat{\theta}_n$: while θ is a fixed *deterministic* unknown quantity, $\hat{\theta}_n$ is a *random* variable, since it is a function of $X_1, \ldots, X_n$ which are random. The estimate $\hat{\theta}_n$ becomes a *realization* of the random variable when we compute its value on the actually observed data. The probability distribution of $\hat{\theta}_n$ is called the *sampling distribution*. The standard deviation of $\hat{\theta}_n$ is called the *standard error*[11],

$$\text{se}[\hat{\theta}_n] = \sqrt{\mathbb{V}[\hat{\theta}_n]}. \tag{3.7}$$

[11] Notice that these definitions mirror the corresponding definitions for $\overline{X}$ that we discussed in the context of survey sampling in Chapter 2.

To quantify how accurate $\hat{\theta}_n$ estimates θ on average, let's define its *bias* as the difference between its expected value and θ:

$$\mathbb{B}[\hat{\theta}_n] = \mathbb{E}[\hat{\theta}_n] - \theta. \tag{3.8}$$

Definition 3.1. An estimate $\hat{\theta}_n$ is called *unbiased* if $\mathbb{E}[\hat{\theta}_n] = \theta$, or, equivalently,

$$\mathbb{B}[\hat{\theta}_n] = 0. \tag{3.9}$$

Unbiasedness is a good property, but its importance should not be overstated: an estimate could be unbiased, but at the same time it could have a very large standard error. In this case, although on average the estimate equals to θ, its particular realizations are likely to be far from θ.

The overall accuracy of a point estimate is often quantified by the *mean squared error* (MSE):

$$\text{MSE}[\hat{\theta}_n] = \mathbb{E}[(\hat{\theta}_n - \theta)^2]. \tag{3.10}$$

It turns out that the MSE can be written in terms of the bias and standard error:

$$
\begin{aligned}
\mathrm{MSE}[\hat{\theta}_n] &= \mathbb{E}[(\hat{\theta}_n - \theta)^2] = \mathbb{E}\Big[\big(\hat{\theta}_n - \mathbb{E}[\hat{\theta}_n] + \underbrace{\mathbb{E}[\hat{\theta}_n] - \theta}_{\mathbb{B}[\hat{\theta}_n]}\big)^2\Big] \\
&= \underbrace{\mathbb{E}\Big[\big(\hat{\theta}_n - \mathbb{E}[\hat{\theta}_n]\big)^2\Big]}_{\mathbb{V}[\hat{\theta}_n]=\mathrm{se}[\hat{\theta}_n]^2} + 2\underbrace{\mathbb{E}\Big[\hat{\theta}_n - \mathbb{E}[\hat{\theta}_n]\Big]}_{0}\,\mathbb{B}[\hat{\theta}_n] + \mathbb{B}[\hat{\theta}_n]^2 \qquad (3.11) \\
&= \mathbb{B}[\hat{\theta}_n]^2 + \mathrm{se}[\hat{\theta}_n]^2.
\end{aligned}
$$

This representation of the MSE as a sum of the squared bias and squared standard error is called the *bias-variance decomposition* of the mean squared error.

Example 3.1. Let us flip a coin n times, and let $X_i = 1$ if we get "heads" on the ith toss, and $X_i = 0$ if we get "tails." If it is unknown whether the coin is fair, then it is reasonable to model the data $X_1, \ldots, X_n$ as an IID sample from the Bernoulli distribution, which is the probability distribution of a random variable which takes two value: 1 with probability p and 0 with probability of $1-p$, where $p \in [0,1]$ is the model parameter[12],

[12] For a fair coin, $p = 1/2$.

$$
X_1, \ldots, X_n \overset{\text{iid}}{\sim} \mathscr{B}(p), \quad X_i = \begin{cases} 1, & \text{with probability } p, \\ 0, & \text{with probability } 1-p. \end{cases} \qquad (3.12)
$$

Suppose we want to estimate p from the data. Let's use the sample mean as a point estimate, $\hat{p}_n = \overline{X}$. This estimate is unbiased,

$$
\mathbb{B}[\hat{p}_n] = \mathbb{E}[\hat{p}_n] - p = \mathbb{E}\left[\frac{1}{n}\sum_{i=1}^n X_i\right] - p = \frac{1}{n}\sum_{i=1}^n \underbrace{\mathbb{E}[X_i]}_{1\cdot p + 0\cdot(1-p)} - p = p - p = 0, \qquad (3.13)
$$

its squared standradrd error is

$$
\begin{aligned}
\mathrm{se}[\hat{p}_n]^2 &= \mathbb{V}[\hat{p}_n] = \mathbb{V}\left[\frac{1}{n}\sum_{i=1}^n X_i\right] = \frac{1}{n^2}\sum_{i=1}^n \mathbb{V}[X_i] \\
&= \frac{1}{n^2}\sum_{i=1}^n \Big(\underbrace{\mathbb{E}[X_i^2]}_{p} - \underbrace{\mathbb{E}[X_i]^2}_{p^2}\Big) = \frac{p(1-p)}{n},
\end{aligned} \qquad (3.14)
$$

and the mean squared error is, therefore,

$$
\mathrm{MSE}[\hat{p}_n] = \mathbb{B}[\hat{p}_n]^2 + \mathrm{se}[\hat{p}_n]^2 = \frac{p(1-p)}{n}. \qquad (3.15)
$$

3.2.2 *Confidence Intervals and Sets*

We have already encountered confidence intervals for the population mean in the context of survey sampling in Section 2.9. Confidence intervals for model parameters are defined similarly.

Definition 3.2. Let $X_1,\ldots,X_n \overset{\text{iid}}{\sim} F(x|\theta) \in \mathscr{F}$. An $100(1-\alpha)\%$ *confidence interval* for parameter θ is a *random* interval I_α constructed from the data, which contains θ with probability $1-\alpha$,

$$\mathbb{P}(\theta \in I_\alpha) = 1-\alpha. \tag{3.16}$$

The value $100(1-\alpha)\%$ is called the *confidence level*[13].

If θ is a vector of parameters, $\theta \in \Theta \subset \mathbb{R}^d$, then the confidence interval I_α is replaced by a confidence set S_α, which can be a random cube, sphere, ellipsoid, or any other random set that traps θ with probability $1-\alpha$.

[13] Usually, confidence levels of 90% ($\alpha = 0.1$) and 95% ($\alpha = 0.05$) are used.

Example 3.2. Suppose that, as in Example 3.1, $X_1,\ldots,X_n \overset{\text{iid}}{\sim} \mathscr{B}(p)$. But now, instead of estimating parameter p, let's construct a confidence interval for p from the data. To this end, we can use Chebyshov's inequality[14], which states that for any random variable Y with mean μ and variance σ^2 and for any $b>0$

$$\mathbb{P}(|Y-\mu| \geq b) \leq \frac{\sigma^2}{b^2}. \tag{3.17}$$

[14] For a quick review of Chebyshov's inequality see Probability Refresher 3.5.

Let's apply this inequality to the sample mean $Y = \overline{X}$. Since, from (3.13) and (3.14), $\mathbb{E}[\overline{X}] = p$ and $\mathbb{V}[\overline{X}] = p(1-p)/n$, we have:

$$\mathbb{P}(|\overline{X}-p| \geq b) \leq \frac{p(1-p)}{nb^2} \leq \frac{1}{4nb^2}. \tag{3.18}$$

In the last inequality we used the fact that if $p \in [0,1]$, then $p(1-p) \leq 1/4$, which becomes obvious if we plot the function $f(p) = p(1-p)$ on the interval $[0,1]$. To find a random interval that contains p with some probability, we need to work with inequality $|\overline{X}-p| < b$, which is equivalent to $\overline{X}-b < p < \overline{X}+b$, instead of $|\overline{X}-p| \geq b$:

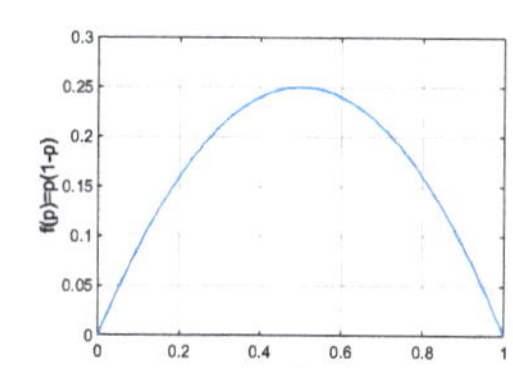

Function $f(p) = p(1-p)$ on the interval $p \in [0,1]$.

$$\mathbb{P}(\overline{X}-b < p < \overline{X}+b) = \mathbb{P}(|\overline{X}-p| < b) = 1-\mathbb{P}(|\overline{X}-p| \geq b) \geq 1-\frac{1}{4nb^2}. \tag{3.19}$$

It remains to set

$$\frac{1}{4nb^2} = \alpha \quad \Leftrightarrow \quad b = \frac{1}{2\sqrt{\alpha n}}, \tag{3.20}$$

and find from (3.19) that

$$\mathbb{P}\left(\overline{X} - \frac{1}{2\sqrt{\alpha n}} < p < \overline{X} + \frac{1}{2\sqrt{\alpha n}}\right) \geq 1-\alpha. \tag{3.21}$$

The interval

$$I_\alpha = \left(\overline{X} - \frac{1}{2\sqrt{\alpha n}},\ \overline{X} + \frac{1}{2\sqrt{\alpha n}}\right) \tag{3.22}$$

is, therefore, *at least* $100(1-\alpha)\%$ confidence interval for p.

As discussed in Probability Refresher 3.5, Chebyshov's inequality is not sharp. This means that although the upper bound in (3.17) is always correct, it is often can be improved and made tighter, that is, smaller. This non-sharpness is the price to pay for the wide applicability of the inequality: it holds for any random variable with finite mean and variance. When we used Chebyshov's inequality for constructing a confidence interval for parameter p, we did not actually used the fact that $X_1,\ldots,X_n$ are independent[15] and bounded, $X_i \in [0,1]$, since the inequality does not need those assumptions. To make use of these properties, let's apply Hoeffding's inequality[16] to the data $X_1,\ldots,X_n \stackrel{\text{iid}}{\sim} \mathscr{B}(p)$:

$$\mathbb{P}(|\overline{X}-p| \geq c) \leq 2\exp\left(-2nc^2\right). \tag{3.23}$$

[15] We used independence for computing the variance $\mathbb{V}[\overline{X}] = p(1-p)/n$, but the same expression can be obtained under a much weaker assumption: pairwise uncorrelatedness.

[16] For a quick review of Hoeffding's inequality see Probability Refresher 3.5.

Therefore,

$$\begin{aligned}\mathbb{P}(\overline{X}-c < p < \overline{X}+c) &= \mathbb{P}(|\overline{X}-p| < c) \\ &= 1-\mathbb{P}(|\overline{X}-p| \geq c) \geq 1-2\exp\left(-2nc^2\right).\end{aligned} \tag{3.24}$$

Now setting

$$2\exp\left(-2nc^2\right) = \alpha \quad \Leftrightarrow \quad c = \sqrt{-\frac{1}{2n}\log\frac{\alpha}{2}}, \tag{3.25}$$

and obtaining that

$$\mathbb{P}\left(\overline{X} - \sqrt{-\frac{1}{2n}\log\frac{\alpha}{2}} < p < \overline{X} + \sqrt{-\frac{1}{2n}\log\frac{\alpha}{2}}\right) \geq 1-\alpha \tag{3.26}$$

allows to conclude that

$$I_\alpha = \left(\overline{X} - \sqrt{-\frac{1}{2n}\log\frac{\alpha}{2}},\ \overline{X} + \sqrt{-\frac{1}{2n}\log\frac{\alpha}{2}}\right) \tag{3.27}$$

is *at least* $100(1-\alpha)\%$ confidence interval for p.

Notice that confidence intervals (3.22) and (3.27) are both centered on the sample mean $\overline{X}$, but have different geometric lengths. The lengths of Chebyshov's I_α^C and Hoeffding's I_α^H intervals are, respectively,

$$|I_\alpha^C| = \frac{1}{\sqrt{\alpha n}} \quad \text{and} \quad |I_\alpha^H| = 2\sqrt{-\frac{1}{2n}\log\frac{\alpha}{2}}. \tag{3.28}$$

Figure 3.1 shows the dependence of the lengths $|I_\alpha^C|$ and $|I_\alpha^H|$ on the sample size n for $\alpha = 0.05$. Both confidence intervals become smaller as n increases. This means that, as expected, the more data we have, the more confident we are about the location of the unknown parameter p. Another important observation is that for any n,

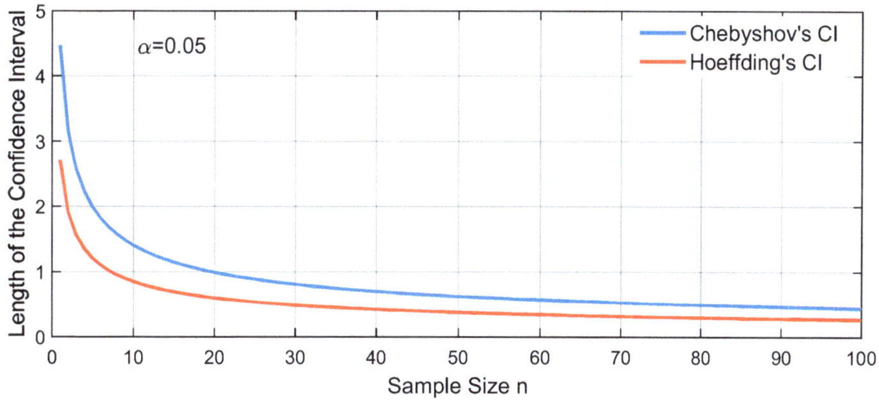

Fig. 3.1 The geometric lengths of Chebyshov's and Hoeffding's confidence intervals for $\alpha = 0.05$.

Hoeffding's interval is shorter than Chebyshov's interval, and, therefore, preferred. This property is a consequence of the fact[17] that Hoeffding's inequality is sharper than Chebyshov's inequality and provides a tighter upper bound. So, although it is less general and has more assumptions, Hoeffding's inequality is more effective for constructing confidence intervals when, like in this example, all the underlying assumptions are satisfied.

[17] Discussed in Probability Refresher 3.5

Confidence intervals for a parameter of interest θ often have the following form:

$$I_\alpha = (\hat{\theta}_n - \mathrm{MOE}_{n,\alpha}, \hat{\theta}_n + \mathrm{MOE}_{n,\alpha}), \tag{3.29}$$

where $\hat{\theta}_n$ is a point estimate of θ and $\mathrm{MOE}_{n,\alpha}$ is the *margin of error* which depends both on α and the sample size n and controls the confidence level of the interval and its geometric size.

For example, quite often point estimates are approximately normally distributed[18]:

[18] Recall the normal approximation for the sample mean in simple random sampling, Section 2.7.

$$\hat{\theta}_n \dot{\sim} \mathscr{N}\left(\theta, \mathrm{se}[\hat{\theta}_n]^2\right) \tag{3.30}$$

This approximation immediately allows to construct an *approximate* $100(1-\alpha)\%$ confidence interval I_α for θ in the from (3.29):

$$I_\alpha = \left(\hat{\theta}_n + z_{\frac{\alpha}{2}} \mathrm{se}[\hat{\theta}_n], \hat{\theta}_n - z_{\frac{\alpha}{2}} \mathrm{se}[\hat{\theta}_n]\right), \tag{3.31}$$

where $z_{\frac{\alpha}{2}}$ is the $(\frac{\alpha}{2})^{\text{th}}$ standard normal quantile, see Fig. 1.7, and $\mathrm{MOE}_{n,\alpha} = -z_{\frac{\alpha}{2}} \mathrm{se}[\hat{\theta}_n]$ is the margin of error. Indeed,

$$\mathbb{P}\left(\hat{\theta}_n + z_{\frac{\alpha}{2}}\mathrm{se}[\hat{\theta}_n] < \theta < \hat{\theta}_n - z_{\frac{\alpha}{2}}\mathrm{se}[\hat{\theta}_n]\right) = \mathbb{P}\Big(z_{\frac{\alpha}{2}} < \underbrace{\frac{\hat{\theta}_n - \theta}{\mathrm{se}[\hat{\theta}_n]}}_{\dot{\sim}\mathcal{N}(0,1)} < -z_{\frac{\alpha}{2}}\Big)$$
$$\approx \Phi(-z_{\frac{\alpha}{2}}) - \Phi(z_{\frac{\alpha}{2}}) = \left(1 - \Phi(z_{\frac{\alpha}{2}})\right) - \Phi(z_{\frac{\alpha}{2}}) = \left(1 - \frac{\alpha}{2}\right) - \frac{\alpha}{2} = 1 - \alpha. \tag{3.32}$$

3.2.3 Hypothesis Testing

While we have seen point estimation and confidence intervals in the context of survey sampling in Chapter 2, hypothesis testing is something completely new for us. In hypothesis testing, we start with the so-called *null hypothesis*, denoted by H_0, that usually represents some "default theory" or "current state of knowledge" that is generally assumed to be true (until data shows otherwise). Then we ask if the observed data provides sufficient evidence to *reject* that default theory. If yes, then we reject the null hypothesis H_0 in favor of an *alternative hypothesis* H_1, also called the *research hypothesis*, which is the opposite of H_0. If not, then we don't reject the null hypothesis or, as often said, we *fail to reject* it. Notice that failing to reject H_0 is logically different from accepting H_0 as being true: the absence of evidence against a hypothesis does not prove that the hypothesis is true[19]. For example, if H_0 states that all cats are black[20] and the data consists of a single observation — a black cat — then the data provides no evidence against the null hypothesis, we fail to reject it based on the data, but it would be wrong to accept H_0. On the contrary, observing a white cat would allow us to safely reject H_0. Since the machinery of hypothesis testing allows to reject the null hypothesis, but does not allow to accept it, often the null hypothesis represents some statement that we want to disprove[21]. We will study hypothesis testing in detail in Chapter 9. Here we just consider a simple example to give a flavor of this important area of statistical inference.

[19] See also Remark 3.4 after Example 3.3.

[20] Which is obviously false: see Fig. 0.1.

[21] Not in a mathematical, but in statistical sense: that is, to show that the observed data is very unlikely to have occurred if the null hypothesis H_0 were true.

Example 3.3. Suppose we want to test if the coin that we flipped in Example 3.1 is fair, that is, the probabilities of observing heads and tails are the same. Let H_0 be the null hypothesis that the coin is fair, and let H_1 be the alternative hypothesis that the coin is not fair. Under the Bernoulli model for the data, $X_1,\dots,X_n \overset{\text{iid}}{\sim} \mathscr{B}(p)$, we can write the hypotheses as follows:

$$H_0 : p = \frac{1}{2} \quad \text{and} \quad H_1 : p \neq \frac{1}{2}. \tag{3.33}$$

Usually, hypothesis testing is performed by constructing a *test statistic* s, a function of the data, whose large values contradict the hull hypothesis. The null hypothesis is then rejected if the observed value of the test statistic is sufficiently large. In this example, it seems reasonable to reject H_0 if the sample mean $\overline{X}$, that, by the law of large numbers, should converge to p as $n \to \infty$, is far from $1/2$. The further away $\overline{X}$ from $1/2$, the more evidence against H_0 the data provides. This suggests the following test statistic

$$s = |\overline{X} - 1/2|. \tag{3.34}$$

If s is larger than some critical threshold, we will reject H_0; otherwise, we will fail to reject the null hypothesis. How large should a test statistic be to reject the null hypothesis? How to quantify the strength of the evidence against the null hypothesis provided by the data? We will discuss these and other important questions of hypothesis testing in Chapter 9.

Remark 3.4. To illustrate the difference between "failing to reject" and "accepting" the null hypothesis, suppose that in Example 3.3 the sample size $n = 10^3$ and the critical threshold for rejecting H_0 is $c = 0.1$. That is, we reject H_0 if and only if $s > 0.1$. Suppose that the observed value of the sample mean is $\overline{X} = 0.7$. Then $s = 0.2 > 0.1$, and we reject H_0. We should feel confident about rejecting H_0, since the sample mean $\overline{X}$, which should be close to p by the LLN, is quite far from the value $p = 0.5$ assumed by the null hypothesis. In this case, the data provides strong evidence against H_0. Suppose now that the observed value of the sample mean is $\overline{X} = 0.55$. Then $s = 0.05 < 0.1$, and we fail to reject H_0. In this case, the evidence provided by the data against the null hypothesis is not strong enough for rejecting the H_0. But can we accept H_0? No, we can't, since the absence of evidence against a hypothesis does not mean the evidence in favor of the hypothesis. In fact, the data with $\overline{X} = 0.55$ is more consistent with the values $p \in (0.525, 0.575)$ rather than with the null value $p = 0.5$.

3.3 Prediction and Regression

Suppose the data contains both inputs $X_1, \ldots, X_n$ and the corresponding responses $Y_1, \ldots, Y_n$, and thus consists of n pairs of observations $(X_1, Y_1), \ldots, (X_n, Y_n)$. The problem of estimating the response Y to a future input X based on the data $(X_1, Y_1), \ldots, (X_n, Y_n)$ and on the observed value of X is called *prediction*. Here, depending on a particular application, "estimating the response" could refer to computing a point estimate of Y, constructing a confidence interval for Y, or testing some hypothesis about Y. The prediction problem is a fundamental problem and it lies at the heart of science and engineering. Over the years, it has been studied by various research communities and, as a result, different terminologies have emerged[22]. The input X is also often called a *covariate*, *predictor*, *regressor*, *feature*, *independent variable*, and *explanatory variable*. The response Y also has several other popular names: *outcome*, *target*, and *dependent variable*.

[22] "A Big Mac is a Big Mac, but they call it, Le Big Mac." Pulp Fiction.

Regression is a collection of techniques for constructing a stochastic model for the relationship between the response variable Y and the input variable X that can be used for prediction. It is based on the so-called *regression function*

$$r(x) = \mathbb{E}[Y | X = x], \tag{3.35}$$

which is the expected value of the response Y given the value of the input $X = x$. The goal of regression is to estimate the regression function $r(x)$ either directly from the data $(X_1,Y_1),\dots,(X_n,Y_n)$, which is called *nonparametric regression*, or by first assuming some parametric model for $r(x)$ and then estimating the model parameters from the data, which is called *parametric regression*. For example, the following parametric model for the regression function

$$r(x) \in \mathscr{F} = \{r(x) = \theta_0 + \theta_1 x, \;\; \theta_0, \theta_1 \in \mathbb{R}\}, \tag{3.36}$$

is called a *simple linear regression* (SLR) model. We will discuss the SLR model and its direct generalization, a *multiple linear regression* (MLR) model, in Chapters 11 and 12, respectively.

3.4 Further Reading

This chapter provided a bird's eye view of the classical statistical inference, main types of stochastic models and inferential problems, and associated fundamental concepts. A thought provoking and stimulating paper by Leo Breiman

- Breiman, L.: Statistical modeling: the two cultures (with comments and a rejoinder by the author). *Statistical Science* 16(3), 199-231 (2001). `https://projecteuclid.org/euclid.ss/1009213726`

compares *stochastic data modeling*, which is a mainstream in statistical research and practice, and which is the main focus of this book, with *algorithmic modeling*, which was originally developed in the fields outside statistics (in particular, in computer science, machine learning, and artificial intelligence) and does not assume any stochastic model for the data. The comments on the paper by prominent statisticians D.R. Cox, B. Efron, B. Hoadley, and E. Parzen as well as the rejoinder by L. Breiman (all are at the end of the aforementioned paper) are also strongly recommended.

3.5 Probability Refresher: Three Fundamental Inequalities

In this section, we will briefly review three probability inequalities that are very useful for statistical inference: Markov's, Chebyshov's, and Hoeffding's inequalities.

The *Markov inequality* provides an upper bound for the probability of the right tail of a non-negative random variable. More precisely, it states that for any non-negative random variable Y and any $a > 0$

$$\mathbb{P}(Y \geq a) \leq \frac{\mathbb{E}[Y]}{a}. \tag{3.37}$$

This inequality is named after the remarkable Russian mathematician Andrey Andreyevich Markov[23] best know for his seminal contributions to probability theory and stochastic processes. In particular, he has discovered and developed the theory of Markov chains, which are special stochastic processes with a very wide range of applications in statistics, physics, biology, finance, scientific computing, information theory, and other fields [6].

[23] An academic great-great-great-grandfather of the author!

It is instructive to prove Markov's inequality for the case where non-negative random variable Y is continuous and has PDF $f(y)$. In this case

$$\begin{aligned}\mathbb{E}[Y] &= \int_0^\infty yf(y)dy = \underbrace{\int_0^a yf(y)dy}_{\geq 0} + \int_a^\infty yf(y)dy \\ &\geq \int_a^\infty \underbrace{y}_{\geq a} f(y)dy \geq \int_a^\infty af(y)dy = a\int_a^\infty f(y)dy = a\mathbb{P}(Y \geq a),\end{aligned} \tag{3.38}$$

Andrey Andreyevich Markov (1856-1922), a Russian mathematician, the discoverer of Markov chains.

which implies (3.37). Notice that we first simply dropped the non-negative integral $\int_0^a yf(y)dy$, and then replaced y in the integral $\int_a^\infty$ with a. This suggests that although the Markov inequality is *general*, i.e. can be applied to *any* non-negative random variable, it is *not sharp*, i.e. the upper bound $\frac{\mathbb{E}[Y]}{a}$ is *not tight*. Indeed, if for instance $a \leq \mathbb{E}[Y]$, then inequality (3.37) becomes $\mathbb{P}(Y \geq a) \leq \frac{\mathbb{E}[Y]}{a} \leq 1$, which is trivial since all probabilities are ≤ 1. Markov's inequality is useful for large values of a.

The *Chebyshov inequality* describes how likely it is for a random variable to be far from its mean value. Namely, for any random variable Y with mean μ and variance σ^2 and for any $b > 0$

$$\mathbb{P}(|Y - \mu| \geq b) \leq \frac{\sigma^2}{b^2}. \tag{3.39}$$

Pafnuty Lvovich Chebyshov (1821-1894), one of the founding fathers of Russian mathematics.

This inequality is named after the outstanding Russian mathematician Pafnuty Lvovich Chebyshov, the teacher of Markov, best known for his seminal contributions to probability theory, number theory, and mechanics.

Remark 3.5. Although his last name is most often transliterated as "Chebyshev" and pronounced with a stress on the first syllable, this is not correct. A more proper transliteration is "Chebyshov" with a stress, according to Pafnuty Lvovich himself, on the last syllable.

The Chebyshov inequality is a straightforward corollary of the Markov inequality. Indeed, by applying Markov's inequality to a non-negative random variable $(Y - \mu)^2$ we have:

$$\mathbb{P}(|Y - \mu| \geq b) = \mathbb{P}((Y - \mu)^2 \geq b^2) \leq \frac{\mathbb{E}[(Y - \mu)^2]}{b^2} = \frac{\mathbb{V}[Y]}{b^2} = \frac{\sigma^2}{b^2}. \tag{3.40}$$

As Markov's inequality, Chebyshov's inequality is also very general: it hold for all random variables with finite mean and variance. As a price for this general applicability, the upper bound $\frac{\sigma^2}{b^2}$ is not tight and useful only if $b > \sigma$.

Wassily Hoeffding (1914-1991), a Finish statistician and probabilist, one of the founders of nonparametric statistics. Reprinted with the permission of the Institute of Mathematical Statistics.

The last probability inequality that we discuss in this section is *Hoeffding's inequality*, which provides an upper bound for the probability that the sample mean of bounded independent random variables deviates from its expected value by more that a given amount. Specifically, let $Y_1, \ldots, Y_n$ be independent random variables such that $Y_i \in [a_i, b_i]$. Then, for any $c > 0$,

$$\mathbb{P}(|\overline{Y} - \mathbb{E}[\overline{Y}]| \geq c) \leq 2\exp\left(-\frac{2n^2c^2}{\sum_{i=1}^n (b_i - a_i)^2}\right). \tag{3.41}$$

This inequality was derived by the remarkable Finish statistician and probabilist Wassily Hoeffding in 1963 [7]. It is often used for constructing confidence intervals as well as for finding the required sample size n needed for obtaining a confidence interval of a given confidence level.

Let's compare Chebyshov's and Hoeffding's upper bounds for the case when both inequalities are applied to the sample mean. For simplicity[24], let's assume that $Y_1, \ldots, Y_n$ are IID and that $[a_i, b_i] = [0, 1]$. Then, according to Chebyshov's inequality,

[24] Similar results will hold for a general case.

$$\mathbb{P}(|\overline{Y} - \mathbb{E}[\overline{Y}]| \geq c) \leq \frac{\mathbb{V}[\overline{Y}]}{c^2} = \frac{\mathbb{V}\left[\frac{1}{n}\sum_{i=1}^n Y_i\right]}{c^2} = \frac{\sum_{i=1}^n \mathbb{V}[Y_i]}{n^2c^2} = \frac{\sigma^2}{nc^2}, \tag{3.42}$$

where $\sigma^2 = \mathbb{V}[Y_i]$ is the variance of Y_i. According to Hoeffding's inequality,

$$\mathbb{P}(|\overline{Y} - \mathbb{E}[\overline{Y}]| \geq c) \leq 2\exp\left(-2nc^2\right). \tag{3.43}$$

Notice that Hoeffding's upper bound in (3.43) is much tighter than Chebyshov's upper bound in (3.42), since the former decays exponentially, but the latter decays only polynomially fast. This result is not surprising since Hoeffding' inequality has stronger assumptions[25] and, as a result, has stronger conclusions.

[25] Hoeffding's inequality assumes that $Y_1, \ldots, Y_n$ are bounded, while Chebyshov's inequality does not.

Problems

3.1. Bias, Standard Error, and Mean Squared Error

This problem will test your basic understanding of the important concepts of bias, standard error, and mean squared error discussed in Section 3.2.1. Suppose that the data $X_1 \ldots, X_n$ is modeled as an IID sample from the uniform distribution on $[0, \theta]$, where $\theta > 0$ is an unknown parameter. That is, $X_1 \ldots, X_n \stackrel{\text{iid}}{\sim} \mathscr{U}[0, \theta]$.

(a) Let $\hat{\theta}_n^{(1)} = 2\overline{X}$, where $\overline{X}$ is the sample mean, be an estimate of the unknown parameter θ. Find the bias, standard error, and MSE of $\hat{\theta}_n^{(1)}$.

(b) Let $\hat{\theta}_n^{(2)} = X_{(n)}$, where $X_{(n)} = \max\{X_1, \dots, X_n\}$ is the largest value in the sample, be another estimate of θ. Find the bias, standard error, and MSE of $\hat{\theta}_n^{(2)}$.

(c) What estimate, $\hat{\theta}_n^{(1)} = 2\overline{X}$ or $\hat{\theta}_n^{(2)} = X_{(n)}$, is more accurate?

3.2. Confidence Interval for the Right Bound of a Uniform Distribution
As above, let $X_1, \dots, X_n \overset{\text{iid}}{\sim} \mathscr{U}[0, \theta]$ be an IID sample from the uniform distribution on $[0, \theta]$, where $\theta > 0$ is unknown. Define $Q = X_{(n)}/\theta$, where $X_{(n)}$ is the nth order statistic, that is, $X_{(n)} = \max\{X_1, \dots, X_n\}$.

(a) Show that the distribution of Q does not depend on θ.[26]

(b) Use Q to construct an $100(1-\alpha)\%$ confidence interval for parameter θ.

[26] A random variable, whose value depends on a parameter, but whose probability distribution does not depend on that parameter, is called a *pivot* or *pivotal quantity*. See Remark 6.3.

References

1. Mumford, D.: The dawning of the age of stochasticity. *Atti della Accademia Nazionale dei Lincei. Classe di Scienze Fisiche, Matematiche e Naturali. Rendiconti Lincei. Matematica e Applicazioni* 11(S1), 107–125 (2000).
`http://eudml.org/doc/289648`
2. Robinson, P.M.: Consistent nonparametric entropy-based testing. *The Review of Economic Studies* 58(3), 437—453 (1991).
`https://doi.org/10.2307/2298005`
3. Hong, Y., White, H.: Asymptotic distribution theory for nonparametric entropy measures of serial dependence. *Econometrica* 73(3), 837—901 (2005).
`http://www.jstor.org/stable/3598868`
4. Cho, J.S., White, H.: Generalized runs tests for the IID hypothesis. *Journal of Econometrics* 162(2), 326–344 (2011).
`https://doi.org/10.1016/j.jeconom.2011.02.001`
5. MacKay, D.J.C.: *Information Theory, Inference, and Learning Algorithms*. Cambridge University Press (2003). Available free online from
`http://www.inference.org.uk/mackay/itila/book.html`
6. Von Hilgers, P., Langville, A.N.: The five greatest applications of Markov chains. *In Proceedings of the Markov Anniversary Meeting*. Boston, MA: Boston Press (2006).
`https://api.semanticscholar.org/CorpusID:17056670`
7. Hoeffding, W.: Probability inequalities for sums of bounded random variables. *Journal of the American Statistical Association* 58(301), 13–30 (1963).
`https://doi.org/10.1080/01621459.1963.10500830`

4
Nonparametric Estimation of the CDF and the Plug-In Principle

No random actions, none not based on underlying principles.
– Marcus Aurelius, *Meditations*

The basic goal of nonparametric inference is to use data $X_1,\ldots,X_n$ to infer an unknown quantity of interest θ, while making as few assumptions about the data as possible. Mathematically, "few assumptions" means that the stochastic model $\mathscr{F}$ used to model the data,

$$X_1,\ldots,X_n \overset{\text{iid}}{\sim} F, \quad F \in \mathscr{F}, \tag{4.1}$$

is large — so large that it can't be parameterized by a finite number of parameters[1].

Let's consider the most general nonparametric model

$$\mathscr{F} = \{\text{all CDFs}\}, \tag{4.2}$$

that assumes only that the data is an IID sample from some probability distribution defined by its CDF F.

In this chapter, we will focus on one of the central problems in nonparametric statistical inference: estimation of a parameter θ of distribution F. But hold on! What do we mean by a "parameter" of $F \in \mathscr{F}$, where $\mathscr{F}$ is a nonparametric model!? So far we have discussed[2] parameters only in the context of parametric models. In general, a parameter of the CDF F is defined as the value of a certain statistical functional on F. Let's discuss this in detail.

[1] In this case, the model $\mathscr{F}$ is often called infinite-dimensional.

[2] In Chapter 3.

4.1 Functionals and Parameters

A *statistical functional* $\mathfrak{f}$ on a stochastic model $\mathscr{F}$ is a map

$$\mathfrak{f} : \mathscr{F} \to \mathbb{R}, \tag{4.3}$$

that takes a CDF $F \in \mathscr{F}$ as an argument and returns a number $\mathfrak{f}(F) \in \mathbb{R}$,

Supplementary Information The online version contains supplementary material available at https://doi.org/10.1007/978-3-032-03848-7_4.

K. M. Zuev, *Fundamentals of Statistical Inference*,
International Series in Operations Research & Management Science 374,
https://doi.org/10.1007/978-3-032-03848-7_4

$$\mathscr{F} \ni F \to \mathfrak{f}(F) \in \mathbb{R}. \tag{4.4}$$

A *parameter* θ of a CDF F is the value of a certain functional $\mathfrak{f}$ on F,

$$\theta = \mathfrak{f}(F). \tag{4.5}$$

Here are several examples of functionals and parameters that are often used:

- Mean or expected value :

$$\mu = \mathfrak{f}(F) = \int x dF(x), \tag{4.6}$$

- Variance:

$$\sigma^2 = \mathfrak{f}(F) = \int (x-\mu)^2 dF(x), \tag{4.7}$$

- Coefficient of variation:

$$\delta = \mathfrak{f}(F) = \frac{\left(\int (x-\mu)^2 dF(x)\right)^{1/2}}{\int x dF(x)} = \frac{\sigma}{\mu}, \tag{4.8}$$

- Median:

$$m = \mathfrak{f}(F) = F^{-1}(1/2), \tag{4.9}$$

- Skewness:

$$\varkappa = \mathfrak{f}(F) = \frac{\int (x-\mu)^3 dF(x)}{\left(\int (x-\mu)^2 dF(x)\right)^{3/2}}, \tag{4.10}$$

- Kurtosis[3]:

$$\kappa = \mathfrak{f}(F) = \frac{\int (x-\mu)^4 dF(x)}{\left(\int (x-\mu)^2 dF(x)\right)^{2}}. \tag{4.11}$$

[3] For statistical interpretation of skewness and kurtosis, see Problem 4.3.

Remark 4.1. The integrals with respect to $dF(x)$ in (4.6)–(4.11) may look confusing, but they are convenient, since they allow to unify the notation for the discrete and continuous cases and consider them simultaneously. So, it is useful to understand what the integral $\int g(x)dF(x)$, where F is a CDF and g is a function, actually means.

If the probability distribution defined by F is continuous with PDF f, then $dF(x) = f(x)dx$ and

$$\int g(x)dF(x) = \int g(x)f(x)dx = \mathbb{E}[g(X)], \quad X \sim F. \tag{4.12}$$

So, the integral $\int g(x)dF(x)$ is simply the expected value of random variable $g(X)$, where X is distributed according to F.

Now let's consider the case where the probability distribution defined by F is discrete with PMF p. In other words, the random variable $X \sim F$ is discrete, takes values in $\{x_1, x_2, \dots\}$, and $\mathbb{P}(X = x_i) = p(x_i)$. Then F can be written in terms of p as follows:

$$F(x) = \mathbb{P}(X \leq x) = \sum_{i: x_i \leq x} \mathbb{P}(X = x_i) = \sum_{i: x_i \leq x} p(x_i) = \sum_i p(x_i) H(x - x_i), \quad (4.13)$$

where $H(x)$ is the Heaviside function (1.15), and so $H(x - x_i) = 1$ if $x_i \leq x$ and $H(x - x_i) = 0$ if $x_i > x$. The Heaviside function $H(x)$ (and, therefore, $F(x)$) is not differentiable in the classical sense, since it is not continuous. The problem, however, occurs only at $x = 0$, where $H(x)$ has a jump from zero to one. This allows to interpret the derivative $H'(x)$ of the Heaviside function as the *Dirac delta function* $\delta(x)$, which can be intuitively defined as a "generalized" function[4] such that

$$\delta(x) = \begin{cases} 0, & \text{if } x \neq 0, \\ \infty, & \text{if } x = 0. \end{cases} \quad \text{and} \quad \int_{-\infty}^{\infty} \delta(x) dx = 1. \quad (4.14)$$

[4] The Dirac delta function is not a function in the classical sense, but it can be rigorously defined as a *measure*.

The delta function has an important property, which is called the *sifting property*[5]: for any function $g(x)$ continuous on (a, b) and any $x_0 \in (a, b)$,

[5] For interested readers, it is proved in Appendix 4.6.

$$\int_{-\infty}^{\infty} g(x) \delta(x - x_0) dx = g(x_0). \quad (4.15)$$

Now, using (4.13), $H'(x) = \delta(x)$, and (4.15), we have:

$$\begin{aligned} \int g(x) dF(x) &= \int g(x) d \sum_i p(x_i) H(x - x_i) = \int g(x) \sum_i p(x_i) dH(x - x_i) \\ &= \int g(x) \sum_i p(x_i) \delta(x - x_i) dx = \sum_i p(x_i) \int g(x) \delta(x - x_i) dx \quad (4.16) \\ &= \sum_i p(x_i) g(x_i) = \mathbb{E}[g(X)], \quad X \sim F. \end{aligned}$$

So, the bottom line is that regardless of whether the probability distribution defined by the CDF F is continuous or discrete, the integral $\int g(x) dF(x)$ equals to the mean of random variable $g(X)$, where X is distributed according to F:

$$\int g(x) dF(x) = \mathbb{E}[g(X)], \quad X \sim F. \quad (4.17)$$

For example, in (4.6), $\int x dF(x) = \mathbb{E}[X]$ is the mean, as expected.

So, returning to our main line of thought, the goal of nonparametric estimation is: given the data $X_1, \ldots, X_n \sim F \in \mathscr{F} = \{\text{all CDFs}\}$, estimate a parameter of interest

$$\theta = \mathfrak{f}(F). \quad (4.18)$$

In this context, the functional $\mathfrak{f}$ is known[6], but the CDF F and, therefore, θ are unknown. Looking at (4.18), the following idea looks rather natural: let's first estimate the CDF $\hat{F} \approx F$, and then estimate the parameter θ by

[6] It is our choice what we want to estimate: the mean, the variance, or some other parameter defined through $\mathfrak{f}$.

$$\hat{\theta} = \mathfrak{f}(\hat{F}). \tag{4.19}$$

Let's follow this path[7].

[7] "It's a dangerous business, Frodo, going out your door. You step onto the road, and if you don't keep your feet, there's no knowing where you might be swept off to." – J.R.R. Tolkien, *The Lord of the Rings*.

4.2 Estimating the CDF

In Section 1.4, we saw[8] that the CDF of the uniform distribution $\mathscr{U}[0,1]$ is accurately approximated by the empirical CDF (eCDF). It turns out that this is not accidental: this is a general result that holds for any probability distribution.

[8] See Fig. 1.5.

Recall that the eCDF $\hat{F}_n$ of sample $X_1,\dots,X_n$ is the CDF that puts probability mass $1/n$ at each data point X_i. More formally,

$$\hat{F}_n(x) = \frac{1}{n}\sum_{i=1}^{n} H(x - X_i), \tag{4.20}$$

where $H(x)$ is the Heaviside step function. The following theorem shows that the eCDF $\hat{F}_n$ is indeed a good estimate of the true CDF that gencrated the data.

Theorem 4.1. *Let $X_1,\dots,X_n \overset{iid}{\sim} F$, then for any x:*

1. $\hat{F}_n(x)$ is an unbiased estimate of $F(x)$,
2. The standard error of $\hat{F}_n(x)$ is

$$\mathrm{se}[\hat{F}_n(x)] = \sqrt{\frac{F(x)(1-F(x))}{n}}, \tag{4.21}$$

3. The mean squared error of $\hat{F}_n(x)$ converges to zero,

$$\mathrm{MSE}[\hat{F}_n(x)] \to 0, \quad \text{as } n \to \infty. \tag{4.22}$$

Proof. Let's look at the definition of the eCDF (4.20): $\hat{F}_n(x)$ can be viewed as the sample mean of quantities $\xi_i(x) = H(x - X_i)$, $i = 1,\dots,n$,

$$\hat{F}_n(x) = \frac{1}{n}\sum_{i=1}^{n} \xi_i(x) = \bar{\xi}(x). \tag{4.23}$$

The quantity $\xi_i(x)$ is a random variable, since X_i is a random variable. The random variable $\xi_i(x)$ takes two values, 0 and 1,

$$\xi_i(x) = H(x - X_i) = \begin{cases} 1, & \text{if } X_i \le x, \\ 0, & \text{if } X_i > x. \end{cases} \tag{4.24}$$

Therefore,

$$\begin{aligned}\mathbb{P}(\xi_i(x)=1) &= \mathbb{P}(X_i \le x) = F(x),\\ \mathbb{P}(\xi_i(x)=0) &= \mathbb{P}(X_i > x) = 1-F(x).\end{aligned} \tag{4.25}$$

This means that for any fixed x, random variables $\xi_1(x),\dots,\xi_n(x)$ are IID from the Bernoulli distribution with parameter $p = F(x)$,

$$\xi_1(x),\dots,\xi_n(x) \overset{\text{iid}}{\sim} \mathscr{B}(F(x)). \tag{4.26}$$

All statements of the theorem follow from the representation of the eCDF as a sample mean of the IID Bernoulli random variables[9]. The expected value is

$$\mathbb{E}[\hat{F}_n(x)] = \mathbb{E}\left[\frac{1}{n}\sum_{i=1}^{n}\xi_i(x)\right] = \frac{1}{n}\sum_{i=1}^{n}\underbrace{\mathbb{E}[\xi_i(x)]}_{F(x)} = F(x), \tag{4.27}$$

and so $\hat{F}_n(x)$ is an unbiased estimate of $F(x)$. The squared standard error is

$$\begin{aligned}\mathrm{se}[\hat{F}_n(x)]^2 = \mathbb{V}[\hat{F}_n(x)] &= \mathbb{V}\left[\frac{1}{n}\sum_{i=1}^{n}\xi_i(x)\right] = \frac{1}{n^2}\sum_{i=1}^{n}\mathbb{V}[\xi_i(x)]\\ &= \frac{1}{n^2}\sum_{i=1}^{n}\Big(\underbrace{\mathbb{E}[\xi_i^2(x)]}_{F(x)} - \underbrace{\mathbb{E}[\xi_i(x)]^2}_{F(x)^2}\Big) = \frac{F(x)(1-F(x))}{n}.\end{aligned} \tag{4.28}$$

Finally, using the bias-variance decomposition (3.11), the mean squared error is

$$\mathrm{MSE}[\hat{F}_n(x)] = \mathbb{B}[\hat{F}_n(x)]^2 + \mathrm{se}[\hat{F}_n(x)]^2 = \frac{F(x)(1-F(x))}{n} \to 0, \quad \text{as } n \to \infty. \tag{4.29}$$

And the theorem is proved. □

This theorem justifies the use of the eCDF as a reasonable point estimate of the data-generating CDF: its mean squared error is small if the sample size is large. Moreover, it turns out that the eCDF $\hat{F}_n(x)$ is a consistent estimate of $F(x)$.

Definition 4.1. A random estimate $\hat{\theta}_n$ of θ is called *consistent* if $\hat{\theta}_n$ converges in probability[10] to θ as $n \to \infty$,

$$\hat{\theta}_n \xrightarrow{\mathbb{P}} \theta. \tag{4.30}$$

If the mean squared error of an estimate $\hat{\theta}_n$ converges to zero, $\mathrm{MSE}[\hat{\theta}_n] \to 0$ as $n \to \infty$, then $\hat{\theta}_n$ is automatically consistent. Indeed, to prove consistency, we need to show that for any $\varepsilon > 0$,

$$\lim_{n\to\infty}\mathbb{P}(|\hat{\theta}_n - \theta| < \varepsilon) = 1. \tag{4.31}$$

But this immediately follows from the Markov inequality[11] applied to the random variable $(\hat{\theta}_n - \theta)^2$:

[9] Compare the following derivations with the derivations in Example 3.1 and notice that they are essentially identical: the role of $X_1,\dots,X_n \overset{\text{iid}}{\sim} \mathscr{B}(p)$ in Example 3.1 is played here by $\xi_1(x),\dots,\xi_n(x) \overset{\text{iid}}{\sim} \mathscr{B}(F(x))$.

[10] For a brief review of convergence in probability, see the beginning of Probability Refresher 2.12.

[11] For a brief review of the Markov inequality, see Probability Refresher 3.5.

$$\mathbb{P}(|\hat{\theta}_n - \theta| < \varepsilon) = 1 - \mathbb{P}(|\hat{\theta}_n - \theta| \geq \varepsilon) = 1 - \mathbb{P}((\hat{\theta}_n - \theta)^2 \geq \varepsilon^2)$$
$$\geq 1 - \frac{\mathbb{E}\left[(\hat{\theta}_n - \theta)^2\right]}{\varepsilon^2} = 1 - \underbrace{\frac{\mathrm{MSE}[\hat{\theta}_n]}{\varepsilon^2}}_{\to 0} \to 1, \quad \text{as } n \to \infty. \tag{4.32}$$

Formally speaking, (4.32) proves that $\lim_{n\to\infty} \mathbb{P}(|\hat{\theta}_n - \theta| < \varepsilon) \geq 1$, but since probabilities can't exceed one, $\lim_{n\to\infty} \mathbb{P}(|\hat{\theta}_n - \theta| < \varepsilon) \leq 1$, and thus (4.31) holds.

According to Theorem 4.1, the mean squared error of $\hat{F}_n(x)$ converges to zero. Therefore, $\hat{F}_n(x) \xrightarrow{\mathbb{P}} F(x)$, and we have the following corollary.

Corollary 4.1. *For any* x*,* $\hat{F}_n(x)$ *is a consistent estimate of* $F(x)$*.*

Intuitively, this means that for any x, if the sample size n is large, then $\hat{F}_n(x)$ is very close to $F(x)$ with large probability.

So, Corollary 4.1 tells us that the eCDF $\hat{F}_n(x)$ converges to the true CDF $F(x)$, but it does not tell us *how fast* this happens. The *Dvoretzky-Kiefer-Wolfowitz* (DKW) inequality[12] does. Namely, the DKW inequality [1] states that for any fixed $\varepsilon > 0$,

[12] Named after Russian-born Israeli mathematician Aryeh Dvoretzky, American statistician Jack Kiefer, and Polish-born American statistician Jacob Wolfowitz.

$$\mathbb{P}\left(\sup_{x\in\mathbb{R}} \left|\hat{F}_n(x) - F(x)\right| > \varepsilon\right) \leq 2\exp\left(-2n\varepsilon^2\right). \tag{4.33}$$

Remark 4.2. The "sup" in (4.33) stands for "supremum." If this technical term is unfamiliar, you can think of it as the "maximum." Intuitively, the DKW inequality states that the largest discrepancy between $\hat{F}_n(x)$ and $F(x)$ shrinks to ε exponentially fast with the sample size n. On the other hand, to reduce the error size by half ($\varepsilon \to \varepsilon/2$) and preserve the same value of the upper bound $2\exp\left(-2n\varepsilon^2\right)$ we need to quadruple the sample size ($n \to 4n$).

Remark 4.3. The proof of the DKW inequality is out of our scope, but if we look at Hoeffding's inequality for the Bernoulli random variables (3.23), and recall that $\hat{F}_n(x)$ can be viewed as the sample mean of the Bernoulli random variables with parameter $p = F(x)$, then the DKW inequality should not be very surprising.

Remark 4.4. The initial upper bound in the DKW inequality was $C\exp(-2n\varepsilon^2)$, where C is some constant. The inequality with $C = 2$ was proved by French statistician Pascal Massart [2]. Therefore, the DKW inequality should have been called the Dvoretzky-Kiefer-Wolfowitz-Massart (DKWM) inequality.

4.2.1 Confidence Set for the CDF

The DKW inequality can be used for constructing a *confidence set* for the CDF, that is, a random set constructed from the data $X_1 \ldots, X_n \overset{\text{iid}}{\sim} F$ that traps F with a specified probability.

To construct a confidence set for F, we need to find two functions F_l, a lower bound, and F_u, an upper bound, such that

$$\mathbb{P}\left(F_l(x) \leq F(x) \leq F_u(x) \text{ for all } x\right) = 1-\alpha \quad (\text{or } \geq 1-\alpha), \tag{4.34}$$

where $\alpha \in (0,1)$ defines the confidence level. As it was mentioned in Section 3.2.2, a confidence interval for a quantity of interest θ often has the form $\hat{\theta}_n \pm \mathrm{MOE}_{n,\alpha}$, where $\hat{\theta}_n$ is a point estimate of θ and $\mathrm{MOE}_{n,\alpha}$ is the margin of error. The eCDF $\hat{F}_n(x)$ is a point estimate of $F(x)$. This motivates to look for $F_l(x)$ and $F_u(x)$ in the following form:

$$F_l(x) = \hat{F}_n(x) - \varepsilon \quad \text{and} \quad F_u(x) = \hat{F}_n(x) + \varepsilon, \tag{4.35}$$

where ε is the margin of error, that should be chosen such that (4.34) is satisfied. The value of ε will probably depend on both n and α. With these lower and upper bounds, using the DKW inequality, we have[13]:

$$\begin{aligned}
&\mathbb{P}\left(F_l(x) \leq F(x) \leq F_u(x), \ \forall x\right) = \mathbb{P}\left(\hat{F}_n(x) - \varepsilon \leq F(x) \leq \hat{F}_n(x) + \varepsilon, \ \forall x\right) \\
&= \mathbb{P}\left(\left|\hat{F}_n(x) - F(x)\right| \leq \varepsilon, \ \forall x\right) = 1 - \mathbb{P}\left(\exists x^* : \ \left|\hat{F}_n(x^*) - F(x^*)\right| > \varepsilon\right) \\
&= 1 - \mathbb{P}\left(\sup_{x\in\mathbb{R}} \left|\hat{F}_n(x) - F(x)\right| > \varepsilon\right) \geq 1 - 2\exp\left(-2n\varepsilon^2\right).
\end{aligned} \tag{4.36}$$

13 "Often, something might seem difficult and dull at first, but then you suddenly grasp it — and everything starts to flow smoothly." – Janusz Korczak, *Rules of Life*.

Now setting

$$2\exp\left(-2n\varepsilon^2\right) = \alpha \quad \Leftrightarrow \quad \varepsilon = \sqrt{-\frac{1}{2n}\log\frac{\alpha}{2}} \tag{4.37}$$

leads to

$$\mathbb{P}\left(\hat{F}_n(x) - \sqrt{-\frac{1}{2n}\log\frac{\alpha}{2}} \leq F(x) \leq \hat{F}_n(x) + \sqrt{-\frac{1}{2n}\log\frac{\alpha}{2}}, \ \forall x\right) \geq 1-\alpha. \tag{4.38}$$

We can slightly refine this confidence set by using the fact the value of the CDF $F(x)$ is always between zero and one. This means that for any random variables A and B

$$\mathbb{P}(A \leq F(x) \leq B) = \mathbb{P}(\max\{A,0\} \leq F(x) \leq \min\{B,1\}). \tag{4.39}$$

Applying this observation to (4.38), we have:

$$\begin{aligned}
&\mathbb{P}\left(F_l(x) \leq F(x) \leq F_u(x), \ \forall x\right) \geq 1-\alpha, \quad \text{where} \\
&F_l(x) = \max\left\{\hat{F}_n(x) - \sqrt{-\frac{1}{2n}\log\frac{\alpha}{2}}, \ 0\right\}, \\
&F_u(x) = \min\left\{\hat{F}_n(x) + \sqrt{-\frac{1}{2n}\log\frac{\alpha}{2}}, \ 1\right\}.
\end{aligned} \tag{4.40}$$

This confidence set for the CDF is called a $(1-\alpha)$ *nonparametric confidence band*[14].

[14] In Problem 4.2, you will construct a 95% confidence band for the CDF of the log return of the S&P 500 index.

4.3 The Plug-In Principle

Since the eCDF $\hat{F}_n$ of a sample $X_1, \ldots, X_n \overset{\text{iid}}{\sim} F$ converges to the underlying data-generating CDF F as the sample size n goes to infinity, and, therefore, $F(x) \approx \hat{F}_n(x)$ for large n, it is natural to expect that any property of F can be accurately approximated by the corresponding property of $\hat{F}_n$. This motivates the following definition.

Definition 4.2. Let $\theta = \mathfrak{f}(F)$ be a parameter of interest, where $\mathfrak{f}$ is a given statistical functional. The estimate

$$\hat{\theta}_n = \mathfrak{f}(\hat{F}_n) \tag{4.41}$$

is called the *plug-in estimate* of θ. In general, plugging in $\hat{F}_n$ instead of F is referred to as the *plug-in principle*.

Suppose that a parameter of interest θ is given by

$$\theta = \mathfrak{f}(F) = \int g(x)dF(x) \overset{\text{Remark 4.1}}{\downarrow}{=} \mathbb{E}[g(X)], \quad X \sim F, \tag{4.42}$$

where $g(x)$ is some function. Then the plug-in estimate of θ is[15]

[15] The following derivations are similar to the derivations in Remark 4.1.

$$\begin{aligned}
\hat{\theta}_n = \mathfrak{f}(\hat{F}_n) &= \int g(x)d\hat{F}_n(x) = \int g(x)d\frac{1}{n}\sum_{i=1}^{n} H(x-X_i) \\
&= \frac{1}{n}\int g(x)\sum_{i=1}^{n} dH(x-X_i) = \frac{1}{n}\int g(x)\sum_{i=1}^{n}\delta(x-X_i)dx \\
&= \frac{1}{n}\sum_{i=1}^{n}\int g(x)\delta(x-X_i)dx = \frac{1}{n}\sum_{i=1}^{n} g(X_i).
\end{aligned} \tag{4.43}$$

Let's consider a specific example.

Example 4.1. The two most often used parameters are the mean and the variance.

(a) Suppose the parameter of interest is the mean,

$$\mu = \int x dF(x). \tag{4.44}$$

In this case, $g(x) = x$, and, therefore, the plug-in estimate

$$\hat{\mu}_n = \frac{1}{n}\sum_{i=1}^{n} X_i = \overline{X} \tag{4.45}$$

is the sample mean of the data. It is an unbiased estimate of the mean:

$$\mathbb{E}[\hat{\mu}_n] = \mathbb{E}[\overline{X}] = \mathbb{E}\left[\frac{1}{n}\sum_{i=1}^{n} X_i\right] = \frac{1}{n}\sum_{i=1}^{n}\underbrace{\mathbb{E}[X_i]}_{\mu} = \mu. \tag{4.46}$$

(b) Suppose the parameter of interest is the variance,

$$\sigma^2 = \int (x-\mu)^2 dF(x) = \int \left(x - \int x dF(x)\right)^2 dF(x) \tag{4.47}$$

The plug-in estimate is then

$$\hat{\sigma}_n^2 = \int \Big(x - \underbrace{\int x d\hat{F}_n(x)}_{\hat{\mu}_n = \overline{X}}\Big)^2 d\hat{F}_n(x) = \int (x-\overline{X})^2 d\hat{F}_n(x). \tag{4.48}$$

In this case, $g(x) = (x-\overline{X})^2$, and, therefore, according to (4.43),

$$\hat{\sigma}_n^2 = \frac{1}{n}\sum_{i=1}^{n}(X_i - \overline{X})^2. \tag{4.49}$$

Unlike the sample mean, the plug-in estimate $\hat{\sigma}_n^2$ is biased[16]. Indeed,

$$\hat{\sigma}_n^2 = \frac{1}{n}\sum_{i=1}^{n} X_i^2 - \frac{2}{n}\overline{X}\underbrace{\sum_{i=1}^{n} X_i}_{n\overline{X}} + \underbrace{\frac{1}{n}\sum_{i=1}^{n}\overline{X}^2}_{\overline{X}^2} = \frac{1}{n}\sum_{i=1}^{n} X_i^2 - \overline{X}^2,$$

$$\begin{aligned}
\mathbb{E}[\hat{\sigma}_n^2] &= \frac{1}{n}\sum_{i=1}^{n}\mathbb{E}[X_i^2] - \mathbb{E}[\overline{X}^2] \\
&= \frac{1}{n}\sum_{i=1}^{n}\big(\underbrace{\mathbb{V}[X_i]}_{\sigma^2} + \underbrace{\mathbb{E}[X_i]^2}_{\mu^2}\big) - \big(\mathbb{V}[\overline{X}] + \underbrace{\mathbb{E}[\overline{X}]^2}_{\mu^2}\big) \\
&= \sigma^2 + \mu^2 - \mathbb{V}\left[\frac{1}{n}\sum_{i=1}^{n} X_i\right] - \mu^2 = \sigma^2 - \frac{1}{n^2}\sum_{i=1}^{n}\underbrace{\mathbb{V}[X_i]}_{\sigma^2} \\
&= \sigma^2 - \frac{\sigma^2}{n} = \frac{n-1}{n}\sigma^2.
\end{aligned} \tag{4.50}$$

So, the plug-in estimate $\hat{\sigma}_n^2$ tends to underestimate σ^2. To construct an unbiased estimate, we simply need to multiply $\hat{\sigma}_n^2$ by the reciprocal of $\frac{n-1}{n}$. Thus,

$$\hat{s}_n^2 = \frac{n}{n-1}\hat{\sigma}_n^2 = \frac{1}{n-1}\sum_{i=1}^{n}(X_i - \overline{X})^2, \tag{4.51}$$

[16] Recall that we have already encountered estimate (4.49) in the context of survey sampling in Section 2.5. In Theorem 2.3, we proved that it is a biased estimate of the population variance. However, in survey sampling, $X_1, \ldots, X_n$ were obtained by simple random sampling from a finite population, while here $X_1, \ldots, X_n$ is an IID sample from distribution F. Notice that the expected value in (2.34) converges to $\frac{n-1}{n}\sigma^2$ derived in (4.50) as the population size $N \to \infty$.

which is sometimes called the *sample variance*, is an unbiased estimate of σ^2. In practice, the sample size n is often large enough so that $\frac{n}{n-1} \approx 1$, and $\hat{\sigma}_n^2$ and $\hat{s}_n^2$ are essentially identical.

As a final remark, it is important to highlight that, although the plug-in estimate $\hat{\theta}_n = \mathfrak{f}(\hat{F}_n)$ is always available and can be used, sometimes more accurate estimates of the parameter of interest $\theta = \mathfrak{f}(F)$ may exist. Especially in the cases, where some additional information about F, other than provided by the data $X_1, \dots, X_n \overset{\text{iid}}{\sim} F$, is known. For example, if distribution F is a member of a certain parametric model $\mathscr{F} = \{F(x|\beta)\}$, then replacing $F(x)$ with $F(x|\hat{\beta})$, where $\hat{\beta}$ is an estimate of the model parameter β, could be better[17] than replacing $F(x)$ with the eCDF $\hat{F}_n(x)$. In other words, the estimate $\hat{\theta} = t(F(x|\hat{\beta}))$ could be more accurate than the plug-in estimate $\hat{\theta}_n = t(\hat{F}_n)$. Let's illustrate this effect with a case study.

[17] "It might be worse, and then again it might be a good deal better." J.R.R. Tolkien, *The Hobbit, or There and Back Again*.

4.4 Case Study: The S&P 500 Index

> *Number one rule of Wall Street: Nobody — I don't care if you're Warren Buffett or Jimmy Buffett — Nobody knows if the stock's going to go up, down, sideways, or in f***ing circles, least of all stockbrokers. It's all a Fugazzi. You know what a Fugazzi is?*
>
> – Martin Scorsese and Terence Winter, *The Wolf of Wall Street* [3]

The Standard & Poor's 500 Index (S&P 500) is a popular stock market index (a mixture of stocks) that consists of the stocks of 500 large companies listed on stock exchanges in the United States. The S&P 500 index is weighted by the market capitalization (or "cap-weighted"). This means that the company's contribution to and influence on the index value is not uniform, but is proportional to the company's market value, which is the total dollar value of the company's outstanding stocks. The list of S&P 500 companies is not fixed, and gets periodically updated based on how companies grow or shrink in value. For example, in June 2020, the top three components of the S&P 500 were the following stocks: Microsoft Corp. (MSFT), Apple Inc. (AAPL), and Amazon.com Inc. (AMZN). Because the index contains essentially all of the largest stocks and accounts for approximately 80% of the total value of the U.S. stock market, the S&P 500 is often considered as an indicator of how well the stock market and the economy are performing overall. The value of the S&P 500 index fluctuates constantly throughout the trading day, and can be monitored on various websites, such as Yahoo! Finance or Google Finance. Figure 4.1 shows the monthly evolution of the adjusted closing price[18] of the S&P 500 between 1927[19] and 2020 together with some recent historical milestones.

Let P_i be the adjusted closing price of the S&P 500 index at time i, where $i = 0, \dots, N = 1112$ corresponds to the first day of each month from Dec 1, 1927 until July 1, 2020. The *return* (or the *simple gross return*) over the holding period from time $i-1$ to time i (one month in our case) is defined as the ratio of the corresponding prices:

[18] The adjusted closing price is an enhancement of the raw closing price that accounts for various corporate actions (splits, dividend distributions, etc.). The adjusted price reflects the true value of the index more accurately and is often used for the performance analysis of the index.

[19] The original index created in 1926 consisted of just 90 stocks. But over the years it expanded, and in 1957, when it started to represent stocks of 500 companies, it was renamed the S&P 500.

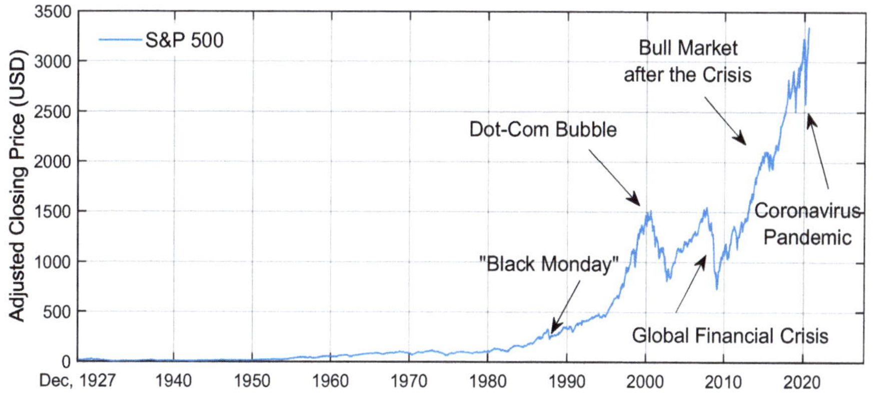

Fig. 4.1 The monthly evolution of the adjusted closing price of the S&P 500 stock market index from December 1, 1927 until July 1, 2020. The historical prices are obtained from `https://finance.yahoo.com/`.

$$R_i = \frac{P_i}{P_{i-1}}, \quad i = 1, \ldots, N. \tag{4.52}$$

For various theoretical and algorithmic reasons, in price series analysis, *log returns* are often used instead of simple returns,

$$r_i = \log R_i = \log\left(\frac{P_i}{P_{i-1}}\right) = \log P_i - \log P_{i-1}, \quad i = 1, \ldots, N. \tag{4.53}$$

Note that if the log return is positive, $r_i > 0$, then the revenue from investing in the S&P 500 index at time $i-1$, holding the investments for one month, and then selling at time i is positive (profit); otherwise, if $r_i < 0$, the revenue is negative (loss). Figure 4.2 shows the monthly log returns $r_1, \ldots, r_N$ that correspond to prices in Fig. 4.1 and the histogram of log returns.

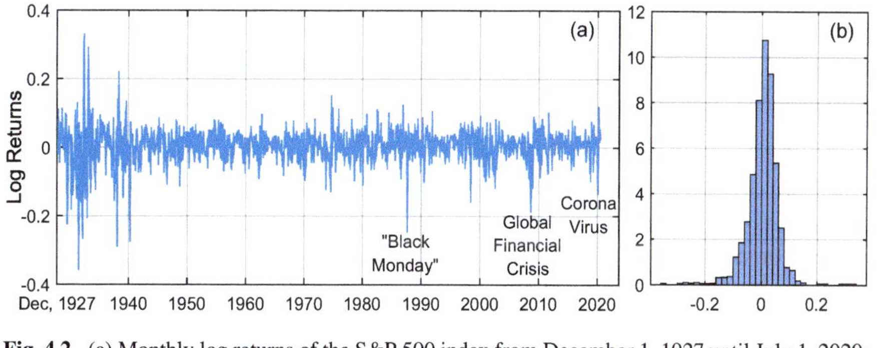

Fig. 4.2 (a) Monthly log returns of the S&P 500 index from December 1, 1927 until July 1, 2020; (b) The histogram of the log returns.

Prices, returns, and log returns are random and cannot be perfectly predicted [4]. Suppose we are interested in the probability that a log return r is below a certain threshold α. So, our parameter of interest is

$$\theta = \mathbb{P}(r < \alpha). \tag{4.54}$$

We can estimate θ from the data $r_1, \ldots, r_N$ both parametrically, by assuming some parametric model for the log returns, and nonparametrically, by using the plug-in estimate. Let's start with a nonparametric approach and compute the plug-in estimate. First, let's rewrite θ in the form (4.5), as the value of a certain statistical functional on the (unknown) CDF F of the log return r:

$$\theta = \mathbb{P}(r < \alpha) = \int_{-\infty}^{\alpha} dF(x) = \int \mathbb{1}_{(-\infty,\alpha)}(x) dF(x), \tag{4.55}$$

where $\mathbb{1}_{(-\infty,\alpha)}$ is the *indicator function* of the interval $(-\infty, \alpha)$, i.e., $\mathbb{1}_{(-\infty,\alpha)}(x) = 1$ if $x < \alpha$ and $\mathbb{1}_{(-\infty,\alpha)}(x) = 0$ if $x \geq \alpha$. Notice that θ has the form (4.42), where $g(x) = \mathbb{1}_{(-\infty,\alpha)}(x)$. Therefore, using (4.43), the plug-in estimate of θ is

$$\hat{\theta}_N = \frac{1}{N} \sum_{i=1}^{N} \mathbb{1}_{(-\infty,\alpha)}(r_i), \tag{4.56}$$

which is simply the proportion of the log returns r_i that are below α.

Now let's derive a parametric estimate of θ. By looking at the histogram on the log returns in Fig. 4.2(b), it seems reasonable, at least at first glance, to assume that the log returns are normally distributed,

$$r_1, \ldots, r_N \overset{\text{iid}}{\sim} \mathcal{N}(\mu, \sigma^2). \tag{4.57}$$

Remark 4.5. This model is called *the log-normal model*, since in this case the simple gross returns $R_1, \ldots, R_N$ are log-normally distributed. It was first introduced back in 1900 by Louis Bachelier in his seminal dissertation *The Theory of Speculation* [5], which anticipated much of modern mathematical finance. If the log returns are normally distributed, then the underlying price process $\{P_i, i = 1, 2, \ldots\}$, where

$$P_i = R_i P_{i-1} = R_i R_{i-1} P_{i-2} = R_i R_{i-1} \ldots R_1 P_0 = P_0 e^{r_1} \ldots e^{r_i} = P_0 e^{r_1 + \ldots + r_i}, \tag{4.58}$$

is called a *log-normal geometric random walk* with parameters μ and σ^2. As the time period between steps $i-1$ and i converges to zero in the appropriate way, the random walk converges to a *geometric Brownian motion*, which is the "standard model" for stock prices used in option pricing [6]. In particular, the famous[20] Black-Scholes-Merton model for option pricing is based on the assumption that the log returns of the underlying asset are normally distributed [7, 8].

[20] Merton and Scholes received the 1997 Nobel Prize in economics for this model. Unfortunately, Black passed away in 1995 and did not receive the prize, since the Nobel Prize is not awarded posthumously.

Under the normality assumption,

$$\theta = \mathbb{P}(r < \alpha) = \mathbb{P}\left(\frac{r-\mu}{\sigma} < \frac{\alpha-\mu}{\sigma}\right) = \Phi\left(\frac{\alpha-\mu}{\sigma}\right), \tag{4.59}$$

where Φ is the standard normal CDF. The model parameters, μ and σ^2, can be estimated from the data as follows[21]:

$$\hat{\mu}_N = \frac{1}{N}\sum_{i=1}^{N} r_i \quad \text{and} \quad \hat{\sigma}_N^2 = \frac{1}{N}\sum_{i=1}^{N}(r_i - \hat{\mu}_N)^2. \tag{4.60}$$

Therefore, the parameter θ can be estimated by

$$\hat{\theta}_N^p = \Phi\left(\frac{\alpha - \hat{\mu}_N}{\hat{\sigma}_N}\right). \tag{4.61}$$

Here, the superscript p indicates that $\hat{\theta}_N^p$ is a parametric estimate.

Given two estimates, the plug-in estimate $\hat{\theta}_N$ and the parametric estimate $\hat{\theta}_N^p$, a natural question is: which estimate is more accurate? Figure 4.3 shows the mean squared errors (MSEs) of the estimates $\hat{\theta}_n$ and $\hat{\theta}_n^p$ as functions of the sample size n for $\alpha = -0.1625$, which corresponds to at least 15% monthly price drop, $P_i < 0.85P_{i-1}$. For each n, estimates $\hat{\theta}_n$ and $\hat{\theta}_n^p$ are based on a sample of size $n < N$ sampled uniformly at random (with replacement) from the data $r_1, \ldots, r_N$, and their MSEs are computed based on 200 independent samples.

[21] These estimates are simply the plug-in estimates (4.45) and (4.49). In Chapter 8, we will show that these estimates are also the *maximum likelihood estimates* (MLEs) of μ and σ^2.

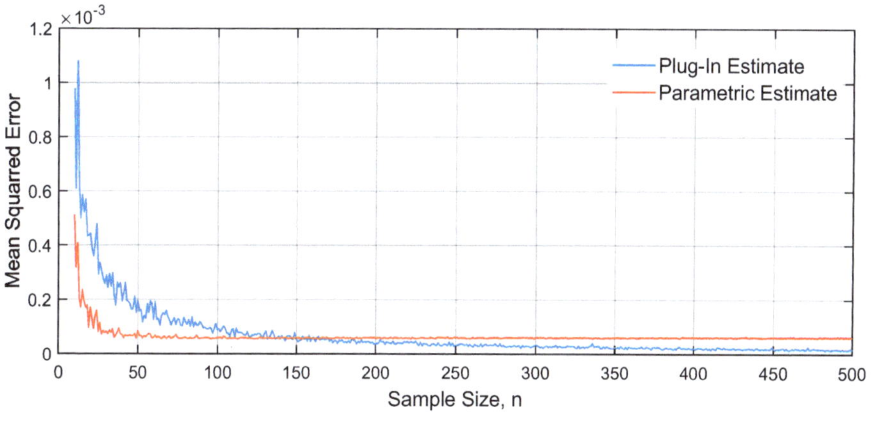

Fig. 4.3 The mean squared errors of the plug-in estimate $\hat{\theta}_n$ and the parametric estimate $\hat{\theta}_n^p$ for $\alpha = -0.1625$. Each estimate is based on a sample of size n obtained from the data $r_1, \ldots, r_N$ by uniform sampling with replacement. For each n, the MSEs are computed based on 200 samples.

The dependence of the MSEs on the sample size illustrates the dynamics of the behavior of the two estimates. If the sample size n is small, then just the data is not enough to accurately estimate θ, and the parametric estimate $\hat{\theta}_n^p$, that uses the normality assumption, outperforms the plug-in estimate $\hat{\theta}_n$. As the sample size n increases, however, the eCDF $\hat{F}_n$ of the log return, as predicted by Theorem 4.1, converges to the true CDF F, and the plug-in estimate, obtained by replacing F with $\hat{F}_n$, becomes more accurate than the parametric estimate.

Why does having more data (as n increases) not help the parametric estimate $\hat{\theta}_n^p$ to be more accurate? The reason is that the normality assumption (4.57) for the log returns of the S&P 500 index is not really correct, especially for the tails of the data, and holds only approximately. Figure 4.4 shows the normal-quantile plot[22] for the data $r_1, \ldots, r_N$, which clearly indicates that the data has long ("heavy") tails that decay slower than the normal tails. Practically, this means that the extreme (large or small) values of log returns are more likely to occur in reality than under the normality assumption. Consequently, instead of normal distribution, heavy-tailed distributions, such as the t-distribution and the generalized hyperbolic distributions, are often used for modeling the log returns of stock indices and other financial assets [9, 10].

[22] See Section 1.5.

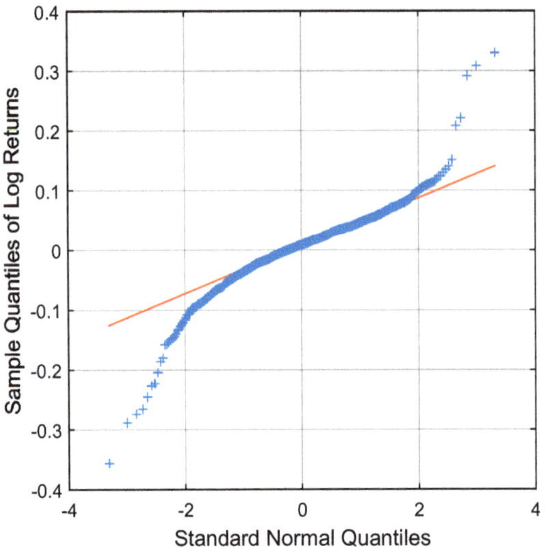

Fig. 4.4 The normal-quantile plot for the S&P 500 log returns $r_1, \ldots, r_N$. The data has long tails.

So, if the sample size n is small ($n < 150$), then the normality assumption (despite the fact that it does not model the tails of the data accurately) helps to compensate for the lack of data and construct a parametric estimate $\hat{\theta}_n^p$, which is more accurate than the nonparametric plug-in estimate $\hat{\theta}_n$. But as the amount of data increases with n, the assumption-free data-driven plug-in estimate becomes more and more accurate, and for $n > 150$ outperforms the parametric estimate.

4.5 Further Reading

This chapter introduced several fundamental notions and important results of nonparametric inference: statistical functionals, parameters, plug-in estimates, and asymptotic properties of the eCDF. For a wonderful overview of the entire field of nonparametric inference, the following excellent book by Larry Wasserman is strongly recommended:

- Wasserman, L.A.: *All of Nonparametric Statistics*. Springer Texts in Statistics (2006). `https://www.springer.com/gp/book/9780387251455`

As the name suggests, the book covers all (or almost all) nonparametric statistics. Of course, it is impossible (and not needed) to cover all topics in detail in a single book. Wasserman's textbook provides a very useful, intuitive, and concise account of most modern topics in nonparametric inference including density estimation, nonparametric regression, minimax estimation, and wavelets.

4.6 Appendix: The Sifting Property of the Dirac Delta Function

The sifting property[23] of the delta function states that for any function $g(x)$ continuous on the interval (a,b) and any $x_0 \in (a,b)$

$$\int_{-\infty}^{\infty} g(x)\delta(x-x_0)dx = g(x_0). \tag{4.62}$$

[23] "The more unnecessary details are put in the more lifelike we take fiction to be." – Tom Shippey, *The Road to Middle-Earth*.

In other words, the integration of the product of a function with $\delta(x-x_0)$ simply "sifts" out the value of the function at x_0. Note that $\delta(x-x_0)$ is just the Dirac delta function (4.14) centered at x_0 instead of zero:

$$\delta(x-x_0) = \begin{cases} 0, & \text{if } x \neq x_0, \\ \infty, & \text{if } x = x_0. \end{cases} \quad \text{and} \quad \int_{-\infty}^{\infty} \delta(x-x_0)dx = 1. \tag{4.63}$$

Intuitively, we can show that (4.62) holds as follows. Since $\delta(x-x_0)$ is zero outside any neighborhood of x_0, for arbitrary small $\varepsilon > 0$

$$\int_{-\infty}^{\infty} g(x)\delta(x-x_0)dx = \int_{x_0-\varepsilon}^{x_0-\varepsilon} g(x)\delta(x-x_0)dx. \tag{4.64}$$

Since $g(x)$ is continuous on the infinitesimal interval $(x_0-\varepsilon, x_0-\varepsilon) \subset (a,b)$, for all $x \in (x_0-\varepsilon, x_0-\varepsilon)$ we have $g(x) \approx g(x_0)$, and so

$$\int_{x_0-\varepsilon}^{x_0-\varepsilon} g(x)\delta(x-x_0)dx \approx g(x_0)\int_{x_0-\varepsilon}^{x_0-\varepsilon} \delta(x-x_0)dx = g(x_0). \tag{4.65}$$

This argument shows that (4.62) is not surprising and sort of expected. More rigorously, we can prove the sifting property as follows:

$$\begin{aligned}\int_{-\infty}^{\infty} g(x)\delta(x-x_0)dx &= \int_{-\infty}^{\infty} \big(g(x_0)+g(x)-g(x_0)\big)\delta(x-x_0)dx \\ &= g(x_0)\int_{-\infty}^{\infty} \delta(x-x_0)dx + \int_{-\infty}^{\infty} \big(g(x)-g(x_0)\big)\delta(x-x_0)dx \\ &= g(x_0) + \int_{x_0-\tau}^{x_0+\tau} \big(g(x)-g(x_0)\big)\delta(x-x_0)dx,\end{aligned} \tag{4.66}$$

where $\tau > 0$ is arbitrary small. It remains to prove that the last integral in (4.66) is zero. Since $g(x)$ is continuous on (a,b), for arbitrary small $\varepsilon > 0$ there exists $\tau > 0$ such that $|g(x)-g(x_0)| < \varepsilon$ whenever $|x-x_0| < \tau$. Therefore,

$$\begin{aligned}\left|\int_{x_0-\tau}^{x_0+\tau} \big(g(x)-g(x_0)\big)\delta(x-x_0)dx\right| &\le \int_{x_0-\tau}^{x_0+\tau} \underbrace{|g(x)-g(x_0)|}_{<\varepsilon}\delta(x-x_0)dx \\ &\le \varepsilon\int_{x_0-\tau}^{x_0+\tau} \delta(x-x_0)dx = \varepsilon.\end{aligned} \tag{4.67}$$

This means that the absolute value of the last integral in (4.66) is less than arbitrary small $\varepsilon > 0$. This is possible only if the integral is zero. This completes the proof of the sifting property.

Problems

4.1. Empirical CDF

Let $X_1,\ldots,X_n \overset{\text{iid}}{\sim} F \in \mathscr{F}$, and let $\hat{F}_n$ be the empirical CDF of the sample. Find the covariance between two random variables $\hat{F}_n(x)$ and $\hat{F}_n(y)$ for $x \neq y$. Compare the derived result with (4.21) for the case where $x = y$.

4.2. Nonparametric Confidence Band for the CDF

In the Case Study in Section 4.4, we discussed how to estimate the probability that a log return of the S&P 500 index will be below a certain threshold. In this problem, you will construct a confidence band for the CDF of the log return. First, download the S&P 500 monthly price data from [11] for the period between Dec 1, 1927 and

July 1, 2020, and compute the log returns $r_1, \ldots, r_N$ given by (4.53) corresponding to the adjusted closing prices, where $N = 1112$. Estimate the CDF F of the log return by the empirical CDF and construct a 95% nonparametric confidence band (4.40) for F. Plot both the estimated CDF and the confidence band.

4.3. Skewness and Kurtosis
Skewness and kurtosis, defined in (4.10) and (4.11), are the parameters that measure the shape of a probability distribution. Skewness $\varkappa$ quantifies the lack of symmetry: if $\varkappa > 0$, then the right tail is heavier than the left tail, and if $\varkappa < 0$, then the left tail is heavier. If the distribution is symmetric (e.g., normal), then $\varkappa = 0$[24].

[24] But $\varkappa = 0$ can also occur if one of the tails is short but fat and the other is long but light, so that overall they have the same "weight" and balance out each other.

For symmetric distributions, the kurtosis κ measures the extent to which probability is concentrated in the tails of the distribution: high κ indicates that the distribution tends to produce more outliers. The kurtosis of any normal distribution $\mathcal{N}(\mu, \sigma^2)$ is $\kappa = 3$. Distribution with $\kappa < 3$ are called *platykurtic* (e.g., the uniform distribution), and distributions with $\kappa > 3$ are called *leptokurtic* (e.g., the Laplace distribution). It is hard to interpret the kurtosis for asymmetric distributions.

Find the plug-in estimates $\hat{\varkappa}_n$ and $\hat{\kappa}_n$ of the skewness and kurtosis. In practice, deviations of $\hat{\varkappa}_n$ and $\hat{\kappa}_n$ from 0 and 3, respectively, are used as an indication of non-normality of the data.

4.4. Estimating the Effectiveness of a New Drug
A pharmaceutical company developed a new drug to treat a certain disease. To check its effectiveness, the following experiment was conducted. A group of $N = 900$ patients with the disease was randomly divided into two equal subgroups. The first subgroup of $n = 450$ patients was given the new drug and the second subgroup of $n = 450$ patients was given an old drug. In the first group, $n_1 = 400$ patients have recovered from the disease, and in the second group, $n_2 = 375$ patients have recovered. Let p_1 and p_2 denote the probabilities of recovery due to the new and old drug respectively. Let $\theta = p_1 - p_2$ be the parameter of interest that quantifies the effectiveness of the new drug compared to the old one: if $\theta > 0$, then the new drug is more effective in treating the disease, and if $\theta < 0$, then the old drug is more effective. Find the plug-in estimate $\hat{\theta}_n$ of θ.

References

1. Dvoretzky, A., Kiefer, J., Wolfowitz, J.: Asymptotic minimax character of the sample distribution function and of the classical multinomial estimator. *Annals of Mathematical Statistics* 27(3), 642-669 (1956).
   `https://doi.org/10.1214/aoms/1177728174`
2. Massart, P.: The tight constant in the Dvoretzky-Kiefer-Wolfowitz inequality. *Annals of Probability* 18(3), 1269-1283 (1990).
   `https://doi.org/10.1214/aop/1176990746`
3. Scorsese, M., Winter, T.: *The Wolf of Wall Street* (2013).
   `https://youtu.be/HEaVXFXm5Wg?si=aJE5IGByPcSV0lrK`
4. Ruppert, D.: *Statistics and Finance: An Introduction*. New York, NY: Springer (2004).
   `https://doi.org/10.1007/978-1-4419-6876-0`

5. Bachelier, L.: *Theory of Speculation*. Gaulthier-Vilar, Paris (1900).
https://www.jstor.org/stable/j.ctt7scn4
6. Hull, J.C.: *Options, Futures, and Other Derivatives*. Pearson (2021).
https://www.pearson.com/en-us/subject-catalog/p/options-futures-and-other-derivatives/P200000005938/9780136939979
7. Black, F., Scholes, M.: The pricing of options and corporate liabilities. *Journal of Political Economy* 81(3), 637–654 (1973).
https://www.jstor.org/stable/1831029
8. Merton, R.C.: Theory of rational option pricing. *The Bell Journal of Economics and Management Science* 4(1), 141–183 (1973).
https://www.jstor.org/stable/3003143
9. Platen, E., Rendek, R.: Empirical evidence on student-t log-returns of diversified world stock indices. *Journal of Statistical Theory and Practice* 2(2), 233-251 (2008).
https://doi.org/10.1080/15598608.2008.10411873
10. Borak, S., Misiorek, A., Weron, R.: Models for heavy-tailed asset returns. In: Cizek, P., Härdle, W., Weron, R. (eds) *Statistical Tools for Finance and Insurance*. Springer, Berlin, Heidelberg (2011).
https://doi.org/10.1007/978-3-642-18062-0_1
11. Yahoo! Finance, S&P 500 Historical Data,
https://finance.yahoo.com/quote/%5EGSPC/history?p=%5EGSPC.
The data set is also available at https://link.springer.com/chapter/10.1007/978-3-032-03848-7_1

5
The Jackknife Method

Guns for show, knives for a pro.
– Guy Ritchie, *Lock, Stock and Two Smoking Barrels* [1]

In the previous chapter, we have learned how to estimate a parameter of interest nonparametrically using the plug-in principle. We have also seen that a plug-in estimate may be biased: for example, the plug-in estimate of the variance is biased. In this chapter, we will learn how to reduce the bias of a biased estimate using the so-called *jackknife method*.

The jackknife method was originally proposed by Maurice Quenouille[1], a British statistician of French ancestry, as a nonparametric method for estimating the bias of an estimate [2, 3]. In 1958, John Tukey extended the method by showing how to use it not only for estimating, but also for reducing the bias, and coined the name "jackknife" [4]. As a pocket knife, this technique can be used as a "quick and dirty" tool for solving a variety of statistical problems.

[1] Despite the remarkable influence of the jackknife method on statistical methodology, I could not find a photo of its inventor on the Internet! Please let me know if you come across one.

5.1 Estimating the Bias

Let $X_1,\dots,X_n$ be data which is modeled as an IID sample from a distribution F, let $\theta = \mathfrak{f}(F)$ be a parameter of interest, where $\mathfrak{f}$ is a given statistical functional, and let

$$\hat{\theta}_n = s(X_1,\dots,X_n) \tag{5.1}$$

be an estimate of θ, where s is a certain statistic (e.g., some function of the data). For example, $\hat{\theta}_n$ can be the plug-in estimate $\hat{\theta}_n = \mathfrak{f}(\hat{F}_n)$, but not necessarily. In practice, many estimates are biased, so that

$$\mathbb{B}[\hat{\theta}_n] = \mathbb{E}[\hat{\theta}_n] - \theta \neq 0. \tag{5.2}$$

This leads to the following natural question: how to estimate the bias $\mathbb{B}[\hat{\theta}_n]$ of the estimate? Let's look at this problem with open naive eyes. We can estimate θ

Supplementary Information The online version contains supplementary material available at https://doi.org/10.1007/978-3-032-03848-7_5.

K. M. Zuev, *Fundamentals of Statistical Inference*,
International Series in Operations Research & Management Science 374,
https://doi.org/10.1007/978-3-032-03848-7_5

in (5.2) simply by $\hat{\theta}_n$, and we can estimate, at least in principle, the expected value $\mathbb{E}[\hat{\theta}_n]$ using the law of large numbers (LLN)[2]. This idea can be implemented in the following way[3]:

[2] It is somewhat a "natural instinct" to estimate expectations by empirical averages using the LLN.

[3] Notice the similarity between (5.3)-(5.5) and the bootstrap method (2.49)-(2.51). The LLN underlies both algorithms.

1. Generate m independent samples, each containing n IID sample units from F:

$$\begin{aligned} \text{Sample } 1: \quad & X_1^{(1)}, \ldots, X_n^{(1)} \overset{\text{iid}}{\sim} F, \\ & \cdots \\ \text{Sample } m: \quad & X_1^{(m)}, \ldots, X_n^{(m)} \overset{\text{iid}}{\sim} F. \end{aligned} \tag{5.3}$$

2. Compute m estimates based on the generated samples:

$$\hat{\theta}_n^{(1)} = s(X_1^{(1)}, \ldots, X_n^{(1)}), \quad \ldots, \quad \hat{\theta}_n^{(m)} = s(X_1^{(m)}, \ldots, X_n^{(m)}). \tag{5.4}$$

   In particular, if $\hat{\theta}_n$ is the plug-in estimate, $\hat{\theta}_n = \mathfrak{f}(\hat{F}_n)$, then $\hat{\theta}_n^{(i)} = \mathfrak{f}(\hat{F}_n^{(i)})$, where $\hat{F}_n^{(i)}$ is the empirical CDF of the sample $X_1^{(i)}, \ldots, X_n^{(i)}$.
3. Estimate the expected value $\mathbb{E}[\hat{\theta}_n]$ by the corresponding empirical average:

$$\mathbb{E}[\hat{\theta}_n] \approx \frac{1}{m} \sum_{i=1}^{m} \hat{\theta}_n^{(i)}. \tag{5.5}$$

The bias can then be estimated as follows:

$$\mathbb{B}[\hat{\theta}_n] = \mathbb{E}[\hat{\theta}_n] - \theta \approx \frac{1}{m} \sum_{i=1}^{m} \hat{\theta}_n^{(i)} - \hat{\theta}_n. \tag{5.6}$$

What is the problem with this approach? A major problem is that we can't really generate m samples from F in Step 1. The only available data from F is our original sample $X_1, \ldots, X_n$. Recall that we faced the same problem in Chapter 2, where we discussed the bootstrap method. The bootstrap approach was used to create multiple copies of $X_1, \ldots, X_n$ to mimic the (finite) target population. The jackknife method also reuses the available data, but adopting a different strategy.

The key idea behind the jackknife method is to emulate (5.3) by cooking up n samples of size $n-1$ from the original data $X_1, \ldots, X_n$ by leaving one data point X_i out at a time[4]:

[4] That is why jackknife is sometimes called a "leave-one-out" method.

1. Form n *jackknife samples* from the data $X_1, \ldots, X_n$:

$$\begin{aligned} \text{Jackknife Sample } 1: \quad & X_2, \ldots, X_n \overset{\text{iid}}{\sim} F, \\ & \cdots \\ \text{Jackknife Sample } i: \quad & X_1, \ldots, X_{i-1}, X_{i+1}, \ldots, X_n \overset{\text{iid}}{\sim} F, \\ & \cdots \\ \text{Jackknife Sample } n: \quad & X_1, \ldots, X_{n-1} \overset{\text{iid}}{\sim} F. \end{aligned} \tag{5.7}$$

2. Compute n *jackknife replications* of $\hat{\theta}_n$ based on the jackknife samples:

$$\begin{aligned}
\text{Jackknife Replication } 1: \quad & \hat{\theta}_n^{(-1)} = s(X_2,\ldots,X_n), \\
& \ldots \\
\text{Jackknife Replication } i: \quad & \hat{\theta}_n^{(-i)} = s(X_1,\ldots,X_{i-1},X_{i+1},\ldots,X_n), \qquad (5.8) \\
& \ldots \\
\text{Jackknife Replication } n: \quad & \hat{\theta}_n^{(-n)} = s(X_1,\ldots,X_{n-1}).
\end{aligned}$$

3. By analogy with (5.5), estimate the expected value $\mathbb{E}[\hat{\theta}_n]$ by the average of jackknife replications $\bar{\theta}_n^J$:

$$\mathbb{E}[\hat{\theta}_n] \approx \frac{1}{n}\sum_{i=1}^{n} \hat{\theta}_n^{(-i)} = \bar{\theta}_n^J. \qquad (5.9)$$

The bias of $\hat{\theta}_n$ can then be estimated by

$$\mathbb{B}[\hat{\theta}_n] = \mathbb{E}[\hat{\theta}_n] - \theta \approx \bar{\theta}_n^J - \hat{\theta}_n. \qquad (5.10)$$

The bad news is that, unfortunately, this estimate is very inaccurate. The good news is that, under certain conditions, it is easy to fix it. Inaccuracy of (5.10) is caused by the inaccuracy of (5.9). Consider two jackknife samples:

$$X_1,\ldots,X_{i-1},X_{i+1},\ldots,X_n \quad \text{and} \quad X_1,\ldots,X_{j-1},X_{j+1},\ldots,X_n. \qquad (5.11)$$

These samples differ only in two sample units: X_i and X_j. The remaining $n-2$ data points are the same in both samples. This means that the corresponding jackknife replications, $\hat{\theta}_n^{(-i)}$ and $\hat{\theta}_n^{(-j)}$, that are based on those samples, are strongly dependent. Therefore, the assumption of the LLN[5], that was supposed to guarantee the accuracy of (5.9) for large n, are not satisfied: although $\hat{\theta}_n^{(-1)},\ldots,\hat{\theta}_n^{(-n)}$ are identically distributed, they are not independent. This is the conceptual reason why (5.10) is not accurate.

It turns out, however, that if the bias $\mathbb{B}[\hat{\theta}_n]$ is not very large, then it is possible to modify the estimate (5.10) and drastically improve its accuracy. Namely, suppose that

$$\mathbb{B}[\hat{\theta}_n] = \frac{a}{n} + \frac{b}{n^2} + O\left(\frac{1}{n^3}\right) \quad \text{as } n \to \infty, \qquad (5.12)$$

where a and b are some constants, and $O\left(1/n^3\right)$ denotes[6] a term that decays to zero not slower than c/n^3 for some constant c. Intuitively, this assumptions means that for large n the bias $\mathbb{B}[\hat{\theta}_n]$ behaves like a/n.

To quantitatively see how inaccurate (5.10) is, that is, how different $\bar{\theta}_n^J - \hat{\theta}_n$ is from $\mathbb{B}[\hat{\theta}_n]$ on average, let's compute the expected value $\mathbb{E}[\bar{\theta}_n^J - \hat{\theta}_n]$ (this will help us to fix the estimate). From (5.12),

$$\mathbb{E}[\hat{\theta}_n] = \theta + \mathbb{B}[\hat{\theta}_n] = \theta + \frac{a}{n} + \frac{b}{n^2} + O\left(\frac{1}{n^3}\right). \qquad (5.13)$$

Equivalently, since $\hat{\theta}_n = s(X_1,\ldots,X_n)$,

[5] For a quick review of the Law of Large Numbers and its assumptions see Probability Refresher 2.12.

[6] This is the so-called *big O notation* often used for describing the limiting behavior of functions.

$$\mathbb{E}[s(X_1,\ldots,X_n)] = \theta + \frac{a}{n} + \frac{b}{n^2} + O\left(\frac{1}{n^3}\right). \tag{5.14}$$

The expected value of the average of jackknife replications $\bar{\theta}_n^J$ is

$$\begin{aligned}\mathbb{E}[\bar{\theta}_n^J] &= \frac{1}{n}\sum_{i=1}^{n}\mathbb{E}\big[\hat{\theta}_n^{(-i)}\big] = \frac{1}{n}\sum_{i=1}^{n}\underbrace{\mathbb{E}\big[s(\overbrace{X_1,\ldots,X_{i-1},X_{i+1},\ldots,X_n}^{\text{sample of size } n-1})\big]}_{\theta+\frac{a}{n-1}+\frac{b}{(n-1)^2}+O\left(\frac{1}{(n-1)^3}\right)} \\ &= \theta + \frac{a}{n-1} + \frac{b}{(n-1)^2} + O\left(\frac{1}{(n-1)^3}\right).\end{aligned} \tag{5.15}$$

Therefore,

$$\begin{aligned}\mathbb{E}[\bar{\theta}_n^J - \hat{\theta}_n] &= \left(\frac{1}{n-1} - \frac{1}{n}\right)a + \left(\frac{1}{(n-1)^2} - \frac{1}{n^2}\right)b + O\left(\frac{1}{n^3}\right) \\ &= \frac{a}{n(n-1)} + \underbrace{\frac{2n+1}{n^2(n-1)^2}b}_{O\left(\frac{1}{n^3}\right)} + O\left(\frac{1}{n^3}\right) = \frac{a}{n(n-1)} + O\left(\frac{1}{n^3}\right).\end{aligned} \tag{5.16}$$

This means that to obtain the leading term $\frac{a}{n}$ of the bias in (5.12), we need to simply "**inflate**" $(\bar{\theta}_n^J - \hat{\theta}_n)$ by multiplying it by $(n-1)$. So, instead of (5.10), let's estimate the bias of $\hat{\theta}_n$ as follows:

$$\mathbb{B}[\hat{\theta}_n] \approx (n-1)(\bar{\theta}_n^J - \hat{\theta}_n) = \hat{\mathbb{B}}_J[\hat{\theta}_n]. \tag{5.17}$$

This estimate is called the *jackknife estimate of the bias*, denoted $\hat{\mathbb{B}}_J[\hat{\theta}_n]$, and the factor $(n-1)$ is called the ***inflation factor***. Under assumption (5.12), the jackknife estimate $\hat{\mathbb{B}}_J[\hat{\theta}_n]$ estimates the bias $\mathbb{B}[\hat{\theta}_n]$ up to order $O\left(\frac{1}{n^2}\right)$:

$$\mathbb{E}\big[\hat{\mathbb{B}}_J[\hat{\theta}_n]\big] = (n-1)\mathbb{E}[\bar{\theta}_n^J - \hat{\theta}_n] = \frac{a}{n} + O\left(\frac{1}{n^2}\right) = \mathbb{B}[\hat{\theta}_n] + O\left(\frac{1}{n^2}\right). \tag{5.18}$$

5.2 Reducing the Bias

A natural way to reduce the bias of an estimate is to subtract the bias (if it is known) or the estimate of the bias (if the bias is unknown but can be estimated) from the estimate. Following this idea, let's define the *jackknifed estimate*[7] $\hat{\theta}_n^J$ as the difference between the original estimate $\hat{\theta}_n$ and the jackknife estimate of the bias $\hat{\mathbb{B}}_J[\hat{\theta}_n]$:

$$\hat{\theta}_n^J = \hat{\theta}_n - \hat{\mathbb{B}}_J[\hat{\theta}_n] = \hat{\theta}_n - (n-1)(\bar{\theta}_n^J - \hat{\theta}_n) = n\hat{\theta}_n - (n-1)\bar{\theta}_n^J. \tag{5.19}$$

[7] Also called the bias-corrected estimate.

Be careful with the notation: $\hat{\theta}_n^J$ is the jackknifed estimate, and $\bar{\theta}_n^J$ is the average of jackknife replication (5.9).

The bias of the jackknifed estimate $\hat{\theta}_n^J$ is an *order magnitude* smaller than the bias of the original estimate $\hat{\theta}_n$ in (5.12), which is $O(1/n)$. Indeed,

$$\begin{aligned}\mathbb{B}\big[\hat{\theta}_n^J\big] &= \mathbb{E}\big[\hat{\theta}_n^J\big] - \theta = \mathbb{E}\big[\hat{\theta}_n - \hat{\mathbb{B}}_J[\hat{\theta}_n]\big] - \theta = \mathbb{E}[\hat{\theta}_n] - \mathbb{E}\big[\hat{\mathbb{B}}_J[\hat{\theta}_n]\big] - \theta \\ &= \mathbb{B}[\hat{\theta}_n] - \mathbb{E}\big[\hat{\mathbb{B}}_J[\hat{\theta}_n]\big] = O\left(\frac{1}{n^2}\right).\end{aligned} \tag{5.20}$$

Moreover, if the original estimate is unbiased for any n, $\mathbb{E}[\hat{\theta}_n] = \theta$, then jackknifing does not destroy that property, and the jackknifed estimate is also unbiased:

$$\begin{aligned}\mathbb{E}\big[\hat{\theta}_n^J\big] &= \mathbb{E}\big[n\hat{\theta}_n - (n-1)\bar{\theta}_n^J\big] = n\mathbb{E}[\hat{\theta}_n] - (n-1)\mathbb{E}[\bar{\theta}_n^J] \\ &= n\theta - (n-1)\frac{1}{n}\sum_{i=1}^{n}\underbrace{\mathbb{E}\big[\hat{\theta}_n^{(-i)}\big]}_{\theta} = n\theta - (n-1)\theta = \theta.\end{aligned} \tag{5.21}$$

Here, the expected value $\mathbb{E}\big[\hat{\theta}_n^{(-i)}\big] = \theta$, because the jackknife replication $\hat{\theta}_n^{(-i)} = s(X_1,\dots,X_{i-1},X_{i+1},\dots,X_n)$ based on a sample of size $(n-1)$ is unbiased, since we assumed that $\mathbb{E}[\hat{\theta}_n] = \mathbb{E}[s(X_1,\dots,X_n)] = \theta$ holds for any sample size n.

Remark 5.1. The formula $\hat{\theta}_n^J = n\hat{\theta}_n - (n-1)\bar{\theta}_n^J$ defines a general jackknifed estimate that was derived under the assumption that the bias of the original estimate $\mathbb{B}[\hat{\theta}_n]$ satisfies (5.12). It can always be used for reducing the bias when this assumption holds. If, however, in addition to (5.12), the exact expression for the bias is known, then it is possible to derive — following exactly the same "jackknifing" steps that we used to derive $\hat{\theta}_n^J$ — a jackknifed estimate with even smaller bias. This is illustrated in Problem 5.1, parts (e) and (f).

Recall that the overall measure of accuracy of an estimate is its mean squared error (MSE), which can be written as the sum of the squared bias and the variance of the estimate (3.11)[8]. If the estimate $\hat{\theta}_n$ is asymptotically unbiased, namely, if its bias satisfies (5.12), then we can reduce the bias even further by using the jackknife method, and adopting $\hat{\theta}_n^J$ instead of $\hat{\theta}_n$. This leads to a natural and important question: *how does jackknifing affect the variance of the estimate?* It can be shown that, under certain technical conditions (which, however, are often hard to check in applications directly), the jackknife method does not strongly affect the variance, that is, the variance remains approximately the same,

[8] This is the bias-variance decomposition of the MSE.

$$\mathbb{V}[\hat{\theta}_n^J] \approx \mathbb{V}[\hat{\theta}_n]. \tag{5.22}$$

Remark 5.2. The jackknife method can be also used for estimating the variance $\mathbb{V}[\hat{\theta}_n]$ or, equivalently, the standard error $\mathrm{se}[\hat{\theta}_n] = (\mathbb{V}[\hat{\theta}_n])^{1/2}$ of the original estimate $\hat{\theta}_n$ using the so-called *pseudo-values*,

$$\tilde{\theta}_n^{(i)} = n\hat{\theta}_n - (n-1)\hat{\theta}_n^{(-i)}, \tag{5.23}$$

which were introduced by John Tukey in [4][9]. The main idea behind the definition of pseudo-values is that these quantities allow to rewrite the jackknifed estimate (5.19) as a sample mean:

$$\begin{aligned}\hat{\theta}_n^J &= n\hat{\theta}_n - (n-1)\bar{\theta}_n^J = \frac{1}{n}\sum_{i=1}^n n\hat{\theta}_n - (n-1)\frac{1}{n}\sum_{i=1}^n \hat{\theta}_n^{(-i)} \\ &= \frac{1}{n}\sum_{i=1}^n \underbrace{\left[n\hat{\theta}_n - (n-1)\hat{\theta}_n^{(-i)}\right]}_{\text{pseudo-value}} = \frac{1}{n}\sum_{i=1}^n \tilde{\theta}_n^{(i)}.\end{aligned} \tag{5.24}$$

[9] According to [5], Tukey named quantities (5.23) "pseudo-values" in an unpublished work.

As an example, if the parameter of interest is the mean $\theta = \mu$ and the estimate is the sample mean of the data $\hat{\theta}_n = \overline{X}$, then the jackknife replication and the corresponding pseudo-value are, respectively,

$$\hat{\theta}_n^{(-i)} = \frac{1}{n-1}\sum_{j\neq i} X_j \quad \text{and} \quad \tilde{\theta}_n^{(i)} = n\overline{X} - (n-1)\frac{1}{n-1}\sum_{j\neq i} X_j = X_i. \tag{5.25}$$

Notice that in this example, $\hat{\theta}_n^J = \hat{\theta}_n = \overline{X}$, that is, jackknifing does not change the original estimate[10], and the pseudo-values are IID, since $X_1,\ldots,X_n$ are IID.

In general, however, the pseudo-values are not IID. Nevertheless, they can be viewed, with a certain degree of accuracy, as "linear approximations" to IID random variables[11]. Indeed, suppose that the statistic s, a multivariate function of the data used for computing the original estimate, can be approximated by the average of univariate functions:

[10] Which is somewhat expected, since the original estimate is unbiased.

[11] This important observation made by John Tukey allows to derive the estimate of $\mathbb{V}[\hat{\theta}_n]$ in (5.28).

$$\hat{\theta}_n = s(X_1,\ldots,X_n) \approx \frac{1}{n}\sum_{i=1}^n \xi(X_i). \tag{5.26}$$

Then, similar to example (5.25)[12], the jackknife replication and the corresponding pseudo-value are

[12] Where $\xi(X_i) = X_i$.

$$\hat{\theta}_n^{(-i)} \approx \frac{1}{n-1}\sum_{j\neq i}\xi(X_j) \quad \text{and} \quad \tilde{\theta}_n^{(i)} \approx \sum_{i=1}^n \xi(X_i) - \sum_{j\neq i}\xi(X_j) = \xi(X_i). \tag{5.27}$$

Since $X_1,\ldots,X_n$ are IID, the pseudo-values $\tilde{\theta}_n^{(1)} \approx \xi(X_1),\ldots,\tilde{\theta}_n^{(n)} \approx \xi(X_n)$ are approximately IID. The accuracy of this approximation depends, of course, on the accuracy of assumption (5.26).

This interpretation of the pseudo-values as approximately IID random variables allows to estimate the variance of $\hat{\theta}_n$ as follows. Assuming that (5.22) holds and treating the pseudo-values *as if* they are IID, we have:

$$\mathbb{V}[\hat{\theta}_n] \approx \mathbb{V}[\hat{\theta}_n^J] = \mathbb{V}\left[\frac{1}{n}\sum_{i=1}^n \tilde{\theta}_n^{(i)}\right] = \frac{1}{n^2}\sum_{i=1}^n \mathbb{V}[\tilde{\theta}_n^{(i)}] = \frac{1}{n}\mathbb{V}[\tilde{\theta}_n^{(i)}]$$
$$\approx \frac{1}{n}\underbrace{\frac{1}{n-1}\sum_{i=1}^n (\tilde{\theta}_n^{(i)} - \hat{\theta}_n^J)^2}_{\text{estimate (4.51) of } \mathbb{V}[\tilde{\theta}_n^{(i)}]} = \hat{\mathbb{V}}_J[\hat{\theta}_n]. \tag{5.28}$$

It can be shown that, under certain conditions, the *jackknife estimate of the variance* $\hat{\mathbb{V}}_J[\hat{\theta}_n]$ is a *consistent*[13] estimate, that is, $\hat{\mathbb{V}}_J[\hat{\theta}_n]$ converges in probability to $\mathbb{V}[\hat{\theta}_n]$,

[13] Recall Definition 4.1.

$$\hat{\mathbb{V}}_J[\hat{\theta}_n] \xrightarrow{\mathbb{P}} \mathbb{V}[\hat{\theta}_n], \quad \text{as } n \to \infty. \tag{5.29}$$

Unfortunately, the technical conditions that guarantee consistency are hard to check in real applications, and there are cases where $\hat{\mathbb{V}}_J[\hat{\theta}_n]$ is not a good estimate. This happens, in particular, when $\hat{\theta}_n = s(X_1, \ldots, X_n)$ is not sufficiently smooth as a function of the data. For instance, if $\hat{\theta}_n = \hat{F}_n^{-1}(1/2)$ is the plug-in estimate of the median or if $\hat{\theta}_n$ depends on order statistics, such as the sample minimum $X_{(1)}$ or the sample maximum $X_{(n)}$, then the jackknife estimate $\hat{\mathbb{V}}_J[\hat{\theta}_n]$ is a poor estimate of the variance. Moreover, it has been shown that the jackknife variance estimate *always* tends to be *biased upwards* [6].

In the next chapter, we will discuss the *bootstrap method*, which is a very popular simulation-based method for estimating standard errors. The bootstrap was inspired by the jackknife and it is a superior technique for estimating standard errors. In some sense the jackknife can be viewed as a linear approximation of the bootstrap [7].

5.3 Example: Jackknifing the Plug-In Estimate of the Variance

Let $X_1, \ldots, X_n \overset{\text{iid}}{\sim} F$ be data, and suppose that the parameter of interest is the variance σ^2 of the distribution F that generated the data. Recall[14] that the plug-in estimate of the variance is

[14] See (4.49) and (4.50).

$$\hat{\sigma}_n^2 = \frac{1}{n}\sum_{i=1}^n (X_i - \overline{X})^2, \tag{5.30}$$

its expected value is

$$\mathbb{E}[\hat{\sigma}_n^2] = \frac{n-1}{n}\sigma^2, \tag{5.31}$$

and, therefore, the bias is

$$\mathbb{B}[\hat{\sigma}_n^2] = \mathbb{E}[\hat{\sigma}_n^2] - \sigma^2 = -\frac{\sigma^2}{n}. \tag{5.32}$$

Let's find the jackknifed estimate (also called the bias-corrected estimate) $\hat{\sigma}_n^{2,J}$ of the variance. First, notice that the bias of the plug-in estimate (5.32) satisfies the assumption (5.12), with $a = -\sigma^2$. By definition (5.19),

$$\hat{\sigma}_n^{2,J} = n\hat{\sigma}_n^2 - (n-1)\bar{\sigma}_n^{2,J}, \tag{5.33}$$

where $\bar{\sigma}_n^{2,J}$ is the average of jackknife replications of $\hat{\sigma}_n^2$,

$$\bar{\sigma}_n^{2,J} = \frac{1}{n}\sum_{i=1}^{n} \hat{\sigma}_n^{2,(-i)}. \tag{5.34}$$

It will be convenient to rewrite the plug-in estimate as in (4.50):

$$\hat{\sigma}_n^2 = \frac{1}{n}\sum_{i=1}^{n} X_i^2 - \overline{X}^2. \tag{5.35}$$

Now let's find the expression for the ith jackknife replication $\hat{\sigma}_n^{2,(-i)}$, which is simply the plug-in estimate of σ^2 based on the ith jackknife sample $X^{(-i)} = (X_1,\dots,X_{i-1},X_{i+1},\dots,X_n)$ of size $(n-1)$:

$$\hat{\sigma}_n^{2,(-i)} = \frac{1}{n-1}\sum_{j\neq i} X_j^2 - \overline{X^{(-i)}}^2 \tag{5.36}$$

This expression is obtained from (5.35) by replacing n with $(n-1)$ and the full sample $X = (X_1,\dots,X_n)$ with the jackknife sample $X^{(-i)} = (X_1,\dots,X_{i-1},X_{i+1},\dots,X_n)$. The mean of the the jackknife sample $X^{(-i)}$ is

$$\overline{X^{(-i)}} = \frac{1}{n-1}\sum_{j\neq i} X_j = \frac{1}{n-1}\Big(\underbrace{\sum_{j=1}^{n} X_j}_{n\overline{X}} - X_i\Big) = \frac{n}{n-1}\overline{X} - \frac{1}{n-1}X_i. \tag{5.37}$$

Substituting this into (5.36) and expanding the square, we have:

$$\begin{aligned}
\hat{\sigma}_n^{2,(-i)} &= \frac{1}{n-1}\Big(\sum_{j=1}^{n} X_j^2 - X_i^2\Big) - \Big(\frac{n}{n-1}\overline{X} - \frac{1}{n-1}X_i\Big)^2 \\
&= \frac{1}{n-1}\sum_{j=1}^{n} X_j^2 \underbrace{- \frac{1}{n-1}X_i^2}_{} - \frac{n^2\overline{X}^2}{(n-1)^2} + \frac{2n\overline{X}X_i}{(n-1)^2} \underbrace{- \frac{1}{(n-1)^2}X_i^2}_{} \\
&= \frac{1}{n-1}\sum_{j=1}^{n} X_j^2 - \frac{n}{(n-1)^2}X_i^2 - \frac{n^2\overline{X}^2}{(n-1)^2} + \frac{2n\overline{X}X_i}{(n-1)^2}.
\end{aligned} \tag{5.38}$$

Substituting this into (5.34) and computing the outer sum, we have:

$$\begin{aligned}\bar{\sigma}_n^{2,J} &= \frac{1}{n}\sum_{i=1}^{n}\left(\frac{1}{n-1}\sum_{j=1}^{n}X_j^2 - \frac{n}{(n-1)^2}X_i^2 - \frac{n^2\overline{X}^2}{(n-1)^2} + \frac{2n\overline{X}X_i}{(n-1)^2}\right)\\
&= \underbrace{\frac{1}{n-1}\sum_{j=1}^{n}X_j^2 - \frac{1}{(n-1)^2}\sum_{i=1}^{n}X_i^2}\underbrace{- \frac{n^2\overline{X}^2}{(n-1)^2} + \frac{2n\overline{X}^2}{(n-1)^2}}\\
&= \frac{n-2}{(n-1)^2}\sum_{i=1}^{n}X_i^2 - \frac{(n^2-2n)\overline{X}^2}{(n-1)^2}.\end{aligned} \tag{5.39}$$

Finally, substituting (5.35) and (5.39) in (5.33), we have:

$$\begin{aligned}\hat{\sigma}_n^{2,J} &= n\left(\frac{1}{n}\sum_{i=1}^{n}X_i^2 - \overline{X}^2\right) - (n-1)\left(\frac{n-2}{(n-1)^2}\sum_{i=1}^{n}X_i^2 - \frac{(n^2-2n)\overline{X}^2}{(n-1)^2}\right)\\
&= \sum_{i=1}^{n}X_i^2 - n\overline{X}^2 - \frac{n-2}{n-1}\sum_{i=1}^{n}X_i^2 + \frac{(n^2-2n)\overline{X}^2}{n-1}\\
&= \frac{1}{n-1}\sum_{i=1}^{n}X_i^2 - \frac{n}{n-1}\overline{X}^2 = \frac{n}{n-1}\underbrace{\left(\frac{1}{n}\sum_{i=1}^{n}X_i^2 - \overline{X}^2\right)}_{\hat{\sigma}_n^2} = \hat{s}_n^2.\end{aligned} \tag{5.40}$$

It turns out that the jackknifed estimate $\hat{\sigma}_n^{2,J}$ of the variance is simply the unbiased sample variance $\hat{s}_n^2$ that we derived in Chapter 4 in (4.51). In this example, the jackknife method helped to reduce the bias of the plug-in estimate to exactly zero.

5.4 Case Study: Alcohol Consumption

Not to waste time on nonsense.

– Marcus Aurelius, *Meditations*

One of the widespread myths (actively supported by marketing and advertising media) is that a little dose of alcohol — a can of beer, a small glass of wine, a shot of whiskey or other spirits — might have a positive effect on health. A recent scientific study, however, has shown that this is not true: no level of alcohol consumption improves health [8]. Moreover, in 2018, the Global Burden of Disease Study[15], which involved more than 500 researchers and academics from more then 40 nations, demonstrated that alcohol contributes substantially, and more than previously estimated, to death, disability, and ill health, globally [9, 10]. In particular, clear and convincing causal links between drinking and premature death, cancer, liver disease, and various cardiovascular problems have been found.

Although alcohol is one of the major causes of death and disability, its global consumption has been steadily increasing, from 5.9 liters of pure alcohol per adult in 1990 to 6.5 liters in 2017 [11, 12]. The main driving force behind this global trend is

[15] The Global Burden of Disease Study, led by the Institute for Health Metrics and Evaluation, is a comprehensive worldwide research program that assesses mortality and disability from major diseases, injuries, and risk factors.

a substantial increase of alcohol consumption in lower-middle income countries. For example, between 2010-2017, the largest increase in drinking occurred in Vietnam (from 4.7 to 8.9 liters) and India (from 4.3 to 5.9 liters). In some countries alcohol intake dropped, with the most significant decrease in Azerbaijan (from 2.9 to 0.5 liters), Russia (from 15.8 to 12.3 liters), and the UK (from 12.3 to 11.4 liters). In the USA and China consumption increased slightly from 9.3 to 9.8 liters and from 7.1 to 7.4 liters, respectively. Unfortunately, global alcohol consumption is expected to grow in the future, and it is predicted that by 2030 half of all world's adults will drink alcohol, almost a quarter will binge drink[16] at least once a month, and per adult consumption will further increase to 7.6 liters. Furthermore, emerging studies (see e.g. [14]) suggest that stress, anxiety, and social isolation caused by the COVID-19 pandemic could increase alcohol drinking and accelerate this global trend.

[16] Binge drinking is a pattern of drinking that brings blood alcohol concentration to 0.08 g/dl or above. This pattern corresponds approximately to consuming 5 or more drinks for males and 4 or more drinks for females, in about 2 hours [13].

In this Case Study, we will look at the correlation between food and alcohol expenditures of British households in 1980-1982 [15]. The data is represented by a $N \times 2$ matrix, where the first column (x) is the budget share for food expenditure, the second column (y) is the budget share for alcohol expenditure, and $N = 1519$ is the number of observations (households). Figure 5.1 shows the scatterplot of the data[17], which is a collection of points (x_i, y_i), $i = 1, \ldots, N$, visually representing the observations.

[17] We will discuss scatterplots in more detail in Chapter 11 in the context of regression analysis.

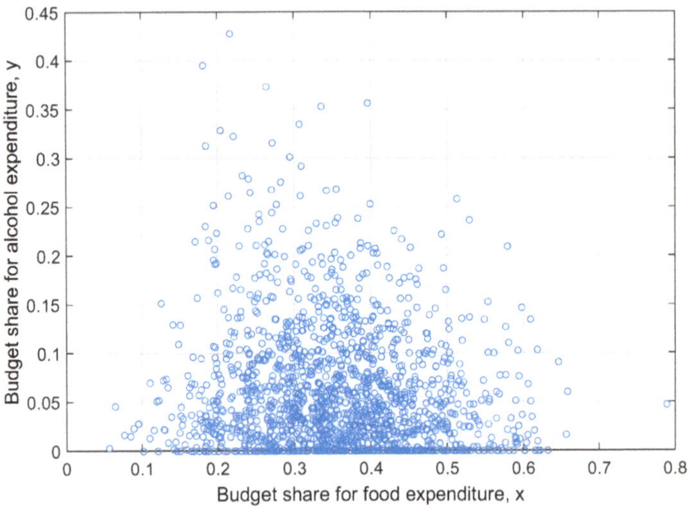

Fig. 5.1 The scatterplot of the budget shares for food and alcohol expenditures in the UK households in 1980-1982. The sample size of the data (number of points) is $N = 1519$.

The scatterplot suggests that, on average, the higher the food expenditure the lower the alcohol expenditure. This qualitative observation can be quantified by the *sample correlation coefficient* (SCC), which is defined as

$$r_{xy} = \frac{\sum_{i=1}^{N}(x_i - \bar{x})(y_i - \bar{y})}{\sqrt{\sum_{i=1}^{N}(x_i - \bar{x})^2 \sum_{i=1}^{N}(y_i - \bar{y})^2}}, \tag{5.41}$$

where $\bar{x}$ and $\bar{y}$ are the sample means of $x_1, \dots, x_N$ and $y_1, \dots, y_N$, respectively.

Remark 5.3. The sample correlation coefficient is a "data analog" of *Pearson's*[18] *correlation coefficient* (PCC) between two random variables X and Y, which is defined as

$$\rho_{XY} = \frac{\text{Cov}(X,Y)}{\sqrt{\mathbb{V}[X]\mathbb{V}[Y]}}. \tag{5.42}$$

The apparent visual similarity between (5.41) and (5.42) is not a coincidence. Namely, it can be shown[19] that the sample correlation coefficient is simply the plug-in estimate of the Pearson correlation coefficient, where samples $x_1, \dots, x_N$ and $y_1, \dots, y_N$ are realizations of random variables X and Y.

The sample correlation coefficient r_{xy} measures the strength and direction of a *linear relationship* between two samples $x = (x_1, \dots, x_N)$ and $y = (y_1, \dots, y_N)$. Its value is always between -1 and 1. If $r_{xy} = \pm 1$, then all data points (x_i, y_i) lie exactly on a line: with a positive slope if $r_{xy} = 1$ or with a negative slope if $r_{xy} = -1$. The value $r_{xy} = 0$ indicates that there is no *linear* relationship between the two samples[20]. If $r_{xy} \in (0,1)$ or $r_{xy} \in (-1,0)$, which is always the case for real data, then points (x_i, y_i) lie approximately on a line: with a positive slope if $r_{xy} > 0$ or with a negative slope if $r_{xy} < 0$. The absolute value $|r_{xy}|$ quantifies the accuracy of approximation of the cloud of data points by a line. Figure 5.2 shows several data sets $(x_1, y_1), \dots, (x_N, y_N)$ and the corresponding sample correlation coefficients[21].

For the alcohol-versus-food data set on Fig. 5.1, the sample correlation coefficient is

$$r = -0.122, \tag{5.43}$$

which is consistent with our previous qualitative observation that the budget share for alcohol expenditure tends to decrease when food's share increases, although the relationship between the two shares is far from being linear.

Suppose now that we don't have access to the entire data set[22]. Instead, we can just obtain a small subset of the data by sampling $(X_1, Y_1), \dots, (X_n, Y_n)$ uniformly at random without replacement from $(x_1, y_1), \dots, (x_N, y_N)$, where the sample size $n = 100$, which is approximately $n/N \times 100\% \approx 6.58\%$ of the original data set. Although we can't now compute the SCC r exactly, we can estimate it from the available data by replacing x_i, $\bar{x}$, y_i, $\bar{y}$, and N in (5.41) with X_i, $\overline{X}$, Y_i, $\overline{Y}$, and n, respectively,

$$\hat{r}_n(X,Y) = \frac{\sum_{i=1}^{n}(X_i - \overline{X})(Y_i - \overline{Y})}{\sqrt{\sum_{i=1}^{n}(X_i - \overline{X})^2 \sum_{i=1}^{n}(Y_i - \overline{Y})^2}}. \tag{5.44}$$

The jackknifed estimate of the SCC r is then

$$\hat{r}_n^J(X,Y) = n\hat{r}_n - (n-1)\bar{r}_n^J, \tag{5.45}$$

[18] In accordance with Arnold's principle (recall Marginal Note 47 in Chapter 1), the correlation coefficient was first defined by French physicist Auguste Bravais.

Karl Pearson (1857-1936), an English mathematician and statistician, who established the discipline of mathematical statistics, practiced law, wrote books on history and religion, and authored more than 500 other publications.

[19] See Problem 6.1(b).

[20] But $r_{xy} = 0$ does not exclude the possibility of existence of a nonlinear relationship between x and y. For examples, see Fig. 5.2

[21] To get a better feel for the concept of correlation coefficient, readers are encouraged to play a guess-the-correlation game at `http://guessthecorrelation.com/`.

[22] For example, we don't have enough resources (time, funding, or both) to collect all $N = 1519$ observations.

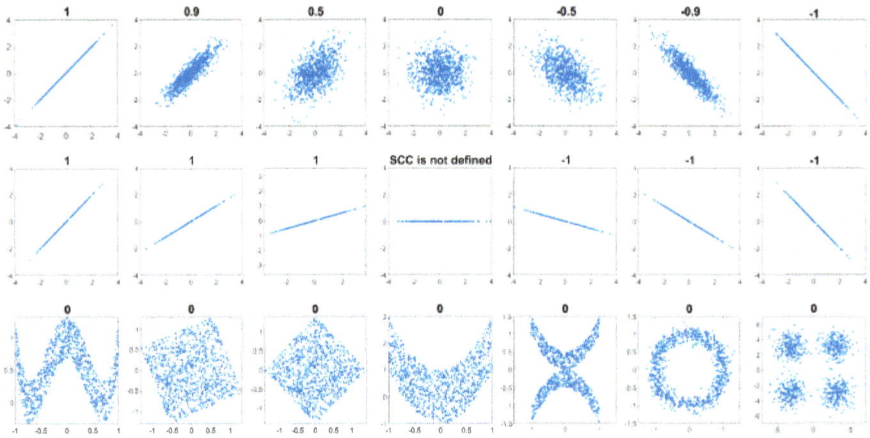

Fig. 5.2 **Clouds of points and the corresponding sample correlation coefficients (SCC). The SCC quantifies the strength and direction of a linear relationship between two samples (top row), i.e. how** closely data points are focused around a line and whether that line is ascending (has positive slope) or descending (has negative slope). But the magnitude of the slope (middle row) and nonlinear relationships (bottom row) are not measured by the SCC. For the cloud for points in the middle, the SCC is not defined since all $y_i = const$, which makes the denominator in (5.41) zero. Clouds of points in the bottom row have different shapes, but the SCC for all of them is zero.

where $\bar{r}_n^J$ is the average of jackknife replications $\hat{r}_n^{(-1)}, \dots, \hat{r}_n^{(-n)}$ of $\hat{r}_n$,

$$\bar{r}_n^J = \frac{1}{n}\sum_{i=1}^{n} \hat{r}_n^{(-i)}, \qquad \hat{r}_n^{(-i)} = \frac{\sum_{j\neq i}\left(X_j - \overline{X^{(-i)}}\right)\left(Y_j - \overline{X^{(-i)}}\right)}{\sqrt{\sum_{j\neq i}\left(X_j - \overline{X^{(-i)}}\right)^2 \sum_{j\neq i}\left(Y_j - \overline{Y^{(-i)}}\right)^2}}, \tag{5.46}$$

where $\overline{X^{(-i)}}$ and $\overline{Y^{(-i)}}$ are the means of the jackknife samples $X^{(-i)}$ and $Y^{(-i)}$,

$$\begin{aligned} X^{(-i)} &= (X_1, \dots, X_{i-1}, X_{i+1}, \dots, X_n), & \overline{X^{(-i)}} &= \frac{1}{n-1}\sum_{j\neq i} X_j, \\ Y^{(-i)} &= (Y_1, \dots, Y_{i-1}, Y_{i+1}, \dots, Y_n), & \overline{Y^{(-i)}} &= \frac{1}{n-1}\sum_{j\neq i} Y_j. \end{aligned} \tag{5.47}$$

To compare statistical properties of two estimates, $\hat{r}_n$ and $\hat{r}_n^J$, we generate $M = 10^4$ independents realizations $(X^{(m)}, Y^{(m)}) = \{(X_1^{(m)}, Y_1^{(m)}), \dots, (X_n^{(m)}, Y_n^{(m)})\}$, $m = 1, \dots, M$, of $(X, Y) = \{(X_1, Y_1), \dots, (X_n, Y_n)\}$ and estimate the bias and the variance of the estimates using the law of large numbers[23] as follows:

[23] See Probability Refresher 2.12.

$$
\begin{aligned}
\mathbb{B}[\hat{r}_n] &= \mathbb{E}[\hat{r}_n] - r \approx \frac{1}{M}\sum_{m=1}^{M} \hat{r}_n(X^{(m)}, Y^{(m)}) - r = 2.1 \times 10^{-3}, \\
\mathbb{B}[\hat{r}_n^J] &= \mathbb{E}[\hat{r}_n^J] - r \approx \frac{1}{M}\sum_{m=1}^{M} \hat{r}_n^J(X^{(m)}, Y^{(m)}) - r = 3.9 \times 10^{-4}, \\
\mathbb{V}[\hat{r}_n] &= \mathbb{E}\left[(\hat{r}_n - \mathbb{E}[\hat{r}_n])^2\right] \\
&\approx \frac{1}{M}\sum_{m=1}^{M}\left(\hat{r}_n(X^{(m)}, Y^{(m)}) - \frac{1}{M}\sum_{k=1}^{M}\hat{r}_n(X^{(k)}, Y^{(k)})\right)^2 = 8.2 \times 10^{-3}, \\
\mathbb{V}[\hat{r}_n^J] &= \mathbb{E}\left[(\hat{r}_n^J - \mathbb{E}[\hat{r}_n^J])^2\right] \\
&\approx \frac{1}{M}\sum_{m=1}^{M}\left(\hat{r}_n^J(X^{(m)}, Y^{(m)}) - \frac{1}{M}\sum_{k=1}^{M}\hat{r}_n^J(X^{(k)}, Y^{(k)})\right)^2 = 8.3 \times 10^{-3}.
\end{aligned}
\tag{5.48}
$$

Consistently with the theoretical results discussed in Section 5.2, the bias of the jackknified estimate $\hat{r}_n^J$ is an order magnitude smaller than the bias of the original estimate $\hat{r}_n$, while the variances of the two estimates are approximately the same.

5.5 Further Reading

In this chapter, we learned how to estimate and reduce the bias of a biased estimate using the jackknife method. If certain assumptions are satisfied, then the jackknife method reduces the bias by an order of magnitude without increasing the variance. A brief description of the jackknife technique together with a summary of its underlying theory, advantages, disadvantages, and its general properties is given in

- Bissell, A.F., Ferguson, R.A.: The Jackknife — Toy, Tool or Two-edged Weapon? *The Statistician* 24(2), 79-100 (1975).
  `https://doi.org/10.2307/2987663`

This paper also presents several case studies that illustrate applications of jackknifing to data sets arising from real problems and demonstrate the misbehavior of the jackknife when it is applied inappropriately. It concludes with a number of practical recommendations for applications of the jackknife method.

Problems

5.1. Jackknife: Theoretical Analysis
Let $X_1, \ldots, X_n$ be data, and suppose that it can be modeled as a sample from the uniform distribution $X_1, \ldots, X_n \overset{\text{iid}}{\sim} \mathscr{U}[0, \theta]$, but the parameter θ is unknown.

(a) Find the plug-in estimate $\hat{\theta}_n$ of θ using the following representation of θ:

$$\theta = \mathfrak{f}(F) = \min\{x \mid F(x) = 1\}, \tag{5.49}$$

where $F(x)$ is the CDF of the uniform distribution. That is, θ is the smallest value of x such that $F(x) = 1$.

(b) Find the bias of the plug-in estimate $\hat{\theta}_n$. Does it satisfy assumption (5.12)?

(c) Find the jackknifed estimate $\hat{\theta}_n^J$ based on the assumption (5.12).

(d) Find the bias of the jackknifed estimate $\hat{\theta}_n^J$. For solving this problem, the following result is useful. If $X_1, \ldots, X_n \sim \mathscr{U}[0, \theta]$ and $X_{(k)}$ is the kth order statistic, then

$$\mathbb{E}[X_{(k)}] = \frac{k\theta}{n+1}. \tag{5.50}$$

(e) [24] Find another jackknifed estimate $\hat{\hat{\theta}}_n^J$ based on the exact expression for the bias derived in part (b) by following the same "jackknifing" steps that we used to derive the general jackknifed estimate $\hat{\theta}_n^J$. How are $\hat{\theta}_n^J$ and $\hat{\hat{\theta}}_n^J$ related?

(f) Find the bias of the jackknifed estimate $\hat{\hat{\theta}}_n^J$ and compare it with the bias of $\hat{\theta}_n^J$.

[24] Parts (e,f) are inspired by the creative solutions of parts (c,d) provided by a Caltech electrical engineering graduate student Sri Aditya Deevi, who took author's course IDS 157 "Statistical Inference" during Spring 2023.

5.2. Jackknife: Practical Implementation

In this problem you will implement the jackknife method. Let $X_1, \ldots, X_n \overset{\text{iid}}{\sim} \mathscr{N}(\mu, \sigma^2)$ be a sample from the normal distribution with variance $\sigma^2 = 1$. Suppose that the parameter of interest is $\theta = e^{\mu}$. The plug-in estimate of θ is then

$$\hat{\theta}_n = e^{\overline{X}}. \tag{5.51}$$

As you will see in part (a), this estimate is biased. The main goal of this problem is to reduce the bias by jackknifing $\hat{\theta}_n$.

(a) Recall that the jackknife method assumes that

$$\mathbb{B}[\hat{\theta}_n] = \frac{a}{n} + \frac{b}{n^2} + O\left(\frac{1}{n^3}\right) \quad \text{as } n \to \infty, \tag{5.52}$$

Check that this assumption is satisfied for $\hat{\theta}_n = e^{\overline{X}}$. For solving this problem, the following result is useful. If $X \sim \mathscr{N}(\mu, \sigma^2)$, then $Y = e^X$ follows the *log-normal distribution*, $Y \sim \ln \mathscr{N}(\mu, \sigma^2)$, and the expected value of Y is

$$\mathbb{E}[Y] = \mathbb{E}[e^X] = e^{\mu + \frac{\sigma^2}{2}}. \tag{5.53}$$

(b) Generate a sample $X_1, \ldots, X_n$ using $\mu = 5$ and $n = 100$. Estimate the bias of $\hat{\theta}_n$ using the jackknife method and compare the estimated value with the exact value obtained in part (a).

(c) In this part, the goal is to empirically observe the theoretical statement that the bias of the jackknifed estimate $\hat{\theta}_n^J$ is smaller than the bias of $\hat{\theta}_n$. First, generate $M = 10^4$ samples $X_1^{(j)}, \ldots, X_n^{(j)}$, $j = 1, \ldots, M$, as in part (b). For each sample compute the plug-in estimate $\hat{\theta}_n^{(j)}$ and the corresponding jackknifed estimate $\hat{\theta}_n^{(j),J}$.

Estimate the biases of $\hat{\theta}_n$ and $\hat{\theta}_n^J$ using the law of large numbers as follows:

$$\mathbb{B}[\hat{\theta}_n] \approx \mathbb{B}_1 = \frac{1}{M}\sum_{j=1}^{M} \hat{\theta}_n^{(j)} - \theta,$$
$$\mathbb{B}[\hat{\theta}_n^J] \approx \mathbb{B}_2 = \frac{1}{M}\sum_{j=1}^{M} \hat{\theta}_n^{(j),J} - \theta. \tag{5.54}$$

Compute both $\mathbb{B}_1$ and $\mathbb{B}_2$. We expect $\mathbb{B}_1$ to be approximately equal to the exact value $\mathbb{B}[\hat{\theta}_n]$ found in part (a) and to be larger (in absolute value) than $\mathbb{B}_2$.

References

1. Ritchie, G.S.: *Lock, Stock and Two Smoking Barrels* (1998). https://youtu.be/Z0mYHjFlh74?si=04m7YIeBsxATUyyx
2. Quenouille, M.H.: Problems in plane sampling. *Annals of Mathematical Statistics* 20(3), 355-375 (1949). https://doi.org/10.1214/aoms/1177729989
3. Quenouille, M.H.: Notes on bias in estimation. *Biometrika* 43(3–4), 353-360 (1956). https://doi.org/10.1093/biomet/43.3-4.353
4. Tukey, J.W.: Bias and confidence in not quite large samples (abstract). *Annals of Mathematical Statistics* 29(2), 614 (1958). https://doi.org/10.1214/aoms/1177706647
5. Miller, R.G.: The jackknife – a review. *Biometrika* 61(1), 1-15 (1974). https://doi.org/10.2307/2334280
6. Efron, B., Stein C.: The jackknife estimate of variance. *The Annals of Statistic* 9(3), 586-596 (1981). https://www.jstor.org/stable/2240822
7. Efron, B.: Bootstrap methods: another look at the jackknife. *The Annals of Statistics* 7(1), 1–26 (1979). https://doi.org/10.1214/aos/1176344552
8. Burton, R., Sheron, N.: No level of alcohol consumption improves health. *The Lancet* 392(10152), 987-988 (2018). https://doi.org/10.1016/S0140-6736(18)31571-X
9. Science Daily: No safe level of alcohol, new study concludes. https://www.sciencedaily.com/releases/2018/08/180824103018.htm.
10. Griswold, M.G., et al.: Alcohol use and burden for 195 countries and territories, 1990–2016: a systematic analysis for the Global Burden of Disease Study 2016. *The Lancet* 392(10152), 1015-1035 (2018). https://doi.org/10.1016/S0140-6736(18)31310-2
11. EurekAlert!: The Lancet: Targets to reduce harmful alcohol use are likely to be missed as global alcohol intake increases. https://www.eurekalert.org/news-releases/658394.
12. Manthey, J., et al.: Global alcohol exposure between 1990 and 2017 and forecasts until 2030: a modelling study. *The Lancet* 393(10190), 2493-2502 (2019). https://doi.org/10.1016/S0140-6736(18)32744-2
13. National Institute of Alcohol Abuse and Alcoholism. NIAAA Newsletter. 2004;3:3. https://www.niaaa.nih.gov/sites/default/files/newsletters/Newsletter_Number3.pdf

14. Grossman, E.R., Benjamin-Neelon, S.E., Sonnenschein, S.: Alcohol Consumption during the COVID-19 Pandemic: A Cross-Sectional Survey of US Adults. *International Journal of Environmental Research and Public Health* 17(24), 9189 (2020). https://doi.org/10.3390/ijerph17249189
15. Blundell, R., Duncan, A., Pendakur, K.: Semiparametric estimation and consumer demand. *Journal of Applied Econometrics*, 13(5), 435–462 (1998). https://www.jstor.org/stable/223219. The data set is also available at https://link.springer.com/chapter/10.1007/978-3-032-03848-7_1

6
The Bootstrap Method

The world stands on absurdities, and without them perhaps nothing at all would happen.

– Fyodor Mikhailovich Dostoevsky, *The Brothers Karamazov*

In the last two chapters we discussed how to estimate a parameter of interest θ nonparametrically using the plug-in principle, $\hat{\theta}_n \approx \theta$, and how to reduce the bias of the estimate $\hat{\theta}_n$ using the jackknife method. In this chapter, we will learn how to estimate the standard error of $\hat{\theta}_n$ and how to construct confidence intervals for θ. Both problems can be solved by the *bootstrap method*[1], which was inspired by the jackknife. The bootstrap, introduced in 1979 by Bradley Efron [1], is a general simulation-based method for measuring the uncertainty of estimates, in particular, for estimating their standard errors and constructing confidence intervals. Its beauty lies in its simplicity and universality: the bootstrap is fully automatic, requires no theoretical calculations, and is always available.

[1] We have already encountered the bootstrap method in Chapter 2 in the context of simple random sampling. Here we will discuss this general simulation method in detail.

6.1 Bootstrapping the Standard Error

As before, let $X_1,\dots,X_n \overset{\text{iid}}{\sim} F \in \mathscr{F}$ be data which is modeled nonparametrically as a sample from a distribution $F \in \mathscr{F}$, where the stochastic model $\mathscr{F} = \{\text{all CDFs}\}$. Let $\hat{\theta}_n = s(X_1,\dots,X_n)$ be an estimate of a parameter of interest $\theta = \mathfrak{f}(F)$ calculated from the data[2], where $\mathfrak{f}$ is a given statistical functional. A fundamental question in statistical inference is *how accurate* $\hat{\theta}_n$ *is*. As discussed in Section 3.2.1, to quantify the overall accuracy of an estimate the mean squared error (MSE) is often used. Using the bias-variance decomposition, the MSE can be written as

$$\text{MSE}[\hat{\theta}_n] = \mathbb{E}[(\hat{\theta}_n - \theta)^2] = \mathbb{B}[\hat{\theta}_n]^2 + \text{se}[\hat{\theta}_n]^2. \tag{6.1}$$

[2] For example, but not necessarily, the plug-in estimate, $\hat{\theta}_n = \mathfrak{f}(\hat{F}_n)$.

Many estimates used in applications, including plug-in estimates, tend to have relatively small biases compared to the magnitudes of their standard errors,

Supplementary Information The online version contains supplementary material available at https://doi.org/10.1007/978-3-032-03848-7_6.

K. M. Zuev, *Fundamentals of Statistical Inference*,
International Series in Operations Research & Management Science 374,
https://doi.org/10.1007/978-3-032-03848-7_6

$$|\mathbb{B}[\hat{\theta}_n]| \ll \mathrm{se}[\hat{\theta}_n]. \tag{6.2}$$

This means that the first term in (6.1) is often negligible compared to the second term, and most of the contribution to the MSE comes from the squared standard error. This observation brings us to the first goal of this chapter: *to learn how to estimate the standard error* $\mathrm{se}[\hat{\theta}_n]$.

By definition, the standard error of $\hat{\theta}_n = s(X_1,\ldots,X_n)$ is

$$\mathrm{se}_F[\hat{\theta}_n] = \mathbb{V}_F[\hat{\theta}_n]^{1/2}, \tag{6.3}$$

where

$$\begin{aligned} \mathbb{V}_F[\hat{\theta}_n] &= \int\cdots\int \left(s(x_1,\ldots,x_n) - \mathbb{E}_F[\hat{\theta}_n]\right)^2 dF(x_1)\ldots dF(x_n), \\ \mathbb{E}_F[\hat{\theta}_n] &= \int\cdots\int s(x_1,\ldots,x_n)dF(x_1)\ldots dF(x_n) \end{aligned} \tag{6.4}$$

are its variance and expected value, respectively. Here we intentionally use the subscript $_F$ to emphasize that the standard error, variance, and expected value of $\hat{\theta}_n$ do depend on F, which is unknown. Since F is unknown, we can't compute $\mathrm{se}_F[\hat{\theta}_n]$ exactly. But we can estimate it using the plug-in principle by replacing F with $\hat{F}_n$. This motivates the following definition.

Definition 6.1. The *ideal bootstrap estimate* of $\mathrm{se}_F[\hat{\theta}_n]$ is the plug-in estimate that uses the empirical CDF $\hat{F}_n$ instead of F:

$$\widehat{\mathrm{se}}_B^{\mathrm{id}}[\hat{\theta}_n] = \mathrm{se}_{\hat{F}_n}[\hat{\theta}_n]. \tag{6.5}$$

Let's consider an example to illustrate this definition.

Example 6.1. Let the parameter of interest $\theta = \mathfrak{f}(F)$ be the mean μ_F of F, and the estimate $\hat{\theta}_n$ be the plug-in estimate, $\hat{\theta}_n = \mathfrak{f}(\hat{F}_n) = \overline{X}$. The standard error of $\hat{\theta}_n$ is then

$$\mathrm{se}_F[\hat{\theta}_n] = \left(\mathbb{V}_F\left[\frac{1}{n}\sum_{i=1}^n X_i\right]\right)^{1/2} = \left(\frac{1}{n^2}\sum_{i=1}^n \mathbb{V}_F[X_i]\right)^{1/2} = \frac{\sigma_F}{\sqrt{n}}, \tag{6.6}$$

where $\sigma_F^2 = \mathbb{V}_F[X_i]$ is the variance of F, which is unknown. The ideal bootstrap estimate of $\mathrm{se}_F[\hat{\theta}_n]$ is, therefore,

$$\widehat{\mathrm{se}}_B^{\mathrm{id}}[\hat{\theta}_n] = \mathrm{se}_{\hat{F}_n}[\hat{\theta}_n] = \frac{\sigma_{\hat{F}_n}}{\sqrt{n}} = \frac{\sqrt{\frac{1}{n}\sum_{i=1}^n (X_i - \overline{X})^2}}{\sqrt{n}} = \frac{1}{n}\left(\sum_{i=1}^n (X_i - \overline{X})^2\right)^{1/2}. \tag{6.7}$$

Here we used (4.49) that provides the expression for the plug-in estimate $\sigma_{\hat{F}_n}$ of the standard deviation σ_F of F.

This example is very special because it is exceptionally simple: the sample mean $\hat{\theta}_n = \overline{X}$ is essentially the only estimate for which explicit analytical calculations in

(6.6), which lead to a closed form expression for $\mathrm{se}_F[\hat{\theta}_n]$, are possible. Usually the ideal bootstrap estimate (6.5) cannot be computed exactly like in (6.7)[3]. Nevertheless, we can always estimate $\widehat{\mathrm{se}}_B^{\mathrm{id}}[\hat{\theta}_n]$ using the law of large numbers[4].

1. Generate B independent random samples of size n from the empirical CDF $\hat{F}_n$:

$$\begin{aligned} \mathscr{S}^{(1)} &= \{X_1^{(1)}, \dots, X_n^{(1)}\} \overset{\text{iid}}{\sim} \hat{F}_n, \\ &\dots \\ \mathscr{S}^{(B)} &= \{X_1^{(B)}, \dots, X_n^{(B)}\} \overset{\text{iid}}{\sim} \hat{F}_n. \end{aligned} \tag{6.8}$$

   Samples $\mathscr{S}^{(1)}, \dots, \mathscr{S}^{(B)}$ are called the *bootstrap samples*.

2. Compute the *bootstrap replications* $\hat{\theta}_n^{(1)}, \dots, \hat{\theta}_n^{(B)}$ of $\hat{\theta}_n = s(X_1, \dots, X_n)$ using the bootstrap samples:

$$\hat{\theta}_n^{(1)} = s(X_1^{(1)}, \dots, X_n^{(1)}), \quad \dots, \quad \hat{\theta}_n^{(B)} = s(X_1^{(B)}, \dots, X_n^{(B)}). \tag{6.9}$$

3. Estimate the standard error

$$\widehat{\mathrm{se}}_B^{\mathrm{id}}[\hat{\theta}_n] = \mathrm{se}_{\hat{F}_n}[\hat{\theta}_n] = \mathbb{V}_{\hat{F}_n}[\hat{\theta}_n]^{1/2} = \mathbb{E}_{\hat{F}_n}\left[\left(\hat{\theta}_n - \mathbb{E}_{\hat{F}_n}[\hat{\theta}_n]\right)^2\right]^{1/2} \tag{6.10}$$

   by the sample standard deviation of the bootstrap replications:

$$\widehat{\mathrm{se}}_B[\hat{\theta}_n] = \left(\frac{1}{B}\sum_{b=1}^{B}\left(\hat{\theta}_n^{(b)} - \frac{1}{B}\sum_{b=1}^{B}\hat{\theta}_n^{(b)}\right)^2\right)^{1/2}. \tag{6.11}$$

   The estimate $\widehat{\mathrm{se}}_B[\hat{\theta}_n]$ is called the *bootstrap estimate* of the standard error.

   The law of large numbers guarantees that

$$\widehat{\mathrm{se}}_B[\hat{\theta}_n] \xrightarrow{\mathbb{P}} \mathrm{se}_{\hat{F}_n}[\hat{\theta}_n], \quad \text{as } B \to \infty, \tag{6.12}$$

and, therefore,

$$\widehat{\mathrm{se}}_B[\hat{\theta}_n] \approx \mathrm{se}_{\hat{F}_n}[\hat{\theta}_n], \quad \text{if } B \gg 1 \ (B \text{ is large}). \tag{6.13}$$

The ideal bootstrap estimate $\mathrm{se}_{\hat{F}_n}[\hat{\theta}_n]$ and its bootstrap estimate $\widehat{\mathrm{se}}_B[\hat{\theta}_n]$ are called *nonparametric* bootstrap estimates[5].

Remark 6.1. This nonparametric bootstrap method differs from the survey sampling bootstrap in Section 2.6 only in Step 1, where bootstrap samples are generated. In the survey sampling bootstrap, we first create the bootstrap population $\mathscr{P}_{boot}$ from samples units $X_1, \dots, X_n$ and then generate bootstrap samples by sampling from $\mathscr{P}_{boot} = \{X_1, \dots, X_1, \dots, X_n, \dots, X_n\}$ uniformly at random *without replacement*[6]. In the nonparametric bootstrap, bootstrap samples are obtained by IID sampling from the empirical CDF $\hat{F}_n$, which is done by sampling from $\{X_1, \dots, X_n\}$ uniformly at random *with replacement*. In both cases, the bootstrap method first approximates

[3] Convince yourself by considering other estimates $\hat{\theta}_n$, e.g. the plug-in estimate of the variance (instead of the mean).

[4] Compare the algorithm presented below with the bootstrap in Section 2.6

[5] There is also a *parametric* bootstrap method that differs from its nonparametric counterpart only in the way the bootstrap samples are generated. We will discuss and compare these two methods in Problem 6.2.

[6] In other words, by using simple random sampling.

the object of interest — the target population $\mathscr{P} \approx \mathscr{P}_{boot}$ or the CDF $F \approx \hat{F}_n$ — and then samples from $\mathscr{P}_{boot}$ or $\hat{F}_n$ using the same sampling method that was used to obtain the original data from $\mathscr{P}$ or F: SRS or IID sampling, respectively.

The bootstrap method for estimating the standard error is schematically illustrated in Fig. 6.1.

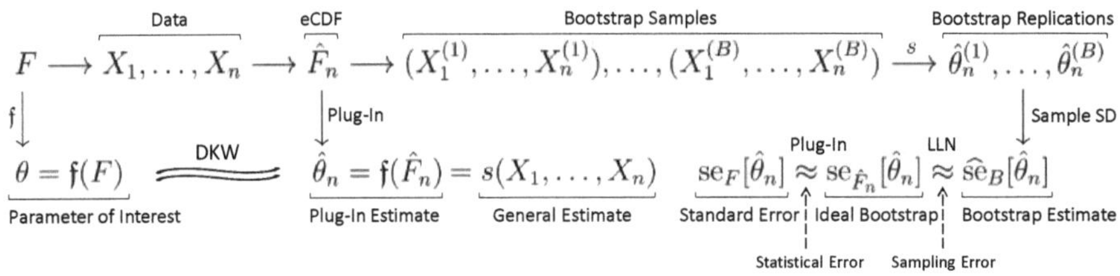

Fig. 6.1 Schematic representation of the nonparametric bootstrap method for estimating the standard error.

There are two types of estimation errors in the bootstrap: *statistical error* and *sampling error*.

- Statistical error occurs in $\text{se}_F[\hat{\theta}_n] \approx \text{se}_{\hat{F}_n}[\hat{\theta}_n]$ because the empirical CDF $\hat{F}_n$ constructed from the data differs from the true CDF F that generated the data. As the sample size increases, $n \to \infty$, the two CDFs get closer, $\hat{F}_n(x) \xrightarrow{\mathbb{P}} F(x)$, but $\hat{F}_n \neq F$ for any finite n. The statistical error is thus controlled by the sample size n, which we can control only if we can get additional data[7].
- Sampling error occurs in $\text{se}_{\hat{F}_n}[\hat{\theta}_n] \approx \widehat{\text{se}}_B[\hat{\theta}_n]$ from using finitely many bootstrap replications $\hat{\theta}_n^{(1)},\dots,\hat{\theta}_n^{(B)}$. If statistic s is cheap to compute, then the sampling error can be made arbitrary small by taking B sufficiently large, since $\widehat{\text{se}}_B[\hat{\theta}_n] \xrightarrow{\mathbb{P}} \text{se}_{\hat{F}_n}[\hat{\theta}_n]$, as $B \to \infty$. But due to the presence of the statistical error, it rarely makes sense to use more than $B = 10^4$ bootstrap replications. In applications that involve complex models, statistics s may be a complicated function of the data which is expensive to compute. In such cases, it becomes important to reduce the number of s-evaluations, in particular, to reduce the number B of bootstrap replications. How small then B can be? This question is thoroughly discussed in [2], where the following rule of thumb is derived: $B = 200$ should be sufficient for most applications.

[7] In some applications this could be time-consuming and expensive.

To sum up, the statistical error is usually larger than the sampling error. While the former arises from a lack of data, which can be difficult to overcome, the latter can often be reduced to a negligible level with brute computing power.

6.2 Bootstrap Confidence Intervals

Bootstrap replications $\hat{\theta}_n^{(1)}, \ldots, \hat{\theta}_n^{(B)}$ in (6.9) can be viewed as approximate[8] realizations of random estimate $\hat{\theta}_n$ and can be used for answering any question about the distribution of $\hat{\theta}_n$. For example, they can be used for estimating the standard error of $\hat{\theta}_n$ as in the previous section, or for constructing confidence intervals.

[8] As the sample size n increases, the quality of this approximation gets better.

Recall[9] that a *confidence interval* for parameter $\theta = \mathfrak{f}(F)$ with the confidence level $100(1-\alpha)\%$ is a *random* interval $I_\alpha = (a,b)$, which contains θ with probability $1-\alpha$,

[9] See Section 3.2.2.

$$\mathbb{P}(a < \theta < b) = 1 - \alpha, \qquad \alpha \in (0,1). \tag{6.14}$$

Let $\hat{\theta}_n = \mathfrak{f}(\hat{F}_n)$ be the plug-in estimate of θ and $\widehat{\text{se}}_B[\hat{\theta}_n]$ be the bootstrap estimate of its standard error. There are several ways to construct a bootstrap confidence intervals for θ. The simplest and most straightforward is the normal interval[10].

[10] Sometimes it is also called the *standard* confidence interval.

6.2.1 Normal Interval

It turns out that for many reasonable distributions F and statistical functionals $\mathfrak{f}$, the plug-in estimate $\hat{\theta}_n = \mathfrak{f}(\hat{F}_n)$ is approximately normally distributed when n is large[11],

[11] Moreover, this large-sample result often holds for general estimates $\hat{\theta}_n = s(X_1, \ldots, X_n)$, not only for plug-in estimates.

$$\hat{\theta}_n \dot{\sim} \mathcal{N}\left(\mathbb{E}[\hat{\theta}_n], \text{se}[\hat{\theta}_n]^2\right), \quad n \gg 1. \tag{6.15}$$

If in addition $\hat{\theta}_n$ is (at least asymptotically) approximately unbiased, $\mathbb{E}[\hat{\theta}_n] \approx \theta$, then

$$\hat{\theta}_n \dot{\sim} \mathcal{N}\left(\theta, \text{se}[\hat{\theta}_n]^2\right), \quad n \gg 1. \tag{6.16}$$

Let's consider a simple example.

Example 6.2. Suppose that the parameter of interest is the mean $\theta = \mu$. The plug-in estimate of θ is then the sample mean $\hat{\theta}_n = \overline{X}$. Thanks to the central limit theorem[12],

[12] See Probability Refresher 2.12.

$$\overline{X} \dot{\sim} \mathcal{N}\left(\mu, \text{se}[\overline{X}]^2 = \frac{\sigma^2}{n}\right), \quad n \gg 1. \tag{6.17}$$

Let' estimate the standard error of $\hat{\theta}_n$ by the bootstrap method, $\text{se}[\hat{\theta}_n] \approx \widehat{\text{se}}_B[\hat{\theta}_n]$. Then (6.16) becomes

$$\hat{\theta}_n \dot{\sim} \mathcal{N}\left(\theta, \widehat{\text{se}}_B[\hat{\theta}_n]^2\right), \quad n \gg 1. \tag{6.18}$$

So, let's assume[13] that

[13] Note that this assumption is equivalent to (6.18).

$$\frac{\hat{\theta}_n - \theta}{\widehat{\text{se}}_B[\hat{\theta}_n]} \dot{\sim} \mathcal{N}(0,1). \tag{6.19}$$

This assumption allows to derive an approximate confidence interval for θ as follows[14]. Since $(\hat{\theta}_n - \theta)/\widehat{\text{se}}_B[\hat{\theta}_n]$ is approximately standard normal,

[14] Compare the derivation below with computations in (3.32)

$$\mathbb{P}\left(z_{\frac{\alpha}{2}} < \frac{\hat{\theta}_n - \theta}{\widehat{\mathrm{se}}_B[\hat{\theta}_n]} < -z_{\frac{\alpha}{2}}\right) \approx \Phi(-z_{\frac{\alpha}{2}}) - \Phi(z_{\frac{\alpha}{2}}) = \left(1 - \frac{\alpha}{2}\right) - \frac{\alpha}{2} = 1 - \alpha, \quad (6.20)$$

where $z_{\frac{\alpha}{2}}$ is the $(\frac{\alpha}{2})^{\text{th}}$ standard normal quantile, see Fig. 1.7. Rewriting the inequalities in terms of θ gives

$$\mathbb{P}\left(\hat{\theta}_n + z_{\frac{\alpha}{2}} \widehat{\mathrm{se}}_B[\hat{\theta}_n] < \theta < \hat{\theta}_n - z_{\frac{\alpha}{2}} \widehat{\mathrm{se}}_B[\hat{\theta}_n]\right) \approx 1 - \alpha. \quad (6.21)$$

Therefore,

$$I_\alpha = \left(\hat{\theta}_n + z_{\frac{\alpha}{2}} \widehat{\mathrm{se}}_B[\hat{\theta}_n], \hat{\theta}_n - z_{\frac{\alpha}{2}} \widehat{\mathrm{se}}_B[\hat{\theta}_n]\right) \quad (6.22)$$

is an approximate $100(1-\alpha)\%$ CI for θ. It is called the *normal confidence interval*. To reassure the reader, the + and − in (6.22) are placed correctly[15].

[15] Recall that $z_{\frac{\alpha}{2}} < 0$.

The normal CI is accurate only under parametric assumption (6.19) that the distribution of estimate $\hat{\theta}_n$ is close to normal. There are cases where this assumption is either not satisfied or it is difficult to check whether it holds or not. This leads to a natural question: *is it possible to construct a confidence interval for θ without making any parametric assumptions?* It turns out that yes, it is possible.

6.2.2 Pivotal Interval

Let $\check{\theta}_n$ be the difference between the estimate $\hat{\theta}_n$ and the parameter of interest θ,

$$\check{\theta}_n = \hat{\theta}_n - \theta, \quad (6.23)$$

and let G be the CDF of random variable $\check{\theta}_n$. Our goal is to find a random interval $I_\alpha = (a, b)$ such that $\mathbb{P}(a < \theta < b) = 1 - \alpha$. Let's rewrite this probability in terms of G:

$$\begin{aligned}\mathbb{P}(a < \theta < b) &= \mathbb{P}(a - \hat{\theta}_n < \theta - \hat{\theta}_n < b - \hat{\theta}_n)\\ &= \mathbb{P}(\hat{\theta}_n - b < \underbrace{\hat{\theta}_n - \theta}_{\check{\theta}_n} < \hat{\theta}_n - a) = G(\hat{\theta}_n - a) - G(\hat{\theta}_n - b). \quad (6.24)\end{aligned}$$

Therefore, we can achieve the desired confidence level $100(1-\alpha)\%$ by requiring

$$\begin{aligned} G(\hat{\theta}_n - a) &= 1 - \frac{\alpha}{2},\\ G(\hat{\theta}_n - b) &= \frac{\alpha}{2}, \end{aligned} \quad (6.25)$$

which is equivalent to

$$
\begin{aligned}
a &= \hat{\theta}_n - G^{-1}\left(1-\frac{\alpha}{2}\right) = \hat{\theta}_n - \mathfrak{g}_{1-\frac{\alpha}{2}}, \\
b &= \hat{\theta}_n - G^{-1}\left(\frac{\alpha}{2}\right) = \hat{\theta}_n - \mathfrak{g}_{\frac{\alpha}{2}},
\end{aligned} \tag{6.26}
$$

where $\mathfrak{g}_{1-\frac{\alpha}{2}} = G^{-1}\left(1-\frac{\alpha}{2}\right)$ and $\mathfrak{g}_{\frac{\alpha}{2}} = G^{-1}\left(\frac{\alpha}{2}\right)$ are the $(1-\frac{\alpha}{2})^{\text{th}}$ and the $(\frac{\alpha}{2})^{\text{th}}$ quantiles of G, respectively.

Remark 6.2. Note that (6.25) is not the only way to achieve the confidence level $100(1-\alpha)\%$. In principle, we can choose any values $G(\hat{\theta}_n - a) = \beta$ and $G(\hat{\theta}_n - b) = \gamma$ as long as $\beta - \gamma = 1-\alpha$. In (6.25), this is done in a balanced way, $\beta = 1-\alpha/2$ and $\gamma = \alpha/2$, so that the tail probabilities, to the left of $\hat{\theta}_n - b$ and to the right of $\hat{\theta}_n - a$, are both equal to $\alpha/2$.

The only obstacle that does not allow us to find a and b from (6.26) is that the CDF G is unknown and, as a result, its quantiles $\mathfrak{g}_q$, $q \in (0,1)$, are also unknown. But, as the name of this chapter suggests, the bootstrap method can help to estimate G, its quantiles, and thus to estimate a and b using (6.26).

If we had a number of realizations of random variable $\check{\theta}_n$, then we could estimate its CDF G by the eCDF constructed from these realizations. But recall that $\check{\theta}_n = \hat{\theta}_n - \theta$ and the bootstrap replications $\hat{\theta}_n^{(1)}, \ldots, \hat{\theta}_n^{(B)}$ of $\hat{\theta}_n$ can be viewed as approximate realizations of $\hat{\theta}_n$. This suggests that we can obtain approximate realizations $\check{\theta}_n^{(1)}, \ldots, \check{\theta}_n^{(B)}$ of $\check{\theta}_n$ by replacing $\hat{\theta}_n$ with its bootstrap replications and estimating θ by the original estimate $\hat{\theta}_n$:

$$
\check{\theta}_n = \hat{\theta}_n - \theta \quad \longrightarrow \quad \check{\theta}_n^{(b)} = \hat{\theta}_n^{(b)} - \hat{\theta}_n, \quad b = 1, \ldots, B. \tag{6.27}
$$

The CDF G can then be estimated as

$$
G(x) \approx \hat{G}_B(x) = \frac{1}{B}\sum_{b=1}^{B} H\big(x - \check{\theta}_n^{(b)}\big), \tag{6.28}
$$

where H, as always, denotes the Heaviside step function (1.15).

Let $\hat{\mathfrak{g}}_\alpha$ denote the αth quantile of $\hat{G}_B$, also called the αth *sample quantile* of $\check{\theta}_n^{(1)}, \ldots, \check{\theta}_n^{(B)}$,

$$
\hat{\mathfrak{g}}_\alpha = \hat{G}_B^{-1}(\alpha). \tag{6.29}
$$

We need to be careful in how we understand the right-hand side of the above formula, since $\hat{G}_B$ is a step function that jumps up by $\frac{1}{B}$ at each $\check{\theta}_n^{(b)}$, $b = 1, \ldots, B$, and therefore strictly speaking, $\hat{G}_B$ is not invertible and $\hat{G}_B^{-1}(\alpha)$ is not defined. There are several ways[16] to make sense of (6.29) and define sample quantiles [3]. One way, illustrated in Fig. 6.2, is to look at the set of all values of x where $\hat{G}_B(x)$ is larger than or equal to α and take the smallest element in that set,

[16] And most of them are implemented in built-in functions in modern programming languages.

$$
\hat{G}_B^{-1}(\alpha) = \min\{x : \hat{G}_B(x) \geq \alpha\}. \tag{6.30}
$$

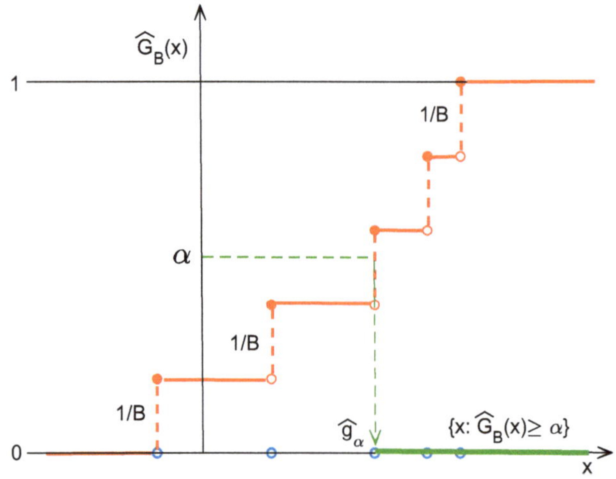

Fig. 6.2 One of possible definitions of the αth sample quantile $\hat{\mathfrak{g}}_\alpha$ of the empirical CDF $\hat{G}_B$.

So, the end points a and b of the confidence interval can be estimated directly from the data as follows:

$$\begin{aligned} a &\approx \hat{a}_B = \hat{\theta}_n - \hat{\mathfrak{g}}_{1-\frac{\alpha}{2}}, \\ b &\approx \hat{b}_B = \hat{\theta}_n - \hat{\mathfrak{g}}_{\frac{\alpha}{2}}. \end{aligned} \tag{6.31}$$

It can be shown that under relatively weak assumptions on F and $\mathfrak{f}$,

$$\mathbb{P}(\hat{a}_B < \theta < \hat{b}_B) \to 1-\alpha, \quad \text{as } n, B \to \infty. \tag{6.32}$$

The interval

$$I_\alpha = (\hat{\theta}_n - \hat{\mathfrak{g}}_{1-\frac{\alpha}{2}}, \hat{\theta}_n - \hat{\mathfrak{g}}_{\frac{\alpha}{2}}) \tag{6.33}$$

is thus an approximate $100(1-\alpha)\%$ CI for θ. It is called the *pivotal confidence interval*.

Remark 6.3. Let $X = (X_1, \ldots, X_n)$ be an IID sample from F and θ be a parameter of F. A random variable $Y(X, \theta)$ that depends on both sample X and parameter θ is called a *pivot* if its distribution does not depend on θ. Once again, the *value* of Y depends on θ, but its *distribution* does not. Intuitively, Y depends on θ both explicitly through θ and implicitly through X, and these two dependences annihilate each other in the distribution of Y. A classical example of a pivot is the so-called *z-score* $Z = \frac{X-\mu}{\sigma}$, where $X \sim \mathcal{N}(\mu, \sigma^2)$. Although Z depends on $\theta = (\mu, \sigma)$, its distribution does not, $Z \sim \mathcal{N}(0,1)$. It can be shown that under weak assumptions on F and $\mathfrak{f}$, the distribution of $\check{\theta}_n = \hat{\theta}_n - \theta$ is not sensitive to $\theta = \mathfrak{f}(F)$, and therefore $\check{\theta}_n$ is an approximate pivot. This explains the name of the confidence interval (6.33).

In this section, we considered two ways to construct a confidence interval for a parameter of interest using the bootstrap method: parametric normal interval and nonparametric pivotal interval. There are also other bootstrap confidence intervals, e.g. the *percentile interval*[17], *studentized interval*[18], BC_a *interval*[19], with many variations. Usually there is a trade-off between the accuracy of an interval and the amount of work needed for its construction. Here we discussed only two bootstrap confidence intervals with the main goal to illustrate the bootstrap method. For more advanced methods, see the references listed below in Section 6.4.

[17] This interval is intuitive, but does not have a solid theoretical support. For "semi-theoretical" justification, see [2].

[18] Also known as "bootstrap-t."

[19] Acronym BC_a stands for "bias-corrected and accelerated."

6.3 Case Study: University Enrollments

An uneducated person accuses others when he is doing badly; a partly educated person accuses himself, an educated person accuses neither some else nor himself.

– Epictetus, *The Encheiridion*

While the importance of education seems to be universally accepted, there is no general consensus on what the main goals of higher education are or should be. Some philosophers, educators, and politicians emphasize "global" society-level purposes, such as shaping students into good citizens and productive members of society, contributing to economical development and growth, and preserving societal cultural values and norms. This line of thought goes back at least to Aristotle, who viewed the primary mission of education in producing good citizens for cities. Other thinkers emphasize "local" individual-level goals of education, namely broadening students' horizons, developing creativity, critical and abstract thinking, general problem solving skills, and positively influencing students' personal development. For example, according to Immanuel Kant, the central goal of education is "the development of character and teaching moral maxims" [4]. For Martin Luther King Jr. the goal of true education is "intelligence plus character" [5]. Although the main goals of higher education is a very interesting (and controversial) topic, it is clearly out of the scope of this book. Instead, in this Case Study, we will look at the change in the total enrollment at large US universities, the main providers of higher education in the United States, between 2000 and 2011.

Figure 6.3 shows a scatterplot of $N = 354$ data points, each corresponding to a large US university. The x and y coordinates of each point are the total enrollments at the corresponding university in year 2000 and 2011, respectively. Only universities with 2011 enrollment larger then 15,000 students are considered.

Let the parameter of interest θ be the ratio of the average enrollments $\bar{y}/\bar{x}$, which can be viewed as a measure of the total change in the university enrollment in the US between 2000 and 2011. Given the data in Fig. 6.3, the value of θ can be computed exactly[20],

$$\theta = \frac{\bar{y}}{\bar{x}} = \frac{\frac{1}{N}\sum_{i=1}^{N} y_i}{\frac{1}{N}\sum_{i=1}^{N} x_i} = 1.41. \tag{6.34}$$

[20] We will use this *true* value as a benchmark.

Suppose now that we don't have access to the full data set, and we can only pick $n < N$ universities uniformly at random from the list of N universities, and record

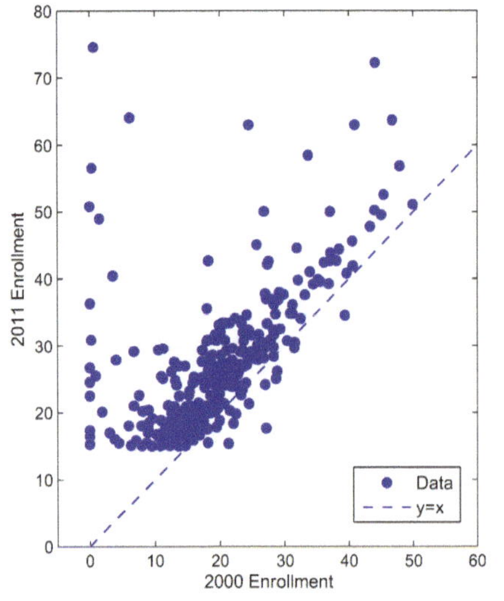

Fig. 6.3 Enrollments (in thousands) of 354 large (with 2011 enrollment larger than 15,000) degree-granting US universities. An "outlier" with large y coordinate, University of Phoenix $(14.8, 307.9)$, is not shown in the Figure. Data source: US Department of Education.

their enrollments $(X_1, Y_1), \ldots, (X_n, Y_n)$. Figure 6.4 shows two such random samples of sizes $n = 10$ (2.82% of the data) and $n = 100$ (28.25% of the data).

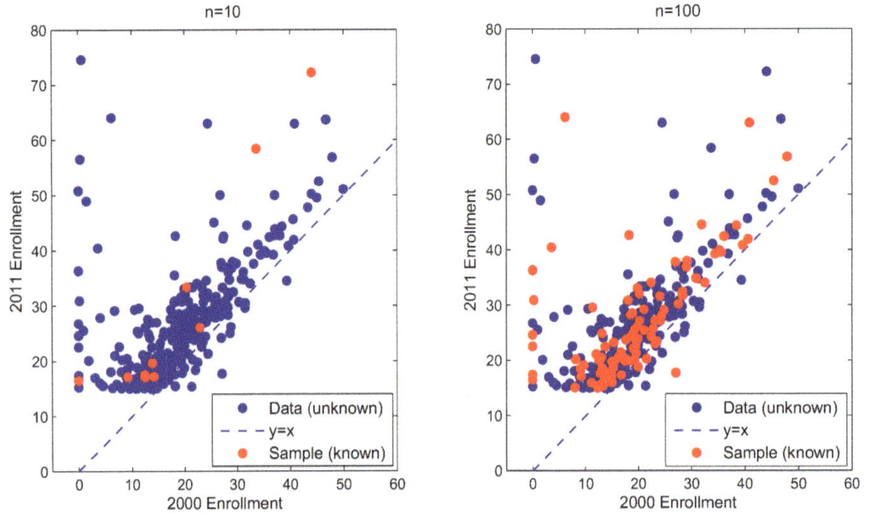

Fig. 6.4 Two random samples (shown in red) of size $n = 10$ (left panel) and $n = 100$ (right panel).

Our goal is to estimate θ, compute the standard error of the estimate using the bootstrap method, and construct normal and pivotal confidence intervals for θ using these samples. The plug-in estimate of θ is given by

$$\hat{\theta}_n = \frac{\overline{Y}}{\overline{X}} = \frac{\frac{1}{n}\sum_{i=1}^n Y_i}{\frac{1}{n}\sum_{i=1}^n X_i}, \tag{6.35}$$

which simply mimics the expression for θ in (6.34). The values of the estimate computed from the samples depicted in Fig. 6.4 are:

$$\hat{\theta}_{10} = 1.61 \quad \text{and} \quad \hat{\theta}_{100} = 1.39. \tag{6.36}$$

As expected, $\hat{\theta}_{100}$ is more accurate and its values is closer to the true value $\theta = 1.41$.

Now let's compute the bootstrap estimate of the standard error of $\hat{\theta}_n$. To make the sampling error completely negligible, we use $B = 10^4$ bootstrap samples

$$(X_1^{(b)}, Y_1^{(b)}), \ldots, (X_n^{(b)}, Y_n^{(b)}) \overset{\text{iid}}{\sim} \hat{F}_n, \qquad b = 1, \ldots, B, \tag{6.37}$$

where $\hat{F}_n$ is the bivariate eCDF that puts probability mass $1/n$ at each pair (X_i, Y_i), $i = 1, \ldots, n$. In other words, each bootstrap sample is obtained by sampling from $(X_1, Y_1), \ldots, (X_n, Y_n)$ uniformly at random with replacement. The corresponding bootstrap replications of $\hat{\theta}_n$ are

$$\hat{\theta}_n^{(b)} = \frac{\overline{Y}^{(b)}}{\overline{X}^{(b)}} = \frac{\frac{1}{n}\sum_{i=1}^n Y_i^{(b)}}{\frac{1}{n}\sum_{i=1}^n X_i^{(b)}}, \qquad b = 1, \ldots, B. \tag{6.38}$$

The values of the bootstrap estimate (6.11) computed from samples shown in Fig. 6.4 are:

$$\widehat{\text{se}}_B[\hat{\theta}_{10}] = 0.14 \quad \text{and} \quad \widehat{\text{se}}_B[\hat{\theta}_{100}] = 0.06. \tag{6.39}$$

As expected, $\widehat{\text{se}}_B[\hat{\theta}_{100}] < \widehat{\text{se}}_B[\hat{\theta}_{10}]$, which agrees with the fact that $\hat{\theta}_{100}$ is a more accurate estimate, since it is based on the sample of a larger size.

Figure 6.5 shows the normal and pivotal confidence intervals for θ constructed from the samples in Fig. 6.4. As expected, both intervals constructed from the small sample ($n = 10$) are longer than the corresponding intervals constructed from the large sample ($n = 100$)[21]. Notice also that for a given sample size n, the normal interval is shorter than the pivotal intervals. This is because in addition to the sample, the normal interval uses a parametric assumption (6.19)[22]. Finally, while the normal intervals are (by definition) symmetric about $\hat{\theta}_n$, the pivotal intervals are not.

It is important to highlight that plug-in estimates (6.36), bootstrap estimates of their standard errors (6.39), and bootstrap confidence intervals in Fig. 6.5 are all computed based on the two *specific* samples shown in Fig. 6.4. The results will of course change if we get different samples. To illustrate the variability of estimates and confidence intervals with respect to different samples, let's repeat all computations for 200 new independent samples from the population of $N = 354$ universities:

[21] The longer a confidence interval is, the more uncertainty is in the value of the parameter.

[22] The more we assume, the more we can infer from the data, provided that our assumptions are correct.

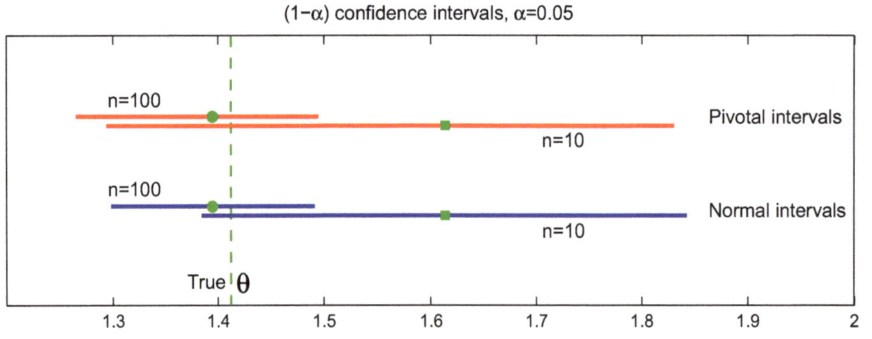

Fig. 6.5 The normal (blue) and pivotal (red) confidence intervals at approximate level 95%. The true value of θ computed in (6.34) is indicated by the dashed line. The estimates $\hat{\theta}_n$ computed in (6.36) are marked by green circles ($n = 100$) and squares ($n = 10$).

100 samples of size $n = 10$ and 100 samples of size $n = 100$. The corresponding simulations results are shown in Fig. 6.6 (estimates) and Fig. 6.7 (confidence intervals).

6.4 Further Reading

The bootstrap method discussed in this chapter is one of the most popular nonparametric methods for estimating the accuracy of random estimates. Soon after it was introduced by Bradley Efron [1], this method and its various modifications became a workhorse in many applications. A clear textbook-length treatment of the bootstrap aimed at scientists and engineers is given in

- Efron, B., Tibshirani, R.: An *Introduction to the Bootstrap*. Chapman & Hall/CRC Monographs on Statistics and Applied Probability, (1994).
  `https://doi.org/10.1201/9780429246593`

A broad coverage of parametric, semiparametric, and nonparametric bootstrap methods with numerous applications, including finite populations, stratified data, linear and nonlinear regression models, and time series, is provided in

- Davison, A.C., Hinkley, D.V.: *Bootstrap Methods and Their Applications*. Cambridge University Press, (1997).
  `https://doi.org/10.1017/CBO9780511802843`

Typical problems with the bootstrap, workable diagnostics, and efficient implementations for computationally involved applications are discussed in

- Canty, A.J., Davison, A.C., Hinkley, D.V., Ventura, V.: Bootstrap diagnostics and remedies. *The Canadian Journal of Statistics*, 34(1), 5-27 (2006).
  `https://www.jstor.org/stable/20445183`

For a more conceptual and somewhat philosophical discussion of the bootstrap the following beautiful essay by Cosma Shalizi is strongly recommended:

- Shalizi, C.: The bootstrap. *American Scientist*, 98(3): 186-190, (2010). `https://doi.org/10.1511/2010.84.186`

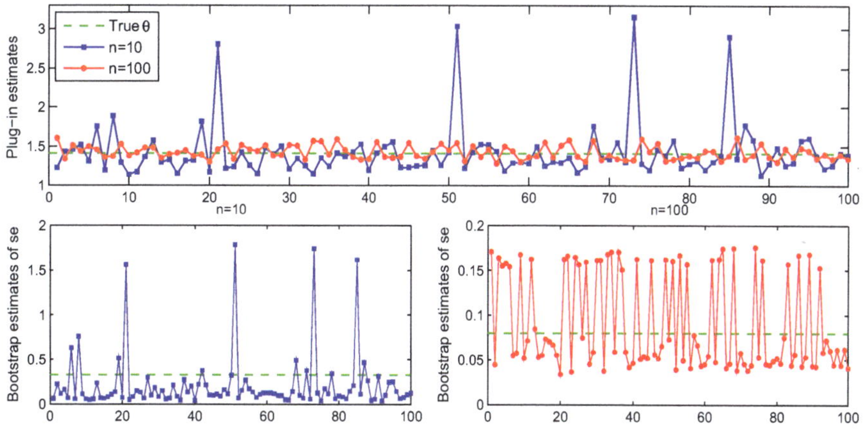

Fig. 6.6 Top panel illustrates the variability in the values of the plug-in estimates $\hat{\theta}_n$. As expected, on average, the estimate for $n = 100$ (red curve) is more accurate. The bottom panel shows the variability of the bootstrap estimates $\widehat{\text{se}}_B[\hat{\theta}_n]$ for $n = 10$ (left) and $n = 100$ (right). Green dashed lines represent the true values of the standard error computed using the full data.

Problems

6.1. Bootstrapping Fiji Earthquakes

In this problem, you will implement the bootstrap method for a) estimating the standard error of an estimate of the correlation between the earthquake magnitude and the number of stations registered that earthquake and b) constructing confidence intervals for that correlation. First, download the data set[23] "Problem6.1_fiji_earthquakes.csv" from `https://link.springer.com/chapter/10.1007/978-3-032-03848-7_1`. The data provides information about the location and size of $n = 1000$ seismic events occurred near Fiji since 1964. The The data has six columns described in Table 6.1.

Let $X_1, \dots, X_n$ be the earthquake magnitudes (5th column) and $Y_1, \dots, Y_n$ be the corresponding numbers of stations that registered that earthquakes (6th column).

(a) Draw a scatterplot of Y (numbers of stations) versus X (magnitudes).

(b) The scatterplot reveals some positive correlation between X and Y. This is expected since the common sense suggests that the larger the magnitude of an

[23] The data was collected by Prof. J.H. Woodhouse of Harvard University (now Emeritus Professor at the University of Oxford).

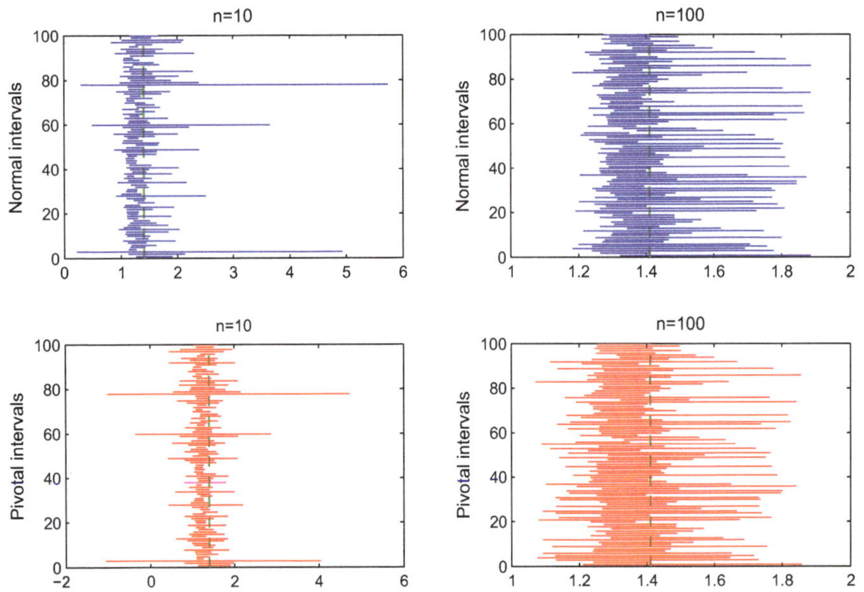

Fig. 6.7 Normal and pivotal confidence intervals at approximate level 95%. For $n = 10$, only 71 out of 100 normal intervals (top left) and 69 pivotal intervals (bottom left) contain the true value of θ indicated by the green dashed line. For $n = 100$, the intervals are more accurate: 83 normal (top right) and 86 pivotal (bottom right) contain the true value of θ.

Column #	Description
1	Event ID
2	Latitude of event
3	Longitude of event
4	Depth (km)
5	Richter Magnitude
6	Number of stations reporting

Table 6.1 Column descriptions of the Fiji earthquakes dataset that contains 6 columns and 1000 rows (earthquakes occurred near Fiji since 1964).

earthquake, the more stations should be able to register it. So, let the parameter of interest be the Pearson correlation coefficient[24] between X and Y,

$$\rho_{XY} = \frac{\text{Cov}(X,Y)}{\sqrt{\mathbb{V}[X]\mathbb{V}[Y]}} \tag{6.40}$$

Find the plug-in estimate $\hat{\rho}_n$ of ρ_{XY} and compute its value using the data.

(c) Find the bootstrap estimate of the standard error of $\hat{\rho}_n$ based on $B = 10^3$ bootstrap samples, and show a histogram of the bootstrap replications $\hat{\rho}_n^{(1)}, \ldots, \hat{\rho}_n^{(B)}$ together with the estimated value $\hat{\rho}_n$.

[24] It was first introduced in Remark 5.3.

(d) Find the normal and pivotal 95% confidence intervals for ρ_{XY}.

6.2. Parametric vs Nonparametric Bootstrap

In this chapter, we have discussed a nonparametric bootstrap method. The bootstrap method can also be used in the context of parametric models[25]. In this case, the method is often referred to as the *parametric bootstrap*. The only difference between parametric and nonparametric bootstraps is in the way the bootstrap samples are generated. In the nonparametric bootstrap, we generate bootstrap samples from the eCDF $\hat{F}_n$ constructed directly from the data $X_1,\dots,X_n \overset{\text{iid}}{\sim} F$, see Fig. 6.1. In the parametric bootstrap, where we assume that $X_1,\dots,X_n \overset{\text{iid}}{\sim} f(x|\theta)$, $\theta \in \Theta$, the bootstrap samples are generated from the estimated PDF $f(x|\hat{\theta}_n)$, where $\hat{\theta}_n$ is an estimate of the model parameter θ[26]. In this problem, you will see that the results obtained by parametric and nonparametric bootstraps can be substantially different.

[25] We will start discussing parametric inference in Chapter 7.

[26] For example, the plug-in estimate, the method of moments estimate (Chapter 7), or the maximum likelihood estimate (Chapter 8)

Let $X_1,\dots,X_n \overset{\text{iid}}{\sim} \mathscr{U}[0,\theta]$ be a sample from the uniform distribution, where θ is a parameter of interest. It seems natural to estimate θ by the largest value in the sample (see Problem 5.1(a)),

$$\hat{\theta}_n = X_{(n)}. \tag{6.41}$$

(a) Find the probability density function (PDF) of $\hat{\theta}_n$ analytically.

(b) Generate a sample $X_1,\dots,X_n \overset{\text{iid}}{\sim} \mathscr{U}[0,\theta]$ of size $n = 50$ with $\theta = 1$. Compare the true distribution of $\hat{\theta}_n$ derived in part (a) with the histograms of $B = 10^4$ parametric and nonparametric bootstrap replications of $\hat{\theta}_n$.

References

1. Efron, B.: Bootstrap methods: another look at the jackknife. *The Annals of Statistics* 7(1), 1–26 (1979). `https://doi.org/10.1214/aos/1176344552`
2. Efron, B., Tibshirani, R.J.: *An Introduction to the Bootstrap*. Chapman and Hall/CRC (1994). `https://doi.org/10.1201/9780429246593`
3. Hyndman, R.J., Fan, Y.: Sample quantiles in statistical packages. *The American Statistician* 50(4), 361–365 (1996). `https://doi.org/10.2307/2684934`
4. Kant, I.: *On Education*. CreateSpace Independent Publishing Platform (2015).
5. King, M.L.Jr.: *The purpose of education*. Published in The Papers of Martin Luther King, Jr. Volume I: Called to Serve, January 1929 - June 1951, University of California Press (1992).

7
The Method of Moments

Some moments are nice, some are nicer, some are even worth writing about.

– Charles Bukowski, *War All the Time*

Nonparametric inference discussed in previous chapters is performed under minimal assumptions on the underlying stochastic model for the given data. On the other hand, usually, the more assumptions we make, the more powerful methods are available for data analysis, and, as a result, the more information and insights we can infer from the data[1]. In this chapter, we will turn our attention to *parametric inference* that makes stronger assumptions and models data $X_1, \ldots, X_n$ as a random sample from a finitely parameterized probability distribution,

[1] Provided that the model assumptions are correct!

$$X_1 \ldots, X_n \overset{\text{iid}}{\sim} F(x|\theta) \in \mathscr{F} = \{F(x|\theta), \quad \theta \in \Theta\}, \tag{7.1}$$

where $F(x|\theta)$ is a CDF, θ is a model parameter, Θ is a parameter space, and $\mathscr{F}$ is a parametric stochastic model for the data. Stochastic model $\mathscr{F}$ is often assumed to be either continuous or discrete[2]. In this case, the probability distribution defined by $F(x|\theta)$ can also be described by the corresponding PDF $f(x|\theta)$ or PMF $p(x|\theta)$, respectively. In general, $\Theta \subset \mathbb{R}^d$ and $\theta = (\theta_1, \ldots, \theta_d)$ is a vector of model parameters. In this parametric framework, the problem of statistical inference reduces then to estimating θ from the data. The main goal of this chapter is to introduce the *method of moments*, one of the first general parametric methods for estimating θ. But before we start talking about the method of moments and its properties, it is important to discuss several conceptual questions concerning parametric inference.

[2] See Remark 3.1.

7.1 Concerning Parametric Inference

The first question which naturally arises after looking at (7.1) is how we can possibly know that $X_1 \ldots, X_n \overset{\text{iid}}{\sim} F(x|\theta)$. In other words, how can we be sure that there exists

Supplementary Information The online version contains supplementary material available at https://doi.org/10.1007/978-3-032-03848-7_7.

K. M. Zuev, *Fundamentals of Statistical Inference*,
International Series in Operations Research & Management Science 374,
https://doi.org/10.1007/978-3-032-03848-7_7

a family of probability distributions $\mathscr{F} = \{F(x|\theta), \theta \in \Theta\}$ and a specific value[3] of θ such that our data $X_1 \ldots, X_n$ is generated exactly from $F(x|\theta)$? Indeed, we cannot. By accepting stochastic model (7.1), we don't really believe that the data is generated *exactly* from $F(x|\theta)$ for some value of θ. Rather we assume (and hope) that there exists a value $\theta_\star$ of θ in Θ such that the distribution $F(x|\theta_\star)$ describes the randomness in the data well enough for all practical purposes[4]. That is,

$$X_1, \ldots, X_n \overset{\text{iid}}{\sim} F(x|\theta_\star), \tag{7.2}$$

and this *approximation* is of sufficient quality to be regarded as exact.

[3] This value $\theta_\star \in \Theta$ is often called the "true" value of the parameter.

[4] It is important to keep in mind that (7.1) is a model, not the underlying reality.

The next important question is how to formulate the stochastic model $\mathscr{F}$, i.e. how to choose a set of probability distributions in (7.1). A parametric stochastic model is usually formulated based on the background knowledge, "physics" of the phenomena that generated the data, and prior experience. For example, it is known that various measures of size (height, weight, length, etc.) calculated on living organisms, certain physical quantities, and measurement errors in experiments are approximately normally distributed; counts of traffic accidents, the numbers of mutations on a strand of DNA per unit length, and the numbers of jumps in a stock price in a given time interval follow a Poisson distribution; inter-spike intervals in neuroscience and sizes of rainfalls and insurance claims are often modeled by the gamma distribution. If such background knowledge and prior experience are not available, then a nonparametric model could be more preferable due to a lower risk of model misspecification[5].

[5] See Marginal Note 10 in Section 3.1.

Finally, is it possible to check the accuracy of assumption (7.1)? It turns out that to some extent, yes, it is possible. First, this assumption can be checked informally by inspecting graphical summaries of the data, such as histograms, boxplots, and Q-Q plots[6]. For example, if the histogram is skewed or bimodal, or if the normal-quantile plot deviates substantially from a straight line, then this is a strong indication that the normal model (3.1) is a bad model for the data. Furthermore, parametric assumption (7.1) can be checked more formally by testing the hypothesis that $X_1 \ldots, X_n \overset{\text{iid}}{\sim} F(x|\theta)$ versus the alternative that this is not true. For example, for testing the normal model (3.1), the Kolmogorov-Smirnov test[7] or the Shapiro-Wilk test[8] can be used. Both formal and informal methods, however, are usually more effective in disproving the model assumption (7.1) rather than in confirming it.

[6] See Chapter 1.

[7] See Section 10.3.

[8] See Remark 10.9.

7.2 Parameters in Parametric Models

Recall[9] that in nonparametric inference, a parameter θ of distribution $F \in \mathscr{F}$, where $\mathscr{F}$ is a nonparametric model, is the value of a certain (known) statistical functional $\mathfrak{f}$ on F, that is $\theta = \mathfrak{f}(F)$. In parametric case (7.1), the notion of a parameter is more intuitively transparent: a parameter of interest is usually θ itself or a component of θ. As the following example illustrates, in some applications, functions of θ can be of interest.

[9] See Section 4.1.

Example 7.1. Consider data $X_1, \dots, X_n$, where X_i is the result of a measles antibody test that measures the number of antibodies in the blood of the ith patient[10]. These antibodies are used by the immune system to identify and neutralize the measles virus. Suppose that we are interested in estimating the proportion π of the target population whose test result X (the number of antibodies) is large than some critical threshold α. We can always estimate π nonparametrically by simply counting the number of X_i with value larger than α and dividing by the sample size,

[10] Antibody tests are often used to check if someone is immune to diseases such as measles, chickenpox, mumps, and rubella.

$$\hat{\pi}_n = \frac{1}{n}\sum_{i=1}^{n} \mathbb{1}_{(\alpha,\infty)}(X_i), \tag{7.3}$$

where $\mathbb{1}_{(\alpha,\infty)}$ is the indicator function: $\mathbb{1}_{(\alpha,\infty)}(x) = 1$ if $x > \alpha$ and $\mathbb{1}_{(\alpha,\infty)}(x) = 0$ if $x \leq \alpha$. We can also estimate π parametrically as follows.

Motivated by an empirical observation that many measurements taken on humans are approximately normally distributed, let us model our data as a sample from a normal distribution,

$$X_1, \dots, X_n \overset{\text{iid}}{\sim} \mathcal{N}(\mu, \sigma^2), \tag{7.4}$$

where $\theta = (\mu, \sigma^2)$ is a vector of model parameters. The proportion π is then the probability that the test result $X \sim \mathcal{N}(\mu, \sigma^2)$ of a randomly chosen member of the population is larger than α,

$$\pi = \mathbb{P}(X > \alpha) = \mathbb{P}\left(\frac{X-\mu}{\sigma} > \frac{\alpha-\mu}{\sigma}\right) = 1 - \Phi\left(\frac{\alpha-\mu}{\sigma}\right), \tag{7.5}$$

where Φ is the standard normal CDF. Note that in this example, we are interested not in the model parameters μ and σ^2 per se, but in the function of model parameter given by (7.5). This function can be estimated by

$$\hat{\pi}_n = 1 - \Phi\left(\frac{\alpha - \hat{\mu}_n}{\hat{\sigma}_n}\right), \tag{7.6}$$

where $\hat{\mu}_n$ and $\hat{\sigma}_n$ are the estimates of μ and σ obtained from the data by some parametric method, for instance, the method of moments (see Example 7.3).

7.3 The Method of Moments

The method of moments (MOM) is the oldest general method for estimating model parameters from the data. It was first introduced in a theoretical context in 1887 by Pafnuty Lvovich Chebyshov in [1], where he used this method for proving his version of the central limit theorem for random variables with zero expectation[11]. For statistical applications, in particular to the theory of evolution, the method of moments was developed and extensively used by Karl Pearson [2].

[11] Chebyshov's CLT was later generalized to the "classical" CLT (see Probability Refresher 2.12) by his student, Aleksandr Lyapunov.

So, suppose that we model data parametrically and, for simplicity, let's assume that the stochastic model $\mathscr{F}$ is continuous. Namely,

$$X_1 \dots, X_n \overset{\text{iid}}{\sim} f(x|\theta) \in \mathscr{F} = \{f(x|\theta), \quad \theta \in \Theta\}, \tag{7.7}$$

where $f(x|\theta)$ is a PDF and $\theta = (\theta_1, \dots, \theta_d)$ is a vector of model parameters that we want to estimate. Recall that the qth *moment* of the probability distribution $f(x|\theta)$ is the expected value of the qth power of $X \sim f(x|\theta)$,

$$m_q(\theta) = \mathbb{E}[X^q] = \int x^q f(x|\theta) dx. \tag{7.8}$$

Since, $f(x|\theta)$ depends on θ, so does the moment $m_q(\theta)$, which is also referred to as the qth *theoretical moment* of $f(x|\theta)$[12].

[12] "Well, here we are, Mr. Pilgrim, trapped in the amber of this moment." – Kurt Vonnegut, *Slaughterhouse-Five*.

Remark 7.1. If the data $X_1, \dots, X_n$ is modeled with a discrete stochastic model

$$X_1 \dots, X_n \overset{\text{iid}}{\sim} p(x|\theta) \in \mathscr{F} = \{p(x|\theta), \quad \theta \in \Theta\}, \tag{7.9}$$

where $p(x|\theta)$ is a PMF on a discrete set $\{x_1, x_2, \dots\}$, then the integral in (7.8) is replaced with the sum:

$$m_q(\theta) = \mathbb{E}[X^q] = \sum_{x_i} x_i^q p(x_i|\theta). \tag{7.10}$$

The definition of the qth theoretical moment for discrete and continuous cases can be unified by using the CDF $F(x|\theta)$ instead of PDF or PMF:

$$m_q(\theta) = \mathbb{E}[X^q] = \int x^q dF(x|\theta). \tag{7.11}$$

For interpretation of the integral in (7.11), see Remark 4.1.

Let's define the qth *sample moment* as follows:

$$\hat{m}_q = \frac{1}{n} \sum_{i=1}^{n} X_i^q. \tag{7.12}$$

It can be shown that the qth sample moment $\hat{m}_q$ is simply the plug-in estimate of the theoretical moment $m_q(\theta)$[13]. By the law of large numbers, applied to random variables $X_1^q, \dots, X_n^q$, the qth sample moment converges to the qth theoretical moment,

[13] To do this, one needs to replace $F(x|\theta)$ in (7.11) with the eCDF $\hat{F}_n$ of sample $X_1 \dots, X_n$ and simply the resulting integral in the spirit of (4.43).

$$\hat{m}_q \xrightarrow{\mathbb{P}} \mathbb{E}[X^q] = m_q(\theta), \quad \text{as } n \to \infty. \tag{7.13}$$

This means that for large sample size n, $\hat{m}_q \approx m_q(\theta)$ with high probability. This approximation lies at the heart of the method of moments.

Definition 7.1. The method of moments (MOM) estimate of $\theta = (\theta_1, \dots, \theta_d)$ is a solution of the following system of d equations with d unknowns:

$$m_q(\theta) = \hat{m}_q, \quad q = 1,\ldots,d. \tag{7.14}$$

In other words, the method of moments tries to match the first d theoretical moments with the corresponding sample moments by finding the appropriate values of $\theta_1,\ldots,\theta_d$. Let's consider a couple of classical examples, one discrete and one continuous.

Example 7.2. Suppose that the data is modeled with the Bernoulli model,

$$X_1,\ldots,X_n \overset{\text{iid}}{\sim} \mathscr{B}(p), \quad p \in [0,1]. \tag{7.15}$$

In this case, we have just one model parameter p, and so $d = 1$. The first theoretical and sample moments are

$$m_1(p) = \mathbb{E}[X] = p \quad \text{and} \quad \hat{m}_1 = \frac{1}{n}\sum_{i=1}^{n} X_i = \overline{X}. \tag{7.16}$$

Matching these two moments $m_1(p) = \hat{m}_1$ gives the MOM estimate of p:

$$\hat{p}_{MOM} = \overline{X}. \tag{7.17}$$

Example 7.3. Suppose that the data is modeled with the normal model,

$$X_1,\ldots,X_n \overset{\text{iid}}{\sim} \mathscr{N}(\mu,\sigma^2), \quad \mu \in \mathbb{R}, \sigma^2 \geq 0. \tag{7.18}$$

In this case, we have two model parameters, μ and σ^2, and so $d = 2$. This means that to find the MOM estimates of μ and σ^2, we need to solve a system of two equations obtained by matching the first two theoretical and samples moments:

$$\begin{aligned} m_1(\mu,\sigma^2) = \hat{m}_1 \quad &\Rightarrow \quad \mathbb{E}[X] = \mu = \frac{1}{n}\sum_{i=1}^{n} X_i = \overline{X}, \\ m_2(\mu,\sigma^2) = \hat{m}_2 \quad &\Rightarrow \quad \mathbb{E}[X^2] = \mathbb{V}[X] + \mathbb{E}[X]^2 = \sigma^2 + \mu^2 = \frac{1}{n}\sum_{i=1}^{n} X_i^2. \end{aligned} \tag{7.19}$$

Solving this systems gives

$$\hat{\mu}_{MOM} = \overline{X} \quad \text{and} \quad \hat{\sigma}^2_{MOM} = \frac{1}{n}\sum_{i=1}^{n} X_i^2 - \left(\frac{1}{n}\sum_{i=1}^{n} X_i\right)^2 = \frac{1}{n}\sum_{i=1}^{n}(X_i - \overline{X})^2. \tag{7.20}$$

In these two examples, the method of moments produces reasonable estimates, which, in fact, coincide with the corresponding plug-in estimates for the mean[14] (4.45) and variance (4.49). This leads to a natural question: are there any examples where the plug-in and MOM estimates are different? Intuitively, the answer is yes (and this is a correct answer), since the plug-in and MOM methods are based on conceptually different ideas.

[14] Both p and μ are the means of the Bernoulli and normal distributions.

Example 7.4. Suppose that the data is modeled with the uniform model,

$$X_1, \ldots, X_n \overset{\text{iid}}{\sim} \mathscr{U}[0, \theta], \tag{7.21}$$

where the right end θ is a model parameter. It can be shown[15] that the plug-in estimate of θ is the maximum value in the sample,

[15] See Problem 5.1(a).

$$\hat{\theta}_n = X_{(n)} = \max\{X_1, \ldots, X_n\}. \tag{7.22}$$

To find the MOM estimate of θ, we need to equate the first theoretical and sample moments,

$$m_1(\theta) = \mathbb{E}[X] = \frac{\theta}{2} \quad \text{and} \quad \hat{m}_1 = \overline{X}, \tag{7.23}$$

which gives an estimate different form (7.22),

$$\hat{\theta}_{MOM} = 2\overline{X}. \tag{7.24}$$

What estimate, $\hat{\theta}_n = X_{(n)}$ or $\hat{\theta}_{MOM} = 2\overline{X}$ is better? This is a good question and it was addressed in Problem 3.1.

In addition to showing that the plug-in and MOM estimates can disagree, Example 7.4 also warns us that we should be careful when using the method of moments, since it could lead to meaningless estimates lying outside of the parameter space. Indeed, consider a special case where the $n = 3$, $X_1 = X_2 = 1$, and $X_3 = 7$. Then the MOM estimate is $\hat{\theta}_{MOM} = 2\overline{X} = 6$, which is an impossible value for θ, since if $X_3 \sim \mathscr{U}[0, \theta]$, then $X_3 \leq \theta$.

It is important to highlight though, that this behavior of the MOM estimate, disturbing at the first glance, is a result of a very small sample size n. If n is large, then the MOM estimate $\hat{\theta}_{MOM}$ is likely to be very close to θ. More formally, under certain (relatively weak) regularity assumptions, $\hat{\theta}_{MOM}$ is a consistent estimate, that is

$$\hat{\theta}_{MOM} \xrightarrow{\mathbb{P}} \theta, \qquad \text{as } n \to \infty. \tag{7.25}$$

Although the general proof of consistency is beyond the scope of this book, it is easy to check that for all considered examples $\hat{\theta}_{MOM}$ is indeed consistent. As an example, let's revisit Example 7.4 and show that the derived estimate $\hat{\theta}_{MOM} = 2\overline{X}$ is consistent. Recall[16] that to prove that an estimate is consistent, it is sufficient to show that the mean squared error (MSE) of the estimate converges to zero as $n \to \infty$. Using the bias-variance decomposition of the MSE (3.11), we have:

[16] See Definition 4.1 and subsequent discussion.

$$\begin{aligned} \text{MSE}[2\overline{X}] &= \mathbb{B}[2\overline{X}]^2 + \text{se}[2\overline{X}]^2 = \left(\mathbb{E}[2\overline{X}] - \theta\right)^2 + \mathbb{V}[2\overline{X}] \\ &= \Big(2\frac{1}{n}\sum_{i=1}^{n}\underbrace{\mathbb{E}[X_i]}_{\theta/2} - \theta\Big)^2 + 4\frac{1}{n^2}\sum_{i=1}^{n}\underbrace{\mathbb{V}[X_i]}_{\theta^2/12} = \frac{\theta^2}{3n} \to 0 \quad \text{as } n \to \infty. \end{aligned} \tag{7.26}$$

One of the main advantages of the method of moments is that MOM estimates are relatively easy to compute, especially for standard stochastic models such as Bernoulli, normal, uniform, etc. On the other hand, MOM estimates are often biased[17] and often more accurate (with smaller mean squared error) estimates exist[18].

[17] For example, $\hat{\sigma}^2_{MOM}$ in Example 7.3 is biased.

[18] For example, the maximum likelihood estimate that we will discuss in the next chapter.

Nevertheless, MOM estimates are often used as initial values for more sophisticated estimation methods that require iterative numerical routines.

7.4 Case Study: The Balding-Nichols Model

> *Alpha children wear grey. They work much harder than we do, because they're so frightfully clever. I'm awfully glad I'm a Beta, because I don't work so hard. And then we are much better than the Gammas and Deltas. Gammas are stupid. They all wear green, and Delta children wear khaki. Oh no, I don't want to play with Delta children. And Epsilons are still worse. They're too stupid to be able to read or write. Besides they wear black, which is such a beastly color. I'm so glad I'm a Beta.*
>
> – Aldous Huxley, *Brave New World*

In this Case Study, we will apply the method of moments for estimating the parameters of the Balding-Nichols model, which is often used in population genetics [3]. The key building block of the Balding-Nichols model is the *beta distribution*[19], which is used for modeling frequencies, proportions, percentages, and other quantities whose range is between 0 and 1. For example, the beta distribution can be used for modeling various survival rates (e.g. for blood cancer), annual interest rates, and click-through rates[20]. It also plays an important role in the theory of order statistics, wavelet analysis, and Bayesian inference.

[19] Also called the beta distribution of the first kind.

[20] A click-through rate is the ratio of users who click on a specific link on a certain web page to the total number of users who view the web page.

A continuous random variable X is said to be distributed according to the beta distribution, denoted $X \sim \mathscr{B}(\alpha, \beta)$, if its PDF is given by

$$f(x \mid \alpha, \beta) = \frac{1}{B(\alpha, \beta)} x^{\alpha-1}(1-x)^{\beta-1}, \qquad 0 < x < 1, \tag{7.27}$$

where $\alpha > 0$ and $\beta > 0$ are the parameters of the distribution, and $B(\alpha, \beta)$ is a normalizing constant that ensures that $f(x \mid \alpha, \beta)$ integrates to 1,

$$B(\alpha, \beta) = \int_0^1 x^{\alpha-1}(1-x)^{\beta-1} dx \qquad \left(\Rightarrow \int_0^1 f(x \mid \alpha, \beta) dx = 1 \right). \tag{7.28}$$

Model parameters α and β control the shape of the beta PDF (7.27). For that reason they are often referred to as the *shape parameters*. Figure 7.1 shows a wide variety of different U-shaped, J-shaped, and bell-shaped beta PDFs corresponding to different values of α and β.

In the Balding-Nichols model [3], the beta distribution is used for modeling the relative frequency of an allele (variant of a gene) in sub-populations of an overall population. The allele frequencies in sub-populations that are more geographically and ancestrally remote tend to differ more. If the average allele frequency in the total population is p, then according to the Balding-Nichols model, the allele frequencies in sub-populations are IID draws from the beta distribution

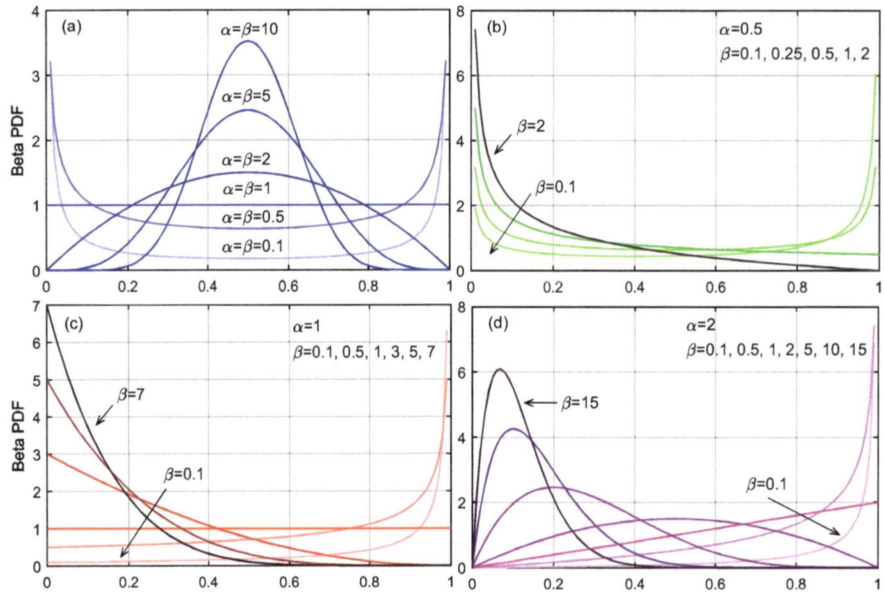

Fig. 7.1 Beta PDFs. In panel (a), $\alpha = \beta$, in this case the beta PDF is symmetric about $x = 1/2$. As the value of $\alpha = \beta$ increases, the shape of the PDF changes from U-shaped, to straight line ($\alpha = \beta = 1$ corresponds to the uniform distribution), and finally to bell-shaped. In panel (b), $\alpha = 0.5$ and as β increases from 0.1 to 2, the shape of the PDF evolves from U-shaped to reverse J-shaped. In panel (c), $\alpha = 1$ and as β increases from 0.1 to 7, the shape of the PDF transforms from J-shaped to reverse J-shaped. Finally, in panel (d), $\alpha = 2$ and as β increases from 0.1 to 15, the shape of the PDF changes from J-shaped, to straight line ($\alpha = 2, \beta = 1$ corresponds to the triangular distribution), and to approximately bell-shaped.

$$\mathscr{B}\left(\frac{1-F}{F}p, \frac{1-F}{F}(1-p)\right), \tag{7.29}$$

where F is the *fixation index*, a commonly used measure for quantifying the genetic differences between populations developed by American geneticist Sewall Wright. The value of the fixation index ranges from 0 (not different) to 1 (completely different), and it strongly depends on the choice of populations. For example, for closely related human populations, such as the Portuguese and the Spaniards, the value of F is significantly below 0.01. On the other hand, for highly unrelated groups, such as the Mbuti population in the Congo region of Africa and the indigenous peoples of New Guinea, the fixation index can be as large as 0.46 [4].

The Balding-Nichols model is widely used in the forensic analysis of DNA profiles[21], genetic epidemiology, and various medical applications. In this Case Study, we will see how the parameters of this model can be estimated using the method of moments. So, suppose that the allele frequencies in sub-populations are modeled as an IID sample from the beta distribution,

[21] No two humans (even "identical" twins) are genetically identical.

$$X_1,\dots,X_n \overset{\text{iid}}{\sim} \mathscr{B}\left(\frac{1-F}{F}p, \frac{1-F}{F}(1-p)\right). \tag{7.30}$$

This model has two parameters: the background allele frequency p and Wright's fixation index F. To find the MOM estimates of p and F, we need to match the first two moments of the beta distribution (7.29) with the corresponding sample moments. As shown in Appendix 7.6, if $X \sim \mathscr{B}(\alpha,\beta)$, then

$$\mathbb{E}[X] = \frac{\alpha}{\alpha+\beta} \quad \text{and} \quad \mathbb{E}[X^2] = \frac{\alpha(\alpha+1)}{(\alpha+\beta)(\alpha+\beta+1)}. \tag{7.31}$$

Substituting the shape parameters

$$\alpha = \frac{1-F}{F}p \quad \text{and} \quad \beta = \frac{1-F}{F}(1-p) \tag{7.32}$$

into (7.31), we can express the first two moments in terms of the model parameters:

$$\begin{aligned}
\mathbb{E}[X] &= \frac{\frac{1-F}{F}p}{\frac{1-F}{F}p+\frac{1-F}{F}(1-p)} = p,\\
\mathbb{E}[X^2] &= \frac{\frac{1-F}{F}p(\frac{1-F}{F}p+1)}{(\frac{1-F}{F}p+\frac{1-F}{F}(1-p))(\frac{1-F}{F}p+\frac{1-F}{F}(1-p)+1)} = p(F(1-p)+p).
\end{aligned} \tag{7.33}$$

The MOM estimate of p and F are then the solutions of the following system of two equations:

$$\begin{aligned}
m_1(p,F) = \hat{m}_1 \quad &\Rightarrow \quad \mathbb{E}[X] = p = \frac{1}{n}\sum_{i=1}^{n} X_i = \overline{X},\\
m_2(p,F) = \hat{m}_2 \quad &\Rightarrow \quad \mathbb{E}[X^2] = p(F(1-p)+p) = \frac{1}{n}\sum_{i=1}^{n} X_i^2.
\end{aligned} \tag{7.34}$$

Solving this systems gives[22]

$$\hat{p}_{MOM} = \overline{X} \quad \text{and} \quad \hat{F}_{MOM} = \frac{\frac{1}{n}\sum_{i=1}^{n}(X_i-\overline{X})^2}{\overline{X}(1-\overline{X})}. \tag{7.35}$$

[22] The MOM estimate $\hat{p}_{MOM} = \overline{X}$ is a natural estimate for the average allele frequency p.

To explore the accuracy of the MOM estimates $\hat{p}_{MOM}$ and $\hat{F}_{MOM}$ numerically, let's generate one hundred samples of size $n = 10$ and one hundred samples of size $n = 100$ from the Balding-Nichols model (7.30) with $p = 0.25$ and $F = 0.1$, and compute the mean and the standard error of the estimates based on these samples:

$$\begin{aligned} n = 10: \quad & \mathbb{E}[\hat{p}_{MOM}] = 0.247, \quad \text{se}[\hat{p}_{MOM}] = 0.043, \\ n = 100: \quad & \mathbb{E}[\hat{p}_{MOM}] = 0.252, \quad \text{se}[\hat{p}_{MOM}] = 0.014, \\ n = 10: \quad & \mathbb{E}[\hat{F}_{MOM}] = 0.089, \quad \text{se}[\hat{F}_{MOM}] = 0.036, \\ n = 100: \quad & \mathbb{E}[\hat{F}_{MOM}] = 0.101, \quad \text{se}[\hat{F}_{MOM}] = 0.014. \end{aligned} \tag{7.36}$$

The MOM estimate $\hat{p}_{MOM}$ of the background allele frequency is unbiased, since

$$\mathbb{E}[\hat{p}_{MOM}] = \mathbb{E}[\overline{X}] = \frac{1}{n}\sum_{i=1}^{n}\mathbb{E}[X_i] = p. \tag{7.37}$$

It is hard to compute $\mathbb{E}[\hat{F}_{MOM}]$ analytically, but the simulation results (7.36) suggest that the MOM estimate $\hat{F}_{MOM}$ of the fixation index is asymptotically unbiased.

Finally, Fig. 7.2 shows the true beta PDF (7.27), where the shape parameters α and β are given by (7.32) with $p = 0.25$ and $F = 0.1$, as well as two estimates of the true PDF based on the MOM estimates $\hat{p}_{MOM}$ and $\hat{F}_{MOM}$ obtained from two samples of sizes $n = 10$ and $n = 100$.

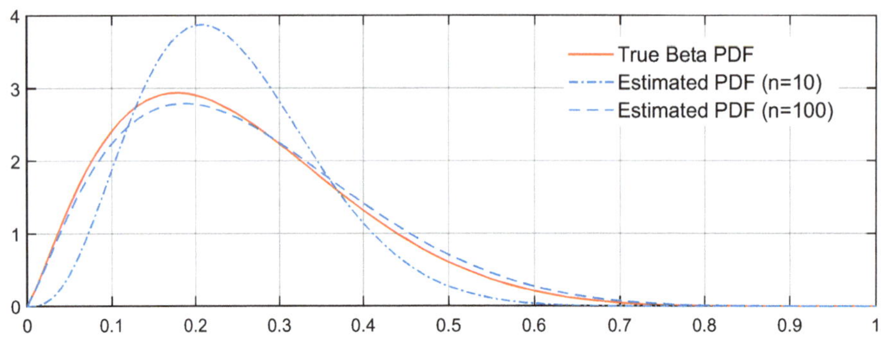

Fig. 7.2 The true beta PDF corresponding to the Balding-Nichols model with $p = 0.25$ and $F = 0.1$ and its MOM-based estimates obtained from samples of sizes $n = 10$ and $n = 100$. The larger the sample size, the more accurately the estimated PDF approximates the true one.

7.5 Further Reading

In this chapter, we learned one of the oldest estimation methods of parametric inference, the methods of moments. In some applications, when the assumed parametric stochastic model is in doubt, MOM estimates can be preferable to other generally more accurate estimates[23]. The main idea behind MOM is to estimate the model parameters by matching theoretical moments of the assumed probability distribution with the corresponding sample moments computed from the data. For complex stochastic models however, theoretical moments (7.8) are often expressed as

[23] Such as the estimates obtained by the method of maximum likelihood, one of the most fundamental and iconic methods of parametric inference that we will discuss in the next chapter.

intractable integrals. In such cases, matching theoretical and samples moments requires solving systems of integral equations. Andrew Gelman[24] in

- Gelman, A.: Method of moments using Monte Carlo simulation. *Journal of Computational and Graphical Statistics* 4(1), 36-54 (1995). `http://stat.columbia.edu/~gelman/research/published/moments.pdf`

presents a computation approach to the method of moments based on Monte Carlo simulation that efficiently resolves this technical problem.

[24] Gelman's blog [5] is also highly recommended.

7.6 Appendix: Moments of the Beta Distribution

The closed-form expressions for the moments of the beta distribution can be derived using properties of the *beta function*[25]

[25] Also known as the Euler integral of the first kind.

$$B(\alpha,\beta) = \int_0^1 x^{\alpha-1}(1-x)^{\beta-1}dx. \tag{7.38}$$

The first important property is that

$$B(\alpha,\beta) = B(\alpha+1,\beta) + B(\alpha,\beta+1). \tag{7.39}$$

Indeed, using the definition of the beta function,

$$\begin{aligned} B(\alpha+1,\beta)+B(\alpha,\beta+1) &= \int_0^1 x^{\alpha}(1-x)^{\beta-1}dx + \int_0^1 x^{\alpha-1}(1-x)^{\beta}dx \\ &= \int_0^1 x^{\alpha-1}(1-x)^{\beta-1}(x+(1-x))dx = B(\alpha,\beta). \end{aligned} \tag{7.40}$$

The second important property is the following identity

$$B(\alpha,\beta+1) = \frac{\beta}{\alpha}B(\alpha+1,\beta). \tag{7.41}$$

It can be proved by integration by parts:

$$\begin{aligned} B(\alpha,\beta+1) &= \int_0^1 x^{\alpha-1}(1-x)^{\beta}dx = \frac{1}{\alpha}\int_0^1 (1-x)^{\beta}dx^{\alpha} \\ &= \underbrace{\frac{1}{\alpha}x^{\alpha}(1-x)^{\beta}\Big|_{x=0}^{x=1}}_{0} - \frac{1}{\alpha}\int_0^1 x^{\alpha}d(1-x)^{\beta} \\ &= \frac{\beta}{\alpha}\int_0^1 x^{\alpha}(1-x)^{\beta-1}dx = \frac{\beta}{\alpha}B(\alpha+1,\beta). \end{aligned} \tag{7.42}$$

Properties (7.39) and (7.41) allow to derive the following key relationship

$$B(\alpha+1,\beta) = \frac{\alpha}{\alpha+\beta} B(\alpha,\beta). \tag{7.43}$$

Indeed,

$$\begin{aligned} B(\alpha,\beta) &= B(\alpha+1,\beta) + B(\alpha,\beta+1) \\ &= B(\alpha+1,\beta) + \frac{\beta}{\alpha} B(\alpha+1,\beta) = \frac{\alpha+\beta}{\alpha} B(\alpha+1,\beta). \end{aligned} \tag{7.44}$$

Now we are ready to compute the moments of the beta distribution. If random variable $X \sim \mathscr{B}(\alpha,\beta)$ with PDF (7.27), then using (7.43)

$$\begin{aligned} \mathbb{E}[X] &= \int_0^1 x f(x|\alpha,\beta)dx = \frac{1}{B(\alpha,\beta)} \int x^{\alpha}(1-x)^{\beta-1}dx \\ &= \frac{B(\alpha+1,\beta)}{B(\alpha,\beta)} = \frac{\alpha}{\alpha+\beta}, \\ \mathbb{E}[X^2] &= \int_0^1 x^2 f(x|\alpha,\beta)dx = \frac{1}{B(\alpha,\beta)} \int x^{\alpha+1}(1-x)^{\beta-1}dx \\ &= \frac{B(\alpha+2,\beta)}{B(\alpha,\beta)} = \frac{\frac{\alpha+1}{\alpha+1+\beta} B(\alpha+1,\beta)}{B(\alpha,\beta)} = \frac{(\alpha+1)\alpha}{(\alpha+\beta+1)(\alpha+\beta)}. \end{aligned} \tag{7.45}$$

Moreover, the recurrence relation (7.43) allows to derive a closed-form expression for the qth moment for any q:

$$\begin{aligned} \mathbb{E}[X^q] &= \int_0^1 x^q f(x|\alpha,\beta)dx = \frac{1}{B(\alpha,\beta)} \int x^{q+\alpha-1}(1-x)^{\beta-1}dx \\ &= \frac{B(\alpha+q,\beta)}{B(\alpha,\beta)} = \frac{\frac{\alpha+q-1}{\alpha+q-1+\beta} B(\alpha+q-1,\beta)}{B(\alpha,\beta)} = \ldots = \\ &= \frac{\frac{\alpha+q-1}{\alpha+q-1+\beta} \frac{\alpha+q-2}{\alpha+q-2+\beta} \cdots \frac{\alpha+1}{\alpha+1+\beta} B(\alpha+1,\beta)}{B(\alpha,\beta)} = \prod_{k=0}^{q-1} \frac{\alpha+k}{\alpha+\beta+k}. \end{aligned} \tag{7.46}$$

Problems

7.1. Exponential Distribution

A random variable X is distributed according to an *exponential distribution*[26] with parameter $\lambda > 0$, denoted $X \sim \mathscr{E}(\lambda)$ if its PDF is

$$f(x|\lambda) = \begin{cases} \lambda e^{-\lambda x}, & \text{if } x \geq 0, \\ 0, & \text{if } x < 0. \end{cases} \tag{7.47}$$

Suppose that data is modeled as an IID sample from the exponential distribution,

[26] The exponential distribution is often used for modeling times between occurrences of uncertain events (failures, successes, earthquakes, customer arrivals, financial defaults, etc.). It plays an important role in the theory of Poisson processes, queuing theory, and reliability engineering.

$$X_1, \ldots, X_n \overset{\text{iid}}{\sim} \mathscr{E}(\lambda). \tag{7.48}$$

Find the method of moments estimate of λ.

7.2. Uniform Distribution with Unknown Support
In Example 7.4, we considered the uniform model on the interval $[0, \theta]$, where only the right end θ is unknown, and found the MOM estimate of θ, which is $\hat{\theta}_{MOM} = \overline{X}$. Suppose now that both endpoints of the interval are unknown. Namely, let

$$X_1, \ldots, X_n \overset{\text{iid}}{\sim} \mathscr{U}[\alpha, \beta], \tag{7.49}$$

where α and β are unknown model parameters, $\alpha < \beta$, and $\mathscr{U}[\alpha, \beta]$ is the uniform distribution on $[\alpha, \beta]$. Find the method of moments estimates of α and β.

7.3. MOM for a Bernoulli-Like Model
To see how the method of moments works in a discrete case, let's consider the following simple Bernoulli-like stochastic model: $X_1, \ldots, X_n \overset{\text{iid}}{\sim} p(x|\theta)$, where $p(x|\theta)$ is a probability mass function defined by

$$p(1|\theta) = \mathbb{P}(X = 1|\theta) = \theta \quad \text{and} \quad p(2|\theta) = \mathbb{P}(X = 2|\theta) = 1 - \theta, \tag{7.50}$$

where $\theta \in (0, 1)$ is the model parameter. Suppose we observe the following sample of size $n = 5$:

$$X_1 = 1, \quad X_2 = 2, \quad X_3 = 1, \quad X_4 = 2, \quad X_5 = 1. \tag{7.51}$$

What is your intuitive estimate of θ? Find the method of moments estimate of θ.

7.4. Estimating the Effectiveness of a New Drug (Revisited)
Recall Problem 4.4[27]. A pharmaceutical company developed a new drug to treat a certain disease. To check its effectiveness, the following experiment was conducted. A group of $N = 900$ patients with the disease was randomly divided into two equal subgroups. The first subgroup of $n = 450$ patients was given the new drug and the second subgroup of $n = 450$ patients was given an old drug. In the first group, $n_1 = 400$ patients have recovered from the disease, and in the second group, $n_2 = 375$ patients have recovered. Let p_1 and p_2 denote the probabilities of recovery due to the new and old drug respectively. Let $\theta = p_1 - p_2$ be the parameter of interest that quantifies the effectiveness of the new drug compared to the old one: if $\theta > 0$, then the new drug is more effective in treating the disease, and if $\theta < 0$, then the old drug is more effective.

In Problem 4.4, we found the the plug-in estimate of θ, $\hat{\theta}_n = (n_1 - n_2)/n$. Now let's use the method of moments. Let

$$\begin{aligned} X_i &= \begin{cases} 1, & \text{if patient } i \text{ in group 1 recovered,} \\ 0, & \text{if patient } i \text{ in group 1 did not recover.} \end{cases} \\ Y_i &= \begin{cases} 1, & \text{if patient } i \text{ in group 2 recovered,} \\ 0, & \text{if patient } i \text{ in group 2 did not recover.} \end{cases} \end{aligned} \tag{7.52}$$

[27] For convenience, it is reminded here.

It makes sense to model samples $X_1,\ldots,X_n$ and $Y_1,\ldots,Y_n$ using the Bernoulli model with parameters p_1 and p_2 respectively,

$$X_1,\ldots,X_n \overset{\text{iid}}{\sim} \mathscr{B}(p_1) \quad \text{and} \quad Y_1,\ldots,Y_n \overset{\text{iid}}{\sim} \mathscr{B}(p_2). \tag{7.53}$$

Find the method of moments estimate $\hat{\theta}_{MOM} = \hat{p}_{1,MOM} - \hat{p}_{2,MOM}$, where $\hat{p}_{1,MOM}$ and $\hat{p}_{2,MOM}$ are the MOM estimates of p_1 and p_2.

References

1. Chebyshev, P.L.: Sur deux théorèèmes relatifs aux probabilitéés. Originally published in Russian in *Zapiski Akademii Nauk* 55 (1887).
2. Pearson, K.: Contributions to the mathematical theory of evolution. *Philosophical Transactions of the Royal Society of London*. A 185 (1994).
`https://www.jstor.org/stable/90667`
3. Balding, D.J., Nichols, R.A.: A method for quantifying differentiation between populations at multi-allelic loci and its implications for investigating identity and paternity. *Genetica* 96(1-2), 3–12 (1995).
`https://doi.org/10.1007/BF01441146`
4. Cavalli-Sforza, L.L., Menozzi, P., Piazza, A.: *The History and Geography of Human Genes*. Princeton University Press, Princeton, New Jersey, 1994.
`https://press.princeton.edu/books/ebook/9780691187266/the-history-and-geography-of-human-genes`
5. Andrew Gelman's Blog "Statistical Modeling, Causal Inference, and Social Science"
`https://statmodeling.stat.columbia.edu/`

8
The Method of Maximum Likelihood

Early hunters and gatherers may not have used the words "method of maximum likelihood" to describe their choice of where and how to hunt and gather, but it is hard to believe they would have been surprised if their method had been described in those terms.
– Stephen M. Stigler, *The Epic Story of Maximum Likelihood [1]*

Maximum likelihood estimation is one of the most popular methods for estimating parameters in parametric models. It was first introduced in 1912 in [2], and then studied and popularized in subsequent works by Ronald Fisher, one of the greatest statisticians of all times [3]. Maximum likelihood estimates are known to be very powerful and have many attractive properties, especially when the sample size is large. In this chapter, we will provide the intuition behind this method, define the likelihood function and the maximum likelihood estimate, consider several classical examples, and discuss the main properties of the method.

Sir Ronald Aylmer Fisher (1890-1962), an English polymath, "a genius who almost single-handedly created the foundations for modern statistical science" [4]. Photo: Courtesy of the University of Adelaide, Special Collections.

8.1 Intuition

Consider a black box that contains $N = 10$ balls of two types: red ● and blue ●. We don't know the exact numbers of red and blue balls in the box, but we do know that there are just two possible cases:

- Case 1: There are three red balls ●●● and seven blue balls ●●●●●●●.
- Case 2: There are seven red balls ●●●●●●● and three blue balls ●●●.

Suppose that we draw a ball from the box at random and see that it is red ●. This observation is our "data". Then, based on this data, Case 2 is intuitively more likely, since it is more *consistent* with the data: in Case 2, the chance of choosing a red ball is higher than in Case 1.

We can describe this example slightly more formally. Let's model our data, i.e. the outcome X of an experiment consisting of randomly drawing a ball from the

Supplementary Information The online version contains supplementary material available at https://doi.org/10.1007/978-3-032-03848-7_8.

K. M. Zuev, *Fundamentals of Statistical Inference*,
International Series in Operations Research & Management Science 374,
https://doi.org/10.1007/978-3-032-03848-7_8

box,

$$X = \begin{cases} 1, & \text{if the chosen ball is red } \bullet, \\ 0, & \text{if the chosen ball is blue } \bullet, \end{cases} \tag{8.1}$$

as a sample from the Bernoulli distribution:

$$X \sim \mathscr{B}(p) \Leftrightarrow X = \begin{cases} 1, & \text{with prob. } p, \\ 0, & \text{with prob. } 1-p, \end{cases} \tag{8.2}$$

where the model parameter p can take only two possible values:

$$\text{Case 1:} \quad p = \frac{3}{10} \quad \text{or} \quad \text{Case 2:} \quad p = \frac{7}{10}. \tag{8.3}$$

Our actually observed data is $X = 1$. Since

$$\mathbb{P}(X = 1 \,|\, p = 3/10) < \mathbb{P}(X = 1 \,|\, p = 7/10), \tag{8.4}$$

the value $p = 7/10$ is more consistent with the observed data.

This example captures the essence of maximum likelihood estimation. If θ is a model parameter of interest (or a vector of parameters), then the method of maximum likelihood aims at finding the value of θ which is most *consistent* with the observed data, where "consistency" of θ with the data is measured by how likely it is for the data to occur given that value of θ. This brings us to the central concept of maximum likelihood estimation: the likelihood function.

8.2 The Likelihood Function and MLE

Let us consider discrete and continuous stochastic models separately. The discrete case is somewhat more straightforward, but at the end we will see that there is no conceptual difference between the two cases[1].

[1] "The creatures outside looked from pig to man, and from man to pig, and from pig to man again; but already it was impossible to say which was which." – George Orwell, *Animal Farm*.

8.2.1 Discrete Stochastic Models

Let $X_1 \ldots, X_n$ be data modeled as an IID sample from a discrete distribution with probability mass function $p(x \,|\, \theta)$,

$$X_1 \ldots, X_n \overset{\text{iid}}{\sim} p(x \,|\, \theta), \quad \theta \in \Theta, \tag{8.5}$$

where Θ is the parameter space, i.e. the set of all possible values of θ. If θ is given, then the probability of observing the data $X_1 \ldots, X_n$ is

$$\mathbb{P}(X_1 \ldots, X_n \mid \theta) \underset{\substack{\uparrow \\ X_1,\ldots,X_n \text{ are independent}}}{=} \prod_{i=1}^{n} \mathbb{P}(X_i \mid \theta) = \prod_{i=1}^{n} p(X_i \mid \theta). \tag{8.6}$$

The value of θ that is most consistent with the data $X_1 \ldots, X_n$ is exactly the value of θ that maximizes the probability (8.6) of observing the data. This motivates the following definition.

Definition 8.1. For a discrete stochastic model (8.5), the *likelihood function* is the joint probability of the data, viewed as a *function of parameter* θ,

$$\mathscr{L}_n(\theta) = \prod_{i=1}^{n} p(X_i \mid \theta). \tag{8.7}$$

Remark 8.1. The subscript $_n$ in the notation for the likelihood function serves as a reminder that the function is obtained based on a sample of size n. This notation will become useful in Section 8.6, where we will study the asymptotic properties of maximum likelihood estimates.

In spite of the fact that the likelihood function $\mathscr{L}_n(\theta)$ is defined in terms of the values of PMF $p(x \mid \theta)$ at $X_1, \ldots, X_n$, the two functions are conceptually different. When we consider the PMF $p(x \mid \theta)$, parameter θ is fixed and x is a variable, which represents possible values of X_i. On the contrary, when we consider the likelihood function $\mathscr{L}_n(\theta)$, the data $X_1, \ldots, X_n$ is fixed (observed) and θ is a variable, which represents different values of the model parameter. The values of the likelihood function $\mathscr{L}_n(\theta_1), \mathscr{L}_n(\theta_2), \ldots$ are the probabilities of observing the fixed data $X_1, \ldots, X_n$ when $\theta = \theta_1, \theta = \theta_2$, etc. The *method of maximum likelihood* searches for the value of θ that makes this probability as large as possible.

Definition 8.2. The *maximum likelihood estimate* (MLE) of θ is the value $\hat{\theta}_{MLE}$ that maximizes the likelihood function,

$$\hat{\theta}_{MLE} = \arg\max_{\theta \in \Theta} \mathscr{L}_n(\theta), \tag{8.8}$$

or, equivalently[2], $\hat{\theta}_{MLE} \in \Theta$ is such that

$$\mathscr{L}_n(\hat{\theta}_{MLE}) \geq \mathscr{L}_n(\theta) \quad \text{for all } \theta \in \Theta. \tag{8.9}$$

[2] Statement (8.9) explains what equation (8.8) means.

To illustrate the method of maximum likelihood, let's consider an example.

Example 8.1. Suppose that data is modeled with the Bernoulli model,

$$X_1, \ldots, X_n \overset{\text{iid}}{\sim} \mathscr{B}(\theta), \quad \theta \in [0,1]. \tag{8.10}$$

To find the MLE estimate $\hat{\theta}_{MLE}$, we need to find the likelihood function and then maximize it with respect to θ. Since the PMF of Bernoulli distribution with parameter θ is

$$p(x|\theta) = \begin{cases} \theta, & \text{if } x = 1, \\ 1-\theta, & \text{if } x = 0. \end{cases} = \theta^x(1-\theta)^{1-x}, \quad x \in \{0,1\}, \tag{8.11}$$

the likelihood function is

$$\mathscr{L}_n(\theta) = \prod_{i=1}^{n} p(X_i|\theta) = \prod_{i=1}^{n} \theta^{X_i}(1-\theta)^{1-X_i} = \theta^{\sum_{i=1}^{n} X_i}(1-\theta)^{n-\sum_{i=1}^{n} X_i}. \tag{8.12}$$

Figure 8.1(a) shows the likelihood function (8.12) for the data $X_1,\ldots,X_n$ generated from the Bernoulli distribution with $\theta = 0.3$ for $n = 10, 100$, and 1000. For convenience of representing all three functions on the same plot[3], they are normalized by dividing by their largest values. In other words, instead of original likelihoods $\mathscr{L}_n(\theta)$, normalized likelihoods $\mathscr{L}_n(\theta)/\max_{\theta\in[0,1]}\mathscr{L}_n(\theta)$ are shown in Fig. 8.1(a). Notice how the likelihood functions become more and more peaked around the true value of the parameter $\theta = 0.3$ as the sample size n increases from $n = 10$ to $n = 1000$.

[3] Their values differ by orders of magnitude.

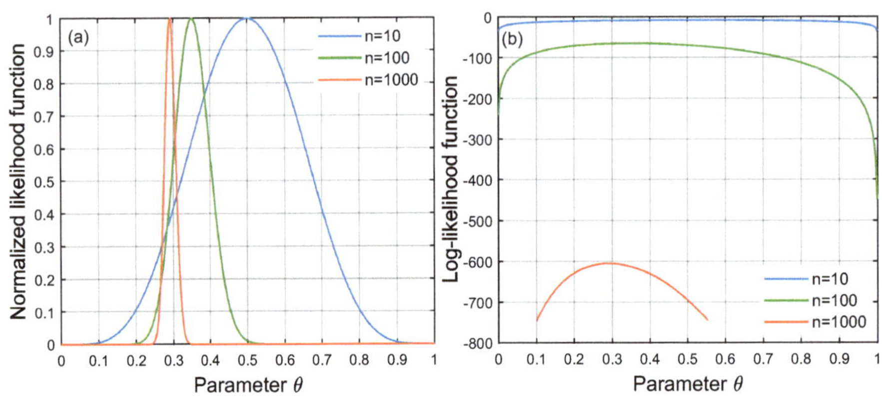

Fig. 8.1 Panel (a) shows three normalized likelihood functions for a sample generated from the Bernoulli model with parameter $\theta = 0.3$ for $n = 10, 100$ and 1000. Panel (b) shows the corresponding log-likelihood functions (8.14).

Now we need to find the value of $\theta \in [0,1]$ that maximizes the likelihood function $\mathscr{L}_n(\theta)$. To this end, we need to solve $\mathscr{L}_n'(\theta) = 0$ for θ (**find the critical point**) and then check that $\mathscr{L}_n''(\theta) < 0$ at the solution (check that this critical point is indeed the global maximizer). But differentiating (8.12) twice, although doable, is a daunting task[4]. It is much easier to differentiate its logarithm, since logarithm converts products into sums, and differentiating sums is easier than differentiating products. Taking the logarithm changes the values of the likelihood, but it does not change the value of the global maximizer, since log is an increasing function. In other words,

$$\arg\max_{\theta\in\Theta} \underbrace{\mathscr{L}_n(\theta)}_{\text{likelihood}} = \arg\max_{\theta\in\Theta} \underbrace{\log\mathscr{L}_n(\theta)}_{\text{log-likelihood}} \tag{8.13}$$

[4] At least for the author.

So, the log-likelihood function is

$$\ell_n(\theta) = \log \mathscr{L}_n(\theta) = \sum_{i=1}^{n} X_i \log\theta + \Big(n - \sum_{i=1}^{n} X_i\Big)\log(1-\theta). \tag{8.14}$$

Figure 8.1(b) shows the log-likelihood functions (8.14) corresponding to the likelihoods shown in Fig. 8.1(a). Notice that the maximizers of likelihoods and log-likelihoods are the same. For example, when $n = 1000$, the maximizer of both (red) functions is approximately $\theta = 0.3$.

The first derivative of the log-likelihood is

$$\ell_n'(\theta) = \frac{\sum_{i=1}^{n} X_i}{\theta} - \frac{n - \sum_{i=1}^{n} X_i}{1-\theta}. \tag{8.15}$$

Solving $\ell_n'(\theta) = 0$ gives a single solution $\hat{\theta} = \frac{1}{n}\sum_{i=1}^{n} X_i = \overline{X}$. Since the second derivative is always negative,

$$\ell_n''(\theta) = -\frac{\sum_{i=1}^{n} X_i}{\theta^2} - \frac{n - \sum_{i=1}^{n} X_i}{(1-\theta)^2} < 0, \tag{8.16}$$

the log-likelihood (8.14) is a concave function[5] and $\hat{\theta}$ is indeed the maximum likelihood estimate of θ,

$$\hat{\theta}_{MLE} = \overline{X}. \tag{8.17}$$

[5] See Fig. 8.1(b).

Notice that the plug-in principle and the method of moments both give the same estimate[6] for the Bernoulli parameter θ, which is the mean of the distribution. The amount of work, however, needed for computing the MLE (8.17) is substantially larger. This is because maximization (8.8) is a non-trivial problem. Now let's extend maximum likelihood estimation to continuous models.

[6] See (4.45) and (7.17).

8.2.2 Continuous Stochastic Models

Suppose that $X_1, \ldots, X_n$ is modeled as an IID sample[7] from a continuous distribution with probability density function $f(x\,|\,\theta)$,

[7] In the Case Study in Section 8.7, we will consider a non IID case.

$$X_1 \ldots, X_n \overset{\text{iid}}{\sim} f(x\,|\,\theta), \quad \theta \in \Theta. \tag{8.18}$$

Unfortunately, we can't mimic the discrete case directly. Indeed, if try to compute the *probability* of observing the data $X_1, \ldots, X_n$ for a given θ, just as we did in (8.6), we will get a disappointing result:

$$\mathbb{P}(X_1 \ldots, X_n\,|\,\theta) = \prod_{i=1}^{n} \underbrace{\mathbb{P}(X_i\,|\,\theta)}_{0} = 0, \tag{8.19}$$

since for any continuous random variable $X \sim f(x)$ with PDF $f(x)$ the probability of observing a particular value $X = x$ is zero:

$$\mathbb{P}(X = x) = \int_x^x f(t)dt = 0. \tag{8.20}$$

Fortunately, there is a remedy for this technical difficulty, which is based on the *probabilistic interpretation* of a probability density function. Recall that if $X \sim f(x)$ is a continuous random variable, then for small $\varepsilon > 0$, $\varepsilon \ll 1$

$$\mathbb{P}(x-\varepsilon < X < x+\varepsilon) = \int_{x-\varepsilon}^{x+\varepsilon} f(x)dx \approx 2\varepsilon f(x). \tag{8.21}$$

The value of the PDF $f(x)$ at point x is thus proportional to the probability of X being in a small neighborhood of x. In other words, $f(x)$ quantifies how likely it is for realizations of X to be near x.

Similarly, if we have n IID continuous random variables $X_1 \dots, X_n \overset{\text{iid}}{\sim} f(x\,|\,\theta)$, then

$$\begin{aligned}\mathbb{P}(x_1-\varepsilon < X_1 < x_1+\varepsilon,\ x_2-\varepsilon < X_2 < x_2+\varepsilon, \dots,\ x_n-\varepsilon < X_n < x_n+\varepsilon)\\ = \prod_{i=1}^{n}\mathbb{P}(x_i-\varepsilon < X_i < x_i+\varepsilon) \approx \prod_{i=1}^{n} 2\varepsilon f(x_i\,|\,\theta) = (2\varepsilon)^n \prod_{i=1}^{n} f(x_i\,|\,\theta).\end{aligned} \tag{8.22}$$

If $\varepsilon \to 0$, then both sides of (8.22) converge to zero, and this equation essentially reduces to (8.19). However, for small but nonzero ε, the value of the *joint PDF* $\prod_{i=1}^{n} f(x_i\,|\,\theta)$ is proportional to the probability of $X = (X_1, \dots, X_n)$ being in a small cubical neighborhood of $x = (x_1, \dots, x_n)$. Here X is a random vector that models the data and x is a particular realization of this vector, the actually observed data. Although in accordance with (8.19) $\mathbb{P}(X = x) = 0$, the probability

$$\mathbb{P}(X \text{ is near } x) \propto \prod_{i=1}^{n} f(x_i\,|\,\theta). \tag{8.23}$$

To maximize the value of this probability, i.e. to make the observed data x as likely to occur under model X as possible, we need to maximize the value of the joint PDF with respect to the model parameter θ. This leads to the definition of the likelihood function for continuous models.

Definition 8.3. For a continuous stochastic model (8.18), the *likelihood function* is the joint density of the data, viewed as a *function of parameter* θ,

$$\mathscr{L}_n(\theta) = \prod_{i=1}^{n} f(X_i\,|\,\theta). \tag{8.24}$$

As in the discrete case (8.7), the data $X_1, \dots, X_n$ in (8.24) represents the actually observed data[8]. The values of the likelihood $\mathscr{L}_n(\theta_1), \mathscr{L}_n(\theta_2), \dots$ are the values of the joint density[9] of the data computed at $X_1, \dots, X_n$ when $\theta = \theta_1$, $\theta = \theta_2$, etc. It is

[8] Using x_i instead of X_i in (8.24) might be more accurate; however, this is not commonly done. Statisticians typically avoid using different notation for similar quantities, preferring instead to infer their meaning from the context. We will adhere to this convention.

[9] Not the probability, as in the discrete case.

important to highlight that although $\mathscr{L}_n(\theta)$ is defined in terms of the values of the joint PDF, it is not a PDF itself[10]. The definition of the MLE remains the same.

[10] In particular, $\int_\Theta \mathscr{L}_n(\theta)d\theta \neq 1$.

Definition 8.4. The *maximum likelihood estimate* (MLE) of θ is the value $\hat{\theta}_{MLE}$ that maximizes the likelihood function,

$$\hat{\theta}_{MLE} = \arg\max_{\theta\in\Theta} \mathscr{L}_n(\theta). \tag{8.25}$$

Let's consider an example of maximum likelihood estimation for a continuous models with two parameters.

Example 8.2. Suppose that data is modeled with the normal model,

$$X_1,\ldots,X_n \overset{\text{iid}}{\sim} \mathscr{N}(\mu,\sigma^2), \quad \mu\in\mathbb{R}, \sigma^2 \geq 0, \tag{8.26}$$

and we want to find the MLE of $\theta = (\theta_1, \theta_2)$, where $\theta_1 = \mu$ and $\theta_2 = \sigma^2$. The likelihood function is[11]

$$\begin{aligned}\mathscr{L}_n(\theta) &= \prod_{i=1}^n \phi(X_i|\mu,\sigma^2) = \prod_{i=1}^n \frac{1}{\sqrt{2\pi}\sigma}\exp\left(-\frac{(X_i-\mu)^2}{2\sigma^2}\right)\\ &\propto \prod_{i=1}^n \frac{1}{\sqrt{\theta_2}}\exp\left(-\frac{(X_i-\theta_1)^2}{2\theta_2}\right) = \theta_2^{-\frac{n}{2}}\exp\left(-\frac{1}{2\theta_2}\sum_{i=1}^n (X_i-\theta_1)^2\right).\end{aligned} \tag{8.27}$$

[11] For convenience, in the second line of this derivation, we drop the constant positive factor of $(2\pi)^{-\frac{n}{2}}$, since its absence does not change the MLE $\hat{\theta}_{MLE}$, which is the global maximizer of $\mathscr{L}_n(\theta)$.

Figure 8.2 shows the likelihood function (8.27) for the data $X_1,\ldots,X_n$ generated from the standard normal distribution $\mathscr{N}(0,1)$ for $n = 10, 100$ and 1000.

As in Example 8.1, it is technically easier to maximize the log-likelihood function, which is given by[12]

$$\ell_n(\theta) = \log\mathscr{L}_n(\theta) = -\frac{n}{2}\log\theta_2 - \frac{1}{2\theta_2}\sum_{i=1}^n (X_i-\theta_1)^2. \tag{8.28}$$

[12] Since we dropped the multiplicative constant $(2\pi)^{-\frac{n}{2}}$ in (8.27), the log-likelihood in (8.28) is given up to an additive constant $-\frac{n}{2}\log 2\pi$.

To find the MLE of θ, we need to calculate the first partial derivatives of $\ell_n(\theta)$ with respect to θ_1 and θ_2, set both derivatives to zero, solve the corresponding system of two equations for θ and find a critical point $\hat{\theta} = (\hat{\theta}_1, \hat{\theta}_2)$, and finally check that the found critical point is indeed the global maximizer of the log-likelihood. Let's implement this plan[13]. The partial derivatives are:

[13] "Gandalf looked curiously and closely at him. 'No, it does not seem right,' he said thoughtfully. 'No, after all I believe your plan is probably the best." – J.R.R. Tolkien, *The Lord of the Rings*.

$$\frac{\partial\ell_n}{\partial\theta_1} = \frac{1}{\theta_2}\sum_{i=1}^n (X_i-\theta_1) \quad \text{and} \quad \frac{\partial\ell_n}{\partial\theta_2} = -\frac{n}{2\theta_2} + \frac{1}{2\theta_2^2}\sum_{i=1}^n (X_i-\theta_1)^2. \tag{8.29}$$

A unique solution of $\partial\ell_n/\partial\theta_1 = 0$, $\partial\ell_n/\partial\theta_2 = 0$ is given by familiar expressions:

$$\hat{\theta}_1 = \overline{X} \quad \text{and} \quad \hat{\theta}_2 = \frac{1}{n}\sum_{i=1}^n (X_i-\overline{X})^2. \tag{8.30}$$

To show that the critical point $\hat{\theta} = (\hat{\theta}_1, \hat{\theta}_2)$ corresponds to the global maximum of log-likelihood, we need to check that the Hessian of $\ell_n(\theta)$, which is the matrix H of

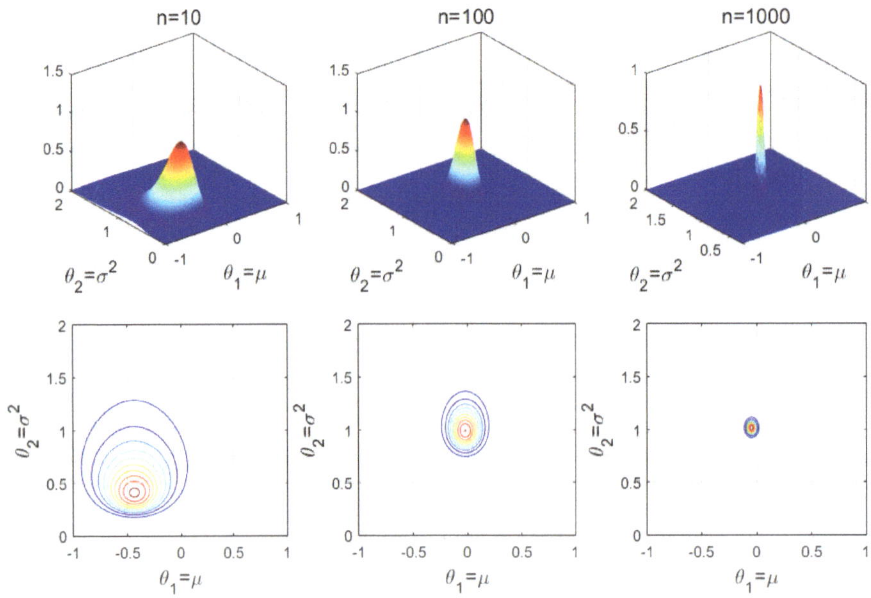

Fig. 8.2 Top row: three likelihood functions for the standard normal data $X_1,\ldots,X_n \overset{\text{iid}}{\sim} \mathcal{N}(0,1)$ for $n = 10, 100$, and 1000. Bottom row: the corresponding contour plots. Red color corresponds to higher values of the likelihood function. Notice how the likelihood function becomes more and more peaked around the true values $\theta_1 = \mu = 0$ and $\theta_2 = \sigma^2 = 1$ as the sample size n increases.

second-order partial derivatives of $\ell_n(\theta)$, is negative definite at $\hat{\theta}$[14],

$$H = \begin{bmatrix} \frac{\partial^2 \ell_n}{\partial \theta_1^2} & \frac{\partial^2 \ell_n}{\partial \theta_1 \partial \theta_2} \\ \frac{\partial^2 \ell_n}{\partial \theta_1 \partial \theta_2} & \frac{\partial^2 \ell_n}{\partial \theta_2^2} \end{bmatrix} \text{ is negative definite at } \hat{\theta}, \tag{8.31}$$

which is equivalent to

$$\left.\frac{\partial^2 \ell_n}{\partial \theta_1^2}\right|_{\theta=\hat{\theta}} < 0 \quad \text{and} \quad \det H|_{\theta=\hat{\theta}} > 0. \tag{8.32}$$

Indeed, the second-order partial derivatives are:

[14] This condition is a high-dimensional analog of concavity $l_n''(\hat{\theta}) < 0$ in one-dimensional case.

$$\frac{\partial^2 \ell_n}{\partial \theta_1^2} = -\frac{n}{\theta_2} < 0 \quad (\text{since } \theta_2 = \sigma^2 > 0),$$

$$\frac{\partial^2 \ell_n}{\partial \theta_2^2} = \frac{n}{2\theta_2^2} - \frac{1}{\theta_2^3}\sum_{i=1}^{n}(X_i - \theta_1)^2,$$

$$\left.\frac{\partial^2 \ell_n}{\partial \theta_2^2}\right|_{\theta=\hat{\theta}} = \frac{1}{\hat{\theta}_2^2}\left(\frac{n}{2} - \frac{\sum_{i=1}^{n}(X_i-\overline{X})^2}{\frac{1}{n}\sum_{i=1}^{n}(X_i-\overline{X})^2}\right) = -\frac{n}{2\hat{\theta}_2^2}, \tag{8.33}$$

$$\frac{\partial^2 \ell_n}{\partial \theta_1 \partial \theta_2} = -\frac{1}{\theta_2^2}\sum_{i=1}^{n}(X_i - \theta_1),$$

$$\left.\frac{\partial^2 \ell_n}{\partial \theta_1 \partial \theta_2}\right|_{\theta=\hat{\theta}} = -\frac{1}{\hat{\theta}_2^2}\sum_{i=1}^{n}(X_i - \overline{X}) = -\frac{1}{\hat{\theta}_2^2}\left(\sum_{i=1}^{n}X_i - n\overline{X}\right) = 0,$$

and, therefore, the determinant of the Hessian at $\theta = \hat{\theta}$ is

$$\det H|_{\theta=\hat{\theta}} = -\frac{n}{\hat{\theta}_2}\left(-\frac{n}{2\hat{\theta}_2^2}\right) = \frac{n^2}{2\hat{\theta}_2^3} > 0 \quad (\text{since } \hat{\theta}_2 > 0). \tag{8.34}$$

So, $\hat{\theta}_1$ and $\hat{\theta}_2$ are the maximum likelihood estimates of the mean μ and variance σ^2,

$$\hat{\mu}_{MLE} = \overline{X} \quad \text{and} \quad \hat{\sigma}^2_{MLE} = \frac{1}{n}\sum_{i=1}^{n}(X_i - \overline{X})^2. \tag{8.35}$$

As in Example 8.1, the method of maximum likelihood does not give us any new estimates: the plug-in and MOM estimates for the mean and the variance of the normal model are exactly the same[15]. At the same time, the amount of work needed to derive the MLEs in (8.35) is again substantially larger. This observations lead to two natural questions: are there any cases where the MLE estimates do not coincide with the plug-in and MOM estimates, and why do we need MLEs, given how relatively hard it is to find them? Before we address these questions in Sections (8.4) and (8.6), it is important to highlight the computational importance of the log-likelihood[16].

[15] See (4.45), (4.49), and (7.20).

[16] Its conceptual importance will become apparent in Section (8.6).

8.3 The Log-Likelihood Function

Examples 8.1 and 8.2 have demonstrated that computationally it is easier to maximize the log-likelihood rather than the original likelihood function. This is not just an artifact of these two examples. This is almost[17] always the case when the likelihood is written as a product of other functions, as in the case where $X_1,\dots,X_n$ are independent (and not necessarily identically distributed). Moreover, many properties of the MLE estimates will be derived in Section 8.6 with the help of the log-likelihood function. Thanks to this universal computational convenience and theoretical importance, the log-likelihood deserves its own formal definition.

[17] See Example 8.3.

Definition 8.5. If $\mathscr{L}_n(\theta)$ is the likelihood, then the log-likelihood function is

$$\ell_n(\theta) = \log \mathscr{L}_n(\theta). \tag{8.36}$$

Finding MLEs requires maximizing $\mathscr{L}_n(\theta)$, or equivalently[18] $\ell_n(\theta)$, which often involves finding the derivatives. The computational convenience of the log-likelihood over the likelihood stems from a simple fact that if $\mathscr{L}_n(\theta)$ is a product, then $\ell_n(\theta)$ is a sum,

[18] From a theoretical point of view.

$$\mathscr{L}_n(\theta) = \prod_{i=1}^{n} f(X_i\,|\,\theta) \quad \Rightarrow \quad \ell_n(\theta) = \sum_{i=1}^{n} \log f(X_i\,|\,\theta), \tag{8.37}$$

and sums are much easier to differentiate than products. Moreover, we can apply the law of large numbers and the central limit theorem to sums. And this is where the theoretic importance of the log-likelihood originates from.

8.4 Plug-In, MOM, and MLE

In both examples of maximum likelihood estimation that we considered so far[19], the MLE estimates fully agree with the corresponding MOM and plug-in estimates. This should not be very surprising, since it is hard to come up with other reasonable estimates for p, μ, and σ^2. For instance, how else could we reasonably estimate the mean μ if not by the sample mean $\overline{X}$? What could be more natural? In other words, all reasonable estimation methods are expected, in some sense, to produce the same natural estimates for p, μ, and σ^2 as those produced by the plug-in, MOM, and MLE methods.

[19] Example 8.1 for the Bernoulli model $\mathscr{B}(p)$ and Example 8.2 for the normal model $\mathscr{N}(\mu, \sigma^2)$.

Nevertheless this raises a question of weather MLE always agrees with MOM and plug-in. Recall[20] that the plug-in method and the method of moments give different results when estimating the upper bound θ of the uniform model $\mathscr{U}[0, \theta]$:

[20] See Problem 5.1(a) and (7.22), and Example 7.4 and (7.24).

$$\hat{\theta}_{plug\text{-}in} = X_{(n)} \quad \text{and} \quad \hat{\theta}_{MOM} = 2\overline{X}. \tag{8.38}$$

Let's investigate what maximum likelihood estimation gives us in this case.

Example 8.3. Suppose that data is modeled with the uniform model,

$$X_1, \ldots, X_n \overset{\text{iid}}{\sim} \mathscr{U}[0, \theta], \tag{8.39}$$

where the right end $\theta > 0$ is a model parameter. The likelihood function is

$$\mathscr{L}_n(\theta) = \prod_{i=1}^{n} f(X_i\,|\,\theta), \tag{8.40}$$

where f is the PDF of the uniform distribution on $[0, \theta]$,

$$f(x\,|\,\theta) = \begin{cases} \frac{1}{\theta}, & \text{if } x \in [0,\theta], \\ 0, & \text{if } x \notin [0,\theta]. \end{cases} \tag{8.41}$$

In this formula, x is the argument of function f and $\theta > 0$ is a fixed parameter. Since in (8.40), $f(X_i\,|\,\theta)$ is viewed as a function of θ and X_i is fixed, it is useful to rewrite (8.41) in an equivalent form, where the value of $f(x\,|\,\theta)$ depends explicitly on θ and $x > 0$ is viewed as a fixed point:

$$f(x\,|\,\theta) = \begin{cases} \frac{1}{\theta}, & \text{if } \theta \geq x, \\ 0, & \text{if } \theta < x. \end{cases} \tag{8.42}$$

Notice that in this example, computing the log-likelihood function is problematic, since $f(X_i\,|\,\theta) = 0$ if $\theta < X_i$ and, therefore, $\log f(X_i\,|\,\theta)$ is not defined for $\theta < X_i$. Luckily we can compute and maximize the likelihood directly. If $\theta < X_{(n)}$, then $f(X_{(n)}\,|\,\theta) = 0$ and $\mathscr{L}_n(\theta) = 0$, since $f(X_{(n)}\,|\,\theta)$ is exactly one of the factors $f(X_1\,|\,\theta), \ldots, f(X_n\,|\,\theta)$ in (8.40). On the other hand, if $\theta \geq X_{(n)}$, then $\theta \geq X_i$ for all $i = 1, \ldots, n$, and all factors $f(X_1\,|\,\theta), \ldots, f(X_n\,|\,\theta)$ equal to $1/\theta$, and $\mathscr{L}_n(\theta) = 1/\theta^n$. So,

$$\mathscr{L}_n(\theta) = \begin{cases} 0, & \text{if } \theta < X_{(n)}, \\ \frac{1}{\theta^n}, & \text{if } \theta \geq X_{(n)}. \end{cases} \tag{8.43}$$

The likelihood function $\mathscr{L}_n(\theta)$ is schematically shown in Fig. 8.3(a). It is a discontinuous functions and its global maximizer, the maximum likelihood estimate of θ, is

$$\hat{\theta}_{MLE} = X_{(n)}. \tag{8.44}$$

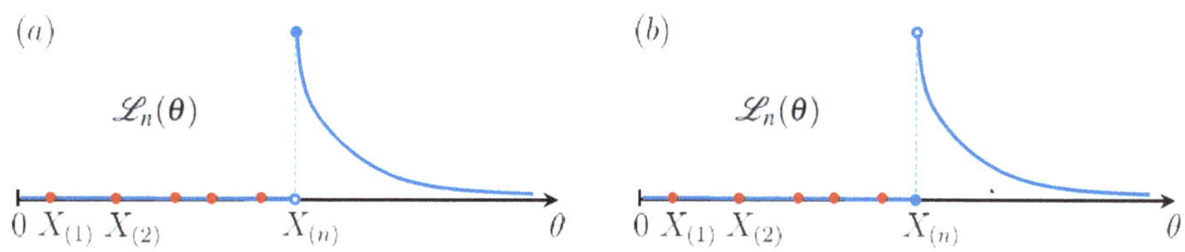

Fig. 8.3 The likelihood functions for stochastic models $\mathscr{U}[0,\theta]$ (panel a) and $\mathscr{U}[0,\theta)$ (panel b). The only difference between the two functions is that in the first case $\mathscr{L}_n(X_{(n)}) = 1/X_{(n)}^n$ and the maximum exists, and in the second case $\mathscr{L}_n(X_{(n)}) = 0$ and the maximum does not exist.

So, in this example, the MLE estimate differs from the MOM estimate but again coincides with the estimate obtained by the plug-in principle. However, the graph of the likelihood in Fig. 8.3(a) suggests how we could slightly change this example to create a new one, where the MLE and plug-in would also disagree.

Example 8.4. Instead of $\mathscr{U}[0,\theta]$, let's consider the uniform model $\mathscr{U}[0,\theta)$, where θ is *not included* into the support of the distribution. In this case, the PDF is

$$f(x|\theta) = \begin{cases} \frac{1}{\theta}, & \text{if } x \in [0,\theta), \\ 0, & \text{if } x \notin [0,\theta). \end{cases} \quad \Leftrightarrow \quad f(x|\theta) = \begin{cases} \frac{1}{\theta}, & \text{if } \theta > x, \\ 0, & \text{if } \theta \leq x. \end{cases} \tag{8.45}$$

and the likelihood function, shown in Fig. 8.3(b), is

$$\mathscr{L}_n(\theta) = \begin{cases} 0, & \text{if } \theta \leq X_{(n)}, \\ \frac{1}{\theta^n}, & \text{if } \theta > X_{(n)}. \end{cases} \tag{8.46}$$

The only difference between the likelihood functions (8.43) and (8.46) is their behavior at $\theta = X_{(n)}$: while in the first case $\mathscr{L}_n(X_{(n)}) = 1/X_{(n)}^n$, in the second case $\mathscr{L}_n(X_{(n)}) = 0$. As a result, (8.46) does not have a global maximizer, and $\hat{\theta}_{MLE}$ does not exist. On the other hand, the plug-in and MOM estimates of θ remain the same as for the model $\mathscr{U}[0,\theta]$, namely $\hat{\theta}_{plug\text{-}in} = X_{(n)}$ and $\hat{\theta}_{MOM} = 2\overline{X}$. This is perhaps the simplest example, where all three estimation methods — plug-in, MOM, and MLE — produce different outcomes[21].

[21] For another example, where the MLE estimate exists and differs substantially from the plug-in estimate, see Problem 8.4.

The last example illustrates that we may encounter somewhat unexpected difficulties when using the method of maximum likelihood. For instance, the MLE may not exist. Before we discuss the good properties of MLEs that have made maximum likelihood estimation one of the most popular parametric estimation methods, let's first outline several potential problems that can arise when finding MLEs.

8.5 Potential Issues with MLEs

All potential difficulties associated with finding MLEs are essentially the difficulties associated with a general problem of finding a global maximum of a function, which is a rather non-trivial task. First, as Example 8.4 illustrates,

1. The MLE *may not exist*. Notice that in Example 8.4 the nonexistence of MLE does not depend on the data: $\hat{\theta}_{MLE}$ does not exist for any $X_1,\ldots,X_n$. This means that this is the "internal" problem of the stochastic model. Usually this can be easily fixed by slightly adjusting the model. For instance, in Example 8.4 replacing $\mathscr{U}[0,\theta)$ with $\mathscr{U}[0,\theta]$ resolves the issue of nonexistence[22].

[22] Example 8.4 becomes Example 8.3.

2. The MLE *may not be unique*. This happens when the likelihood function has more than one global maximizer. This is schematically shown in Fig. 8.4. Non-uniqueness can be usually resolved by adding more data and recomputing the likelihood function.
3. In many applications, it is virtually *impossible to calculate* the MLE *analytically*. In this sense, Examples 8.1, 8.2, and 8.3 are exceptions, since the parametric models considered in these examples are simple, and the corresponding likelihood and log-likelihood functions have relatively simple expressions, which are

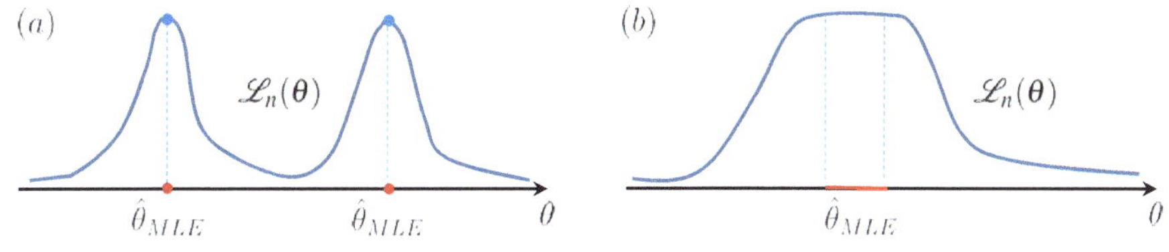

Fig. 8.4 Non-uniqueness of the MLE. In panel (a), there are two MLEs; in panel (b) there is a continuum of MLEs.

easy to maximize analytically. Most stochastic models used in real-world applications are more involved, have many parameters, and often contain Bernoulli, normal, uniform, and other standard probability distributions as their building blocks. The resulting (log-)likelihood functions are complex and are typically maximized by numerical or simulation-based optimization methods[23].

4. The likelihood function may have *many local maxima*. This is illustrated in Fig. 8.5. In this case, a maximizer found by an optimization method may be a local maximizer and, therefore, not the MLE. This issue is especially troublesome when θ is a high-dimensional vector of parameters and it is impossible to plot the graph of the likelihood function and visually inspect whether the found maximizer is the global one. A simple and popular strategy for dealing with this problem is to run the optimization algorithm several times independently, every time starting from a new point in the parameters space Θ, and choose the maximizer that corresponds to the largest value of the likelihood function[24].

[23] For example, see the Case Study in Section 8.7.

[24] This strategy, of course, does not guarantee that the found maximizer is indeed the MLE, but it sheds light on whether the likelihood has local maxima and increases the chances of finding the MLE.

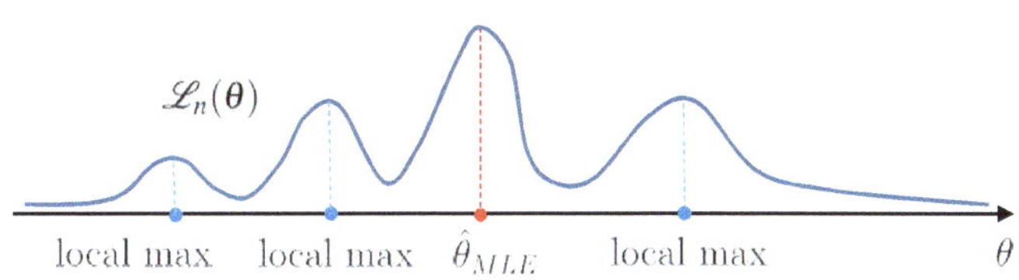

Fig. 8.5 A likelihood function with three local maxima and one global maximum that corresponds to the MLE.

5. The MLE may be *sensitive* to small perturbations in the data. This means that small changes in the data, $X_i \to \tilde{X}_i \approx X_i$, could lead to a substantial change in the value of the MLE: $\hat{\theta}_{MLE} \to \tilde{\hat{\theta}}_{MLE} \not\approx \hat{\theta}_{MLE}$. There are three scenarios when this could happen. First, if the parametric stochastic model is not adequate for describing the randomness in the data, then the likelihood function $\mathscr{L}_n(\theta)$ could be approximately flat over a relatively large domain $\Theta_0 \subset \Theta$ containing the MLE, $\hat{\theta}_{MLE} \in \Theta_0$, so that all values of $\theta \in \Theta_0$ are approximately equally non-consistent with the data[25]. In this case, small changes in the data resulting into a small change in the likelihood, $\mathscr{L}_n(\theta) \to \widetilde{\mathscr{L}_n}(\theta) \approx \mathscr{L}_n(\theta)$, could lead to a large change in the value of the MLE, see Fig. 8.6.

[25] All values of $\theta \in \Theta_0$ make the data equally unlikely to occur.

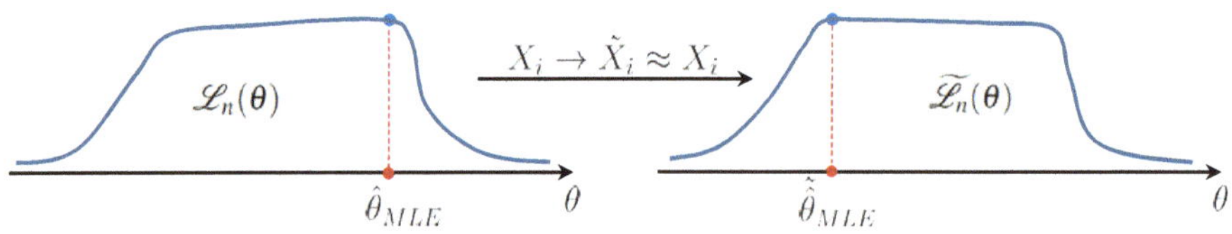

Fig. 8.6 Sensitivity of the MLE with respect to small perturbations in the data. Even small changes could make the elephant change his orientation inside the boa.

"My drawing was not of a hat. It was a boa constrictor digesting an elephant." – Antoine de Saint-Exupéry, *The Little Prince*.

Second, if stochastic model $\mathscr{F} = \{F(x|\theta), \theta \in \Theta\}$ is adequate for the data but not sensitive to the values of $\theta \in \Theta_0 \subset \Theta$, that is $F(x|\theta_1) \approx F(x|\theta_2)$ for any $\theta_1, \theta_2 \in \Theta_0$, then the likelihood function will be also approximately flat over Θ_0. If Θ_0 contains the MLE, then small perturbations in the data could lead to the behavior illustrated in Fig. 8.6, although in this case, all values of $\theta \in \Theta_0$ will be approximately equally well consistent with the data. In this scenario, high sensitivity of the MLE is a consequence of a bad parameterization of a good model[26]. Finally, the MLE may be sensitive to small perturbations in the data if the sample size n is small. In this case, the amount of data is not enough for the likelihood function to nail down the true value of the parameter accurately. This kind of sensitivity is demonstrated in [6] for the binomial model $X_1, \ldots, X_n \overset{\text{iid}}{\sim} \mathscr{B}(m, p)$, where the sample size $n = 5$, the success probability $p = 0.32$, and the number of trials m is the single model parameter. If $X_1 = 16, X_2 = 18, X_3 = 22, X_4 = 25, X_5 = 27$ (this data was generated from the model with $m = 75$), then $\hat{m}_{MLE} = 99$. It turns out that if we increase the last data point just by one, $X_5 = 28$, then the value of MLE will increase to $\hat{m}_{MLE} = 190$.

26 Models where the mapping $\theta \mapsto F(x|\theta)$ is *not one-to-one* are called *unidentifiable*.

After discussing the potential problems associated with MLEs, it is a good time to enhance our appreciation of and confidence in the method of maximum likelihood by focusing on the attractive properties of MLEs.

8.6 Good Properties of MLEs

Popularity of the method of maximum likelihood stems from the nice asymptotic properties of MLEs: if the parametric model $\mathscr{F} = \{F(x|\theta), \theta \in \Theta\}$ satisfies certain regularity conditions, then the MLE is consistent, asymptotically normal, asymptotically unbiased, asymptotically efficient, and equivariant. Let's discuss these properties in detail. For definiteness, we focus on continuous stochastic models, but the same results will hold for discrete models as well[27]. So, suppose that

$$X_1 \ldots, X_n \overset{\text{iid}}{\sim} f(x|\theta), \quad \theta \in \Theta, \tag{8.47}$$

is a correct model for the data and let $\theta_\star \in \Theta$ denote the true value[28] of θ.

27 Roughly speaking, in the subsequent formulas PDFs and integrals should be replaced with PMFs and sums.

28 See Marginal Notes 9 in Section 3.1.1 and 3 in Section 7.1.

8.6.1 Consistency

Consistency is a desirable and in some sense "expected" property of any reasonable estimate. As the method of moments estimate, the MLE is also a *consistent estimate*:

$$\hat{\theta}_{MLE} \xrightarrow{\mathbb{P}} \theta_\star, \qquad \text{as } n \to \infty. \tag{8.48}$$

Intuitively, this means that if the parametric model is specified correctly and we have a lot of data, then the MLE will estimate the true value of the parameter accurately: if $n \gg 1$, then $\hat{\theta}_{MLE} \approx \theta_\star$ with large probability.

The proof of consistency is based on the important concept of Kullback-Leibler distance introduced by Solomon Kullback and Richard Leibler in [7], which is used not only in statistical inference, but also in probability theory, machine learning, and information theory (both classical and quantum).

Definition 8.6. Let g and h be two probability density functions[29]. The *Kullback-Leibler (KL) distance* is a measure of how h is different from g, defined as follows:

$$KL(g,h) = \int g(x) \log \frac{g(x)}{h(x)} dx. \tag{8.49}$$

Remark 8.2. It can be shown that the KL distance is always non-negative, $KL(g,h) \geq 0$, and $KL(g,h) = 0$ if and only if $g(x) = h(x)$ almost everywhere[30]. But the KL distance is not symmetric, $KL(g,h) \neq KL(h,g)$, and it does not satisfy the triangle inequality. So, it is a "distance" is the statistical, not geometric, sense of the word. It is also called the KL divergence, relative entropy, and information gain.

Let's now prove that (8.48) holds and the MLE is indeed a consistent estimate[31]. To find the MLE, we need to maximize the log-likelihood function,

$$\ell_n(\theta) = \sum_{i=1}^{n} \log f(X_i \mid \theta) \to \max. \tag{8.50}$$

Since $X_1, \ldots, X_n$ are IID, random variables $\log f(X_1 \mid \theta), \ldots, \log f(X_n \mid \theta)$ are also IID[32]. Therefore, by the law of large numbers,

$$\frac{1}{n}\ell_n(\theta) = \frac{1}{n}\sum_{i=1}^{n} \log f(X_i \mid \theta) \xrightarrow{\mathbb{P}} \mathbb{E}\left[\log f(X \mid \theta)\right], \qquad X \sim f(x \mid \theta_\star). \tag{8.51}$$

The expected value is given by the integral

$$\mathbb{E}\left[\log f(X \mid \theta)\right] = \int f(x \mid \theta_\star) \log f(x \mid \theta) dx, \tag{8.52}$$

which looks almost like the KL distance between densities $f(x \mid \theta_\star)$ and $f(x \mid \theta)$. It is not quite the KL distance, but it is easy to cook it up:

[29] If g and h are probability mass functions that define discrete distributions, then the integral in definition (8.49) must be replaced with a sum.

[30] This means $g(x) = h(x)$ for all x except perhaps a "negligible" set that has zero probability (e.g. a set of finite number of points).

[31] To keep things intuitive, we will provide a "sketch of proof" rather than a formal rigorous proof. For a rigorous proof as well as all underlying assumptions, see [8].

[32] They may have a complicated distribution, but they are still independent and identically distributed.

$$
\begin{aligned}
\mathbb{E}[\log f(X\,|\,\theta)] &= \int f(x\,|\,\theta_\star)\log\left(\frac{f(x\,|\,\theta)}{f(x\,|\,\theta_\star)}f(x\,|\,\theta_\star)\right)dx \\
&= \int f(x\,|\,\theta_\star)\log\left(\frac{f(x\,|\,\theta)}{f(x\,|\,\theta_\star)}\right)dx + \underbrace{\int f(x\,|\,\theta_\star)\log f(x\,|\,\theta_\star)dx}_{\xi(\theta_\star),\text{ some function of }\theta_\star} \\
&= -\underbrace{\int f(x\,|\,\theta_\star)\log\left(\frac{f(x\,|\,\theta_\star)}{f(x\,|\,\theta)}\right)dx}_{KL(f(x\,|\,\theta_\star),f(x\,|\,\theta))} + \xi(\theta_\star) = -KL(\theta_\star,\theta)+\xi(\theta_\star),
\end{aligned}
\tag{8.53}
$$

where, for brevity, $KL(f(x\,|\,\theta_\star), f(x\,|\,\theta))$ is denoted by $KL(\theta_\star,\theta)$. So, by the LLN,

$$\frac{1}{n}\ell_n(\theta) \xrightarrow{\mathbb{P}} -KL(\theta_\star,\theta)+\xi(\theta_\star), \tag{8.54}$$

and, therefore, for large n

$$\ell_n(\theta) \approx -nKL(\theta_\star,\theta)+n\xi(\theta_\star). \tag{8.55}$$

In maximizing the log-likelihood with respect to θ, the second term $n\xi(\theta_\star)$ plays no role, since it does not depend on θ. It follows that maximization of $\ell_n(\theta)$ is equivalent to the minimization of the KL distance $KL(\theta_\star,\theta)$, which is always non-negative and equals to zero if and only if $f(x\,|\,\theta) = f(x\,|\,\theta_\star)$, that is if[33] $\theta = \theta_\star$. So,

$$\hat{\theta}_{MLE} = \arg\max_{\theta\in\Theta} \ell_n(\theta) \approx \arg\min_{\theta\in\Theta} KL(\theta_\star,\theta) = \theta_\star, \tag{8.56}$$

where the approximation holds for large n with large probability, gets more accurate as n increases, and becomes exact in the limit as $n\to\infty$. This means $\hat{\theta}_{MLE} \xrightarrow{\mathbb{P}} \theta_\star$ as $n\to\infty$ and, thus, $\hat{\theta}_{MLE}$ is consistent.

[33] Here we implicitly assume that model (8.47) is ***identifiable***: the mapping $\theta \mapsto f(x\,|\,\theta)$ is *one-to-one* and, therefore, $KL(\theta_1,\theta_2)=0 \Leftrightarrow \theta_1=\theta_2$. Compare with Marginal Note 26.

8.6.2 Asymptotic Normality

The consistency of the maximum likelihood estimate tells us that if the sample size is large, then the estimate will be close to the true value of the parameter, $\hat{\theta}_{MLE} \approx \theta_\star$, but it does not tell us anything about the limiting probability distribution of the MLE: how exactly is $\hat{\theta}_{MLE}$ distributed around $\theta_\star$ for $n \gg 1$? It turns out that, under certain regularity conditions, the MLE is *asymptotically normal*[34],

$$\frac{\hat{\theta}_{MLE}-\theta_\star}{\mathrm{se}[\hat{\theta}_{MLE}]} \xrightarrow{\mathbb{D}} \mathcal{N}(0,1), \qquad \text{as } n\to\infty. \tag{8.57}$$

Intuitively, this means that if n is large, then $\hat{\theta}_{MLE}$ is approximately normally distributed centered at the true value $\theta_\star$,

[34] For simplicity, in this section, we consider the case, where θ is a single parameter. But asymptotic normality of the MLE holds even when θ is a vector of parameters.

$$\hat{\theta}_{MLE} \stackrel{\cdot}{\sim} \mathcal{N}(\theta_\star, \mathrm{se}[\hat{\theta}_{MLE}]^2), \qquad \text{for } n \gg 1. \tag{8.58}$$

Moreover, the variance $\mathbb{V}[\hat{\theta}_{MLE}] = \mathrm{se}[\hat{\theta}_{MLE}]^2$ of the maximum likelihood estimate

$$n\mathbb{V}[\hat{\theta}_{MLE}] \to \frac{1}{\mathscr{I}(\theta_\star)}, \qquad \text{as } n \to \infty, \tag{8.59}$$

and, therefore, for large n can be approximated as

$$\mathbb{V}[\hat{\theta}_{MLE}] \approx \frac{1}{n\mathscr{I}(\theta_\star)}, \qquad \text{for } n \gg 1. \tag{8.60}$$

Here $\mathscr{I}(\theta_\star)$ is the Fisher information[35], a fundamental and important concept related to the KL distance, which often appears in different contexts in statistical inference, Bayesian statistics, information theory, and machine learning.

Definition 8.7. Let X be a random variable distributed according to PDF $f(x\,|\,\theta_\star)$, that belongs to a family of PDFs parameterized by $\theta \in \Theta$,

$$X \sim f(x\,|\,\theta_\star) \in \mathscr{F} = \{f(x\,|\,\theta),\ \ \theta \in \Theta\}. \tag{8.61}$$

Then the *Fisher information* associated with $X \sim f(x\,|\,\theta_\star)$ is[36]

$$\mathscr{I}(\theta_\star) = \mathbb{E}\left[\left(\left.\frac{\partial \log f(X\,|\,\theta)}{\partial \theta}\right|_{\theta=\theta_\star}\right)^2\right]. \tag{8.62}$$

Remark 8.3. It is important to stress that the Fisher information $\mathscr{I}(\theta_\star)$ is defined for a random variable whose distribution is a *member* of a smoothly parameterized *family* of distributions $\mathscr{F}$. It is not defined for a single random variable considered in isolation, since (8.62) assumes differentiation with respect to θ, and to compute this derivative we should be able to "vary" θ in the neighborhood of $\theta_\star$. So, without the family of distributions $\mathscr{F}$ the definition does not make sense. If X is a discrete random variable, then all PDFs in the above definition are replaced with PMFs. A discrete case is considered in Example 8.5.

Remark 8.4. It turns out that for the Fisher information to be meaningful and useful, the stochastic model $\mathscr{F}$ in Definition 8.7 must be *regular*, that is:

1. The support of $f(x\,|\,\theta)$, that is $\{x : f(x\,|\,\theta) \neq 0\}$, must not depend on $\theta \in \Theta$.
2. The derivative $\partial \log f(x\,|\,\theta)/\partial\theta$ must exist for all x and θ.
3. Integration with respect to x and differentiation with respect to θ must commute:

$$\int g(x)\frac{\partial}{\partial\theta} f(x\,|\,\theta)dx = \frac{\partial}{\partial\theta}\int g(x) f(x\,|\,\theta)dx. \tag{8.63}$$

Most stochastic models used in applications, such as normal, exponential, beta, Bernoulli, Poisson, and Cauchy models, are regular. An example of a non-regular

[35] "We systematically overestimate the value of access to information and underestimate the value of access to each other." – Clay Shirky.

[36] Don't be scared away by the dangerously looking expression in (8.62). We will figure out what it means.

model is the uniform model $\mathscr{U}[0,\theta]$, since the first property is violated: the support depends on θ. In what follows, we assume that stochastic model $\mathscr{F}$ is regular.

To derive an intuitive interpretation of the Fisher information, to understand what $\mathscr{I}(\theta_\star)$ actually measures, and what it tells us about the distribution $f(x\,|\,\theta_\star)$ within the family $\{f(x\,|\,\theta), \theta \in \Theta\}$, let's consider the main "ingredient" of (8.62), the partial derivative, which is called the *score function*:

$$s(x\,|\,\theta_\star) = \left.\frac{\partial \log f(x\,|\,\theta)}{\partial \theta}\right|_{\theta=\theta_\star} = \frac{\left.\frac{\partial f(x\,|\,\theta)}{\partial \theta}\right|_{\theta=\theta_\star}}{f(x\,|\,\theta_\star)}. \tag{8.64}$$

The numerator $\left.\frac{\partial f(x\,|\,\theta)}{\partial \theta}\right|_{\theta=\theta_\star}$ is the rate of change of PDF f at x as the value of θ varies around $\theta_\star$. The score function $s(x\,|\,\theta_\star)$ is, therefore, the *relative rate of change* of f at x as θ varies around $\theta_\star$, see Fig. 8.7.

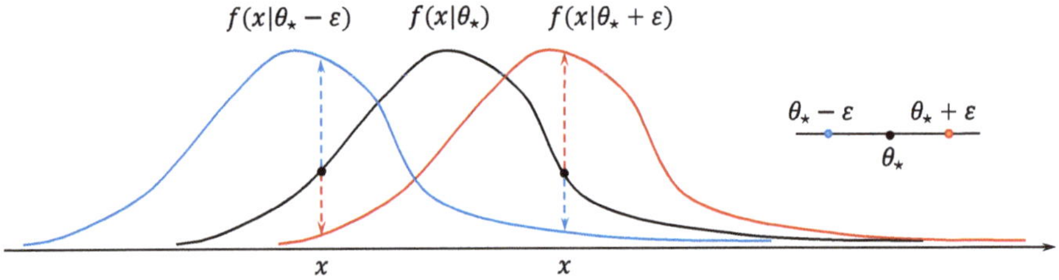

Fig. 8.7 The distribution $f(x\,|\,\theta_\star)$ (in the middle) corresponds to $\theta = \theta_\star$. An infinitesimal perturbation of the parameter $\theta_\star \to \theta_\star \pm \varepsilon$, $\varepsilon \approx 0$, results into a perturbation of the distribution $f(x\,|\,\theta_\star) \to f(x\,|\,\theta_\star \pm \varepsilon)$. For each x, the value of the score function $s(x\,|\,\theta_\star)$ quantifies how strongly and in what direction the value of f at x changes as a result of this perturbation.

Depending on the sign of the derivative $\left.\frac{\partial f(x\,|\,\theta)}{\partial \theta}\right|_{\theta=\theta_\star}$, for some values of x the relative rate of change is negative and for some x it is positive. It turns out that the *average* relative rate of change over all possible values of x is zero,

$$\mathbb{E}[s(X\,|\,\theta_\star)] = 0. \tag{8.65}$$

This should not be surprising. For any θ, $f(x\,|\,\theta)$ is a PDF, and therefore the area under the curve[37] equals to one for any θ. If f increases at some values of x as a result of perturbing θ, then it must decrease at some other values of x to keep the total area equal to one[38]. And so the relative rate of change of f should be zero on average. More formally,

[37] The integral of $f(x\,|\,\theta)$ with respect to x.

[38] See Fig. 8.7.

$$\begin{aligned}\mathbb{E}[s(X\,|\,\theta_\star)] &= \int s(x\,|\,\theta_\star)f(x\,|\,\theta_\star)dx = \int \frac{\left.\frac{\partial f(x\,|\,\theta)}{\partial \theta}\right|_{\theta=\theta_\star}}{f(x\,|\,\theta_\star)} f(x\,|\,\theta_\star)dx \\ &= \int \left.\frac{\partial f(x\,|\,\theta)}{\partial \theta}\right|_{\theta=\theta_\star} dx = \left.\frac{\partial}{\partial \theta}\right|_{\theta=\theta_\star} \underbrace{\int f(x\,|\,\theta)dx}_{1} = 0.\end{aligned} \tag{8.66}$$

This happens because positive and negative values of $s(x\,|\,\theta_\star)$, resulting from increases and decreases of f at x, balance each other and get canceled on average.

The Fisher information $\mathscr{I}(\theta_\star)$ is the average *squared* relative rate of change of PDF f as the value of θ varies around $\theta_\star$,

$$\mathscr{I}(\theta_\star) = \mathbb{E}\left[s(X\,|\,\theta_\star)^2\right]. \tag{8.67}$$

By squaring the score function, we ignore the sign of the relative rate of change and focus on the *magnitude* of change. So, the Fisher information $\mathscr{I}(\theta_\star)$ is a measure of how fast the entire distribution $f(x\,|\,\theta)$ changes as θ is perturbed around $\theta_\star$.

Is it good or bad to have large $\mathscr{I}(\theta_\star)$ for estimating θ whose true value is $\theta_\star$? If $\mathscr{I}(\theta_\star)$ is large, then $f(x\,|\,\theta)$ changes quickly when θ varies near $\theta_\star$. This means that $f(x\,|\,\theta_\star)$ is very different from its neighboring distributions $f(x\,|\,\theta)$, where $\theta \in (\theta_\star - \varepsilon, \theta_\star) \cup (\theta_\star, \theta_\star + \varepsilon)$, and therefore it will be easy to estimate $\theta_\star$ from the data.

So, the upshot of this discussion is the following: *the larger the Fisher information $\mathscr{I}(\theta_\star)$ is, the easier it is to estimate the true value $\theta_\star$ of θ from the data.* This conclusion is in full agreement with (8.60): the larger $\mathscr{I}(\theta_\star)$ is, the smaller is the variance of the MLE.

While definition (8.62) allowed us to derive an intuitive interpretation of the Fisher information, it is not convenient for computations. It can be shown[39] that the Fisher information can be equivalently written as follows:

[39] Problem 8.6 will ask you to do this.

$$\mathscr{I}(\theta_\star) = -\mathbb{E}\left[\left.\frac{\partial^2 \log f(X\,|\,\theta)}{\partial \theta^2}\right|_{\theta=\theta_\star}\right], \quad \text{where } X \sim f(x\,|\,\theta_\star). \tag{8.68}$$

Remark 8.5. This formula is not only more convenient for computation, but also provides a relationship between the Fisher information $\mathscr{I}(\theta_\star)$ and the KL distance $KL(\theta_\star, \theta)$ between distributions $f(x\,|\,\theta_\star)$ and $f(x\,|\,\theta)$. Let's compute the second derivative of $KL(\theta_\star, \theta)$ at $\theta = \theta_\star$:

$$\begin{aligned}\left.\frac{\partial^2}{\partial \theta^2}\right|_{\theta=\theta_\star} KL(\theta_\star, \theta) &= \left.\frac{\partial^2}{\partial \theta^2}\right|_{\theta=\theta_\star} \int f(x\,|\,\theta_\star) \log\left(\frac{f(x\,|\,\theta_\star)}{f(x\,|\,\theta)}\right) dx \\ &= \left.\frac{\partial^2}{\partial \theta^2}\right|_{\theta=\theta_\star} \int f(x\,|\,\theta_\star)(\log f(x\,|\,\theta_\star) - \log f(x\,|\,\theta))dx \\ &= -\int f(x\,|\,\theta_\star) \left.\frac{\partial^2}{\partial \theta^2}\right|_{\theta=\theta_\star} \log f(x\,|\,\theta)dx = \mathscr{I}(\theta_\star).\end{aligned} \tag{8.69}$$

To get a better feel for the Fisher information, let's consider a couple of concrete examples: one discrete and one continuous.

Example 8.5. Suppose that data is modeled with the Bernoulli model,

$$X_1,\dots,X_n \overset{\text{iid}}{\sim} \mathscr{B}(\theta), \quad \theta \in [0,1]. \tag{8.70}$$

Before we start any computations, let's try to predict the *qualitative behavior* of the Fisher information $\mathscr{I}(\theta_\star)$ as a function of the true value $\theta_\star \in [0,1]$. If $\theta_\star = 0$, then just one observation ($n=1$) is sufficient for estimating the true value exactly. Indeed, in this case $X_i = 0$ for all $i = 1,\dots,n$, and $\hat{\theta}_{MLE} = \overline{X} = 0 = \theta_\star$. Similarly, if $\theta_\star = 1$, then $\hat{\theta}_{MLE} = \theta_\star = 1$ for any sample size n. Therefore, if $\theta_\star$ is close 0 or 1, then it will be easy to estimate it accurately, and we should expect the Fisher information $\mathscr{I}(\theta_\star)$ to be large. On the other hand, if $\theta_\star = 1/2$, then we may need a lot of observations (large n) before the sample mean $\hat{\theta}_{MLE} = \overline{X}$ gets close to $1/2$ with large probability. This means that $\theta_\star = 1/2$ is relatively difficult to estimate accurately, and we should expect $\mathscr{I}(\theta_\star)$ to be small around $\theta_\star = 1/2$. Finally, thanks to the symmetry of the binary Bernoulli model[40], it should be equally easy/difficult to accurately estimate $\theta_\star$ and $1-\theta_\star$. In terms of the Fisher information, this means that $\mathscr{I}(\theta_\star)$ should be symmetric about $\theta_\star = 1/2$, that is $\mathscr{I}(\theta_\star) = \mathscr{I}(1-\theta_\star)$. The expected qualitative behavior of the Fisher information is shown in Fig. 8.8.

[40] If θ is replaced with $1-\theta$, then $X_i = 0$ essentially plays the role of $X_i = 1$ and vice versa.

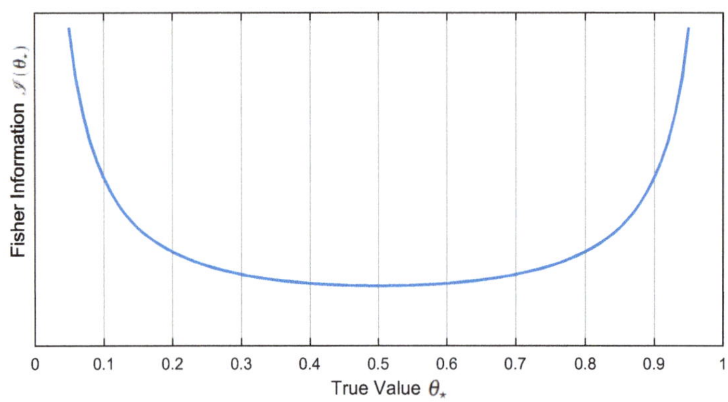

Fig. 8.8 The expected qualitative behavior of the Fisher information for the Bernoulli model.

Now let's do some computations with the hope that they will confirm our expectations. Since the Bernoulli PMF is $p(x|\theta) = \theta^x(1-\theta)^{1-x}$, where $x \in \{0,1\}$, its logarithm is

$$\log p(x|\theta) = x\log\theta + (1-x)\log(1-\theta). \tag{8.71}$$

To use (8.68), we need to compute the second derivative of $\log p(x|\theta)$ at $\theta = \theta_\star$:

$$\frac{\partial \log p(x|\theta)}{\partial\theta} = \frac{x}{\theta} - \frac{1-x}{1-\theta}, \qquad \left.\frac{\partial^2 \log p(x|\theta)}{\partial\theta^2}\right|_{\theta=\theta_\star} = -\frac{x}{\theta_\star^2} - \frac{1-x}{(1-\theta_\star)^2}. \tag{8.72}$$

So, the Fisher information is

$$\begin{aligned}\mathscr{I}(\theta_\star) &= -\mathbb{E}\left[\left.\frac{\partial^2 \log p(X\,|\,\theta)}{\partial\theta^2}\right|_{\theta=\theta_\star}\right] = \mathbb{E}\left[\frac{X}{\theta_\star^2} + \frac{1-X}{(1-\theta_\star)^2}\right] \\ &= \frac{1}{(1-\theta_\star)^2}\underbrace{\mathbb{P}(X=0)}_{1-\theta_\star} + \frac{1}{\theta_\star^2}\underbrace{\mathbb{P}(X=1)}_{\theta_\star} = \frac{1}{\theta_\star(1-\theta_\star)}.\end{aligned} \tag{8.73}$$

The derived expression fully agrees with the predicted qualitative behavior shown in Fig. 8.8: $\mathscr{I}(\theta_\star)$ becomes (infinitely) large as $\theta_\star \to 0$ or $\theta_\star \to 1$, it has a global minimum at $\theta_\star = 1/2$, and it is symmetric about $\theta_\star = 1/2$.

Example 8.6. Suppose that data is modeled with the normal model,

$$X_1,\ldots,X_n \overset{\text{iid}}{\sim} \mathscr{N}(\mu,\sigma^2), \quad \mu \in \mathbb{R}, \tag{8.74}$$

where σ^2 is known and μ is the single model parameter. As in the previous example, let's try to understand what kind of function $\mathscr{I}(\mu_\star)$ is before attempting any formal calculations. Given the variance σ^2, are there any values of $\mu_\star$ that would be easier or more difficult to estimate from the data compared to other values? It seems no, for as long as the width of the normal distribution is fixed, the sample mean $\hat{\mu}_{MLE} = \overline{X}$ should estimate any true value $\mu_\star$ with the same accuracy. This means that the Fisher information, as a function of $\mu_\star$, should be constant, $\mathscr{I}(\mu_\star) = c$. Furthermore, the smaller the variance σ^2, the less $X_1,\ldots,X_n$ are spread around $\mu_\star$, and the more accurately $\hat{\mu}_{MLE} = \overline{X}$ should estimate $\mu_\star$. Consequently, the constant (with respect to $\mu_\star$) c should be a decreasing function of σ^2.

Now let's perform the formal calculation. The logarithm of the normal PDF[41] is

[41] See (1.33).

$$\log\phi(x\,|\,\mu,\sigma^2) = -\log\sqrt{2\pi}\sigma - \frac{(x-\mu)^2}{2\sigma^2}. \tag{8.75}$$

The derivatives are given by

$$\frac{\partial \log\phi(x\,|\,\mu,\sigma^2)}{\partial\mu} = \frac{x-\mu}{\sigma^2}, \qquad \left.\frac{\partial^2 \log\phi(x\,|\,\mu,\sigma^2)}{\partial\mu^2}\right|_{\mu=\mu_\star} = -\frac{1}{\sigma^2}. \tag{8.76}$$

The Fisher information is thus indeed a constant

$$\mathscr{I}(\mu_\star) = \frac{1}{\sigma^2}, \tag{8.77}$$

that decreases as the variance σ^2 increases.

For a detailed discussion of the Fisher information and its applications to confidence intervals, hypothesis tests, Bayesian inference, and measuring model complexity, see an accessible and concrete tutorial article [9].

Let's now come back to the main topic of this section, the asymptotic normality of the maximum likelihood estimate,

$$\hat{\theta}_{MLE} \dot{\sim} \mathcal{N}\left(\theta_\star, \frac{1}{n\mathscr{I}(\theta_\star)}\right), \tag{8.78}$$

and provide a sketch of proof of this somewhat surprising and beautiful result[42].

Recall that $\hat{\theta}_{MLE}$ is the global maximizer of the log-likelihood function

$$\hat{\theta}_{MLE} = \arg\max_{\theta\in\Theta} \ell_n(\theta), \qquad \ell_n(\theta) = \sum_{i=1}^{n} \log f(X_i \,|\, \theta). \tag{8.79}$$

To get a relationship between $\hat{\theta}_{MLE}$ and $\theta_\star$, let's Taylor-expand the derivative of the log-likelihood $\ell_n'(\theta)$ about $\theta_\star$. Ignoring the higher order terms,

$$\ell_n'(\theta) \approx \ell_n'(\theta_\star) + (\theta - \theta_\star)\ell_n''(\theta_\star). \tag{8.80}$$

If we set $\theta = \hat{\theta}_{MLE}$ in the above approximation and note that $\ell_n'(\hat{\theta}_{MLE}) = 0$, since $\hat{\theta}_{MLE}$ is a critical point of $\ell_n(\theta)$, then we get

$$0 \approx \ell_n'(\theta_\star) + (\hat{\theta}_{MLE} - \theta_\star)\ell_n''(\theta_\star) \;\Rightarrow\; \hat{\theta}_{MLE} - \theta_\star \approx -\frac{\ell_n'(\theta_\star)}{\ell_n''(\theta_\star)} = -\frac{\frac{1}{n}\ell_n'(\theta_\star)}{\frac{1}{n}\ell_n''(\theta_\star)}. \tag{8.81}$$

Here we sprinkled the factor of $\frac{1}{n}$ in both the numerator and denominator in order to convert sums[43] into averages, so that we can use the central limit theorem and the law of large numbers[44]. To derive the asymptotic distribution of $\hat{\theta}_{MLE}$, let's use the CLT to find the asymptotic distribution of $\frac{1}{n}\ell_n'(\theta_\star)$ and use the LLN to find approximate value of $\frac{1}{n}\ell_n''(\theta_\star)$ for large n.

Using the central limit theorem, the definition of the score function (8.64), and the derived facts[45] that $\mathbb{E}[s(X\,|\,\theta_\star)] = 0$ and $\mathbb{E}\left[s(X\,|\,\theta_\star)^2\right] = \mathscr{I}(\theta_\star)$, we have:

$$\begin{aligned}
\frac{1}{n}\ell_n'(\theta_\star) &= \frac{1}{n}\sum_{i=1}^{n} \left.\frac{\partial \log f(X_i\,|\,\theta)}{\partial\theta}\right|_{\theta=\theta_\star} = \frac{1}{n}\sum_{i=1}^{n} s(X_i\,|\,\theta_\star) \\
&\dot{\sim} \mathcal{N}\Bigg(\underbrace{\mathbb{E}[s(X\,|\,\theta_\star)]}_{0}, \frac{\overbrace{\mathbb{V}[s(X\,|\,\theta_\star)]}^{\mathbb{E}[s(X\,|\,\theta_\star)^2]=\mathscr{I}(\theta_\star)}}{n}\Bigg) = \mathcal{N}\left(0, \frac{\mathscr{I}(\theta_\star)}{n}\right).
\end{aligned} \tag{8.82}$$

Using the law of large numbers and the equivalent definition of the Fisher information (8.68), we have:

$$\begin{aligned}
\frac{1}{n}\ell_n''(\theta_\star) &= \frac{1}{n}\sum_{i=1}^{n} \left.\frac{\partial^2 \log f(X_i\,|\,\theta)}{\partial^2\theta}\right|_{\theta=\theta_\star} \\
&\approx \mathbb{E}\left[\left.\frac{\partial^2 \log f(X\,|\,\theta)}{\partial^2\theta}\right|_{\theta=\theta_\star}\right] = -\mathscr{I}(\theta_\star).
\end{aligned} \tag{8.83}$$

Finally, combining (8.81), (8.82), and (8.83),

[42] Again, our goal here is to be as intuitive and conceptually clear as possible. For a rigorous proof and all regularity conditions required for asymptotic normality (in addition to those listed in Remark 8.4), see [10].

[43] Since ℓ_n is a sum (8.79), so are ℓ_n' and ℓ_n''.

[44] See Probability Refresher 2.12 for a brief review of CLT and LLN.

[45] See (8.65) and (8.67).

$$\hat{\theta}_{MLE} - \theta_\star \dot{\sim} \frac{\mathcal{N}\left(0, \frac{\mathcal{I}(\theta_\star)}{n}\right)}{\mathcal{I}(\theta_\star)} = \mathcal{N}\left(0, \frac{1}{n\mathcal{I}(\theta_\star)}\right), \tag{8.84}$$

which is equivalent to (8.78).

The asymptotic normality of the maximum likelihood estimate

$$\sqrt{n\mathcal{I}(\theta_\star)}(\hat{\theta}_{MLE} - \theta_\star) \xrightarrow{\mathbb{D}} \mathcal{N}(0,1), \quad \text{as } n \to \infty \tag{8.85}$$

is a beautiful theoretical result, but it is hard to use in applications since usually the Fisher information $\mathcal{I}(\theta_\star)$ is unknown, since the true value $\theta_\star$ is unknown[46]. It can be shown[47] that if we replace $\mathcal{I}(\theta_\star)$ with its "plug-in" estimate $\mathcal{I}(\hat{\theta}_{MLE})$, then the asymptotic normality will still hold:

$$\sqrt{n\mathcal{I}(\hat{\theta}_{MLE})}(\hat{\theta}_{MLE} - \theta_\star) \xrightarrow{\mathbb{D}} \mathcal{N}(0,1), \quad \text{as } n \to \infty. \tag{8.86}$$

[46] Unless $\mathcal{I}(\theta_\star)$ is a constant and does not depend on $\theta_\star$, as for example in the normal case (8.77).

[47] Using Slutzky's theorem and the continuous mapping theorem, for details see [11].

This result is more application-friendly and can be used for constructing *approximate confidence intervals* for the true value $\theta_\star$ and, as we will see in the next chapter, for testing hypotheses. If n is large, then the random variable on the left hand side of (8.86) is approximately distributed according to the standard normal distribution, and, therefore,

$$\mathbb{P}\left(z_{\frac{\alpha}{2}} < \sqrt{n\mathcal{I}(\hat{\theta}_{MLE})}(\hat{\theta}_{MLE} - \theta_\star) < -z_{\frac{\alpha}{2}}\right) \approx 1 - \alpha, \tag{8.87}$$

where $z_{\frac{\alpha}{2}}$ is the $(\frac{\alpha}{2})^{\text{th}}$ standard normal quantile, see Fig. 1.7. Rearranging the terms gives

$$\mathbb{P}\left(\hat{\theta}_{MLE} + \frac{z_{\frac{\alpha}{2}}}{\sqrt{n\mathcal{I}(\hat{\theta}_{MLE})}} < \theta_\star < \hat{\theta}_{MLE} - \frac{z_{\frac{\alpha}{2}}}{\sqrt{n\mathcal{I}(\hat{\theta}_{MLE})}}\right) \approx 1 - \alpha, \tag{8.88}$$

which means that

$$I_\alpha = \left(\hat{\theta}_{MLE} + \frac{z_{\frac{\alpha}{2}}}{\sqrt{n\mathcal{I}(\hat{\theta}_{MLE})}}, \hat{\theta}_{MLE} - \frac{z_{\frac{\alpha}{2}}}{\sqrt{n\mathcal{I}(\hat{\theta}_{MLE})}}\right) \tag{8.89}$$

is an approximate $100(1-\alpha)\%$ confidence interval for $\theta_\star$. This interval is called the ***MLE based confidence interval***.

Example 8.7. Suppose that data is modeled with the Bernoulli model,

$$X_1, \ldots, X_n \overset{\text{iid}}{\sim} \mathcal{B}(p), \quad p \in [0,1]. \tag{8.90}$$

Then $\hat{p}_{MLE} = \overline{X}$, the Fisher information[48] is $\mathcal{I}(p_\star) = 1/p_\star(1-p_\star)$, and the MLE based confidence interval is

[48] See Example 8.5.

$$I_\alpha = \left(\overline{X} + z_{\frac{\alpha}{2}} \sqrt{\frac{\overline{X}(1-\overline{X})}{n}},\ \overline{X} - z_{\frac{\alpha}{2}} \sqrt{\frac{\overline{X}(1-\overline{X})}{n}} \right). \tag{8.91}$$

Compare this confidence interval with intervals (3.22) and (3.27) derived from the Chebyshov's and Hoeffding's inequalities.

Example 8.8. Suppose that data is modeled with the normal model,

$$X_1,\ldots,X_n \overset{\text{iid}}{\sim} \mathscr{N}(\mu,\sigma^2), \quad \mu \in \mathbb{R}, \tag{8.92}$$

where σ^2 is known and μ is the single model parameter. Then $\hat{\mu}_{MLE} = \overline{X}$, the Fisher information[49] is $\mathscr{I}(\mu_\star) = 1/\sigma^2$, and the MLE based confidence interval is

[49] See Example 8.6.

$$I_\alpha = \left(\overline{X} + z_{\frac{\alpha}{2}} \frac{\sigma}{\sqrt{n}},\ \overline{X} - z_{\frac{\alpha}{2}} \frac{\sigma}{\sqrt{n}} \right). \tag{8.93}$$

Remark 8.6. In this section, we discussed the asymptotic normality of the MLE for the case when θ is a single model parameter. But this property also holds if $\theta = (\theta_1,\ldots,\theta_d)$ is a vector of model parameters. Namely, it can be shown [12] that

$$\sqrt{n\mathscr{I}(\theta_\star)}(\hat{\theta}_{MLE} - \theta_\star) \xrightarrow{\mathbb{D}} \mathscr{N}(0,\mathbb{I}_d), \qquad \text{as } n \to \infty, \tag{8.94}$$

where $\mathbb{I}_d$ is the $d \times d$ identity matrix, $\mathscr{N}(0,\mathbb{I}_d)$ is the d-dimensional standard normal distribution, and $\mathscr{I}(\theta_\star)$ is the *Fisher information matrix* of size $d \times d$ whose elements $\mathscr{I}_{ij}(\theta_\star)$ are defined, by analogy with (8.62) and (8.68), as

$$\begin{aligned} \mathscr{I}_{ij}(\theta_\star) &= \mathbb{E}\left[\left.\frac{\partial \log f(X\,|\,\theta)}{\partial \theta_i}\right|_{\theta=\theta_\star} \left.\frac{\partial \log f(X\,|\,\theta)}{\partial \theta_j}\right|_{\theta=\theta_\star} \right], \\ \mathscr{I}_{ij}(\theta_\star) &= -\mathbb{E}\left[\left.\frac{\partial^2 \log f(X\,|\,\theta)}{\partial \theta_i \partial \theta_j}\right|_{\theta=\theta_\star} \right]. \end{aligned} \tag{8.95}$$

The Fisher information matrix can be viewed as a Riemannian metric on the d-dimensional model parameter space and is used in *information geometry*, an interdisciplinary field that studies stochastic models by methods of differential geometry. But this is a different story [13].

8.6.3 Asymptotic Unbiasedness

A direct corollary of asymptotic normality is *asymptotic unbiasedness* of the MLE. As the sample size n increases, the distribution of $\hat{\theta}_{MLE}$ converges to the normal distribution with mean $\theta_\star$. Therefore, the expected value of $\hat{\theta}_{MLE}$ converges to the true value,

$$\mathbb{E}[\hat{\theta}_{MLE}] \to \theta_\star, \qquad \text{as } n \to \infty. \tag{8.96}$$

Remark 8.7. At first glance it may seem that asymptotic unbiasedness (8.96) also follows form consistency, $\hat{\theta}_{MLE} \xrightarrow{\mathbb{P}} \theta_\star$ as $n \to \infty$, but this is not true. Somewhat counterintuitively, convergence of a random variable in probability[50], $X_n \xrightarrow{\mathbb{P}} a$, does not imply $\mathbb{E}[X_n] \to a$. As a counterexample, let

[50] For convergence in probability, see Probability Refresher 2.12.

$$X_n = \begin{cases} 0, & \text{with probability } 1 - \frac{1}{n}, \\ n, & \text{with probability } \frac{1}{n}. \end{cases} \tag{8.97}$$

Then $X_n \xrightarrow{\mathbb{P}} 0$ and $n \to \infty$. Indeed, for any $\varepsilon > 0$,

$$\lim_{n\to\infty} \mathbb{P}(|X_n - 0| < \varepsilon) = \lim_{n\to\infty} \mathbb{P}(X_n = 0) = \lim_{n\to\infty} \left(1 - \frac{1}{n}\right) = 1. \tag{8.98}$$

On the other hand,

$$\mathbb{E}[X_n] = 0 \times \mathbb{P}(X_n = 0) + n \times \mathbb{P}(X_n = n) = 1, \qquad \text{for any } n. \tag{8.99}$$

Let's revisit examples 8.1, 8.2, and 8.3, and check that the derived maximum likelihood estimates are indeed asymptotically unbiased.

Example 8.9. Suppose that data is modeled with the Bernoulli model,

$$X_1, \ldots, X_n \overset{\text{iid}}{\sim} \mathscr{B}(\theta), \quad \theta \in [0,1], \tag{8.100}$$

and $\theta_\star$ is the true value of θ. The MLE of θ is $\hat{\theta}_{MLE} = \overline{X}$, and its expectation is

$$\mathbb{E}[\hat{\theta}_{MLE}] = \mathbb{E}[\overline{X}] = \frac{1}{n} \sum_{i=1}^{n} \underbrace{\mathbb{E}[X_i]}_{\theta_\star} = \theta_\star. \tag{8.101}$$

In this case, $\hat{\theta}_{MLE}$ is not just asymptotically unbiased, it is unbiased for any n[51].

[51] Exactly unbiased estimates are, of course, also asymptotically unbiased.

Example 8.10. Suppose that data is modeled with the normal model,

$$X_1, \ldots, X_n \overset{\text{iid}}{\sim} \mathscr{N}(\mu, \sigma^2), \quad \mu \in \mathbb{R}, \sigma^2 \geq 0, \tag{8.102}$$

and $\mu_\star$ and $\sigma_\star^2$ are the true values of model parameters. The MLEs are $\hat{\mu}_{MLE} = \overline{X}$ and $\hat{\sigma}^2_{MLE} = \frac{1}{n}\sum_{i=1}^{n}(X_i - \overline{X})^2$, and their expected values are[52]

[52] The expected value of $\hat{\sigma}^2_{MLE} = \frac{1}{n}\sum_{i=1}^{n}(X_i - \overline{X})^2$ is computed in (4.50).

$$\begin{aligned} \mathbb{E}[\hat{\mu}_{MLE}] &= \mathbb{E}[\overline{X}] = \frac{1}{n} \sum_{i=1}^{n} \underbrace{\mathbb{E}[X_i]}_{\mu_\star} = \theta_\star, \\ \mathbb{E}[\hat{\sigma}^2_{MLE}] &= \frac{n-1}{n}\sigma_\star^2 \to \sigma_\star^2, \qquad \text{as } n \to \infty. \end{aligned} \tag{8.103}$$

So, $\hat{\mu}_{MLE}$ is unbiased for any n, and $\hat{\sigma}^2_{MLE}$ is asymptotically unbiased.

Example 8.11. Suppose that data is modeled with the uniform model,

$$X_1,\dots,X_n \overset{\text{iid}}{\sim} \mathscr{U}[0,\theta], \tag{8.104}$$

and $\theta_\star$ is the true value of θ. The MLE of θ is $\hat{\theta}_{MLE} = X_{(n)}$, which is the largest value in the sample. Its expected value is computed in[53] (81):

$$\mathbb{E}[\hat{\theta}_{MLE}] = \mathbb{E}[X_{(n)}] = \frac{n}{n+1}\theta_\star \to \theta_\star, \qquad \text{as } n \to \infty. \tag{8.105}$$

In this example, the maximum likelihood estimate is not asymptotically normal[54] since the model $\mathscr{U}[0,\theta]$ is not regular[55], but it is still asymptotically unbiased.

[53] See also (5.50).

[54] The PDF of $X_{(n)}$ is derived in (80).

[55] See Remark 8.4.

8.6.4 Asymptotic Efficiency

Recall[56] that the overall accuracy of a point estimate $\hat{\theta}_n$ is usually quantified by the mean squared error (MSE), which, using the bias-variance decomposition (3.11), can be written as

$$\text{MSE}[\hat{\theta}_n] = \mathbb{B}[\hat{\theta}_n]^2 + \mathbb{V}[\hat{\theta}_n]. \tag{8.106}$$

Asymptotic unbiasedness of the MLE discussed in the previous section implies that the bias of the MLE converges to zero as the sample size increases,

$$\mathbb{B}[\hat{\theta}_{MLE}] = \mathbb{E}[\hat{\theta}_{MLE}] - \theta_\star \to 0, \qquad \text{as } n \to \infty. \tag{8.107}$$

Therefore, if n is large, then the MLE is approximately unbiased[57], i.e. its bias is approximately zero, $\mathbb{B}[\hat{\theta}_{MLE}] \approx 0$. This means that the largest contribution to the MSE of the MLE comes from the variance of the MLE. This observation leads to the following natural question: what is the *smallest* possible value of the *variance* of an unbiased estimate? The answer is given by the *Cramér-Rao inequality*, also called the *information inequality*.

Let $\hat{\theta}_n$ be any unbiased estimate of parameter θ constructed from the data $X_1,\dots,X_n \overset{\text{iid}}{\sim} f(x|\theta), \theta \in \Theta$, and let $\theta_\star$ be the true value of θ. Then, under the regularity conditions listed in Remark 8.4, we have[58],

$$\mathbb{V}[\hat{\theta}_n] \geq \frac{1}{n\mathscr{I}(\theta_\star)}. \tag{8.108}$$

Interestingly, the Cramér-Rao inequality from a conceptual point of view is a corollary of the Cauchy-Schwarz inequality, which is one of the most fundamental and widely used inequalities in mathematics. A sketch of proof of (8.108) is given in Appendix 8.9.

Remark 8.8. The Cramér-Rao inequality is named after Swedish statistician Harald Cramér and Indian-American statistician Calyampudi Radhakrishna Rao, who independently proved the inequality in 1946 and 1945, respectively. This is yet another manifestation of Arnold's principle[59], since the inequality was first proved by New Zealand statisticians Alexander Aitken and Harold Silverstone in 1942 [15].

[56] See Section 3.2.1.

[57] Or exactly unbiased in some special cases, as we saw in Examples 8.9 and 8.10.

[58] A rigorous proof of (8.108), and its generalizations are discussed in [14].

[59] See Marginal Note 47 in Chapter 1.

Moreover, the result was apparently obtained in 1939 in Silverstone's doctoral dissertation under Aitken's supervision.

The Cramér-Rao inequality provides a fundamental lower bound for the variance (and, therefore, for the mean squared error) of an unbiased estimate of a parameter. This lower bound is the best accuracy we could hope to achieve with an unbiased estimate constructed from an IID sample of size n. This motivates the following definition.

Definition 8.8. An unbiased estimate $\hat{\theta}_n$ is called *efficient* if it achieves the Cramér-Rao lower bound, that is if

$$\mathbb{V}[\hat{\theta}_n] = \frac{1}{n\mathscr{I}(\theta_\star)}. \tag{8.109}$$

Remark 8.9. The term "efficient" is overused in the statistical literature. In general, an estimate is efficient (or more efficient) if it is the best (or better) in some sense. For example, we often say that an estimate $\hat{\theta}_n$ is more efficient than another estimate $\tilde{\theta}_n$ if $\mathrm{MSE}[\hat{\theta}_n] < \mathrm{MSE}[\tilde{\theta}_n]$. In Definition 8.8, we consider only a class of unbiased estimates, and we say that an estimate it that class is efficient if it has the smallest possible variance, or equivalently, the smallest possible MSE. But this does not exclude a possibility of existing a biased estimate with the MSE less than in (8.109).

Note that the Cramér-Rao lower bound $1/n\mathscr{I}(\theta_\star)$ is exactly the asymptotic variance of the MLE (8.60). Combining this with the asymptotic unbiasedness of the MLE implies that the MLE is *asymptotically efficient*. Intuitively, this means that if n is large, then for well-behaved stochastic models, the MLE is approximately the most accurate estimate among all unbiased estimates,

$$\mathbb{B}[\hat{\theta}_{MLE}] \approx 0 \quad \text{and} \quad \mathbb{V}[\hat{\theta}_{MLE}] \approx \frac{1}{n\mathscr{I}(\theta_\star)}. \tag{8.110}$$

Let's illustrate this property with the Bernoulli and normal stochastic models.

Example 8.12. Suppose that data is modeled with the Bernoulli model,

$$X_1,\dots,X_n \overset{\text{iid}}{\sim} \mathscr{B}(\theta), \quad \theta \in [0,1], \tag{8.111}$$

and $\theta_\star$ is the true value of θ. The MLE $\hat{\theta}_{MLE} = \overline{X}$ is unbiased[60], the Fisher information[61] is $\mathscr{I}(\theta_\star) = 1/\theta_\star(1-\theta_\star)$, and the variance of the MLE is

[60] See Example 8.9.

[61] See Example 8.5.

$$\mathbb{V}[\hat{\theta}_{MLE}] = \frac{1}{n^2}\sum_{i=1}^{n}\mathbb{V}[X_i] = \frac{1}{n^2}\sum_{i=1}^{n}\theta_\star(1-\theta_\star) = \frac{1}{n\frac{1}{\theta_\star(1-\theta_\star)}} = \frac{1}{n\mathscr{I}(\theta_\star)}. \tag{8.112}$$

In this example, the Cramér-Rao lower bound is achieved exactly (not just asymptotically), and the MLE is efficient for any sample size n.

Example 8.13. Suppose that data is modeled with the normal model,

$$X_1,\dots,X_n \overset{\text{iid}}{\sim} \mathscr{N}(\mu,\sigma^2), \quad \mu \in \mathbb{R}, \tag{8.113}$$

where σ^2 is known, μ is the model parameter, and $\mu_\star$ is its the true value. The MLE $\hat{\mu}_{MLE} = \overline{X}$ is unbiased[62], the Fisher information[63] is $\mathscr{I}(\mu_\star) = 1/\sigma^2$, and the variance of the MLE is

[62] See Example 8.10.

[63] See Example 8.6.

$$\mathbb{V}[\hat{\mu}_{MLE}] = \frac{1}{n^2}\sum_{i=1}^{n} \mathbb{V}[X_i] = \frac{1}{n^2}\sum_{i=1}^{n} \sigma^2 = \frac{\sigma^2}{n} = \frac{1}{n\frac{1}{\sigma^2}} = \frac{1}{n\mathscr{I}(\mu_\star)}. \tag{8.114}$$

As in the previous example, $\hat{\mu}_{MLE}$ is efficient for any sample size n.

In general, the MLE may not be efficient for finite n. For example, the MLE of the variance in the normal model, $\hat{\sigma}^2_{MLE} = \frac{1}{n}\sum_{i=1}^{n}(X_i - \overline{X})^2$, is biased for any n, and therefore, can't be efficient by definition. Moreover, even if the MLE is efficient, there may still exist a slightly biased estimate with a smaller mean squared error.

8.6.5 Equivariance

Equivariance is a nice non-asymptotic[64] "bonus" to the asymptotic properties discussed in the previous sections. Suppose we are interested in estimating ψ, which is a function of the model parameter θ, that is $\psi = g(\theta)$. It turns out that if $\hat{\theta}_{MLE}$ is the MLE of θ, then the MLE of ψ is simply

[64] Valid for any n.

$$\hat{\psi}_{MLE} = g(\hat{\theta}_{MLE}). \tag{8.115}$$

In other words, maximum likelihood estimation commutes with any function,

$$\widehat{g(\theta)}_{MLE} = g(\hat{\theta}_{MLE}), \tag{8.116}$$

which can be represented by the following commutative diagram:

$$\begin{array}{ccc} \theta & \xrightarrow{\;g\;} & \psi \\ {\scriptstyle MLE}\downarrow & & \downarrow{\scriptstyle MLE} \\ \hat{\theta}_{MLE} & \xrightarrow{\;g\;} & \hat{\psi}_{MLE} \end{array} \tag{8.117}$$

This property is called *equivariance* or *functional invariance*[65] of the MLE.

[65] Although "functional commutativity" would be a better name for this property.

Example 8.14. Let's revisit Example 7.1, where

$$X_1, \ldots, X_n \overset{\text{iid}}{\sim} \mathscr{N}(\mu, \sigma^2) \tag{8.118}$$

are the results of a measles antibody test on n patients, and the quantity of interest is the proportion π of the target population whose test result is larger than a certain critical threshold α. Then[66],

[66] See (7.5) for derivation.

$$\pi = 1 - \Phi\left(\frac{\alpha - \mu}{\sigma}\right), \tag{8.119}$$

and, thanks to equivariance, the MLE of π is

$$\hat{\pi}_{MLE} = 1 - \Phi\left(\frac{\alpha - \hat{\mu}_{MLE}}{\hat{\sigma}_{MLE}}\right) = 1 - \Phi\left(\frac{\alpha - \overline{X}}{\sqrt{\frac{1}{n}\sum_{i=1}^{n}(X_i - \overline{X})^2}}\right). \tag{8.120}$$

The equivariance property (8.115) holds for any function $g : \Theta \to \Psi$, where Θ is the original model parameter space and Ψ is the set of possible values of $\psi = g(\theta)$. If g is a one-to-one correspondence, then equivariance is relatively easy to prove[67]. In this case, there exists the inverse function $g^{-1} : \Psi \to \Theta$, $\theta = g^{-1}(\psi)$, and we can re-parameterize the stochastic model $\mathscr{F}$ using ψ instead of θ:

[67] A proof for a general case is given in [14].

$$\begin{aligned} f(x\,|\,\theta) &= f(x\,|\,g^{-1}(\psi)) = \tilde{f}(x\,|\,\psi), \\ \mathscr{F} &= \{f(x\,|\,\theta),\ \ \theta \in \Theta\} = \{\tilde{f}(x\,|\,\psi),\ \ \psi \in \Psi\}. \end{aligned} \tag{8.121}$$

Let $\widetilde{\mathscr{L}_n}$ be the likelihood function in the ψ-parameterization. Then, for any $\theta \in \Theta$

$$\widetilde{\mathscr{L}_n}(g(\theta)) = \widetilde{\mathscr{L}_n}(\psi) = \prod_{i=1}^{n} \tilde{f}(X_i\,|\,\psi) = \prod_{i=1}^{n} f(X_i\,|\,\theta) = \mathscr{L}_n(\theta). \tag{8.122}$$

Therefore, for any $\psi \in \Psi$,

$$\widetilde{\mathscr{L}_n}(\psi) = \mathscr{L}_n(\theta) \le \mathscr{L}_n(\hat{\theta}_{MLE}) = \widetilde{\mathscr{L}_n}(g(\hat{\theta}_{MLE})), \tag{8.123}$$

which means that $g(\hat{\theta}_{MLE})$ is the global maximizer of the likelihood $\widetilde{\mathscr{L}_n}(\psi)$ and hence is the MLE of ψ.

8.7 Case Study: Geometric Preferential Attachment

Maybe, maybe... But all these curved spaces smell of mathematical mysticism.

– Arkady and Boris Strugatsky, *The Land of Crimson Clouds*

A network is a collection of points, called nodes or vertices, some of which are joined by lines, called edges or links. Nodes represent "things" and links represent certain relationships between them. Networks are often used as models of complex systems that capture the pattern of connection between system components. Complex systems that can be represented as networks are everywhere, they are tightly woven into the fabric of the modern world, form a skeleton of our civilization, and guide our daily lives. Examples include technological networks (transportation networks, power grids, the Internet), social networks (friendship networks, online social networks, financial networks), information networks (the world wide web, recommendation systems, citation networks), and biological networks (neural networks, metabolic networks, food webs). Complex networks, their structure and dynamics, and processes on complex networks are studied in *network science*, an emerging interdisciplinary field at the intersection of mathematics, statistics, and computer science [16].

In this Case Study, we will consider a stochastic model for growing complex networks, called the *geometric preferential attachment* (GPA) model [17][68], and will see how one of its parameters can be estimated from the network data by the method of maximum likelihood.

The basic idea underlying the GPA model is that when a node joins a network it is likely to connect to "popular" nodes[69] as well as to "similar" nodes[70]. The time $t \in \mathbb{N}$ a node joins the network is used as a proxy for node's popularity: all other things being equal, the older the node (the smaller t), the more popular it is. The similarity feature of node t is modeled by a random variable ψ_t distributed over a unit circle that represents the "similarity space." While ψ_t itself does not have any physical meaning, the angular distance $\Delta\psi_{st} = \pi - |\pi - |\psi_s - \psi_t||$ between ψ_s and ψ_t quantifies the similarity between two nodes. Upon joining the network, a new node t connects to an already existing node $s < t$ if s is both popular and similar to t, that is if $s^\alpha \Delta\psi_{st}$ is sufficiently small, where $\alpha \in [0,1]$ is a model parameter that controls the relative importance of popularity and similarity. The network connections are therefore formed by a trade-off between popularity and similarity.

[68] The author could not resist the temptation to base this case study on his own research.

[69] "The rich gets risher."

[70] "Birds of feather flock together."

The described mechanism for link formation allows a geometric interpretation which is very useful for theoretical (and philosophical) analysis of the GPA model. Let $r_s(t)$ denote the radial coordinate of node s at time $t > s$, and let's define

$$r_s(t) = \alpha \log s + (1-\alpha)\log t, \qquad \alpha \in [0,1]. \tag{8.124}$$

Then at time t, when node t joins the network, nodes $s < t$ and t can be represented by points on the plane with polar coordinates $(r_s(t), \psi_s)$ and $(r_t(t) = \log t, \psi_t)$, respectively. It turns out that if we interpret the plane not as the usual flat Euclidean plane, but as the *curved hyperbolic plane* [18], then the hyperbolic distance d_{st} between nodes s and t is approximately

$$\begin{aligned} d_{st} &= r_s(t) + r_t(t) + \log\frac{\Delta\psi_{st}}{2} \\ &= \alpha\log s + (1-\alpha)\log t + \log t + \log\frac{\Delta\psi_{st}}{2} \\ &= (2-\alpha)\log t + \log\frac{s^\alpha \Delta\psi_{st}}{2}. \end{aligned} \tag{8.125}$$

Therefore, for a given t, the minimization of $s^\alpha \Delta\psi_{st}$ with respect to s is equivalent to the minimization of d_{st}. This means that new nodes simply connect to the hyperbolically closest existing nodes.

How does a new node t find its position ψ_t on the unit circle (similarity space)? The GPA models assumes that the similarity space has "hot" regions, and that the hotter the region, the more attractive it is for a new node. The hotness of a region is quantified by the node density in that region. So, the higher the node density in a particular region of the unit circle, the higher the probability that a new node is placed in that region. This general idea leads to the following algorithm for generating a growing network[71].

[71] For more technical details and intuition behind the GPA model, see [17].

<u>GPA Model</u>

0. Initially the network is empty. New nodes t appear one a time $t = 1, 2, \ldots$. For each t:
1. The angular coordinate $\psi_t \in [0, 2\pi]$ of a new node t is defined as follows:
   (a) Sample $\phi_1, \ldots, \phi_t \stackrel{\text{iid}}{\sim} \mathcal{U}[0, 2\pi]$ uniformly at random. Points with polar coordinates $(r_t(t), \phi_1), \ldots, (r_t(t), \phi_t)$, where $r_t(t) = \log t$, are the "candidate" positions for newborn node t in the hyperbolic plane.
   (b) Define the attractiveness $A_t(\phi_i)$ of the ith candidate $(r_t(t), \phi_i)$ as the number of existing nodes that lie within hyperbolic distance $r_t(t)$ from it.
   (c) Set $\psi_t = \phi_i$ with probability

$$p_t(\phi_i \mid \phi_1, \ldots, \phi_{i-1}, \phi_{i+1}, \ldots, \phi_t) = \frac{A_t(\phi_i) + \lambda}{\sum_{j=1}^{t} (A_t(\phi_j) + \lambda)}, \tag{8.126}$$

   where $\lambda \geq 0$ is a parameter, called the *initial attractiveness*.
2. The radial coordinate of node t is set to $r_t(t) = \log t$. The radial coordinates of all existing nodes $s < t$ are updated to $r_s(t) = \alpha \log s + (1 - \alpha) \log t$.
3. Node t connects to m hyperbolically closest existing nodes (if $t \leq m$, then t connects to all existing nodes).

The GPA model has three parameters: the number of links m established by every new node upon its birth, the speed of radial expansion α, and the initial attractiveness λ. Model parameter m controls the average node degree of the network, which equals to $2m$. The GPA model generates scale-free networks (i.e. with the power-law degree distribution), and α determines the power-law exponent. Finally, λ controls the heterogeneity of the distribution of angular coordinates: the heterogeneity is a decreasing function of λ, and as $\lambda \to \infty$, the distribution converges to the uniform distribution on the circle. Figure 8.9 shows three networks generated by the GPA model with $m = 3$, $\alpha = 2/3$, and $\lambda = 0.1, 1$, and 10.

In this Case Study, our focus is on the estimation of initial attractiveness λ from the network data, i.e. from the list of n nodes and links between them. While the other two model parameters, m and α, can be estimated directly form the network average degree and the degree distribution, estimating λ is more involved and based on the method of maximum likelihood. The initial attractiveness λ controls the distribution of angular coordinates $\psi_1, \ldots, \psi_n$ of the nodes. The HyperMap method developed in [19] allows to infer the node coordinates $(r_1, \psi_1), \ldots, (r_n, \psi_n)$ in the hyperbolic plane from the network data[72]. So, given the inferred node coordinates, the likelihood function is

[72] Also using maximum likelihood estimation.

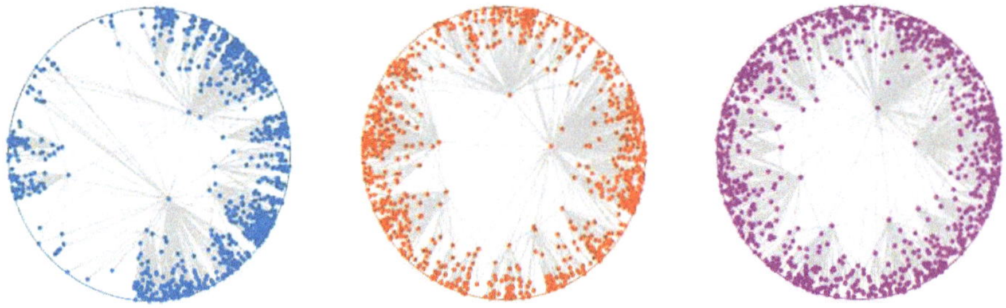

Fig. 8.9 Complex networks with $n = 10^3$ nodes generated according to the GPA model with $m = 3$, $\alpha = 2/3$, and $\lambda = 0.1, 1$, and 10. The smaller the value of λ, the more heterogeneous the distribution of angular coordinates $\psi_1, \ldots, \psi_n$ of nodes.

$$\mathscr{L}_n(\lambda) = f(\psi_1, \ldots, \psi_n \,|\, \lambda), \tag{8.127}$$

where $f(\psi_1, \ldots, \psi_n \,|\, \lambda)$ is the join density of the angular coordinates. Note that the angular coordinates are not IID, and the joint density cannot be factorized into the product of marginal densities, as it was the case in all previous examples. That is $f(\psi_1, \ldots, \psi_n \,|\, \lambda) \neq \prod_{i=1}^n f(\psi_i \,|\, \lambda)$. Instead, using conditioning,

$$\mathscr{L}_n(\lambda) = f(\psi_1 \,|\, \lambda) \times f(\psi_2 \,|\, \lambda, \psi_1) \times \ldots \times f(\psi_n \,|\, \lambda, \psi_1, \ldots, \psi_{n-1}), \tag{8.128}$$

where $f(\psi_1 \,|\, \lambda)$ is the density of the angular coordinate ψ_1 of the first node,

$$f(\psi_1 \,|\, \lambda) = \frac{1}{2\pi}, \qquad \psi_1 \in [0, 2\pi], \tag{8.129}$$

since $\psi_1 \sim \mathscr{U}[0, 2\pi]$ is uniformly distributed on the circle (see Step 1 of the GPA algorithm); $f(\psi_2 \,|\, \lambda, \psi_1)$ is the conditional density of the angular coordinate ψ_2 of the second node,

$$\begin{aligned} f(\psi_2 \,|\, \lambda, \psi_1) &\propto \int_0^{2\pi} f(\phi_1) p_2(\psi_2 \,|\, \phi_1) d\phi_1 \\ &= \frac{1}{2\pi} \int_0^{2\pi} \frac{A_2(\psi_2) + \lambda}{A_2(\psi_2) + A_2(\phi_1) + 2\lambda} d\phi_1, \end{aligned} \tag{8.130}$$

where $f(\phi_1) = 1/2\pi$ is the density of $\phi_1 \sim \mathscr{U}[0, 2\pi]$ and $p_2(\psi_2 \,|\, \phi_1)$ is the probability in (8.126) with $t = 2$; and, finally, $f(\psi_n \,|\, \lambda, \psi_1, \ldots, \psi_{n-1})$ is the conditional density of the angular coordinate ψ_n of the last n^{th} node[73],

$$\begin{aligned} &f(\psi_n \,|\, \lambda, \psi_1, \ldots, \psi_{n-1}) \\ &\propto \int_0^{2\pi} \cdots \int_0^{2\pi} f(\phi_1) \ldots f(\phi_{n-1}) p_n(\psi_n \,|\, \phi_1, \ldots, \phi_{n-1}) d\phi_1 \ldots d\phi_{n-1} \\ &= \frac{1}{(2\pi)^{n-1}} \int_0^{2\pi} \cdots \int_0^{2\pi} \frac{A_n(\psi_n) + \lambda}{A_n(\psi_n) + \sum_{j=1}^{n-1} A_n(\phi_j) + n\lambda} d\phi_1 \ldots d\phi_{n-1}. \end{aligned} \tag{8.131}$$

[73] "It gave him the greatest joy to watch lines of formulas emerge from his pen — almost always very complex, yet usually elegant, beautiful, and full of deep meaning." – Arkady and Boris Strugatsky, *The Land of Crimson Clouds*.

The likelihood function is thus proportional to

$$\mathscr{L}_n(\lambda) \propto \int_0^{2\pi} \frac{(A_2(\psi_2)+\lambda)d\phi_1}{A_2(\psi_2)+A_2(\phi_1)+2\lambda} \times \ldots \times \int_0^{2\pi} \ldots \int_0^{2\pi} \frac{(A_n(\psi_n)+\lambda)d\phi_1 \ldots d\phi_{n-1}}{A_n(\psi_n)+\sum_{j=1}^{n-1} A_n(\phi_j)+n\lambda}. \tag{8.132}$$

The log-likelihood is therefore (up to an additive constant)

$$\ell_n(\lambda) = \sum_{t=2}^{n} \log \int_0^{2\pi} \ldots \int_0^{2\pi} \frac{(A_t(\psi_t)+\lambda)d\phi_1 \ldots d\phi_{t-1}}{A_t(\psi_t)+\sum_{j=1}^{t-1} A_n(\phi_j)+t\lambda}. \tag{8.133}$$

Multiple integrals in (8.133) cannot be calculated analytically, since the attractiveness function A_t cannot be written in closed from. Nevertheless, the log-likelihood function $\ell_n(\lambda)$ can be accurately and efficiently estimated by the Monte Carlo method (see [17] for details). The MLE of λ can then be found by discretizing an interval $[0, \lambda_{\max}]$ for some large $\lambda_{\max}$, computing the the log-likelihood $\ell_n(\lambda)$ at all discretization points, and then finding the point with the largest value of $\ell_n(\lambda)$. Figure 8.10 shows six log-likelihood functions computed from six networks generated by the GPA model using six different values of the initial attractiveness λ. In all cases, the global maximizer of the log-likelihood $\hat{\lambda}_{MLE}$ coincides with the true value of λ with high accuracy. Furthermore, the smaller the true value of λ, the more peaked the log-likelihood is around $\hat{\lambda}_{MLE}$, and the easier it is to estimate the MLE accurately[74].

[74] The initial attractiveness is expected to be small in real complex networks, since small values of λ correspond to heterogeneous distributions of angular node coordinates, which lead to the formation of node communities (see the left network in Fig. 8.9), whose presence is one of the universal properties of real networks [16]. For example, it is estimated that the initial attractiveness for the Internet is $\lambda_{Int} = 0.7$ [17].

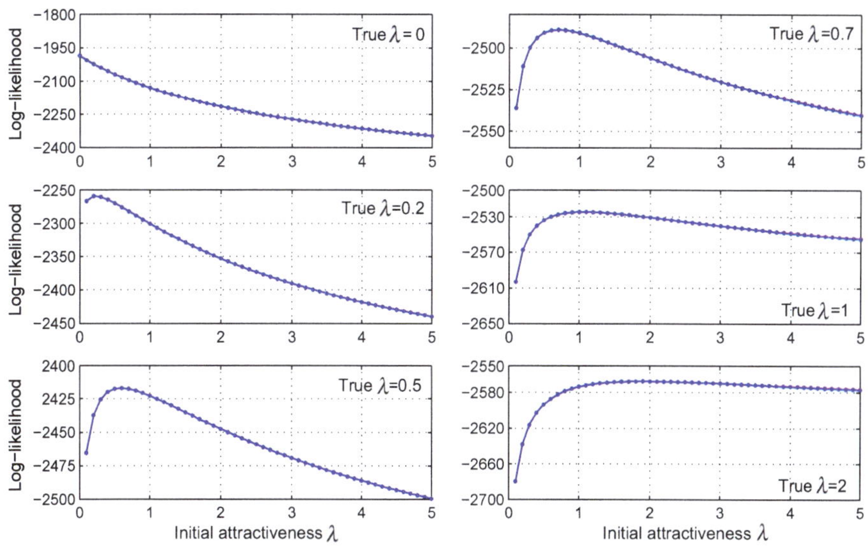

Fig. 8.10 The estimated log-likelihood functions (8.133) for networks of size $n = 10^3$ generated by the GPA model with $\lambda = 0, 0.2, 0.5, 0.7, 1$, and 2.

8.8 Further Reading

In this chapter, we learned one of the most popular and widely used parametric estimation methods, the method of maximum likelihood. The basic idea behind this method is to find the value of the model parameter which is as consistent with the observed data as possible. This idea leads to an optimization problem, where the goal is to find the global maximizer of the likelihood function, which is called the MLE (maximum likelihood estimate). For simple stochastic models, this maximization problem can be solved analytically and a closed-form expression for the MLE can be derived[75]. For more complex models, however, maximizing the likelihood function is more involved and often requires certain numerical or simulation methods. We discussed both potential difficulties associated with likelihood maximization and attractive asymptotic properties of MLEs that contribute significantly to the popularity of the method. The method of maximum likelihood is often used in machine learning algorithms and its applications can be found in all data-related disciplines. A tutorial exposition of maximum likelihood estimation with a good conceptual explanation of the method and instructive examples is provided in

[75] But even for simple stochastic models, derivations can be quite involved.

- Myung, J.: Tutorial on maximum likelihood estimation. *Journal of Mathematical Psychology* 47, 90-100 (2003). `https://doi.org/10.1016/S0022-2496(02)00028-7`

8.9 Appendix: The Cramér-Rao Inequlity

A starting point in the proof of the Cramér-Rao inequlity (8.108) is the fundamental Cauchy-Schwarz inequality[76], which states that for any two vectors u and v of an inner product space V

$$|\langle u,v\rangle|^2 \leq \langle u,u\rangle \cdot \langle v,v\rangle, \tag{8.134}$$

where $\langle\cdot,\cdot\rangle$ is the inner product defined on V. If V is the vector space of random variables equipped with the inner product

[76] The Cauchy-Schwarz inequality is one of the most fundamental inequalities, appearing in various contexts and with applications throughout all of mathematics [20].

$$\langle X,Y\rangle = \mathbb{E}[XY], \tag{8.135}$$

then the Cauchy-Schwarz inequality becomes

$$|\mathbb{E}[XY]|^2 \leq \mathbb{E}[X^2]\mathbb{E}[Y^2]. \tag{8.136}$$

As a corollary, the covariance between any random variables X and Y satisfies the so-called *covariance inequality*:

$$\begin{aligned} |\mathrm{Cov}(X,Y)|^2 &= |\mathbb{E}[(X-\mathbb{E}[X])(Y-\mathbb{E}[Y])]|^2 \\ &\leq \mathbb{E}\left[(X-\mathbb{E}[X])^2\right]\mathbb{E}\left[(Y-\mathbb{E}[Y])^2\right] = \mathbb{V}[X]\mathbb{V}[Y]. \end{aligned} \tag{8.137}$$

Remark 8.10. The covariance inequality implies, in particular, that Pearson's correlation coefficient (5.42) is always between -1 and $+1$, $|\rho_{XY}| \leq 1$.

To prove the Cramér-Rao inequlity, let's apply the covariance inequality to the following two random variables: $\hat{\theta}_n(X_1,\dots,X_n)$ and $\sum_{i=1}^n s(X_i\,|\,\theta_\star)$, where s is the score function (8.64),

$$\left|\mathrm{Cov}\left(\hat{\theta}_n, \sum_{i=1}^n s(X_i\,|\,\theta_\star)\right)\right|^2 \leq \mathbb{V}[\hat{\theta}_n]\mathbb{V}\left[\sum_{i=1}^n s(X_i\,|\,\theta_\star)\right] \tag{8.138}$$

Since $X_1,\dots,X_n$ are IID, so are random variables $s(X_1\,|\,\theta_\star),\dots,s(X_n\,|\,\theta_\star)$. Therefore,

$$\begin{aligned}\mathbb{V}\left[\sum_{i=1}^n s(X_i\,|\,\theta_\star)\right] &= \sum_{i=1}^n \mathbb{V}[s(X_i\,|\,\theta_\star)] \\ &= \sum_{i=1}^n \Big(\underbrace{\mathbb{E}[s(X_i\,|\,\theta_\star)^2]}_{\mathscr{I}(\theta_\star),\text{ see (8.67)}} - \underbrace{\mathbb{E}[s(X_i\,|\,\theta_\star)]^2}_{0,\text{ see (8.65)}}\Big) = n\mathscr{I}(\theta_\star).\end{aligned} \tag{8.139}$$

The covariance is[77] (buckle up!)

$$\begin{aligned}
\mathrm{Cov}\left(\hat{\theta}_n, \sum_{i=1}^n s(X_i\,|\,\theta_\star)\right) &= \mathbb{E}\left[\hat{\theta}_n \sum_{i=1}^n s(X_i\,|\,\theta_\star)\right] - \mathbb{E}\left[\hat{\theta}_n\right]\mathbb{E}\left[\sum_{i=1}^n s(X_i\,|\,\theta_\star)\right] \\
&= \mathbb{E}\left[\hat{\theta}_n \sum_{i=1}^n s(X_i\,|\,\theta_\star)\right] - \mathbb{E}\left[\hat{\theta}_n\right]\sum_{i=1}^n \underbrace{\mathbb{E}[s(X_i\,|\,\theta_\star)]}_{0,\text{ see (8.65)}} = \mathbb{E}\left[\hat{\theta}_n \sum_{i=1}^n s(X_i\,|\,\theta_\star)\right] \\
&= \int\cdots\int \hat{\theta}_n(x_1,\dots,x_n)\sum_{i=1}^n s(x_i\,|\,\theta_\star)\;\underbrace{\prod_{j=1}^n f(x_j\,|\,\theta_\star)}_{\text{joint PDF of }X_1,\dots,X_n}\;dx_1\dots dx_n \\
&= \int\cdots\int \hat{\theta}_n(x_1,\dots,x_n)\sum_{i=1}^n \frac{\left.\frac{\partial f(x_i\,|\,\theta)}{\partial\theta}\right|_{\theta=\theta_\star}}{f(x_i\,|\,\theta_\star)}\prod_{j=1}^n f(x_j\,|\,\theta_\star)dx_1\dots dx_n \\
&= \int\cdots\int \hat{\theta}_n(x_1,\dots,x_n)\sum_{i=1}^n \left.\frac{\partial f(x_i\,|\,\theta)}{\partial\theta}\right|_{\theta=\theta_\star}\prod_{j=1, j\neq i}^n f(x_j\,|\,\theta_\star)dx_1\dots dx_n \\
&= \int\cdots\int \hat{\theta}_n(x_1,\dots,x_n)\left.\frac{\partial}{\partial\theta}\right|_{\theta=\theta_\star}\prod_{j=1}^n f(x_j\,|\,\theta)dx_1\dots dx_n \\
&= \left.\frac{\partial}{\partial\theta}\right|_{\theta=\theta_\star}\int\cdots\int \hat{\theta}_n(x_1,\dots,x_n)\prod_{j=1}^n f(x_j\,|\,\theta)dx_1\dots dx_n \\
&= \left.\frac{\partial}{\partial\theta}\right|_{\theta=\theta_\star}\mathbb{E}[\hat{\theta}_n] = \left.\frac{\partial}{\partial\theta}\right|_{\theta=\theta_\star}\theta = 1.
\end{aligned} \tag{8.140}$$

[77] Recall that $\mathrm{Cov}(X,Y) = \mathbb{E}[XY] - \mathbb{E}[X]\mathbb{E}[Y]$.

"Keep looking at the book!" he hissed urgently.
"What?"
"Don't Panic."
"I'm not panicking!"
"Yes you are."
"Alright so I'm panicking, what else is there to do?"
– Douglas Adams, *The Hitchhiker's Guide to the Galaxy*.

In the last line, we used the assumption that $\hat{\theta}_n$ is an unbiased estimate, and therefore $\mathbb{E}[\hat{\theta}_n] = \theta$ for all $\theta \in \Theta$. Finally, combining (8.138), (8.139), and (8.140) gives the desired result

$$\mathbb{V}[\hat{\theta}_n] \geq \frac{1}{n\mathscr{I}(\theta_\star)} \tag{8.141}$$

and hence completes the proof.

Problems

8.1. MLE for a Bernoulli-Like Model

Let's revisit the Bernoulli-like model introduced in Problem 7.3. Suppose that data is modeled by $X_1, \dots, X_n \overset{\text{iid}}{\sim} p(x\,|\,\theta)$, where $p(x\,|\,\theta)$ is a PMF defined by

$$p(1\,|\,\theta) = \mathbb{P}(X=1\,|\,\theta) = \theta \quad \text{and} \quad p(2\,|\,\theta) = \mathbb{P}(X=2\,|\,\theta) = 1-\theta, \tag{8.142}$$

where $\theta \in (0,1)$ is the model parameter. Suppose we observe the following sample of size $n=5$:

$$X_1 = 1, \quad X_2 = 2, \quad X_3 = 1, \quad X_4 = 2, \quad X_5 = 1. \tag{8.143}$$

What is your intuitive estimate of θ? Find the maximum likelihood estimate of θ.

8.2. Poisson Distribution

The *Poisson distribution*[78] is one of the most important discrete probability distributions, which is used for modeling various natural and social phenomena. For example, the Poisson distribution can be used for modeling the number of mutations in a strand of DNA or the number of car accidents at a road intersection. It also plays an important role in modeling the risk of large earthquakes and even in the causal set approach to quantum gravity! In general, the Poisson distribution is used to for modeling the uncertain number of certain events occurring in a fixed period of time or in a given volume of space.

[78] As one may already expect, obeying Arnold's principle (see Marginal Note 47 in Chapter 1), the Poisson distribution was originally discovered not by Poisson. Indeed, it was first discovered (approximately 125 years before Poisson) by Abraham de Moivre.

A random variable X is distributed according to the Poisson distribution with parameter $\lambda > 0$, denoted $X \sim \mathscr{P}(\lambda)$, if its PMF is

$$p(x\,|\,\lambda) = e^{-\lambda}\frac{\lambda^x}{x!}, \quad x \in \{0,1,2,\dots\}. \tag{8.144}$$

If $X \sim \mathscr{P}(\lambda)$, then its mean and variance are $\mathbb{E}[X] = \mathbb{V}[X] = \lambda$. Suppose that data is modeled as an IID sample from the Poisson distribution,

$$X_1, \dots, X_n \overset{\text{iid}}{\sim} \mathscr{P}(\lambda), \tag{8.145}$$

and let $\lambda_\star$ denote the true value of λ.

(a) Find the maximum likelihood estimate $\hat{\lambda}_{MLE}$ of λ.

(b) Show that $\hat{\lambda}_{MLE}$ a consistent estimate of λ.

(c) Find the Fisher information $\mathscr{I}(\lambda_\star)$.

(d) Find the variance of $\hat{\lambda}_{MLE}$ and check that $\mathbb{V}[\hat{\lambda}_{MLE}] \approx 1/n\mathscr{I}(\lambda_\star)$ for $n \gg 1$, as predicted by (8.60).

(e) Find the asymptotic distribution of $\hat{\lambda}_{MLE}$ as $n \to \infty$. Does your result agree with the general result (8.78)?

(f) Is $\hat{\lambda}_{MLE}$ asymptotically unbiased?

(g) Is $\hat{\lambda}_{MLE}$ efficient?

8.3. Uniform Distribution with Unknown Support (Revisited)
In Example 8.3, we considered the uniform model on the interval $[0,\theta]$, where only the right end θ is unknown, and found the MLE estimate of θ, which is $\hat{\theta}_{MLE} = X_{(n)}$, the maximum value in the sample. Suppose now that both endpoints of the interval are unknown. Namely, let

$$X_1,\ldots,X_n \overset{\text{iid}}{\sim} \mathscr{U}[\alpha,\beta], \tag{8.146}$$

where α and β are unknown model parameters, $\alpha < \beta$, and $\mathscr{U}[\alpha,\beta]$ is the uniform distribution on $[\alpha,\beta]$. In Problem 7.2, we found the MOM estimates of the model parameters. Find the MLE estimates of α and β.

8.4. MLE vs Plug-In Estimate
As in Problem 8.3, let's assume that data is modeled as an IID sample from the uniform distribution

$$X_1,\ldots,X_n \overset{\text{iid}}{\sim} \mathscr{U}[\alpha,\beta], \tag{8.147}$$

where α and β are unknown model parameters, $\alpha < \beta$. Suppose that our parameter of interest is the mean of the distribution,

$$\mu = \frac{\alpha+\beta}{2}. \tag{8.148}$$

(a) Find the maximum likelihood estimate $\hat{\mu}_{MLE}$ of μ.

(b) The plug-in estimate of μ is the sample mean $\hat{\mu}_n = \overline{X}$ (see Example 4.1(a)). Let's compare the accuracy of $\hat{\mu}_{MLE}$ and $\hat{\mu}_n$ in a specific example. Let the true values of parameters be $\alpha_\star = 1$, $\beta_\star = 3$, and let the sample size be $n = 10$. Find the mean squared error of $\hat{\mu}_n$ analytically using the bias-variance decomposition and a fact that if $X \sim \mathscr{U}[\alpha,\beta]$, then its variance is $\mathbb{V}[X] = (\beta-\alpha)^2/12$. Find the mean squared error of $\hat{\mu}_{MLE}$ by simulation. Namely, generate $m = 10^4$ samples $X_1^{(i)},\ldots,X_n^{(i)} \overset{\text{iid}}{\sim} \mathscr{U}[\alpha_\star,\beta_\star]$, $i = 1,\ldots,m$, and use them to estimate $\text{MSE}[\hat{\mu}_{MLE}]$ exploiting the law of large numbers. Which estimate is more accurate?

8.5. MLE and Confidence Interval
Suppose that data is modeled with the normal model with unit variance $\sigma^2 = 1$,

$$X_1,\ldots,X_n \overset{\text{iid}}{\sim} \mathscr{N}(\mu,1). \tag{8.149}$$

For each sample unit X_i, let's defined an indicator-like random variable

$$Y_i = \begin{cases} 1, & \text{if } X_i > 0, \\ 0, & \text{if } X_i \leq 0, \end{cases} \qquad i = 1, \ldots, n. \tag{8.150}$$

Since $X_1, \ldots, X_n$ are IID, so are $Y_1, \ldots, Y_n$. Let ψ be their common mean, $\psi = \mathbb{E}[Y_i]$, and let $\psi_\star$ denote its true value, which depends on the true value $\mu_\star$ of μ.

(a) Express ψ in terms on μ and find the MLE of ψ based on the data $X_1, \ldots, X_n$.

(b) Construct an approximate 95% confidence interval for $\psi_\star$ from the data $X_1, \ldots, X_n$.

8.6. Alternative Formula for the Fisher Information
Definition 8.7 defines the Fisher information associated with a random variable $X \sim f(x\,|\,\theta_\star) \in \{f(x\,|\,\theta), \theta \in \Theta\}$ as follows

$$\mathscr{I}(\theta_\star) = \mathbb{E}\left[\left(\left.\frac{\partial \log f(X\,|\,\theta)}{\partial \theta}\right|_{\theta=\theta_\star}\right)^2\right]. \tag{8.151}$$

This formula allowed us to obtain an intuitive interpretation of the Fisher information in Section 8.6.2. However, it is not convenient for calculations. Show that the Fisher information can be defined equivalently as follows:

$$\mathscr{I}(\theta_\star) = -\mathbb{E}\left[\left.\frac{\partial^2 \log f(X\,|\,\theta)}{\partial \theta^2}\right|_{\theta=\theta_\star}\right]. \tag{8.152}$$

References

1. Stigler, S.M.: The epic story of maximum likelihood. *Statistical Science* 22(4), 598-620 (2007).
`https://doi.org/10.1214/07-STS249`
2. Fisher, R.A.: On an absolute criterion for fitting frequency curves. *Messenger of Mathematics* 41, 155-160 (1912).
`https://www.jstor.org/stable/2246266`
3. Aldrich, J.: R.A. Fisher and the making of maximum likelihood 1912-1922. *Statistical Science* 12(3), 162-176 (1997).
`https://doi.org/10.1214/ss/1030037906`
4. Hald, A.: *A History of Mathematical Statistics from 1750 to 1930*. New York: Wiley (1998).
5. Aldrich, J.: *A Guide to R.A. Fisher.*
`http://www.economics.soton.ac.uk/staff/aldrich/fisherguide/rafframe.htm`.
6. Ingram Olkin, I., Petkau, A.J., Zidek, J.V.: A comparison of n estimators for the binomial distribution. *Journal of the American Statistical Association* 76(375), 637-642 (1981).
`https://doi.org/10.2307/2287523`
7. Kullback, S., Leibler, R.A.: On information and sufficiency. *The Annals of Mathematical Statistics* 22(1), 79-86 (1951).
`https://doi.org/10.1214/aoms/1177729694`

8. Wald, A.: Note on the consistency of the maximum likelihood estimate. *The Annals of Mathematical Statistics* 20(4), 595-601 (1949).
https://www.jstor.org/stable/2236315
9. Ly, A., Marsman, M., Verhagen, J., Grasman., R., Wagenmakers, E.-J.: A tutorial on Fisher information. *Journal of Mathematical Psychology* 80, 40-55 (2017).
https://doi.org/10.1016/j.jmp.2017.05.006
10. Lehmann, E.L.: *Elements of Large-Sample Theory*. Springer, New York, NY (1999).
https://doi.org/10.1007/b98855
11. Wasserman, L.A.: *All of Statistics*. Springer, New York, NY (2004).
https://doi.org/10.1007/978-0-387-21736-9
12. van der Vaart, A.W.: *Asymptotic Statistics*. Cambridge University Press, Cambridge (1998).
https://doi.org/10.1017/CBO9780511802256
13. Nielsen, F.: An elementary introduction to information geometry. *Entropy* 22(10), 1100 (2020).
https://doi.org/10.3390/e22101100
14. Casella, G., Berger, R.L.: *Statistical Inference*. 2nd edition, Duxbury Advanced Series, Pacific Grove, CA, (2002).
15. Shenton, L.R.: The so-called Cramer-Rao inequality. *The American Statistician* 24(2), 36 (1970).
https://www.jstor.org/stable/2681931
16. Newman, M.E.J.: *Networks: An Introduction*. Oxford: Oxford University Press, 2nd edition (2018).
https://global.oup.com/academic/product/networks-9780198805090
17. Zuev K.M., Boguñá, M., Bianconi, G., Krioukov, D.: Emergence of soft communities from geometric preferential attachment. *Scientific Reports* 5, 9421 (2015).
https://doi.org/10.1038/srep09421
18. Bonahon, F.: *Low-Dimensional Geometry*. American Mathematical Society (2009).
https://bookstore.ams.org/stml-49
19. Papadopoulos, F., Psomas, C., Krioukov, D: Network mapping by replaying hyperbolic growth. *IEEE/ACM Transactions on Networking* 23(1), 198-211 (2015).
https://doi.org/10.1109/TNET.2013.2294052
20. Steele, J.M.: *The Cauchy-Schwarz Master Class*. Cambridge University Press (2004).
https://doi.org/10.1017/CBO9780511817106

9
Hypothesis Testing: A General Framework

By my nature, I am incapable of refraining from constructing hypotheses about everything that surrounds me. I don't like being without hypotheses, I can't do without them.
– Arkady and Boris Strugatsky, *Burdened with Evil, or Forty Years Later*

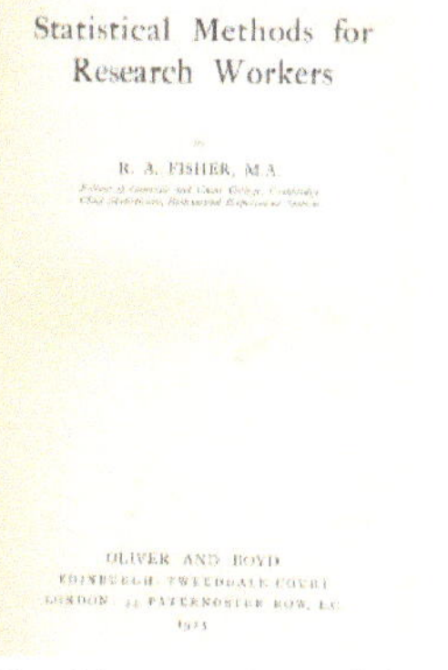

The title page of one of the most influential books on statistical methods.

In previous chapters, we learned how to estimate model parameters and other quantities of interest in both parametric and nonparametric settings. In many applications, however, researchers are interested in checking certain statements about parameters rather than estimating their values per se. Consider, for example, a pharmaceutical company that developed a new drug for reducing blood pressure. Let θ denote the average change in blood pressure after taking the drug. The principal question is whether the drug works or not. So, the main goal is to test

$$H_0 : \theta = 0 \quad \text{versus} \quad H_1 : \theta \neq 0. \tag{9.1}$$

The hypothesis H_0 is called the *null hypothesis*. It states that, on average, the new drug has zero effect[1] on blood pressure. The *alternative hypothesis* H_1 (also called the *research hypothesis*) states that there is some effect. In this example, testing H_0 versus H_1 is a primary problem. Even if we reject the null hypothesis and decide that the drug affects the blood pressure, that is $\theta \neq 0$, estimating the exact value of θ is an important, yet secondary problem. A part of statistical inference that deals with this kind of "either/or" problems is called *statistical hypothesis testing*.

[1] Hence the name "null."

The development of modern theory of statistical hypothesis testing began in 1908, when William Gosset[2] introduced the t-test [1]. Ronald Fisher generalized the t-test, extended its applicability, and created a new paradigm for hypothesis testing in his famous book "Statistical Methods for Research Workers" [2], which became very influential. Jerzy Neyman and Egon Pearson[3] further developed hypothesis testing by introducing key concepts and proving important theoretical results in their seminal papers [3, 4, 5].

[2] Who wrote under the pseudonym "Student," see Section 10.2.

[3] A son of Karl Pearson, see Remark 5.3.

In this chapter, we will develop a general framework for testing statistical hypotheses and discuss in detail an important but often misused and misunderstood concept of the p-value.

Supplementary Information The online version contains supplementary material available at https://doi.org/10.1007/978-3-032-03848-7_9.

K. M. Zuev, *Fundamentals of Statistical Inference*,
International Series in Operations Research & Management Science 374,
https://doi.org/10.1007/978-3-032-03848-7_9

9.1 Alice, Bob, and Two Coins

To get started, let's consider a simple example of a testing problem that will help us to illustrate all main notions and concepts introduced down the road.

Alice has two coins: a fair coin with the probability of heads $p_0 = 0.5$ and a biased coin with the probability of heads $p_1 = 0.7$. Alice chooses one of the coins, tosses it $n = 10$ times and tells Bob the total number of heads k, but does not tell him which coin was tossed. Based on the value of k, Bob has to infer which coin was tossed.

Intuitively, it is clear that the larger k is, the more likely it was a biased coin. If Alice tossed coin i, where $i = 0$ corresponds to the fair (unbiased) coin and $i = 1$ to the biased coin, then the total number of heads is a random variable K that follows the binomial distribution with parameters $n = 10$ and $p = p_i$, $K \sim \mathscr{B}(n, p_i)$. The probability of getting exactly k heads is then

$$\mathbb{P}(K = k \,|\, p = p_i) = \binom{n}{k} p_i^k (1 - p_i)^{n-k}, \qquad k = 0, \ldots, n, \;\; i = 0, 1. \tag{9.2}$$

Figure 9.1 shows the values of this probability for both coins and different k.

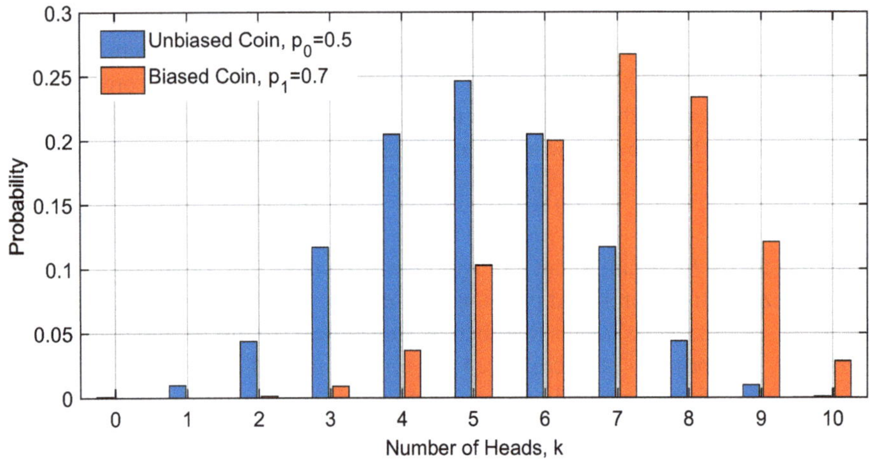

Fig. 9.1 Probabilities (9.2) of getting k heads in $n = 10$ tosses for biased and unbiased coins.

For example, if the number of heads $k = 2$, then

$$\frac{\mathbb{P}(K = 2 \,|\, p = p_1)}{\mathbb{P}(K = 2 \,|\, p = p_0)} \approx \frac{1}{30}, \tag{9.3}$$

which means that the fair coin is approximately 30 times more likely to produce 2 heads than the biased coin. Although both probabilities are very small (less than 0.05) and thus observing 2 heads is an unlikely event, given that this event has occurred, Bob should conclude that Alice tossed the fair coin. On the other hand, if the number of heads $k = 8$, then

$$\frac{\mathbb{P}(K = 8 \mid p = p_1)}{\mathbb{P}(K = 8 \mid p = p_0)} \approx 5, \tag{9.4}$$

and, therefore, it is approximately 5 times more likely that the biased coin was tossed. In general, if the number of heads is k, then it seems reasonable for Bob to conclude that Alice tossed the biased coin if and only if

$$\mathbb{P}(K = k \mid p = p_1) > \mathbb{P}(K = k \mid p = p_0), \tag{9.5}$$

which is equivalent to the number of heads being in the following range[4]:

[4] See Fig. 9.1.

$$k \in \{7, 8, 9, 10\}. \tag{9.6}$$

Otherwise, if $k < 7$, then $\mathbb{P}(K = k \mid p = p_1) < \mathbb{P}(K = k \mid p = p_0)$, and Bob should guess that the fair coin was tossed.

In what follows, we will refer to this relatively simple testing example involving Alice, Bob, and two Coins, as the *ABC example*.

9.2 General Testing Framework

Let $X_1, \ldots, X_n$ be our data which is modeled as an IID sample from a distribution F,

$$X_1, \ldots, X_n \overset{\text{iid}}{\sim} F \in \mathscr{F}, \tag{9.7}$$

where the stochastic model $\mathscr{F}$ can be parametric or nonparametric. Let θ be a parameter of interest, and Θ be the parameter space, i.e. the set of all possible values of θ. Furthermore, let $\Theta = \Theta_0 \sqcup \Theta_1$ be a partition of the parameter space into two disjoint subsets[5]. Suppose we want to test the *null hypothesis* that θ is in Θ_0 versus the *alternative hypothesis* that θ is in Θ_1,

[5] Recall that $\Theta = \Theta_0 \sqcup \Theta_1$ means that $\Theta = \Theta_0 \cup \Theta_1$ and $\Theta_0 \cap \Theta_1 = \varnothing$.

$$H_0 : \theta \in \Theta_0 \quad \text{vs} \quad H_1 : \theta \in \Theta_1. \tag{9.8}$$

Let $\mathscr{X}$ be the *sample space*, that is the set of all possible values of the data,

$$\mathscr{X} = \{X = (X_1, \ldots, X_n)\} \subset \mathbb{R}^n. \tag{9.9}$$

We test hypotheses by constructing an appropriate subset $\mathscr{R} \subset \mathscr{X}$ of the sample space, called the *rejection region*:

- If $X = (X_1, \ldots, X_n) \in \mathscr{R}$, then we reject H_0,
- If $X = (X_1, \ldots, X_n) \notin \mathscr{R}$, then we fail to reject H_0.

The rejection region $\mathscr{R}$ is constructed in such a way that $X \in \mathscr{R}$ means evidence against the null hypothesis, which is strong enough for rejecting H_0. If the data does not provide sufficient evidence against H_0, that is if $X \notin \mathscr{R}$, then we do not reject H_0, or, in other words, fail to reject it. At the same time, we do not accept H_0 as being true, since the absence of evidence against a hypothesis does not imply that the hypothesis is true[6]. The logical difference between "failing to reject" and "accepting" the null hypothesis was discussed in detail in Section 3.2.3 and vividly illustrated in Remark 3.4.

[6] For example, the fact that you are reading this book right now does not provide any evidence against a hypothesis that all people in the word are reading this book right now. But this does not mean the hypothesis is true (unfortunately for the author).

Remark 9.1. Nevertheless, the term "accept the null hypothesis" is often used in the literature and practice, especially when a *decision* must be made based on the testing results. For example, the pharmaceutical company must make a decision whether to manufacture the new drug based on testing (9.1). In this case, accepting the null hypothesis does not mean that we believe it is true, it simply means that we will act as if it were true. If the data does not provide evidence against the null hypothesis that states that the new drug does not work, the company will not manufacture it.

The rejection region $\mathscr{R}$ usually has the following form:

$$\mathscr{R} = \{X \in \mathscr{X} \mid s(X) \geq c\}, \tag{9.10}$$

where $s(X)$ is a *test statistic* and c is a *critical value*. The test statistic, which in principle can be any function of the data $X = (X_1, \ldots, X_n)$, is a "lens" through which we look at the data and test hypotheses[7]. It defines the *shape* of the rejection region. The test statistic should be chosen such that its large values contradict the null hypothesis. The critical value is a certain threshold, exceeding which leads to the rejection of the null hypothesis. It defines the *size* of the rejection region. Figure 9.2 schematically illustrates the described testing framework.

[7] "Dear friend, do you not see
That all that is visible to us
Is but a glimmer, only shadows
Of things invisible to eyes?"
– V.S. Solovyov (1892)

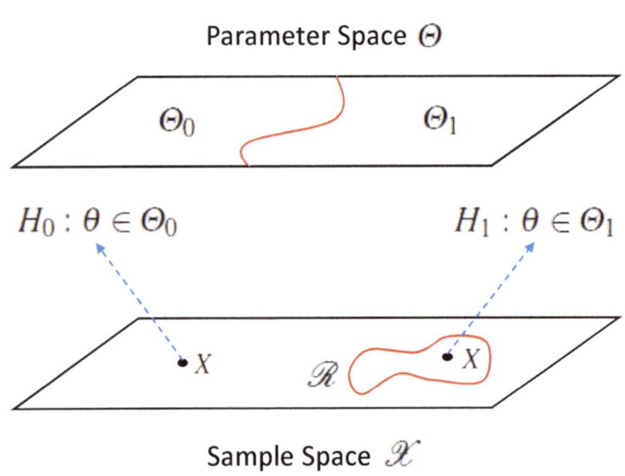

Fig. 9.2 A big picture of testing framework. If the data belongs to the rejection region, $X \in \mathscr{R}$, then we reject the null hypothesis H_0. Otherwise, if $X \notin \mathscr{R}$, then we fail to reject H_0 (or accept it). Constructing a test is equivalent to defining a rejection region $\mathscr{R}$ in the sample space $\mathscr{X}$.

Example 9.1. Let's see how the ABC example from the previous section fits the describe general framework for hypothesis testing. Our data is the total number of heads K in $n = 10$ tosses of a coin, which is modeled as a sample from the binomial distribution,

$$K \sim \mathscr{B}(n, p). \tag{9.11}$$

The parameter of interest θ is the the probability of heads p, which belongs to the parameter space consisting of just two points, $\Theta = \{0.5, 0.7\}$. The partition of the parameter space into two disjoint subsets is $\Theta = \Theta_0 \sqcup \Theta_1$, where $\Theta_0 = \{0.5\}$ and $\Theta_1 = \{0.7\}$. The null and alternative hypotheses are

$$H_0 : \theta \in \Theta_0 \Leftrightarrow p = 0.5 \quad \text{and} \quad H_1 : \theta \in \Theta_1 \Leftrightarrow p = 0.7. \tag{9.12}$$

The sample space, which is the set of all possible values of K, is $\mathscr{X} = \{0, 1, \dots, 10\}$, and the rejection region is $\mathscr{R} = \{7, 8, 9, 10\} \subset \mathscr{X}$. The test statistics is the ratio of probabilities

$$s(k) = \frac{\mathbb{P}(K = k \,|\, p = 0.7)}{\mathbb{P}(K = k \,|\, p = 0.5)}, \tag{9.13}$$

and the corresponding critical value that defines $\mathscr{R}$ is $c = 1$.

9.3 The Null and Alternative Hypotheses

From a formal point of view, it may appear that the null and alternative hypotheses play absolutely symmetric roles. In practice, however, the roles of H_0 and H_1 are often conceptually different[8]. In many applications, the *null hypothesis* — as its name suggests — states that "there is no effect" (e.g. of a drug on a disease), "there is no relationship" (e.g. between variables) or "there is no difference" (e.g. between two populations). The null hypothesis H_0 is generally assumed to be true until data indicates otherwise. The *alternative* (or *research*) *hypothesis*, on the other hand, claims the opposite: "there is an effect," "there is a relationship," "there is a difference," etc. The rejection region, which is a set of data that provides evidence against the null hypothesis, is then a place where "discovery" lives. The alternative hypothesis H_1 is usually a statement that carries the burden of proof. It is accepted instead of the null hypothesis only if it is strongly supported by the data.

[8] Why this is the case will become more clear after we discuss statistical errors in the next section.

In some cases, it is useful to think about hypothesis testing as a criminal trial. According to the presumption of innocence, a person accused of a crime is considered innocent until proven guilty (beyond the shadow of a doubt). The null hypothesis in this context is H_0: "the person is not guilty." It will be rejected in favor of the alternative, H_1: "the person is guilty," only if the data provides strong evidence against the null hypothesis.

It is important to highlight that the null and alternative hypotheses must be formulated without using the data which is intended for testing these hypotheses. Testing hypotheses suggested by the data on the same data is a form of circular reasoning.

If a dataset suggests that a certain property holds, then testing whether the property indeed holds using the same dataset is meaningless: the hypothesis that the property holds will most likely be accepted, but this would not provide any evidence that the property holds more broadly. For example, if last month's financial reports indicate that "investing in cryptocurrency is profitable," then this (research) hypothesis must be tested on the data from other months. Generating hypotheses based on the data and then testing them on the same data is sometimes called *post-hoc theorizing* and should be avoided.

9.4 Two Types of Statistical Errors

> *Suppose a stochastic model M is advanced to represent (or to explain) certain phenomena P. The question immediately arises whether the representation is satisfactory. The observations provide a certain number, frequently only a small number, of data, $x_1, \ldots, x_n$ and these data must be used to decide whether to accept or to reject the model M. In situations of this kind the possibility of erroneous decisions cannot be eliminated, and the best one can do is to seek methods of making decisions that, in a sense, minimize the risk of mistakes.*
>
> – Jerzy Neyman, *Statistics – Servant of All Sciences* [6]

Can we guarantee that we make no errors when drawing conclusions from the data? No, in most cases not. Usually, the data provides some, but not full, information about the quantity of interest. It helps to reduce the uncertainty, but does not eliminate it completely. The errors are, therefore, conceptually unavoidable. For instance, in the ABC example, the biased coin may produce five heads[9], in which case Bob will make an error by guessing that the coin was unbiased.

[9] The probability of this event is about 0.1.

There are two types of statistical errors in hypothesis testing:

- *Type I error*: rejecting the null hypothesis H_0 when it is true,
- *Type II error*: failing to reject the null hypothesis H_0 when it is false,

which are schematically shown in Figure 9.3.

		State of the World	
		H_0 is false	H_0 is true
Decision based on the Data	Reject H_0	☺	Type I Error
	Fail to Reject H_0	Type II Error	☺

Fig. 9.3 Two types of statistical errors: reject H_0 when it is true (type I error) and fail to reject H_0 when it is not true (type II error).

Purely mathematically[10], making both errors is equally bad. In practice, however, type I and type II errors often have drastically different consequences and making a type I error is usually much worse than making a type II error. For example, convicting an innocent person (type I error) is much worse that acquitting a guilty person (type II error); using a drug that does not work (type I error) is worse than discarding a working drug (type II error). In Pascal's wager[11], the French mathematician and philosopher avoids the type I error: don't believe in God while He exists.

[10] That is when we focus on abstractions and ignore the context the abstractions came from.

[11] "Either God exists or He doesn't. Either I believe in God or I don't. Of the four possibilities, only one is to my disadvantage. To avoid that possibility, I believe in God." – Blaise Pascal.

Remark 9.2. A type I error is also called a *false positive* (we falsely decide that there is an effect when in fact there is not), and a type II error is also called a *false negative* (we falsely decide that there is no effect when, in fact, there is). Although this terminology is rather widespread in the literature, in some cases it can be confusing. For example, for the hypotheses defined in (9.14), it is more natural to call the type I error a false negative and the type II error a false positive.

As we will see later in this chapter, if the sample size is fixed, then it is generally impossible to make the probabilities of both errors arbitrarily small. The machinery of hypothesis testing allows us to control only one of the two errors — it is chosen to be the type I error — and make its probability smaller than any specified threshold[12]. Understanding this can help in some cases, where making one error is much worse than making the other, to understand which of the two competing hypotheses that we want to test should be the null hypothesis and which the alternative: H_0 and H_1 should be chosen such that rejecting H_0 when it is true is worse than failing to reject H_0 when it is false.

[12] Reducing the probability of one error usually increases the chance of the other.

Example 9.2. Consider an oncologist who wants to test whether a patient has cancer based on the result of a biopsy[13]. Which of the two conflicting hypotheses, "there is cancer" and "there is no cancer," should be the null hypothesis? To make the type I error the worst out of the two errors, the hypotheses should be defined as follows:

$$H_0 : \text{there is cancer} \quad \text{vs} \quad H_1 : \text{there is no cancer}. \tag{9.14}$$

[13] A biopsy is a surgical procedure that involves extraction of sample cells or a piece of tissue for determining whether a disease is present.

Indeed, concluding mistakenly that there is no cancer (type I error) is worse than concluding mistakenly that there is cancer (type II error). While in the former case, the patient will not get required treatment immediately, and precious time could be lost, in the latter case, the subsequent analysis will show that there is no cancer.

Example 9.3. Consider an engineer who has designed a new earthquake-resistant building and, before constructing the building, wants to test whether the probability p_c of building collapse under an earthquake excitation is smaller than a certain acceptable threshold $p_c^\star$ or not[14]. Which of the two competing hypotheses, $p_c \leq p_c^\star$ and $p_c > p_c^\star$, should be the null hypothesis? To make the type I error the worst of the two errors, the hypotheses should be defined as follows:

$$H_0 : p_c > p_c^\star \text{ (not safe)} \quad \text{vs} \quad H_1 : p_c \leq p_c^\star \text{ (safe)}. \tag{9.15}$$

[14] To test, using building and earthquake models, and then, based on the results, decide whether to construct the real building.

Indeed, constructing a building that is not safe (type I error) is much worse than deciding not to construct a safe building (type II error).

9.5 The Power Function

Probabilities of both type I and type II errors can be computed using the so-called *power function*. If $\mathscr{R}$ is the rejection region, then the probability of a type I error is

$$\mathbb{P}(\text{Type I error}) = \mathbb{P}(\text{Reject } H_0 \,|\, H_0 \text{ is true}) = \mathbb{P}(X \in \mathscr{R} \,|\, \theta \in \Theta_0). \tag{9.16}$$

The probability of a type II error is

$$\begin{aligned}\mathbb{P}(\text{Type II error}) &= \mathbb{P}(\text{Fail to reject } H_0 \,|\, H_1 \text{ is true}) = \mathbb{P}(X \notin \mathscr{R} \,|\, \theta \in \Theta_1)\\ &= 1 - \mathbb{P}(X \in \mathscr{R} \,|\, \theta \in \Theta_1).\end{aligned} \tag{9.17}$$

From the last two equations, we see that both probabilities can be expressed in terms of the probability of event $X \in \mathscr{R}$ given the value of θ. This motivates the following important definition.

Definition 9.1. The *power function* of a hypothesis test with rejection region $\mathscr{R}$ is a function $\beta : \Theta \to [0,1]$ defined by

$$\beta(\theta) = \mathbb{P}(X \in \mathscr{R} \,|\, \theta). \tag{9.18}$$

In terms of the error probabilities,

$$\beta(\theta) = \begin{cases} \mathbb{P}(\text{Type I error}), & \text{if } \theta \in \Theta_0,\\ 1 - \mathbb{P}(\text{Type II error}), & \text{if } \theta \in \Theta_1. \end{cases} \tag{9.19}$$

For an "ideal" test, where both error probabilities equal to zero[15], the power function is therefore zero on Θ_0 and one on Θ_1. The ideal power function is schematically shown in Fig. 9.4. The closer the power function of a test is to the ideal power function, the better the test performs.

[15] It is impossible to construct the ideal test in practice, but it is a useful theoretical concept.

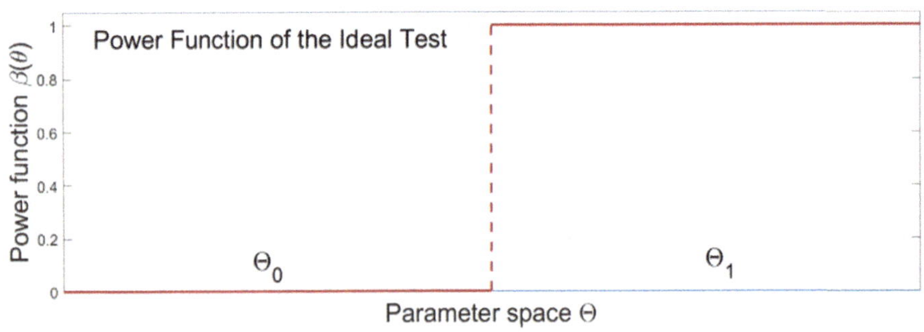

Fig. 9.4 The ideal power function of the ideal test, where the probabilities of type I and type II errors are zero, $\beta(\theta) = 0$ if $\theta \in \Theta_0$ and $\beta(\theta) = 1$ if $\theta \in \Theta_1$.

Example 9.4. The power function of the test constructed in the ABC example is defined on $\Theta = \Theta_0 \sqcup \Theta_1 = \{0.5, 0.7\}$ and is equal to

$$\begin{aligned}\beta(p) &= \mathbb{P}(K \in \mathscr{R} \,|\, p) = \mathbb{P}(K \in \{7,8,9,10\} \,|\, p) = \sum_{k=7}^{10} \mathbb{P}(K = k \,|\, p) \\ &= \sum_{k=7}^{10} \binom{10}{k} p^k (1-p)^{10-k} \approx \begin{cases} 0.172, & \text{if } p = 0.5, \\ 0.650, & \text{if } p = 0.7. \end{cases}\end{aligned} \tag{9.20}$$

The probabilities of type I and type II errors are therefore

$$\begin{aligned}\mathbb{P}(\text{Type I error}) &= \beta(0.5) = 0.172, \\ \mathbb{P}(\text{Type II error}) &= 1 - \beta(0.7) = 0.35.\end{aligned} \tag{9.21}$$

Although this test is far from being ideal (probability of errors are not zero), as we will see, in some sense, this is the best test available.

It is natural to expect from a reasonable test that it is more likely to reject the null hypothesis when it is false ($\theta \in \Theta_1$) then when it is true ($\theta \in \Theta_0$). That is,

$$\mathbb{P}(X \in \mathscr{R} \,|\, \theta_1) \geq \mathbb{P}(X \in \mathscr{R} \,|\, \theta_0), \qquad \text{for all } \theta_0 \in \Theta_0, \theta_1 \in \Theta_1, \tag{9.22}$$

or, in terms of the power function,

$$\beta(\theta_1) \geq \beta(\theta_0), \qquad \text{for all } \theta_0 \in \Theta_0, \theta_1 \in \Theta_1. \tag{9.23}$$

A test that satisfies (9.23) is called *unbiased*. For instance, the test constructed in the ABC example is unbiased, since $\beta(0.7) > \beta(0.5)$.

9.6 The Size of a Test

It is impossible to control both type I and type II errors and make their probabilities arbitrarily small[16]. The reason behind this is the following. As equation (9.16) indicates, to reduce the probability of type I error, we need to shrink the rejection region $\mathscr{R}$. But shrinking $\mathscr{R}$ leads, according to equation (9.17) to the increase of the probability of type II error. Similarly, to reduce the probability of type II error, we need to expand the rejection region, but this will result into the increase of the probability of type I error. So, there is a *trade-off* between type I and type II errors: reducing the probability of one of the two errors increases the probability of the other[17].

As an extreme example, we can completely eliminate a possibility of type I error by taking $\mathscr{R} = \varnothing$, that is by never rejecting the hull hypothesis. In this case, $\mathbb{P}(\text{Type I error}) = 0$, but $\mathbb{P}(\text{Type II error}) = 1$. Analogously, we can make the type II error impossible by taking $\mathscr{R} = \mathscr{X}$, that is by always failing to reject the null hypothesis. In this case, $\mathbb{P}(\text{Type II error}) = 0$, but $\mathbb{P}(\text{Type I error}) = 1$.

[16] "Work with what you have control of and you'll have your hands full." – James Stockdale, *Courage Under Fire*.

[17] Neyman and Pearson suggested that "how the balance [between the two types of error] should be struck must be left to the investigator," [4], p. 296.

Since the type I error is the worst of the two errors, controlling its probability is more important. This leads to the following key definition.

Definition 9.2. The *size* of a test with power function $\beta(\theta)$ is

$$\alpha = \sup_{\theta\in\Theta_0} \beta(\theta) = \sup_{\theta\in\Theta_0} \mathbb{P}(\text{Type I error}). \tag{9.24}$$

Remark 9.3. The "sup" in the above equation stands for "supremum," which is the least upper bound of a set. If this technical terms is unfamiliar, you can think of it as the "maximum." The size of a test is thus the maximum possible probability of type I error. This is schematically illustrated in Fig. 9.5.

Remark 9.4. A related notion, which is often confused and used interchangeably with the test size, is the test level. We say that a test has *level* α if $\sup_{\theta\in\Theta_0} \beta(\theta) \le \alpha$, i.e. its size is less than or equal to α.

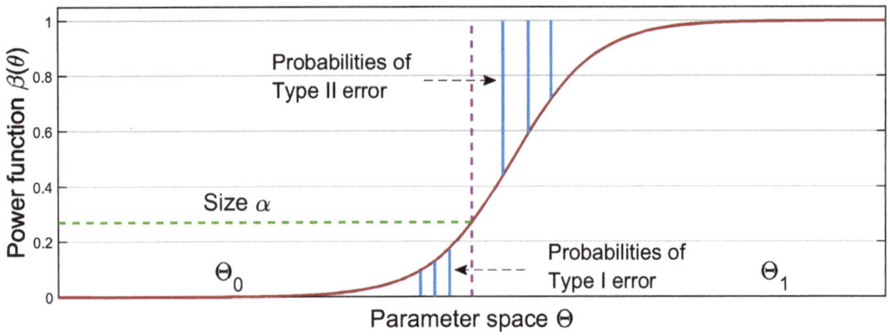

Fig. 9.5 The power function of a test with size α. The vertical dashed purple line represents the boundary between Θ_0 and Θ_1. The vertical solid blue lines are the probabilities of type I and type II errors for different values of θ. The maximum possible probability of type I error is α. Here, it is achieved at the boundary between Θ_0 and Θ_1.

The concept of size allows to formulate the following conceptual strategy for controlling testing errors. Let's fix an admissible value of α[18], and consider the set $\mathscr{T}_\alpha$ of all tests of size α or less. In other words, $\mathscr{T}_\alpha$ is the set of all rejections regions $\mathscr{R}$ satisfying

$$\sup_{\theta\in\Theta_0} \underbrace{\mathbb{P}(X\in\mathscr{R}|\theta)}_{\beta(\theta)} \le \alpha. \tag{9.25}$$

Whatever test (rejection region) from $\mathscr{T}_\alpha$ we end up using for testing our hypotheses, we are guaranteed that the probability of making type I error is at most α. A test $t \in \mathscr{T}_\alpha$ will also have a small probability of type II error if its power function $\beta(\theta)$ is large when $\theta \in \Theta_1$. So, among all tests in $\mathscr{T}_\alpha$, let's find a test $t^\dagger \in \mathscr{T}_\alpha$ with the power function $\beta^\dagger(\theta)$ that satisfies

$$\beta^\dagger(\theta) \ge \beta(\theta), \qquad \text{for all } \theta \in \Theta_1, \tag{9.26}$$

18 Typical values used in applications are $\alpha = 0.01$, $\alpha = 0.05$ (advocated by Fisher [2]) and $\alpha = 0.1$. The choice depends on how strongly we want to reduce a possibility of type I error.

and every $\beta(\theta)$ that is a power function of a test in $\mathscr{T}_\alpha$. If such test $t^\dagger \in \mathscr{T}_\alpha$ exists, it is called a *uniformly*[19] *most powerful* (UMP) test at *level*[20] α.

In many cases, UMP tests do not exist, since requirement (9.26) is quite strong. Even when a UMP test exists, it is often hard to construct and use it in practice. In the next chapter, we will consider several very popular tests which, in general, are not UMP, but still have high power, perform reasonably well, and are often used in various applications.

[19] Here "uniformly" means that (9.26) holds for *all* $\theta \in \Theta_1$.

[20] "Level" and not "size" because $t^\dagger$ is allowed to have size less than α.

As it follows from Example 9.4, the size of the test constructed in the ABC example is $\alpha = \beta(0.5) = 0.172$. Let's consider a more complex example where we construct a test of an arbitrary size $\alpha \in (0,1)$ and then use the sample size n to control the probability of a type II error.

Example 9.5. Suppose that data is modeled with the normal model,

$$X_1,\dots,X_n \overset{\text{iid}}{\sim} \mathscr{N}(\mu,\sigma^2), \quad \mu \in \mathbb{R}, \tag{9.27}$$

where σ^2 is known and μ is the single parameter. And suppose we want to test

$$H_0 : \mu \leq 0 \quad \text{vs} \quad H_1 : \mu > 0. \tag{9.28}$$

In this case, the parameter space is $\Theta = (-\infty,\infty)$, and its partition into two disjoint subsets is $\Theta = \Theta_0 \sqcup \Theta_1$, where $\Theta_0 = (-\infty, 0]$ and $\Theta_1 = (0,\infty)$. To construct a test, we need to define a test statistic $s(X)$ whose large values contradict the hull hypothesis. Intuitively, it seems reasonable to use the sample mean[21],

[21] In Problem 10.6, we will see that this is indeed a good choice that leads to the so-called generalized likelihood ratio test.

$$s(X) = \overline{X} = \frac{1}{n}\sum_{i=1}^{n} X_i, \tag{9.29}$$

and reject H_0 whenever $X = (X_1,\dots,X_n)$ is in the rejection region

$$\mathscr{R} = \{X \in \mathscr{X} \mid \overline{X} \geq c\}, \tag{9.30}$$

where c is some critical value. The power function of this test is

$$\beta(\mu) = \mathbb{P}(X \in \mathscr{R} \mid \mu) = \mathbb{P}(\overline{X} \geq c \mid \mu). \tag{9.31}$$

Given μ, the sample mean $\overline{X}$ is normally distributed with parameters μ and σ^2/n,

$$\overline{X} \sim \mathscr{N}\left(\mu, \frac{\sigma^2}{n}\right) \quad \iff \quad \frac{\overline{X}-\mu}{\sigma/\sqrt{n}} \sim \mathscr{N}(0,1), \tag{9.32}$$

and, therefore, the power function is

$$\beta(\mu) = \mathbb{P}\left(\frac{\overline{X}-\mu}{\sigma/\sqrt{n}} \geq \frac{c-\mu}{\sigma/\sqrt{n}} \,\middle|\, \mu\right) = 1 - \Phi\left(\frac{c-\mu}{\sigma/\sqrt{n}}\right), \tag{9.33}$$

where Φ is the standard normal CDF. The power function $\beta(\mu)$ is a monotonically increasing function of μ, and it approaches 0 and 1 as μ goes to minus and plus infinity,

$$\mu \uparrow \Rightarrow \beta \uparrow, \quad \mu \to -\infty \Rightarrow \beta \to 0, \quad \mu \to +\infty \Rightarrow \beta \to 1. \tag{9.34}$$

Note that equation (9.33) defines not just a single power function, but a family of power functions parameterized by the critical value c. For any fixed μ, the power function is a monotonically decreasing function of c, and it approaches 1 and 0 as c goes to minus and plus infinity,

$$c \uparrow \Rightarrow \beta \downarrow, \quad c \to -\infty \Rightarrow \beta \to 1, \quad c \to +\infty \Rightarrow \beta \to 0. \tag{9.35}$$

Figure 9.6 shows the power function in (9.33) for $\sigma = 1$, $n = 10$, and three different values of the critical value c. Notice that c controls the size α of the test: as c increases (the rejection region (9.30) shrinks and the test rejects H_0 less often), the size α decreases (the type I error becomes less likely to occur).

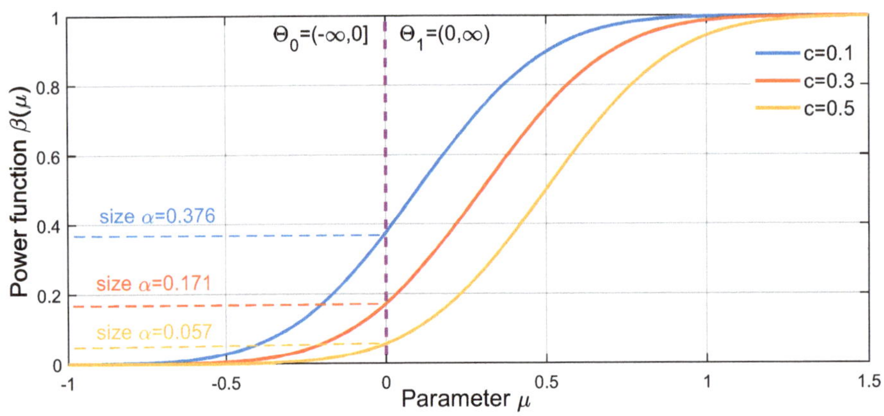

Fig. 9.6 The power function (9.33) for $\sigma = 1$, $n = 10$, and $c = 0.1$ (blue), $c = 0.3$ (red), and $c = 0.5$ (yellow). The corresponding sizes are $\alpha = 0.376$, $\alpha = 0.171$, and $\alpha = 0.057$, respectively. The vertical dashed purple line represents the boundary between $\Theta_0 = \{\mu \le 0\}$ and $\Theta_1 = \{\mu > 0\}$.

To construct a test of a given size α, we need to find c such that

$$\alpha = \sup_{\mu \in \Theta_0} \beta(\mu) = \sup_{\mu \le 0} \beta(\mu) = \beta(0) = 1 - \Phi\left(\frac{c\sqrt{n}}{\sigma}\right). \tag{9.36}$$

Here $\sup_{\mu \le 0} \beta(\mu) = \beta(0)$ since β is an increasing function of μ. Solving the above equation for c gives

$$c = \frac{\sigma}{\sqrt{n}} \Phi^{-1}(1-\alpha) = \frac{\sigma}{\sqrt{n}} z_{1-\alpha}. \tag{9.37}$$

So, the size α test based on the rejection region (9.30) has the following form:

$$\text{Reject } H_0 : \mu \le 0 \quad \Longleftrightarrow \quad \overline{X} \ge \frac{\sigma}{\sqrt{n}} z_{1-\alpha}. \tag{9.38}$$

Suppose now that we can also control the sample size n[22]. For any n, however small, test (9.38) guarantees that the probability of type I error does not exceed α. But can we use n to control the probability of type II error? Let's investigate.

To make the probability of type II error small, we need to make the power function $\beta(\mu)$ large for $\mu \in \Theta_1$, i.e. for $\mu > 0$[23]. So, let's first find the power function of test (9.38). To this end, we need to substitute the value of c in (9.37) into (9.33):

$$\beta(\mu) = 1 - \Phi\left(\frac{\frac{\sigma}{\sqrt{n}} z_{1-\alpha} - \mu}{\sigma/\sqrt{n}}\right) = 1 - \Phi\left(z_{1-\alpha} - \frac{\mu\sqrt{n}}{\sigma}\right). \tag{9.39}$$

If $\mu > 0$, then the power function β increases as n increases. If $\mu < 0$, then the power function β decreases as n increases. And in the limit, as $n \to \infty$, it converges to the ideal power function[24], except at $\mu = 0$, where, by construction, $\beta(0) = \alpha$:

$$\lim_{n\to\infty} \beta(\mu) = \begin{cases} 0, & \text{if } \mu < 0, \\ \alpha, & \text{if } \mu = 0, \\ 1, & \text{if } \mu > 0. \end{cases} \tag{9.40}$$

Since for any finite n, the power function $\beta(\mu)$ is a continuous function of μ and $\beta(0) = \alpha$, in a neighborhood of zero $\beta(\mu) \approx \alpha \ll 1$. Therefore, we cannot make $\beta(\mu)$ arbitrary close to one for all $\mu > 0$. In other words, with a finite sample size, we cannot make the probability of type II error arbitrary close to zero for all $\mu > 0$. But we can step away from $\mu = 0$ by $\varepsilon > 0$, $\varepsilon \ll 1$ and require the power function to be large at $\mu = \varepsilon$,

$$\beta(\varepsilon) = 1 - \gamma, \qquad \gamma > 0, \gamma \ll 1. \tag{9.41}$$

The constant γ plays the same role for the type II error as α does for the type I error. The above equation leads to the equation on the sample size:

$$1 - \gamma = 1 - \Phi\left(z_{1-\alpha} - \frac{\varepsilon\sqrt{n}}{\sigma}\right). \tag{9.42}$$

Solving for n gives

$$\sqrt{n} = \frac{\sigma}{\varepsilon}\left(z_{1-\alpha} - z_\gamma\right). \tag{9.43}$$

As expected, the sample size is a decreasing function of both ε and γ and it blows up at $\varepsilon = 0$ and at $\gamma = 0$,

$$\varepsilon \downarrow \Rightarrow n \uparrow, \quad \gamma \downarrow \Rightarrow n \uparrow, \quad \varepsilon \to 0 \text{ or } \gamma \to 0 \Rightarrow n \to \infty. \tag{9.44}$$

So, if the sample size n is given by (9.43), then the size α test (9.38) guarantees that the probability of type II error is at most γ if $\mu \in [\varepsilon, \infty) \subset \Theta_1$. If $\mu \in (0, \varepsilon)$, then the probability of type II error is larger than γ. Figure 9.7 shows the power function (9.39) for $\sigma = 1$, $\alpha = 0.1$, $\gamma = 0.1$, $\varepsilon = 0.1$, and samples size (9.43).

Let's now briefly summarize the general testing framework that we have developed so far. To construct a test, we need to define a rejection region

[22] For example, we design an experiment, and try to determine what sample size is appropriate.

[23] See (9.19).

[24] See Fig. 9.4.

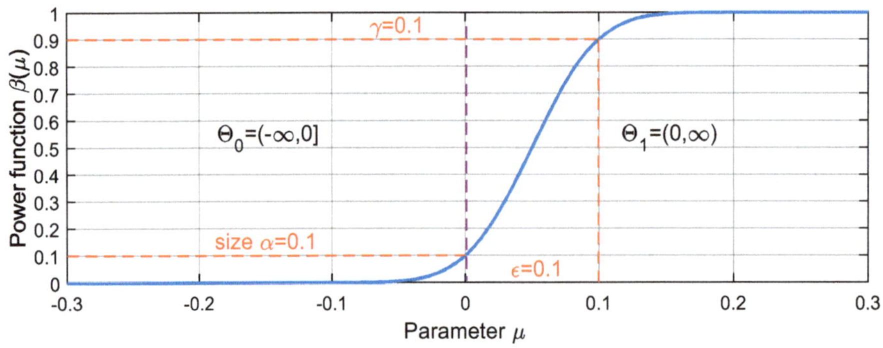

Fig. 9.7 The power function (9.39) for $\sigma = 1$, $\alpha = 0.1$, $\gamma = 0.1$, $\varepsilon = 0.1$. The samples size $n = 657$ is given by (9.43). For this test, $\mathbb{P}(\text{Type I error}) \leq \alpha$ and $\mathbb{P}(\text{Type II error}) \leq \gamma$ if $\mu \geq \varepsilon$.

$$\mathscr{R} = \{X \in \mathscr{X} \mid s(X) \geq c\} \tag{9.45}$$

by choosing a test statistic $s(X)$ and critical value c. Choosing $s(X)$ is an art, but often reasonable candidates exist. The guiding principle for choosing $s(X)$ is that its large values should contradict the null hypothesis H_0[25]. The critical value c controls the size α of the test, which is a desired upper bound for the probability of type I error. After $s(X)$ has been chosen and α has been specified, to find c we need to solve the following equation:

[25] See also Remark 10.12.

$$\sup_{\theta \in \Theta_0} \beta(\theta) = \alpha \quad \Longleftrightarrow \quad \sup_{\theta \in \Theta_0} \mathbb{P}(s(X) \geq c \,|\, \theta) = \alpha. \tag{9.46}$$

Finally, if we can control the sample size n, then we can try to bound the probability of type II error in the spirit of Example 9.5 by exploiting the dependence of the power function $\beta(\theta)$ on n.

9.7 The Concept of the P-value

The p-value is one of the most iconic and often used (and sometimes misused [7]) concepts of hypothesis testing. Let's start with a motivation for this concept.

9.7.1 Motivation

In Example 9.5, we derived a size α test that rejects the null hypothesis $H_0 : \mu \leq 0$ in favor of the alternative $H_1 : \mu > 0$ whenever the sample mean of $X_1, \ldots, X_n \overset{\text{iid}}{\sim} \mathscr{N}(\mu, \sigma^2)$ is sufficiently large, namely

$$\overline{X} \geq c_\alpha = \frac{\sigma}{\sqrt{n}} z_{1-\alpha}. \tag{9.47}$$

Suppose that we have two samples, $X^{(1)} = (X_1^{(1)}, \dots, X_n^{(1)})$ and $X^{(2)} = (X_1^{(2)}, \dots, X_n^{(2)})$, both sample means are larger than c_α, and the second sample mean is larger than the first one,

$$\overline{X}^{(2)} = \frac{1}{n}\sum_{i=1}^{n} X_i^{(2)} > \overline{X}^{(1)} = \frac{1}{n}\sum_{i=1}^{n} X_i^{(1)} > c_\alpha. \tag{9.48}$$

This situation is schematically shown in Fig. 9.8.

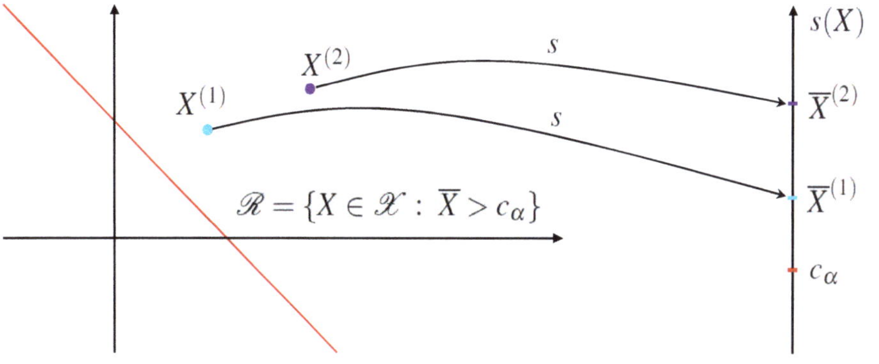

Fig. 9.8 Both samples, $X^{(1)}$ and $X^{(2)}$, belong to the rejection region $\mathscr{R}$, which is a hyperplane in $\mathbb{R}^n$ whose boundary $\partial\mathscr{R} = \{X \in \mathscr{X} \mid \overline{X} = c_\alpha\}$ is represented by the red line. The test statistic $s : X \mapsto \overline{X}$ maps samples to their sample means $\overline{X}^{(1)}$ and $\overline{X}^{(2)}$. The second sample $X^{(2)}$ lies deeper in the rejection region than the first sample $X^{(1)}$, and, as a result, $\overline{X}^{(2)} > \overline{X}^{(1)} > c_\alpha$.

The test will reject H_0 based on both samples, since $X_i^{(1)} > c_\alpha$ and $X_i^{(2)} > c_\alpha$. However, the second sample $X^{(2)}$ provides *more evidence* against H_0, since it lies *deeper* in the rejection region (further away from its boundary), which is manifested by the inequality $\overline{X}^{(2)} > \overline{X}^{(1)}$. So, we should be more confident in rejecting H_0 based on $X^{(2)}$ than based on $X^{(1)}$. Nevertheless, the testing theory that we developed so far *does not distinguish* between $X^{(1)}$ and $X^{(2)}$, and regardless of which of the two samples we observe, the verdict of testing will be the same: the test rejects H_0.

This leads to a natural general question: can we, rather than simply reporting that a size α test rejects or fails to reject H_0, be *more informative*? Is it possible to *quantify* that one sample provides more evidence against H_0 (with respect to the chosen test statistic) than another? It turns out that yes, and the concept of the p-value allows to do that. Let's now formally define the p-value.

9.7.2 Formal Definition

Consider an arbitrary test of size α with the rejection region

$$\mathscr{R}_\alpha = \{X \in \mathscr{X} \mid s(X) \geq c_\alpha\}. \tag{9.49}$$

Here the subscript $_\alpha$ explicitly indicates the dependence of the critical value and, therefore, the rejection region on α. Recall that the critical value c_α is determined from α by solving the following equation:

$$\alpha = \underbrace{\sup_{\theta \in \Theta_0} \mathbb{P}(s(X) \geq c \,|\, \theta)}_{\text{decreasing function of } c}. \tag{9.50}$$

How does the rejection region $\mathscr{R}_\alpha$ depend on the size α? The key observation for answering this question is that the right-hand side of (9.50) is a decreasing function of c. Indeed, as c increases, it becomes less likely for event $s(X) \geq c$ to occur. So, if α increases, then c_α decreases and therefore $\mathscr{R}_\alpha$ expands[26],

$$\alpha \uparrow \implies c_\alpha \downarrow \implies \mathscr{R}_\alpha \uparrow. \tag{9.51}$$

This means that the family of rejection regions parameterized by $\alpha \in (0,1)$ is a collection of nested subsets,

$$\alpha' > \alpha \implies \mathscr{R}_\alpha \subset \mathscr{R}_{\alpha'}. \tag{9.52}$$

In particular, if the size α test rejects H_0 based on sample X, then for any $\alpha' > \alpha$ the size α' test also rejects H_0 based on X,

$$X \in \mathscr{R}_\alpha \implies X \in \mathscr{R}_{\alpha'} \quad \text{for all } \alpha' > \alpha. \tag{9.53}$$

Therefore, given the observed data $X \in \mathscr{X}$, there exists the *smallest* size $\alpha_{\min}$ such that the size $\alpha_{\min}$ test rejects H_0. This smallest test size is exactly the p-value.

Definition 9.3. The *p-value* of X is the smallest size $\alpha_{\min}$ at which the test rejects the null hypothesis H_0 based on the data X,

$$\mathfrak{p}(X) = \alpha_{\min} = \min_{\alpha \in (0,1)} \{\alpha \,|\, X \in \mathscr{R}_\alpha\}. \tag{9.54}$$

Remark 9.5. It is important to highlight that the p-value[27] $\mathfrak{p}(X)$ depends not only on the data X, but also on the test statistic s, which is a "lens" through which we look at the data when testing hypotheses. When we need to explicitly indicate the s-dependence, we will write $\mathfrak{p}_s(X)$. So, $\mathfrak{p}_{s_1}(X) \neq \mathfrak{p}_{s_2}(X)$ for different statistics $s_1 \neq s_2$.

[26] That is, the *test size* α literally controls the *geometric size* of the rejection region $\mathscr{R}_\alpha$.

[27] We will use $\mathfrak{p}(X)$ and $\alpha_{\min}$ as two alternative notations for the p-value: $\mathfrak{p}(X)$ better reflects the name "p-value," and $\alpha_{\min}$ better reflects the essence of the p-value (the smallest size).

9.7.3 Geometric Interpretation

Geometrically, the p-value $\alpha_{\min}$ defines the smallest rejection region $\mathscr{R}_{\alpha_{\min}}$ that contains X. More precisely, if $\partial\mathscr{R}_\alpha$ is the boundary of the rejection region,

$$\partial\mathscr{R}_\alpha = \{X \in \mathscr{X} \mid s(X) = c_\alpha\}, \tag{9.55}$$

then $\alpha_{\min}$ defines the rejection region whose *boundary* contains X,

$$X \in \partial\mathscr{R}_{\alpha_{\min}}. \tag{9.56}$$

This is schematically illustrated in Fig. 9.9.

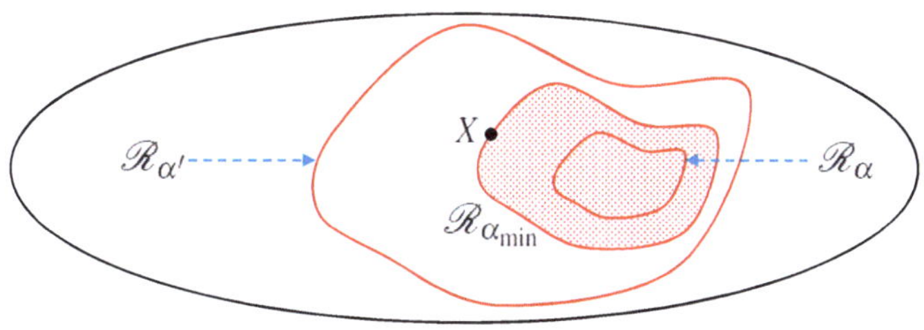

Fig. 9.9 Geometric interpretation of the p-value $\mathfrak{p}(X)$. If the test size α is too small, then the data $X \notin \mathscr{R}_\alpha$ and we fail to reject H_0. If α' is sufficiently large, then $X \in \mathscr{R}_{\alpha'}$ and we reject H_0. By gradually increasing the size from α to α', we expand the rejection region until we reach the value $\alpha_{\min} \in (\alpha, \alpha')$ for which $X \in \partial\mathscr{R}_{\alpha_{\min}}$. This test size is the p-value of X, $\mathfrak{p}(X) = \alpha_{\min}$.

Therefore, to find the p-value of X, we need to solve for $\alpha_{\min}$ the following equation:

$$s(X) = c_{\alpha_{\min}}. \tag{9.57}$$

9.7.4 Qualitative Interpretation

What is the intuitive *qualitative* interpretation of the p-value? Consider two samples, X and $\widetilde{X}$, and suppose that $\mathfrak{p}(\widetilde{X}) < \mathfrak{p}(X)$ or, equivalently, $\tilde{\alpha}_{\min} < \alpha_{\min}$. For example, X is as shown in Fig. 9.9 and $\widetilde{X}$ is somewhere inside $\mathscr{R}_\alpha$ in the same figure[28]. Then, since c_α is a decreasing function of α,

$$s(\widetilde{X}) = c_{\tilde{\alpha}_{\min}} > c_{\alpha_{\min}} = s(X). \tag{9.58}$$

The larger the value of the test statistic s, the stronger the data contradicts the null hypothesis. Therefore, $s(\widetilde{X}) > s(X)$ means that sample $\widetilde{X}$ provides more evidence against the null hypothesis than sample X. Thus, we have shown that

[28] Please feel free to draw a dot in Fig. 9.9 representing $\widetilde{X} \in \mathscr{R}_\alpha$.

$$\mathfrak{p}(\widetilde{X}) < \mathfrak{p}(X) \implies \widetilde{X} \text{ provides more evidence against } H_0 \text{ than } X. \tag{9.59}$$

So, qualitatively, the p-value is a *measure of evidence against the null hypothesis* provided by the data and chosen test statistic.

Let's compute the p-value for the example considered in Section 9.7.1 that we used for motivating the concept and compare $\mathfrak{p}(X^{(1)})$ and $\mathfrak{p}(X^{(2)})$.

Example 9.6. Suppose that data is modeled with the normal model,

$$X_1, \ldots, X_n \overset{\text{iid}}{\sim} \mathscr{N}(\mu, \sigma^2), \quad \mu \in \mathbb{R}, \tag{9.60}$$

where σ^2 is known and μ is the single parameter. And suppose we want to test

$$H_0 : \mu \leq 0 \quad \text{vs} \quad H_1 : \mu > 0. \tag{9.61}$$

In Example 9.5, we derived the size α test with the rejection region

$$\mathscr{R}_\alpha = \left\{ X \in \mathscr{X} \mid \overline{X} \geq c_\alpha = \frac{\sigma}{\sqrt{n}} z_{1-\alpha} \right\}. \tag{9.62}$$

If α increases, then $z_{1-\alpha}$ decreases, c_α decreases, and the hyperplane $\mathscr{R}_\alpha$ expands. Thus, to find the p-value of X, we need to find $\alpha_{\min}$ such that $X \in \partial \mathscr{R}_{\alpha_{\min}}$. So, we need to solve for $\alpha_{\min}$ the following equation:

$$\overline{X} = c_{\alpha_{\min}} = \frac{\sigma}{\sqrt{n}} z_{1-\alpha_{\min}}, \tag{9.63}$$

which gives[29]

$$\alpha_{\min} = \mathfrak{p}(X) = 1 - \Phi\left(\frac{\sqrt{n}}{\sigma} \overline{X} \right). \tag{9.64}$$

[29] Recall that $\Phi(z_\alpha) = \alpha$.

Now if $X^{(1)} = (X_1^{(1)}, \ldots, X_n^{(1)})$ and $X^{(2)} = (X_1^{(2)}, \ldots, X_n^{(2)})$ are two samples and the second sample lies deeper in the rejection region, $\overline{X}^{(2)} > \overline{X}^{(1)}$, see Fig. 9.8, then

$$\overline{X}^{(2)} > \overline{X}^{(1)} \implies \mathfrak{p}(X^{(2)}) < \mathfrak{p}(X^{(1)}). \tag{9.65}$$

As we discussed above on an intuitive level, the second sample provides more evidence against the null hypothesis than the first sample, since it lies deeper in the rejection region. The p-value confirms and quantifies this intuition.

9.7.5 Quantitative Interpretation

While (9.59) provides a qualitative interpretation of the p-value, in applications we need to know how to interpret p-values *quantitatively*, i.e. how to interpret their specific numerical values. Table 9.1 contains a popular conventional scheme for the interpretation of the p-values which is often used in practice.

P-Value	Conventional Interpretation
$\mathfrak{p}(X) < 0.01$	very strong evidence against H_0
$0.01 < \mathfrak{p}(X) < 0.05$	strong evidence against H_0
$0.05 < \mathfrak{p}(X) < 0.1$	weak evidence against H_0
$\mathfrak{p}(X) > 0.1$	no evidence against H_0

Table 9.1 A widely used yet often criticized scheme for the interpretation of values of the p-value.

It is important to highlight that there is nothing fundamental in the cutoff values 0.01, 0.05, and 0.1 listed in the table[30]. Their usage is purely historical and conventional, and the cutoffs can be changed if the underlying application requires this.

Instead of reporting that a test of a *certain* size α rejects or fails to reject H_0, we can report the corresponding p-value $\mathfrak{p}(X)$ of the data X. This is *more informative*, since the p-value tells us what decision the test will make for *any* size α: the test will reject H_0 for all $\alpha > \mathfrak{p}(X)$ and will fail to reject H_0 for all $\alpha < \mathfrak{p}(X)$.

Remark 9.6. If $\mathfrak{p}(X) < \alpha$, then this result is often called "statistically significant" at the level α, where α is typically 0.05. Intuitively, data X is statistically significant if it is unlikely to have occurred when the null hypothesis is true. If the result is declared as statistically significant, then the null hypothesis is rejected. Unfortunately, in the literature and in the media, the term "statistical significance" is overused, often loosely used, and sometime confused with "scientific" or "practical significance."[31] As it was explained above, it is more informative to report the exact value of the p-value than to state that the result is statistically significant at a certain level.

To get a better feel for the interpretation scheme in Tab. 9.1 and to consider a case where the sample space $\mathscr{X}$ and the rejection region $\mathscr{R}$ are discrete sets[32], let's compute the p-value for the ABC example.

Example 9.7. The rejection region in the ABC example has the following form

$$\mathscr{R} = \{k \in \{0, \dots, n\} \mid s(k) \geq c\}, \tag{9.66}$$

where the test statistic is the ratio of probabilities,

$$s(k) = \frac{\mathbb{P}(K = k | p = p_1)}{\mathbb{P}(K = k | p = p_0)} = \frac{\binom{n}{k} p_1^k (1-p_1)^{n-k}}{\binom{n}{k} p_0^k (1-p_0)^{n-k}} = \left(\frac{p_1}{p_0}\right)^k \left(\frac{1-p_1}{1-p_0}\right)^{n-k}, \tag{9.67}$$

the sample size $n = 10$, and the values of parameter p (the probability of heads) corresponding to the null and alternative hypotheses are $p_0 = 0.5$ and $p_1 = 0.7$. The "natural" value $c = 1$ of the critical value used in Example 9.1 leads to the rejection region $\mathscr{R} = \{7, 8, 9, 10\}$. When computing the p-value, we don't fix c, which fully determines the rejection region; instead, we need to consider the family of rejection regions (9.66), and then find the smallest region that contains the actually observed data $K = k$. The size of this smallest rejection region is the p-value.

Parameterization of the family of discrete rejection regions (9.66) with continuously varying c is not convenient, since different but close values of c define the

[30] For that reason, these cutoff values have been criticized, especially in recent years. For instance, it was suggested in [8] to replace the iconic 0.05 value (originally proposed by Fisher merely for definiteness and convenience for decision-making) with 0.005. We will discuss the unresolved debate about the p-value in more detail in Section 9.7.8.

[31] For example, if a new drug reduces blood pressure, but by a very small amount, then the result is statistically significant (there is effect), but practically insignificant (the effect size is too small for being useful).

[32] As opposed to continuous sets as in Example 9.6.

[33] For instance, any critical value $c \in (0.98, 2.25)$ leads to $\mathscr{R} = \{7, 8, 9, 10\}$.

[34] Large values of both k and $s(k)$ contradict H_0.

same rejection region[33] due to the discreteness of the set of all possible values of the test statistics: $s(k)$ has only $n+1$ different values, for $k = 0, \ldots, n$. These values increase with k, $s(k+1) > s(k)$. This is intuitively expected[34], evident from Fig. 9.1, and can be proved directly by using the expression for $s(k)$ in (9.67):

$$\frac{s(k+1)}{s(k)} = \frac{p_1}{p_0}\frac{1-p_0}{1-p_1} > 1. \tag{9.68}$$

If $k \in \mathscr{R} \Leftrightarrow s(k) \geq c$, then $s(k+1) > s(k) \geq c$ and, thus, $k+1 \in \mathscr{R}$. So, if k belongs to the rejection region $\mathscr{R}$, then so do $k+1, k+2, \ldots, n$. This means that all rejection regions have the following simple form:

$$\mathscr{R} = \{n_{\min}, \ldots, n\}, \tag{9.69}$$

where $n_{\min} = 0, \ldots, n+1$ is a discrete parameter that parameterizes the family of rejection regions. If $n_{\min} = 0$, then $\mathscr{R} = \mathscr{X}$, if $n_{\min} = n+1$, then $\mathscr{R} = \varnothing$.

If $K = k$ is the actually observed number of heads, then the smallest rejection region that contains k is $\mathscr{R}_{\min} = \{k, \ldots, n\}$. The p-value of k is the size of the test with rejection region $\mathscr{R}_{\min}$, i.e. the probability that $K \sim \mathscr{B}(n, p)$ is in $\mathscr{R}_{\min}$, given that $p = p_0$:

$$\begin{aligned}\mathfrak{p}(k) &= \mathbb{P}(K \in \mathscr{R}_{\min} \,|\, p = p_0) = \mathbb{P}(K \geq k \,|\, p = p_0) \\ &= \sum_{i=k}^{n} \mathbb{P}(K = i \,|\, p = p_0) = \sum_{i=k}^{n} \binom{n}{i} p_0^i (1-p_0)^{n-i}.\end{aligned} \tag{9.70}$$

Figure 9.10 shows the p-value $\mathfrak{p}(k)$ for $k = 0, \ldots, n$.

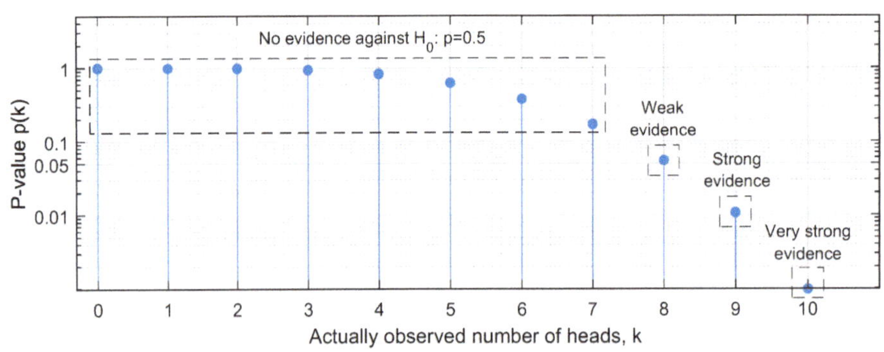

Fig. 9.10 The p-value $\mathfrak{p}(k)$ as a function of the observed number of heads k in the ABC example. According to the conventional scheme for the interpretation of the numerical values of the p-values summarized in Table 9.1, any $k \leq 7$ provides no evidence against the null hypothesis that the coin is unbiased, $H_0 : p = 0.5$, $k = 8$ provides week evidence, $k = 9$ provides strong evidence, and, finally, $k = 10$ (all tosses are heads) provides very strong evidence against H_0.

9.7.6 Probabilistic Interpretation

By Definition 9.3, the p-value is a certain size of a hypothesis test, and the size of a test is the supremum of the corresponding power function[35], which is the probability of a certain event[36]. This general observation suggests that the p-value should have a purely *probabilistic interpretation*. To derive this probabilistic interpretation[37], it is important to use different notations for the actually observed data, $x = (x_1, \dots, x_n)$, and its model $X = (X_1, \dots, X_n) \stackrel{\text{iid}}{\sim} F \in \mathscr{F}$. Just like we did in Example 9.7, where we used k to denote the actually observed number of heads and $K \sim \mathscr{B}(n, p)$ to denote its model. So, if x is the actually observed data (deterministic realization of random vector X), then the p-value of x is the size of the test with rejection region $\mathscr{R}_{\alpha_{\min}}$:

[35] See Definition 9.2.

[36] See Definition 9.1.

[37] Which would justify the "p" in the name "p-value."

$$\begin{aligned} \mathfrak{p}(x) = \alpha_{\min} = \sup_{\theta\in\Theta_0} \beta(\theta) &= \sup_{\theta\in\Theta_0} \mathbb{P}(X \in \mathscr{R}_{\alpha_{\min}} \,|\, \theta) \\ &= \sup_{\theta\in\Theta_0} \mathbb{P}(s(X) \geq c_{\alpha_{\min}} \,|\, \theta). \end{aligned} \tag{9.71}$$

On the other hand, according to (9.57), $\alpha_{\min}$, the p-value of x, satisfies $s(x) = c_{\alpha_{\min}}$. Therefore,

$$\mathfrak{p}(x) = \sup_{\theta\in\Theta_0} \mathbb{P}(s(X) \geq s(x) \,|\, \theta). \tag{9.72}$$

If the hull hypothesis is a *simple hypothesis*, that is, if it has the following form

$$H_0 : \theta = \theta_0, \tag{9.73}$$

where $\theta_0 \in \Theta$ is just a single point in the parameter space, then the supremum in (9.72) disappears and the formula simplifies to

$$\mathfrak{p}(x) = \mathbb{P}(s(X) \geq s(x) \,|\, \theta = \theta_0) = \mathbb{P}(s(X) \geq s(x) \,|\, H_0). \tag{9.74}$$

So, if H_0 is simple, then the p-value is the probability, under H_0, to observe a value of the test statistic $s(X)$ which is at least as extreme as the value $s(x)$ that has been actually observed. This probabilistic interpretation, which is schematically illustrated in Fig. 9.11, often allows to quickly compute the p-value when the null hypothesis is simple.

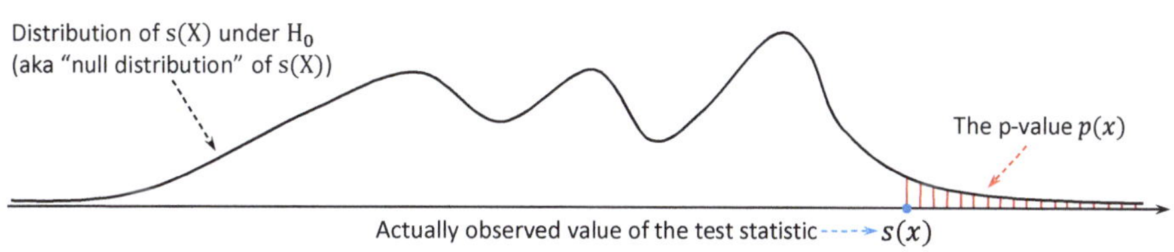

Fig. 9.11 Probabilistic interpretation of the p-value for the case, where the null hypothesis H_0 is simple. The p-value of the observed data $\mathfrak{p}(x) = \mathbb{P}(s(X) \geq s(x) \,|\, H_0)$ is the (red-shaded) area under the null distribution of the test statistic to the right of the observed value $s(x)$ of the test statistic. If $\mathfrak{p}(x)$ is small, then $s(x)$ is large, and, thus, x provides evidence against H_0.

For instance, let's revisit Example 9.7, where we computed the p-value for the ABC example. Applying (9.74) immediately gives[38]

$$\mathfrak{p}(k) = \mathbb{P}(s(K) \geq s(k) \,|\, p = p_0) = \mathbb{P}(K \geq k \,|\, p = p_0), \tag{9.75}$$

which is exactly (9.70). This method is much faster than the one we used in Example 9.7, where we had to consider the family of rejection regions and then change their parameterization from continuous to discrete.

If the null hypothesis $H_0 : \theta \in \Theta_0$ is not simple, then the probabilistic interpretation of the p-value is slightly more complicated and given by (9.72): the p-value is the maximum possible (under H_0) probability to observe the value of the test statistic which is the same or larger than the value that has been actually observed. If maximizing the probability over Θ_0 is not difficult, then this probabilistic interpretation often also allows to compute the p-value relatively quickly, compared to other methods[39].

[38] Here we use the established fact that $s(k)$ is an increasing function of k, and, therefore, $s(K) \geq s(k)$ is equivalent to $K \geq k$.

[39] Problem 9.2 will ask you to revisit Example 9.6 and compute the p-value using formula (9.72).

9.7.7 Misinterpretations

The p-value is a non-trivial concept. We discussed a motivation behind this concept, a rather involved formal definition, and several different interpretations: geometric, qualitative, quantitative, and probabilistic. It is now important to discuss three main misinterpretations of the p-value, which sometimes appear even in academic papers published in respected journals. For more common misinterpretations, not only of p-values, but also confidence intervals and statistical tests in general, see a highly recommended guide[40] [9].

1. *The p-value is not the probability that H_0 is true.*
   The p-value does have a probabilistic interpretation, but it is not the probability that the null hypothesis is true given the observed data, i.e. $\mathfrak{p}(x) \neq \mathbb{P}(H_0 \,|\, X = x)$. If fact, the p-value is computed under the assumption that H_0 is true.
2. *A statistically significant result*[41] *does not necessarily mean that H_0 is false.*
   A small p-value can occur not only when H_0 is false, but also when it is true, but there was an error in the data, or if some test assumption was violated, or even by random chance. For instance, in the ABC example, even if H_0 is true (the chosen coin is unbiased), Alice can still get $k = 9$ heads out of $n = 10$ tosses. In that case, the p-value will be small, $\mathfrak{p}(k = 9) < 0.05$ (see Fig. 9.10), and the observed data $k = 9$ will provide strong evidence against H_0 despite it being true. Since the p-value is a measure of evidence against H_0, not a measure of whether H_0 is true, a small p-value should serve as a "red flag" that the observed data is unusual (unexpected, surprising) if H_0 is true (and all test assumptions are satisfied).
3. *A large p-value does not provide evidence in favor of H_0.*
   Recall that the p-value depends not only of the data, but also on the test statistic[42], which is a "lens" through which we look at the data and test hypotheses. A large

[40] It has 18 various misinterpretations of p-values!

[41] Such as $\mathfrak{p}(x) < 0.05$.

[42] See Remark 9.5.

p-value can occur not only when H_0 is true, but also when it is false, but the test is statistic s is poorly chosen, so that it does not discriminate well between H_0 and H_1. This situation is illustrated in Fig. 9.12. Suppose that H_0 is false and we

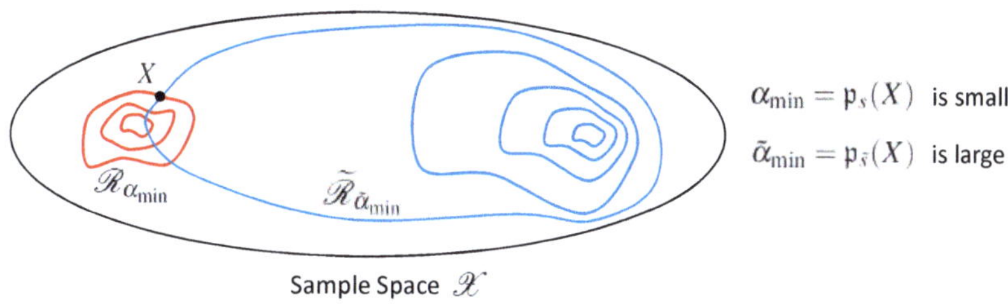

Fig. 9.12 The red contours on the left and the blue contours on the right represent the families of rejection regions $\mathscr{R}$ and $\tilde{\mathscr{R}}$ with respect to test statistics s and $\tilde{s}$, respectively. The data $X \in \partial\mathscr{R}_{\alpha_{\min}}$ and $X \in \partial\tilde{\mathscr{R}}_{\tilde{\alpha}_{\min}}$, where $\alpha_{\min} = \mathfrak{p}_s(X)$ and $\tilde{\alpha}_{\min} = \mathfrak{p}_{\tilde{s}}(X)$ are the p-values of X computed with respect to s and $\tilde{s}$.

have two different test statistics: a good statistic s, which discriminates between H_0 and H_1 well, and a bad statistic $\tilde{s}$, which does not. Since H_0 is false and s is good, the p-value $\mathfrak{p}_s(X)$ computed with respect to s will be small, which is evidence against H_0. Since $\tilde{s}$ is bad, the small size rejection regions $\tilde{R}_{\tilde{\alpha}}$ defined by $\tilde{s}$ (small blue regions in the figure) will be far from the data X. To "capture" the data, $X \in \partial\tilde{\mathscr{R}}_{\tilde{\alpha}_{\min}}$, we will have to expand the rejection region substantially, and this will result in the large p-value $\tilde{\alpha}_{\min} = \mathfrak{p}_{\tilde{s}}(X)$ with respect to $\tilde{s}$. But we can't interpret the large p-value $\mathfrak{p}_{\tilde{s}}(X)$ as evidence in favor of H_0. Conceptually: the absence of evidence against a hypothesis does not provide evidence in favor of the hypothesis.

To provide a less abstract and more concrete example, suppose we have two samples, $X_1 \dots, X_n \overset{\text{iid}}{\sim} F_X$ and $Y_1, \dots, Y_n \overset{\text{iid}}{\sim} F_Y$, from two distributions, F_X and F_Y, and we want to test the null hypothesis that the two distributions are the same versus the alternative hypothesis that they are different[43],

$$H_0 : F_X = F_Y \quad \text{vs} \quad H_1 : F_X \neq F_Y. \tag{9.76}$$

To test these hypotheses, we need to choose a test statistic such that its large values contradict H_0. Let's use the absolute value of the difference of the sample means,

$$s(X,Y) = |\overline{X} - \overline{Y}|. \tag{9.77}$$

According to (9.57), the p-value $\alpha_{\min}$ satisfies the following equation:

$$s(X,Y) = c_{\alpha_{\min}}, \tag{9.78}$$

where the critical value c_α is a decreasing function of the size α[44]. If the p-value $\alpha_{\min}$ is small, then $c_{\alpha_{\min}}$ and $s(X,Y)$ are large, and this provides evidence against H_0, since $|\overline{X} - \overline{Y}|$ being large is not compatible with $F_X = F_Y$. On the

[43] For example, $X_1 \dots, X_n$ and $Y_1, \dots, Y_n$ could represent, respectively, the changes in blood pressure after taking a new drug and after placebo. The null hypothesis states that the drug has zero effect, and the alternative hypothesis states that the effect exists.

[44] See (9.51).

other hand, if the p-value $\alpha_{\min}$ is large, then $c_{\alpha_{\min}}$ and $s(X,Y)$ are small, but this does not provide evidence in favor of H_0, since $|\overline{X}-\overline{Y}|$ being small is perfectly compatible with $F_X \neq F_Y$. Distributions can have the same means but be different. The chosen test statistic (9.77) measures a certain aspect of discrepancy between F_X and F_Y, but it does not discriminate between H_0 and H_1 well. A much better statistic for testing (9.76) will be discussed in Section 10.4.

9.7.8 The P-value Debate

The smart way to keep people passive and obedient is to strictly limit the spectrum of acceptable opinion, but allow very lively debate within that spectrum — even encourage the more critical and dissident views.
– Noam Chomsky, *The Common Good*

The complexity of the concept and, as a result, its frequent misuses and misinterpretations on the one hand, and the popularity of the concept and the ubiquity of applications where it is used on the other, have sparked a hot debate about the p-value. In 2015, the editors of the journal *Basic and Applied Social Psychology* argued that p-values are more misleading than informative and made a radical decision to completely ban the use of p-values in their journal [10]. In 2016, the *American Statistical Association* (ASA) released a statement [11], where it recommended to stop using statistical significance[45], i.e. to stop making dichotomous decisions (accept or reject) based only on whether the p-value passes a specified threshold, such as $p < 0.05$. Nevertheless, the ASA did not recommend to stop using the p-values per se. A call to retire statistical significance [12] — to ban not the p-values, but their use for threshold-based categorical decision-making — was signed by 854 scientists from 52 countries. Many researchers, instead of abandoning statistical significance, proposed to lower the significance threshold from 0.05 to 0.005 [8, 13], but many others disagreed with this suggestion for different and sometimes opposite reasons[46] [15, 16]. In the context of this debate, several alternatives to the p-value have been proposed [17], but none of them have seen a widespread use. On the other side of the debate are scientists and researchers who recommend to keep using the p-values and not to abandon statistical significance, e.g. [16, 18, 19]. They assert that inferences are unavoidably dichotomous, that statistical significance has a crucial practical importance, and that misuses and misinterpretations of p-values made by some researchers is not a good reason for banning the use of p-values altogether.

[45] See Remark 9.6

[46] "Hearing that p-values are terrible and that, by the way, they are not low enough recalls the vacationer's complaint that "The food was horrible – and the portions were so small!" The two complaints nullify each other." [14].

What can I recommend, in the face of this unresolved debate, to the reader who has just learned about hypothesis testing, p-values, and statistical significance? Three things:

1. Refine your understanding of the p-value. Even if you will not use it, other researchers will, and when you see a statement about the p-value in a paper, poster, or presentation, it is important to understand what it means.

2. When using the p-value, always report its computed value, not just whether it is below or above a certain threshold. Be clear whether the computed value is exact or estimated[47]. Check whether the data which is used for testing satisfies the assumptions of the statistical test which is used for computing the p-value[48].
3. Before declaring statistical significance or non-significance and using it for rejecting or accepting the null hypothesis, read the papers cited above. This should help to avoid potential pitfalls and mistakes. In particular, the significance threshold (be it the traditional 0.05 or something else[49]) must be set before the p-value is computed, and statistical significance should not be confused with practical or scientific significance[50]. Finally, keep in mind that the p-value is just a single-number measure of the evidence against the null hypothesis provided by the data and subjectively chosen test statistic. Don't give it more importance than it actually deserves. Keep calm if the p-value is less than 0.05.

[47] As we will see in the next chapter, in many cases it is hard or impossible to compute the p-value exactly and we will have to estimate it.

[48] For example, (9.64) was computed under the normality assumption (9.60).

[49] As Fisher noted, "No scientific worker has a fixed level of significance at which from year to year, and in all circumstances, he rejects hypotheses; he rather gives his mind to each particular case in the light of his evidence and his ideas." [2].

[50] See Marginal Note 31.

9.8 Further Reading

This chapter described a general theoretical framework for testing statistical hypotheses and illustrated it with elementary examples. We discussed the conceptual difference between the null and alternative hypotheses, two types of possible errors made in testing, and the key notions of power function, size of a test, and p-value. One of the most comprehensive books on hypothesis testing, which can also be used as a reference book, is

- Lehmann, E.L., Romano, J.P.: *Testing Statistical Hypotheses*. New York: Springer, 3rd Edition (2005). `https://doi.org/10.1007/0-387-27605-X`

In some applications, there is a need to perform *multiple testing*, that is, to simultaneously test several hypotheses. For instnace, this happens when the efficacy of a new drug is tested on different sub-populations, or hypotheses about gene expression levels are tested for thousands of genes using data from DNA microarrays, or one wants to test which coefficients in a multiple regression model[51] are significantly different from zero. Even if for each individual test the probability of type I error α is small, the probability of making *at least one* type I error in N tests can be very large. Indeed, assuming that all tests a independent,

[51] See Chapter 12.

$$\begin{aligned}&\mathbb{P}(\text{at least one type I error in } N \text{ tests}) = 1 - \mathbb{P}(\text{no type I errors in } N \text{ tests})\\&= 1 - \mathbb{P}(\text{no type I error in test } 1, \ldots, \text{no type I error in test } N)\\&= 1 - \prod_{i=1}^{N} \underbrace{\mathbb{P}(\text{no type I error in test } i)}_{1-\alpha} = 1 - (1-\alpha)^N.\end{aligned} \tag{9.79}$$

For example, if $\alpha = 0.05$ and $N = 100$, then the probability of at least one type I error is approximately 0.994. Similarly, if p-values are used for measuring statistical significance, then some of them are likely to fall below a chosen significance threshold purely by chance, and the corresponding test results will subsequently be declared statistically significant even if they are not. This problem and its solutions are discussed in detail in Chapter 9 of Lehmann and Romano's book, but perhaps

- Chen, S.Y., Feng, Z., Yi., X.: A general introduction to adjustment for multiple comparisons. *The Journal of Thoracic Disease* 9(6), 1725-1729 (2017). `https://doi.org/10.21037/jtd.2017.05.34`

can provide a less theoretical, but more practical and reader-friendly introduction to the problem, along with an overview of available solutions.

Problems

9.1. ABC Example: Test Sizes and Error Probabilities
Consider the ABC example introduced in Section 9.1 and studied in Examples 9.1, 9.4, and 9.7. Recall[52] that the rejection region has the following form:

[52] See (9.69) and the preceding discussion.

$$\mathscr{R}_{n_{\min}} = \{n_{\min}, \dots, n\}, \tag{9.80}$$

where $n_{\min} = 0, \dots, n+1$ and $n = 10$. Find the size of the test with rejection region $\mathscr{R}_{n_{\min}}$ and the corresponding probabilities of Type I and Type II errors, and plot these probabilities versus $n_{\min}$.

9.2. Probabilistic Interpretation of the P-value
Suppose that data is modeled with the normal model,

$$X_1, \dots, X_n \overset{\text{iid}}{\sim} \mathscr{N}(\mu, \sigma^2), \quad \mu \in \mathbb{R}, \tag{9.81}$$

where σ^2 is known and μ is the single parameter. And suppose we want to test

$$H_0 : \mu \leq 0 \quad \text{vs} \quad H_1 : \mu > 0. \tag{9.82}$$

In Example 9.6, we used the sample mean $s(X) = \overline{X}$ as a test statistic and found the corresponding p-value of $X = x$,

$$\mathfrak{p}(x) = 1 - \Phi\left(\frac{\sqrt{n}}{\sigma}\,\overline{x}\right), \tag{9.83}$$

using a "geometric" approach based on (9.57). Find the p-value of $X = x$ using a probabilistic approach derived in Section 9.7.6.

9.3. Testing the Uniform Distribution
Suppose that data is modeled with the uniform model,

$$X_1, \dots, X_n \overset{\text{iid}}{\sim} \mathscr{U}[0, \theta], \tag{9.84}$$

where the right end θ is a model parameter. And suppose we want to test

$$H_0 : \theta = \theta_0 \quad \text{vs} \quad H_1 : \theta > \theta_0, \tag{9.85}$$

where θ_0 is some positive constant. Consider $X_{(n)} = \max\{X_1, \ldots, X_n\}$, the largest value in the sample. We expect $X_{(n)}$ to be smaller under H_0 than under H_1. It seems natural to reject H_0 if $X_{(n)}$ is large: the larger $X_{(n)}$ is, the more evidence against H_0 it provides. For example, if $X_{(n)} > \theta_0$, then H_0 is definitely false. So, let's use a test with the test statistic $s(X) = X_{(n)}$ and rejection region

$$\mathscr{R} = \{X \mid X_{(n)} \geq c\}, \tag{9.86}$$

where c is a critical value.

(a) Find the power function $\beta(\theta)$ of this test.

(b) Construct a test of size α by choosing c appropriately.

(c) Compute the p-value of $X_{(n)} = x_{(n)} = 0.48$ if the sample size $n = 20$ and $\theta_0 = 1/2$.

9.4. Testing the Normal Distribution
Suppose that data is modeled with the normal model,

$$X_1, \ldots, X_n \overset{\text{iid}}{\sim} \mathscr{N}(\mu, \sigma^2), \quad \mu \in \mathbb{R}, \tag{9.87}$$

where σ^2 is known and μ is a model parameter. And suppose that, instead of (9.82), we now want to test

$$H_0 : \mu = \mu_0 \quad \text{vs} \quad H_1 : \mu = \mu_1, \tag{9.88}$$

where $\mu_0 < \mu_1$. Intuitively, the larger the sample mean $\overline{X}$ is, the more evidence against H_0 the data provides. So, let's consider a test with the rejection region

$$\mathscr{R} = \{X \mid \overline{X} \geq c\}. \tag{9.89}$$

(a) Construct a test of size α.

(b) Find its power under H_1 and explore its behavior if $\mu_1 \approx \mu_0$ and if $\mu_1 \to \infty$.

(c) Find the limit of the power function of the size α test as the sample size $n \to \infty$.

(d) Compute the p-value of the actually observed data $\overline{X} = \bar{x}$ for $\mu_0 = -1$, $\mu_1 = 1$, $\sigma = 10$, and $n = 1000$, and plot it is a function of $\bar{x}$ for $\bar{x} \in [-2, 2]$.

References

1. Student: The probable error of a mean. *Biometrika* 6(1), 1-25 (1908). https://doi.org/10.2307/2331554
2. Fisher, R.A.: *Statistical Methods for Research Workers*. Edinburgh: Oliver & Boyd (1925). https://doi.org/10.1007/978-1-4612-4380-9_6
3. Neyman, J., Pearson, E.S.: On the use and interpretation of certain test criteria for purposes of statistical inference. *Biometrika* 20A (1/2), 175-250 (1928). https://doi.org/10.2307/2331945

4. Neyman, J., Pearson, E.S.: On the problem of the most efficient tests of statistical hypotheses. *Philosophical Transactions of the Royal Society of London*, Ser. A, 231, 289-337 (1933). https://www.jstor.org/stable/91247
5. Neyman, J., Pearson, E.S.: The testing of statistical hypotheses in relation to probabilities a priori. *Proceedings of the Cambridge Philosophical Society* 29(4), 492-510 (1933). https://doi.org/10.1017/S030500410001152X
6. Neyman, J.: Statistics – servant of all sciences. *Science* 122(3126), 401-406 (1955). https://doi.org/10.1126/science.122.3166.401
7. Aschwanden, C.: Statisticians found one thing they can agree on: it's time to stop misusing p-values. (2016). http://fivethirtyeight.com/features/statisticians-found-one-thing-they-can-agree-on-its-time-to-stop-misusing-p-values/
8. Benjamin, D.J., Berger, J.O., Johnson, V.E., et al.: Redefine statistical significance. *Nature Human Behaviour* 2, 6-10 (2018). https://doi.org/10.1038/s41562-017-0189-z
9. Greenland, S., Senn, S.J., Rothman, K.J., Carlin, J.B., Poole, C., Goodman, S.N., Altman, D.G.: Statistical tests, P values, confidence intervals, and power: a guide to misinterpretations. *European Journal of Epidemiology* 31, 337-350 (2016). https://doi.org/10.1007/s10654-016-0149-3
10. Trafimow, D., Marks, M.: Editorial. *Basic and Applied Social Psychology* 37, 1-2 (2015). https://doi.org/10.1080/01973533.2015.1012991
11. Wasserstein, R.L., Lazar, N.A.: The ASA Statement on p-values: context, process, and purpose. *The American Statistician* 70(2), 129-133 (2016). https://doi.org/10.1080/00031305.2016.1154108
12. Amrhein, V., Greenland, S., McShane, B.: Scientists rise up against statistical significance. *Nature* 567, 305-307 (2019). https://doi.org/10.1038/d41586-019-00857-9
13. Ioannidis, J.P.A.: The proposal to lower p value thresholds to .005. *The Journal of the American Medical Association* 319(14), 1429-1430 (2018). https://doi.org/10.1001/jama.2018.1536
14. Krueger, J.I., Heck, P.R.: The heuristic value of p in inductive statistical inference. *Frontiers in Psychology* 8, 908 (2017). https://doi.org/10.3389/fpsyg.2017.00908
15. Trafimow, D., Amrhein, V., Areshenkoff., C.N. et al: Manipulating the alpha level cannot cure significance testing. *Frontiers in Psychology* 9, 699 (2018). https://doi.org/10.3389/fpsyg.2018.00699
16. Di Leo, G., Sardanelli, F.: Statistical significance: p value, 0.05 threshold, and applications to radiomics-reasons for a conservative approach. *European Radiology Experimental* 4, 18 (2020). https://doi.org/10.1186/s41747-020-0145-y
17. Wasserstein, R.L., Schirm, A.L., Lazar, N,A,: Moving to a world beyond "p<0.05". *American Statistician* 73, 1-19 (2019). https://doi.org/10.1080/00031305.2019.1583913
18. Ioannidis, J.P.A.: The importance of predefined rules and prespecified statistical analyses: do not abandon significance. *The Journal of the American Medical Association* 321(21), 2067-2068 (2019). https://doi.org/10.1001/jama.2019.4582
19. Leek, J.T., Peng, R.D.: Statistics: p-values are just the tip of the iceberg. *Nature* 520, 612 (2015). https://doi.org/10.1038/520612a

10
Popular Statistical Tests

What is right is not always popular and what is popular is not always right.

– Albert Einstein

In the previous chapter, we discussed a general framework for hypothesis testing. Our next goal is to consider several specific statistical tests that are often used in practice. In this chapter, we will learn the Wald test and the t-test for testing simple hypotheses, the Kolmogorov-Smirnov test for testing the equality of probability distributions, the likelihood ratio test, which, thanks to its power, is one of the most widely used tests in parametric inference, and an elegant nonparametric permutation test for testing whether two samples were generated by the same data generating process. To illustrate how these tests can be used in applications, we will consider two case studies: we will compare the cholesterol levels in men and women and will statistically test Mendel's theory of inheritance.

10.1 Wald Test

Suppose that data $X_1, \ldots, X_n$ is modeled as an IID sample from a distribution F,

$$X_1, \ldots, X_n \overset{\text{iid}}{\sim} F \in \mathscr{F}, \tag{10.1}$$

where $\mathscr{F}$ is a parametric or nonparametric stochastic model. Let $\theta \in \mathbb{R}$ be a parameter of interest and suppose that we want to test

$$H_0 : \theta = \theta_0 \quad \text{vs} \quad H_1 : \theta \neq \theta_0, \tag{10.2}$$

where $\theta_0 \in \mathbb{R}$ is a hypothesized value. A hypothesis of the form $H_0 : \theta = \theta_0$, where θ_0 is a single point in the parameter space, is called a *simple hypothesis*. In practice, the hypothesized value is often set to zero, $\theta_0 = 0$, which corresponds to "no effect," "no relationship," or "no difference."[1]

[1] Recall the conceptual role of the null hypothesis described in Section 9.3.

Supplementary Information The online version contains supplementary material available at https://doi.org/10.1007/978-3-032-03848-7_10.

K. M. Zuev, *Fundamentals of Statistical Inference*,
International Series in Operations Research & Management Science 374,
https://doi.org/10.1007/978-3-032-03848-7_10

The *Wald test* for testing (10.2) bridges the gap between three different types of inferential problems: point estimation, confidence intervals, and hypothesis testing. Let $\hat{\theta}_n$ be an estimate of θ and $\widehat{\text{se}}[\hat{\theta}_n]$ be an estimate of the standard error $\text{se}[\hat{\theta}_n]$ of $\hat{\theta}_n$. Assume that $\hat{\theta}_n$ is normally distributed,

Abraham Wald (1902-1950), a Hungarian mathematician, who made important contributions to statistics and economics, and was known for "his gentleness, and extraordinary strength with which he attacked his problems." [1]. Photo: Archives of the Mathematisches Forschungsinstitut Oberwolfach.

$$\frac{\hat{\theta}_n - \theta_\star}{\widehat{\text{se}}[\hat{\theta}_n]} \sim \mathcal{N}(0,1), \tag{10.3}$$

where $\theta_\star$ is the true value of θ. It can be hard to check whether this assumption holds for a general estimate $\hat{\theta}_n$, which, in principle, can be any function of the data $X_1, \ldots, X_n$. However, if $\mathcal{F}$ is a parametric model and $\hat{\theta}_n$ is the maximum likelihood estimate, $\hat{\theta}_n = \hat{\theta}_{MLE}$, then its standard error can be estimated as

$$\text{se}[\hat{\theta}_{MLE}] \approx \widehat{\text{se}}[\hat{\theta}_{MLE}] = 1/\sqrt{n\mathcal{I}(\hat{\theta}_{MLE})}, \tag{10.4}$$

where $\mathcal{I}$ is the Fisher information, and it is asymptotically normal[2],

$$\sqrt{n\mathcal{I}(\hat{\theta}_{MLE})}(\hat{\theta}_{MLE} - \theta_\star) \xrightarrow{\mathbb{D}} \mathcal{N}(0,1), \qquad \text{as } n \to \infty. \tag{10.5}$$

[2] See Section 8.6.2

This means that for the MLE estimate, the normality assumption (10.3) holds, at least approximately, for large n. This assumption is also likely to hold approximately for an unbiased (or asymptotically unbiased) estimate $\hat{\theta}_n$ that can be expressed as a sum of IID terms. In this case, the central limit theorem can be used to argue that $\hat{\theta}_n$ is approximately normally distributed centered at the true value $\theta_\star$.

If the normality assumption (10.3) holds and the null hypotheses H_0 is true, then the hypothesized value θ_0 is the true value of the parameter and the quantity

$$W = \left|\frac{\hat{\theta}_n - \theta_0}{\widehat{\text{se}}[\hat{\theta}_n]}\right| \quad \text{is expected to be not far from zero.} \tag{10.6}$$

In other words, large values of this quantity will contradict the null hypothesis. The Wald test uses W as the test statistic. Namely,

$$\text{The Wald test rejects } H_0 \quad \Longleftrightarrow \quad W = \left|\frac{\hat{\theta}_n - \theta_0}{\widehat{\text{se}}[\hat{\theta}_n]}\right| \geq c. \tag{10.7}$$

To construct the Wald test of size α (so that the maximum possible probability of type I error is α), we need[3] to solve for c the following equation

[3] Recall (9.46).

$$\sup_{\theta \in \Theta_0} \mathbb{P}(W \geq c \,|\, \theta) = \alpha. \tag{10.8}$$

Since the null hypotheses is simple, $\Theta_0 = \{\theta_0\}$, we have

$$\alpha = \sup_{\theta \in \Theta_0} \mathbb{P}(W \geq c \,|\, \theta) = \mathbb{P}(W \geq c \,|\, \theta_0) = \mathbb{P}\left(\left| \underbrace{\frac{\hat{\theta}_n - \theta_0}{\widehat{\mathrm{se}}[\hat{\theta}_n]}}_{\sim \mathcal{N}(0,1)} \right| \geq c \,\middle|\, \theta_0 \right) = 2\Phi(-c) \quad (10.9)$$

Therefore, the critical value is $c = -\Phi^{-1}(\alpha/2) = -z_{\alpha/2} = z_{1-\alpha/2}$, and

$$\text{The size } \alpha \text{ Wald test rejects } H_0 \quad \Longleftrightarrow \quad W = \left| \frac{\hat{\theta}_n - \theta_0}{\widehat{\mathrm{se}}[\hat{\theta}_n]} \right| \geq z_{1-\frac{\alpha}{2}}. \quad (10.10)$$

Now, let's find the power function of this test and explore its properties. Recall[4] that the power function $\beta(\theta)$ is the probability of rejecting the null hypotheses given that the true value of the parameter is θ,

[4] See Definition 9.1.

$$\beta(\theta) = \mathbb{P}\left(\left| \frac{\hat{\theta}_n - \theta_0}{\widehat{\mathrm{se}}[\hat{\theta}_n]} \right| \geq z_{1-\frac{\alpha}{2}} \,\middle|\, \theta \right). \quad (10.11)$$

Note that, in general, $(\hat{\theta}_n - \theta_0)/\widehat{\mathrm{se}}[\hat{\theta}_n]$ is not distributed according to the standard normal distribution[5]. Given that θ is the true value, $(\hat{\theta}_n - \theta)/\widehat{\mathrm{se}}[\hat{\theta}_n] \sim \mathcal{N}(0,1)$. To use this, first, we need to get rid of the absolute value in (10.11):

[5] It is distributed according to $\mathcal{N}(0,1)$ only if H_0 is true, that is, $\theta = \theta_0$.

$$\begin{aligned}
\beta(\theta) &= \mathbb{P}\left(\frac{\hat{\theta}_n - \theta_0}{\widehat{\mathrm{se}}[\hat{\theta}_n]} \geq z_{1-\frac{\alpha}{2}} \,\middle|\, \theta \right) + \mathbb{P}\left(\frac{\hat{\theta}_n - \theta_0}{\widehat{\mathrm{se}}[\hat{\theta}_n]} \leq \underbrace{-z_{1-\frac{\alpha}{2}}}_{z_{\frac{\alpha}{2}}} \,\middle|\, \theta \right) \\
&= \mathbb{P}\left(\underbrace{\frac{\hat{\theta}_n - \theta}{\widehat{\mathrm{se}}[\hat{\theta}_n]}}_{\sim \mathcal{N}(0,1)} \geq z_{1-\frac{\alpha}{2}} + \frac{\theta_0 - \theta}{\widehat{\mathrm{se}}[\hat{\theta}_n]} \,\middle|\, \theta \right) + \mathbb{P}\left(\underbrace{\frac{\hat{\theta}_n - \theta}{\widehat{\mathrm{se}}[\hat{\theta}_n]}}_{\sim \mathcal{N}(0,1)} \leq z_{\frac{\alpha}{2}} + \frac{\theta_0 - \theta}{\widehat{\mathrm{se}}[\hat{\theta}_n]} \,\middle|\, \theta \right) \quad (10.12) \\
&= 1 - \Phi\left(z_{1-\frac{\alpha}{2}} + \frac{\theta_0 - \theta}{\widehat{\mathrm{se}}[\hat{\theta}_n]} \right) + \Phi\left(z_{\frac{\alpha}{2}} + \frac{\theta_0 - \theta}{\widehat{\mathrm{se}}[\hat{\theta}_n]} \right).
\end{aligned}$$

Let's derive several basic properties of this function. First, since $\beta(\theta)$ is the power function of the Wald test of size α, we should have $\beta(\theta_0) = \alpha$. Indeed,

$$\beta(\theta_0) = 1 - \Phi(z_{1-\frac{\alpha}{2}}) + \Phi(z_{\frac{\alpha}{2}}) = 1 - (1 - \alpha/2) + \alpha/2 = \alpha. \quad (10.13)$$

Next, if $\theta \to \pm\infty$, then the power function converges to 1,

$$\begin{aligned}
\lim_{\theta \to +\infty} \beta(\theta) &= 1 - \Phi(-\infty) + \Phi(-\infty) = 1, \\
\lim_{\theta \to -\infty} \beta(\theta) &= 1 - \Phi(+\infty) + \Phi(+\infty) = 1.
\end{aligned} \quad (10.14)$$

It can be also checked that it is symmetric about θ_0, that is,

$$\beta(\theta_0 - x) = \beta(\theta_0 + x), \quad \text{for any } x \in \mathbb{R}. \quad (10.15)$$

For many reasonable estimates, the standard error $\mathrm{se}[\hat{\theta}_n] \approx \widehat{\mathrm{se}}[\hat{\theta}_n] \to 0$ as the sample size increases[6]. In this case, the power function of the size α Wald test converges to the ideal power function, except at θ_0, where it equals to α by definition[7]:

[6] See, for instance, (10.4).

[7] We have already seen this behavior of the power function in Example 9.5.

$$\lim_{\widehat{\mathrm{se}}[\hat{\theta}_n]\to 0} \beta(\theta) = \begin{cases} 1 - \Phi(+\infty) + \Phi(+\infty) = 1, & \text{if } \theta < \theta_0, \\ \alpha, & \text{if } \theta = \theta_0, \\ 1 - \Phi(-\infty) + \Phi(-\infty) = 1, & \text{if } \theta > \theta_0. \end{cases} \tag{10.16}$$

The power function of the Wald test of size α is shown schematically in Fig. 10.1.

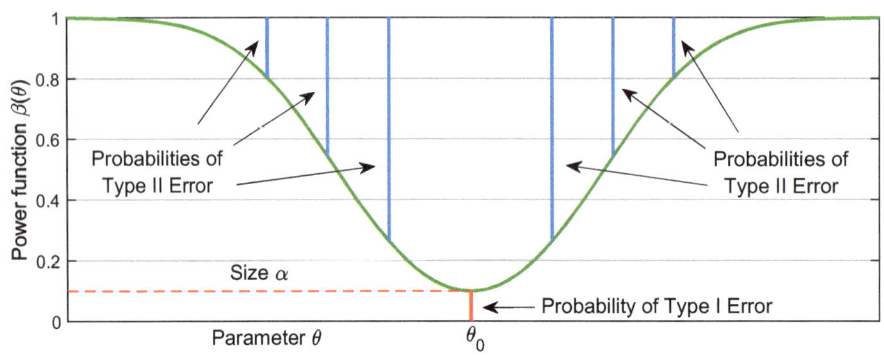

Fig. 10.1 The power function of the size α Wald test defined in (10.10). The probability of Type I error, represented by the vertical red line, coincides with the size and equals to α. The probability of the Type II error, represented by vertical blue lines, is large if the true value of the parameter is close to the hypothesized value θ_0 and decreases as the true value goes away from θ_0. If $\widehat{\mathrm{se}}[\hat{\theta}_n] \to 0$ as $n \to \infty$, then we can reduce the type II error probability by increasing the sample size.

The normality assumption (10.3) allows to effortlessly construct a confidence interval for the true value $\theta_\star$ of the parameter. Namely, if (10.3) holds, then

$$I_\alpha^W = \left(\hat{\theta}_n - \widehat{\mathrm{se}}[\hat{\theta}_n] z_{1-\frac{\alpha}{2}}, \hat{\theta}_n + \widehat{\mathrm{se}}[\hat{\theta}_n] z_{1-\frac{\alpha}{2}}\right) \tag{10.17}$$

is an $100(1-\alpha)\%$ confidence interval for $\theta_\star$. Indeed,

$$\begin{aligned} \mathbb{P}(\theta_\star \in I_\alpha^W) &= \mathbb{P}\left(\hat{\theta}_n - \widehat{\mathrm{se}}[\hat{\theta}_n] z_{1-\frac{\alpha}{2}} < \theta_\star < \hat{\theta}_n + \widehat{\mathrm{se}}[\hat{\theta}_n] z_{1-\frac{\alpha}{2}}\right) \\ &= \mathbb{P}\left(\underbrace{-z_{1-\frac{\alpha}{2}}}_{z_{\frac{\alpha}{2}}} < \underbrace{\frac{\hat{\theta}_n - \theta_\star}{\widehat{\mathrm{se}}[\hat{\theta}_n]}}_{\sim\mathcal{N}(0,1)} < z_{1-\frac{\alpha}{2}}\right) = \Phi(z_{1-\frac{\alpha}{2}}) - \Phi(z_{\frac{\alpha}{2}}) = 1 - \alpha. \end{aligned} \tag{10.18}$$

Note that the size α Wald test derived in (10.10) rejects $H_0 : \theta = \theta_0$ if and only if the hypothesized value θ_0 does not belong to the confidence interval I_α^W for $\theta_\star$,

$$\text{The size } \alpha \text{ Wald test rejects } H_0 \quad \Longleftrightarrow \quad \theta_0 \notin I_\alpha^W. \tag{10.19}$$

This means that using the size α Wald test for testing $H_0 : \theta = \theta_0$ is equivalent to checking whether θ_0 belongs to the $100(1-\alpha)\%$ confidence interval I_α^W.

To compute the p-value $\mathfrak{p}(x)$ of the actually observed data $x = (x_1, \dots, x_n)$, it is convenient to use its probabilistic interpretation[8]: $\mathfrak{p}(x)$ is the probability of observing a value of the Wald statistic $W(X)$ that is at least as large as the value $W(x)$ that has been actually observed, under the assumption that H_0 is true:

$$\begin{aligned}\mathfrak{p}(x) = \mathbb{P}(W(X) \geq W(x) | H_0) &= \mathbb{P}\left(\left|\underbrace{\frac{\hat{\theta}_n - \theta_0}{\widehat{\text{se}}[\hat{\theta}_n]}}_{\sim\mathcal{N}(0,1)}\right| \geq W(x) \middle| \theta_0\right) \\ &= 2\Phi(-W(x)).\end{aligned} \tag{10.20}$$

Let's finish our discussion of the Wald test with an example, where the test is used for comparing the mean values of two populations.

Example 10.1. Suppose we have two data sets, $X_1, \dots, X_n$ and $Y_1, \dots, Y_m$, which are modeled as IID samples from distributions F_X and F_Y, respectively[9]. Let μ_X and μ_Y be the mean values of F_X and F_Y, and suppose we want to test whether $\mu_X = \mu_Y$, or equivalently,

$$H_0 : \Delta\mu = 0 \quad \text{vs} \quad H_1 : \Delta\mu \neq 0, \tag{10.21}$$

where $\Delta\mu = \mu_X - \mu_Y$. A natural estimate of $\Delta\mu$ is the difference of the sample means:

$$\widehat{\Delta\mu} = \overline{X} - \overline{Y}. \tag{10.22}$$

Its (squared) standard error is

$$\text{se}^2[\widehat{\Delta\mu}] = \mathbb{V}[\widehat{\Delta\mu}] = \mathbb{V}[\overline{X}] + \mathbb{V}[\overline{Y}] = \frac{\sigma_X^2}{n} + \frac{\sigma_Y^2}{m}, \tag{10.23}$$

where σ_X^2 and σ_Y^2 are the variances of distributions F_X and F_Y. We can estimate the standard error by estimating the variances σ_X^2 and σ_Y^2 with the corresponding sample variances[10]:

$$\widehat{\text{se}}[\widehat{\Delta\mu}] = \sqrt{\frac{\hat{s}_{X,n}^2}{n} + \frac{\hat{s}_{Y,m}^2}{m}}, \tag{10.24}$$

where

$$\hat{s}_{X,n}^2 = \frac{1}{n-1}\sum_{i=1}^{n}(X_i - \overline{X})^2 \quad \text{and} \quad \hat{s}_{Y,m}^2 = \frac{1}{m-1}\sum_{i=1}^{m}(Y_i - \overline{Y})^2. \tag{10.25}$$

If sample sizes n and m are sufficiently large, then, thanks to the central limit theorem, $\widehat{\Delta\mu}$ is approximately normally distributed, its mean equals to the true value of $\Delta\mu$, and, thanks to the law of large numbers, its standard deviation $\text{se}[\widehat{\Delta\mu}] \approx \widehat{\text{se}}[\widehat{\Delta\mu}]$. In this case, the normality assumption (10.3) is approximately satisfied and

[8] See Section 9.7.6 in general and equation (9.74) in particular.

[9] For example, the two samples are the heights of randomly chosen n Hobbits and m Dwarves or the numbers of Quidditch goals scored by randomly chosen n Gryffindor and m Slytherin students.

[10] Recall that the sample variance, defined in (4.51), is an unbiased estimate of the variance.

$$\text{The size } \alpha \text{ Wald test rejects } H_0 \quad \iff \quad \frac{|\overline{X}-\overline{Y}|}{\sqrt{\frac{\hat{s}^2_{X,n}}{n}+\frac{\hat{s}^2_{Y,m}}{m}}} \geq z_{1-\frac{\alpha}{2}}. \tag{10.26}$$

The p-value of the actually observed data $x=(x_1,\ldots,x_n)$ and $y=(y_1,\ldots,y_m)$ is

$$\mathfrak{p}(x,y)=2\Phi\left(\frac{-|\bar{x}-\bar{y}|}{\sqrt{\frac{\hat{s}^2_{x,n}}{n}+\frac{\hat{s}^2_{y,m}}{m}}}\right). \tag{10.27}$$

As intuitively expected, the larger the difference $|\bar{x}-\bar{y}|$, the smaller the p-value, and the stronger the evidence against the null hypothesis that the means are the same.

10.2 Student's T-Test

The *t-test* was introduced in 1908 by English statistician, research chemist, and experimental brewer William Sealy Gosset [2]. Gosset spent his entire career at Guinness Brewery in Dublin, the world's largest brewery at that time. For proprietary reasons, Guinness required Gosset to publish his work under the pseudonym "Student," so that readers would not know he was a scientist at the brewery. Student discovered the t-test empirically by analyzing patterns in the data. That Student's results are correct was rigorously proved later by Fisher. Gosset was a humble, modest, and tolerant man with a good sense of humor [3]. As the story goes, when Gosset was thanked for his contributions to statistics, he replied: "Oh, that's nothing. Fisher would have discovered it all anyway."

William Sealy Gosset (1876-1937), to this day still better known under the pseudonym "Student," who was described by Ronald Fisher as "the Faraday of statistics" for his extraordinary ability to solve important practical statistical problems despite relatively weak mathematical ability.

Unlike the Wald test that does not make any distributional assumptions for the data, the t-test assumes that the data is normally distributed:

$$X_1,\ldots,X_n \overset{\text{iid}}{\sim} \mathcal{N}(\mu,\sigma^2), \tag{10.28}$$

where μ and σ^2 are unknown model parameters. Suppose we want to test

$$H_0:\mu=\mu_0 \quad \text{vs} \quad H_1:\mu\neq\mu_0, \tag{10.29}$$

where μ_0 is a hypothesized value. Since the null hypothesis is simple, we can try to use the Wald test. To this end, we need to construct an estimate $\hat{\mu}_n$ of μ and an estimate $\widehat{\text{se}}[\hat{\mu}_n]$ of the standard error $\text{se}[\hat{\mu}_n]$ of $\hat{\mu}_n$, and then check that the Wald normality assumption (10.3) is satisfied. Let's estimate the mean μ by the sample mean,

$$\hat{\mu}_n=\overline{X}. \tag{10.30}$$

Its (squared) standard error is

$$\mathrm{se}^2[\hat{\mu}_n] = \mathbb{V}[\overline{X}] = \mathbb{V}\left[\frac{1}{n}\sum_{i=1}^{n} X_i\right] = \frac{1}{n^2}\sum_{i=1}^{n}\underbrace{\mathbb{V}[X_i]}_{\sigma^2} = \frac{\sigma^2}{n}. \tag{10.31}$$

Estimating the variance σ^2 with the sample variance $\hat{s}_n^2 = \frac{1}{n-1}\sum_{i=1}^{n}(X_i - \overline{X})^2$ gives

$$\widehat{\mathrm{se}}[\hat{\mu}_n] = \frac{\hat{s}_n}{\sqrt{n}}. \tag{10.32}$$

Thanks to the central limit theorem, the Wald normality assumption (10.3) is *approximately* satisfied *if* the sample size n is *large*:

$$\frac{\hat{\mu}_n - \mu_\star}{\widehat{\mathrm{se}}[\hat{\mu}_n]} = \frac{\overline{X} - \mu_\star}{\hat{s}_n/\sqrt{n}} \mathrel{\dot\sim} \mathcal{N}(0,1) \quad \text{if } n \text{ is large}, \tag{10.33}$$

where $\mu_\star$ is the true value of μ. Therefore, if n is large, we can use the Wald test.

But what if the sample size n is small[11]? In this case, the normal approximation in (10.33) will be inaccurate and the Wald test will not be applicable. Small samples often occur in the analyses of medical, chemical, biological, and agricultural data.

[11] In his work at the brewery, Gosset analyzed very small samples.

Fortunately, its turns out that it is possible to derive the *exact* distribution of random variable

$$T = \frac{\overline{X} - \mu_\star}{\hat{s}_n/\sqrt{n}}. \tag{10.34}$$

Namely, it can be shown that for *any n*, random variable T is distributed according to the *t-distribution* with $(n-1)$ degrees of freedom[12], denoted

$$T \sim \mathcal{T}(n-1). \tag{10.35}$$

[12] For the purpose of this discussion, "degrees of freedom" is just the name of the parameter of the t-distribution.

The probability density function $f_T(x|n-1)$ of Student's t-distribution with $(n-1)$ degrees of freedom has a rather complicated expression[13]

$$f_T(x|n-1) = \frac{\Gamma\left(\frac{n}{2}\right)}{\sqrt{(n-1)\pi}\,\Gamma\left(\frac{n-1}{2}\right)}\left(1 + \frac{x^2}{n-1}\right)^{-\frac{n}{2}}, \tag{10.36}$$

where Γ is a special function, called the gamma function, that extends the factorial function to complex numbers. In general, the degrees of freedom $(n-1)$ can be any positive number, not necessarily integer. In other words, (10.36) is well defined for any $n > 1$. But if $n > 2$ is integer, as in our case, where n represents the sample size, then the PDF (10.36) admits an alternative expression without the gamma function:

$$f_T(x|n-1) = \begin{cases} \frac{2\cdot 4 \ldots (n-4)(n-2)}{3\cdot 5 \ldots (n-5)(n-3)\sqrt{n-1}\pi}\left(1 + \frac{x^2}{n-1}\right)^{-\frac{n}{2}}, & \text{if } n \text{ is even}, \\ \frac{3\cdot 5 \ldots (n-4)(n-2)}{2\cdot 4 \ldots (n-5)(n-3)\sqrt{n-1}2}\left(1 + \frac{x^2}{n-1}\right)^{-\frac{n}{2}}, & \text{if } n \text{ is odd}. \end{cases} \tag{10.37}$$

[13] Student miraculously derived this expression by fitting known distributions to the data. Fisher rigorously proved that Student's formula is correct by re-deriving it using n-dimensional geometry. In his letter to Karl Pearson (asking to check Fisher's proof), Gosset wrote: "I don't feel at home in more than three dimensions even if I could understand it otherwise." [3].

Remark 10.1. If $n = 2$, then the t-distribution reduces to the standard *Cauchy distribution* with the PDF $f_T(x|1) = f_C(x) = 1/\pi(1+x^2)$. The sample size $n = 2$ is, of course, pathologically small, although Gosset has studied the chemical properties of barley with samples of size as small as $n = 3$.

Student's[14] t-distribution has two important properties: it is symmetric about zero and it converges to the standard normal distribution as the number of degrees of freedom n increases,

$$\mathscr{T}(n) \to \mathscr{N}(0,1), \quad \text{as } n \to \infty. \tag{10.38}$$

Moreover, if $n \geq 30$, then the approximation $\mathscr{T}(n) \approx \mathscr{N}(0,1)$ is often considered sufficiently accurate for practical purposes. Figure 10.2 shows the PDF of the t-distribution for different values of n together with the standard normal PDF.

[14] Confirming Arnold's principle (recall Marginal Note 47 in Chapter 1), Student's t-distribution was first discovered in 1876, the year Gosset was born, by German mathematician Jacob Lüroth [4].

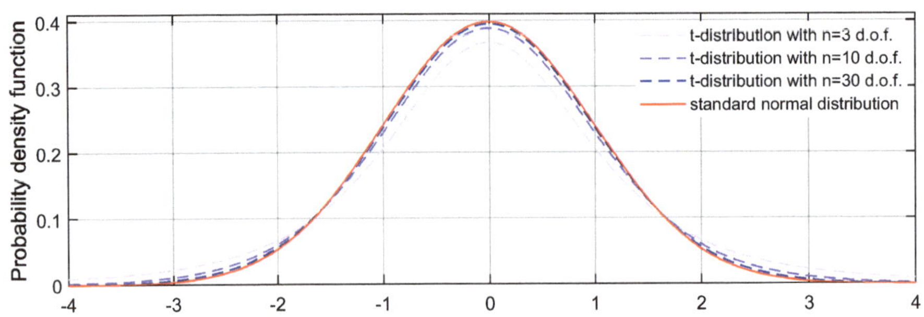

Fig. 10.2 The probability density functions of the t-distribution with $n = 3, 10$, and 30 degrees of freedom (dashed blue lines) and their convergence to the standard normal PDF (red solid line).

The distributional result (10.35), which holds for any sample size n, allows to construct a new test for testing (10.29), called *Student's t-test*, and use it instead of the Wald test, which is inapplicable (inaccurate) for small n. If the null hypothesis H_0 is true, then the hypothesized value μ_0 is the true value of the parameter and

$$\frac{\overline{X} - \mu_0}{\hat{s}_n/\sqrt{n}} \sim \mathscr{T}(n-1). \tag{10.39}$$

Therefore, under H_0, the quantity

$$G = \left| \frac{\overline{X} - \mu_0}{\hat{s}_n/\sqrt{n}} \right| \quad \text{is expected to be not far from zero.} \tag{10.40}$$

In other words, large values of G will contradict the null hypothesis. Student's test uses G[15] as the test statistic. Namely,

[15] "G" stands for Gosset.

$$\text{Student's t-test rejects } H_0 \quad \Longleftrightarrow \quad G = \left| \frac{\overline{X} - \mu_0}{\hat{s}_n/\sqrt{n}} \right| \geq c. \tag{10.41}$$

The t-test of size α is derived in the same way as the Wald test of size α (10.10), with the only difference being that instead of the standard normal distribution that underlies the Wald test, the t-test is based on the t-distribution:

$$\text{The size } \alpha \text{ Student's t-test rejects } H_0 \iff G = \left| \frac{\overline{X} - \mu_0}{\hat{s}_n / \sqrt{n}} \right| \geq t_{n-1,1-\frac{\alpha}{2}}, \tag{10.42}$$

where $t_{n,q}$ is the qth quantile of the t-distribution with n degrees of freedom. That is, $t_{n,q}$ is a number such that

$$\int_{-\infty}^{t_{n,q}} f_T(x|n)dx = q \quad \text{or, equivalently,} \quad F_T(t_{n,q}|n) = q, \tag{10.43}$$

where $F_T(x|n)$ is the CDF of the t-distribution with n degrees of freedom.

Just as the Wald test, the size α t-test can be written in terms of an $100(1-\alpha)\%$ confidence interval for the true value $\mu_\star$, which is given by[16]

[16] Compare with (10.17).

$$I_\alpha^T = \left(\overline{X} - \frac{\hat{s}_n}{\sqrt{n}} t_{n-1,1-\frac{\alpha}{2}}, \overline{X} + \frac{\hat{s}_n}{\sqrt{n}} t_{n-1,1-\frac{\alpha}{2}} \right). \tag{10.44}$$

Indeed, what should be by now a familiar computation shows that[17]

[17] Compare with (10.18).

$$\begin{aligned}
\mathbb{P}(\mu_\star \in I_\alpha^T) &= \mathbb{P}\left(\overline{X} - \frac{\hat{s}_n}{\sqrt{n}} t_{n-1,1-\frac{\alpha}{2}} < \mu_\star < \overline{X} + \frac{\hat{s}_n}{\sqrt{n}} t_{n-1,1-\frac{\alpha}{2}} \right) \\
&= \mathbb{P}\left(\underbrace{-t_{n-1,1-\frac{\alpha}{2}}}_{t_{n-1,\frac{\alpha}{2}}} < \underbrace{\frac{\overline{X} - \mu_\star}{\hat{s}_n / \sqrt{n}}}_{\sim \mathscr{T}(n-1)} < t_{n-1,1-\frac{\alpha}{2}} \right) \\
&= F_T(t_{n-1,1-\frac{\alpha}{2}} | n-1) - F_T(t_{n-1,\frac{\alpha}{2}} | n-1) = 1 - \alpha.
\end{aligned} \tag{10.45}$$

It remains to notice that test (10.42) rejects $H_0 : \mu = \mu_0$ if and only if the hypothesized value for the mean μ_0 does not belong to the confidence interval I_α^T for $\mu_\star$,

$$\text{The size } \alpha \text{ Student's t-test rejects } H_0 \iff \mu_0 \notin I_\alpha^T. \tag{10.46}$$

The p-value $\mathfrak{p}(x)$ of the actually observed data $x = (x_1, \ldots, x_n)$ is given by[18]

[18] Compare with (10.20)

$$\begin{aligned}
\mathfrak{p}(x) = \mathbb{P}(G(X) \geq G(x) | H_0) &= \mathbb{P}\left(\left| \underbrace{\frac{\overline{X} - \mu_0}{\hat{s}_n / \sqrt{n}}}_{\sim \mathscr{T}(n-1)} \right| \geq G(x) \,\middle|\, \mu_0 \right) \\
&= 2F_T(-G(x) | n-1).
\end{aligned} \tag{10.47}$$

Finally, it is important to reiterate and highlight that

$$\mathscr{T}(n) \to \mathscr{N}(0,1), \quad F_T(x|n) \to \Phi(x), \quad t_{n,q} \to z_q, \quad \text{as } n \to \infty, \tag{10.48}$$

and, therefore, Student's t-test converges to the (10.33)-based Wald test as the sample size n increases. In particular, if $n \geq 30$, then the two test are essentially identical.

10.2.1 Two-Sample T-Test

Suppose now, as in Example 10.1, that we have two samples $X = (X_1, \ldots, X_n)$ and $Y = (Y_1, \ldots, Y_m)$, and we want to test whether the underlying probability distributions that generated X and Y have the same mean values:

$$H_0 : \Delta\mu = 0 \quad \text{vs} \quad H_1 : \Delta\mu \neq 0, \tag{10.49}$$

where $\Delta\mu = \mu_X - \mu_Y$. But unlike Example 10.1, where no distributional assumptions were made[19], let's assume that both samples are normally distributed:

$$X_1, \ldots, X_n \overset{\text{iid}}{\sim} \mathcal{N}(\mu_X, \sigma_X^2) \quad \text{and} \quad Y_1, \ldots, Y_m \overset{\text{iid}}{\sim} \mathcal{N}(\mu_Y, \sigma_Y^2). \tag{10.50}$$

This testing problem, called the *Behrens-Fisher problem*, is one the classical problems in statistics that appear in many applications.

Recall that the two-sample Wald test (10.26) is based on the assumption that, under H_0, the distribution of statistic

$$\frac{\overline{X} - \overline{Y}}{\sqrt{\frac{\hat{s}_{X,n}^2}{n} + \frac{\hat{s}_{Y,m}^2}{m}}} \tag{10.51}$$

is standard normal[20]. This assumption holds approximately if sample sizes n and m are sufficiently large. But if n or m are small, then the assumption does not hold and the Wald test is not applicable.

The exact distribution of statistic (10.51) under H_0 is unknown. Nevertheless, it can be shown [5] that if both samples X and Y are normally distributed and $\Delta\mu = 0$, then for any n and m, statistic (10.51) is approximately distributed according to the t-distribution with d degrees of freedom

$$\frac{\overline{X} - \overline{Y}}{\sqrt{\frac{\hat{s}_{X,n}^2}{n} + \frac{\hat{s}_{Y,m}^2}{m}}} \mathrel{\dot{\sim}} \mathcal{T}(d), \tag{10.52}$$

where d is given by the *Welch-Satterthwaite equation*[21] [5, 6]:

$$d = \frac{\left(\frac{\hat{s}_{X,n}^2}{n} + \frac{\hat{s}_{Y,m}^2}{m}\right)^2}{\frac{1}{n-1}\left(\frac{\hat{s}_{X,n}^2}{n}\right)^2 + \frac{1}{m-1}\left(\frac{\hat{s}_{Y,m}^2}{m}\right)^2}. \tag{10.53}$$

[19] The only assumption was that $X_1, \ldots, X_n$ are IID and $Y_1, \ldots, Y_m$ are IID.

[20] This assumption stems from the Wald normality assumption (10.3).

[21] Also known as the Satterthwaite formula and the Welch degrees of freedom.

Remark 10.2. Although approximation (10.52) is valid for any $n > 1$ and $m > 1$, in practice, it is recommended to use it when the sample sizes are not extremely small, for example when $n > 5$ and $m > 5$ [7].

A statistical test for comparing the means (10.49), which is based on the distributional result (10.52), (10.53) is called the *two-sample t-test* or the *Welch t-test*. The two-sample t-test of a given size α is derived by analogy with the corresponding Wald test and it has the following familiar form[22]:

$$\text{The size } \alpha \text{ two-sample t-test rejects } H_0 \quad \Longleftrightarrow \quad \frac{|\overline{X}-\overline{Y}|}{\sqrt{\frac{\hat{s}^2_{X,n}}{n}+\frac{\hat{s}^2_{Y,m}}{m}}} \geq t_{d,1-\frac{\alpha}{2}}. \quad (10.54)$$

[22] Compare with (10.26).

The p-value of the actually observed data $x = (x_1, \ldots, x_n)$ and $y = (y_1, \ldots, y_m)$ is

$$\mathfrak{p}(x,y) = 2F_T\left(\frac{-|\bar{x}-\bar{y}|}{\sqrt{\frac{\hat{s}^2_{x,n}}{n}+\frac{\hat{s}^2_{y,m}}{m}}}\,\middle|\, d\right). \quad (10.55)$$

Note that the two-sample Wald and Welch t-test are approximate, since they are both based on statistic (10.51), whose exact distribution under the null hypothesis is unknown and is approximated by the standard normal and t-distribution, respectively. As the sample sizes increase $n,m \to \infty$, the Welch degrees of freedom $d \to \infty$, the t-distribution converges to the standard normal $\mathscr{T}(d) \to \mathscr{N}(0,1)$, and the two tests become asymptotically exact and equivalent. For finite (and especially small) sample sizes, the Welch t-test is often recommended as the best test. A comprehensive review of the existing statistical tests for solving the Behrens-Fisher problem (and analogous problems for non-normal samples) together with practical recommendations is provided in [8].

10.3 Kolmogorov-Smirnov Test

Suppose that data $X_1, \ldots, X_n$ is modeled as an IID sample from a distribution F,

$$X_1, \ldots, X_n \overset{\text{iid}}{\sim} F \in \mathscr{F}, \quad (10.56)$$

where $\mathscr{F}$ is a stochastic model. If we are interested in testing whether $\theta = \theta_0$, where θ is some parameter and θ_0 is its hypothesized value, then we can use the Wald test. In addition, if $\mathscr{F} = \{\mathscr{N}(\mu,\sigma^2), \mu \in \mathbb{R}, \sigma^2 \geq 0\}$ is the normal model and the parameter of interest θ is the mean μ, then (especially if the sample size n is small) we can use the t-test.

But what if, instead of just a single parameter, we are interested in the full probability distribution F? In other words, suppose that, instead of testing $\theta = \theta_0$ versus $\theta \neq \theta_0$, we want to test

$$H_0 : F = F_0 \quad \text{vs} \quad H_1 : F \neq F_0, \tag{10.57}$$

where F_0 is some specified distribution[23]. To visually compare the data distribution with any given probability distribution, we can use the quantile-quantile (Q-Q) plot[24]: the closer the Q-Q plot resembles the line $y = x$, the more accurately the reference distribution F_0 describes the distribution of the data. If there is a clear strong discrepancy between the Q-Q plot and the line $y = x$, then we can safely reject H_0 without any further quantitative analysis. However, if the discrepancy is not strong and it is not clear whether it occurred due to the random variation in the data or due to the inadequacy of F_0, then a formal statistical test is need for testing (10.57).

[23] Distribution F_0 is also called the "reference" or "null" distribution. Note that F_0 must be specified without using the data. This is discussed in more detail below.

[24] See Section 1.5.

Andrey Nikolaevich Kolmogorov (1903-1987), a Soviet mathematician, one of the greatest mathematicians of the XX century, one of the founders of the modern probability theory. Photo: Archives of the Mathematisches Forschungsinstitut Oberwolfach.

Nikolai Vasilyevich Smirnov (1900-1966), a Soviet mathematician, one of the founders of nonparametric mathematical statistics. Photo: Archives of the Steklov Mathematical Institute.

The *Kolmogorov-Smirnov test* (KS test) is a general nonparametric test, developed by Andrey Kolomogorov and Nikolai Smirnov, that allows to quantitatively assess whether a given sample was drawn from a specified probability distribution. If the null hypothesis H_0 is true and $X_1, \ldots, X_n$ is indeed an IID sample from F_0, then[25] the empirical CDF (eCDF) $\hat{F}_n(x)$ of the sample is a *consistent estimate* of the reference CDF $F_0(x)$. That is, for any x

$$\frac{1}{n}\sum_{i=1}^{n} H(x = X_i) = \hat{F}_n(x) \xrightarrow{\mathbb{P}} F_0(x), \quad \text{as } n \to \infty. \tag{10.58}$$

Moreover[26], the mean squared error of $\hat{F}_n(x)$ converges to zero,

$$\text{MSE}[\hat{F}_n(x)] = \mathbb{E}\left[(\hat{F}_n(x) - F_0(x))^2\right] \to 0, \quad \text{as } n \to \infty. \tag{10.59}$$

Intuitively, these properties mean that if H_0 is true and the sample size n is large, then the difference between $\hat{F}_n(x)$ and $F_0(x)$ is likely to be small for all x.

Consider the following quantity, called the *Kolmogorov-Smirnov statistic* (KS statistic), which measures the largest absolute difference between $\hat{F}_n$ and F_0:

$$D = \sup_{x \in \mathbb{R}} \left|\hat{F}_n(x) - F_0(x)\right|. \tag{10.60}$$

The KS statistics, illustrated in Fig. 10.3, quantifies the discrepancy between the empirical and hypothesized CDFs of the data. If the null hypothesis H_0 is true, then, thanks to (10.58) and (10.59),

$$D = \sup_{x \in \mathbb{R}} \left|\hat{F}_n(x) - F_0(x)\right| \quad \text{is expected to be not far from zero.} \tag{10.61}$$

Large values of D, on the other hand, will indicate a large discrepancy between F_0 and $\hat{F}_n$ (which converges to the true distribution) and, therefore, will contradict H_0.

The Kolmogorov-Smirnov test is based on the statistic (10.60). Namely,

$$\text{The KS test rejects } H_0 \quad \Longleftrightarrow \quad D = \sup_{x \in \mathbb{R}} \left|\hat{F}_n(x) - F_0(x)\right| \geq c. \tag{10.62}$$

To construct the KS test of size α, we need to find the critical value c such that[27]

[25] See Corollary 4.1.

[26] See Theorem 4.1.

[27] Recall that the size of a test is the maximum possible probability of the type I error, see Remark 9.3. Since $H_0 : F = F_0$ is simple, the size is simply the probability of type I error.

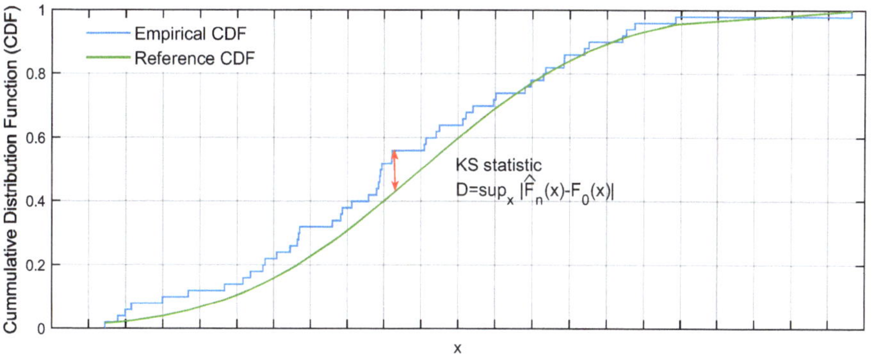

Fig. 10.3 The Kolmogorov-Smirnov statistic measures the distance (illustrated with the red double arrow) between the empirical CDF $\hat{F}_n$ and the reference CDF F_0 by computing the largest absolute difference between their values.

$$\alpha = \mathbb{P}(\text{Type I error}) = \mathbb{P}(\text{Reject } H_0 | H_0 \text{ is true}) = \mathbb{P}(D \geq c | H_0 \text{ is true}). \quad (10.63)$$

Inverting the inequality sign, $\mathbb{P}(D \geq c | H_0) = 1 - \mathbb{P}(D < c | H_0)$, and rearranging the terms lead to the following equation on c:

$$\mathbb{P}(D < c | H_0) = 1 - \alpha. \quad (10.64)$$

The DKW inequality (4.33) provides only a lower bound for the probability on the left-hand side,

$$\mathbb{P}(D < c | H_0) \geq 1 - 2\exp\left(-2nc^2\right), \quad (10.65)$$

which can't be used for finding c. The exact asymptotic distribution of the KS statistic D under the null hypothesis H_0 was derived by Andrey Kolmogorov in [9]. He proved that if the reference distribution F_0 is *continuous*, then the probability $\mathbb{P}(D < c | H_0)$ does not depend on F_0, and for any $x > 0$

$$\mathbb{P}(\sqrt{n}D < x | H_0) \to F_K(x) = \sum_{k=-\infty}^{+\infty} (-1)^k \exp\left(-2k^2x^2\right), \quad \text{as } n \to \infty. \quad (10.66)$$

In other words, the continuous random variable $\sqrt{n}D$ converges in distribution to a positive random variable K whose CDF is given by F_K,

$$\text{Under } H_0: \quad \sqrt{n}D \xrightarrow{\mathbb{D}} K, \quad \text{as } n \to \infty. \quad (10.67)$$

The probability distribution $\mathscr{K}$ of random variable K defined by the CDF F_K is called the *Kolmogorov distribution* and the asymptotic result (10.67) is called the *Kolmogorov theorem*.

Remark 10.3. Although it is not relevant for the Kolmogorov-Smirnov test, it is still interesting to mention that the mean and variance of the Kolmogorov distribution can be computed exactly [10],

$$\mu_K = \sqrt{\frac{\pi}{2}} \log 2 \approx 0.8687 \quad \text{and} \quad \sigma_K^2 = \frac{\pi^2}{12} - \frac{\pi}{2}(\log 2)^2 \approx 0.0678. \tag{10.68}$$

Remark 10.4. If x is small, then the series in (10.66) converges very slowly. In this case, the following theoretically equivalent [11, 12], but numerically more convenient expression for $F_K(x)$ can be used:

$$F_K(x) = \frac{\sqrt{2\pi}}{x} \sum_{k=1}^{\infty} \exp\left(-\frac{(2k-1)^2\pi^2}{8x^2}\right). \tag{10.69}$$

By keeping just the first term corresponding to $k = 1$, we can obtain a relatively simple and accurate approximation[28] for $F_K(x)$ for small x,

$$F_K(x) \approx \frac{\sqrt{2\pi}}{x} \exp\left(-\frac{\pi^2}{8x^2}\right). \tag{10.70}$$

[28] The asymptotic formula (10.70) was first given in the original paper by Kolmogorov [9].

The probability density function of the Kolmogorov distribution is shown in Fig. 10.4.

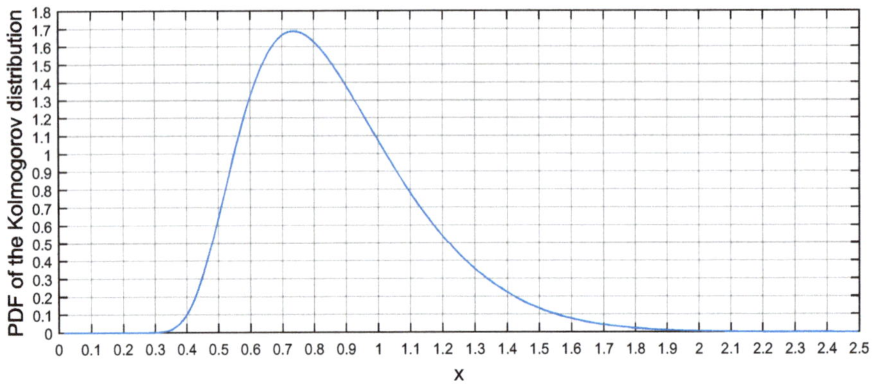

Fig. 10.4 The probability density function $f_K(x) = F_K'(x)$ of the Kolmogorov distribution.

Returning to equation (10.64) and using the fact that, under H_0 and for large n, the multiple of the KS statistic $\sqrt{n}D$ is distributed approximately according to the Kolmogorov distribution, $\sqrt{n}D \mathrel{\dot\sim} \mathscr{K}$, we have:

$$1 - \alpha = \mathbb{P}(D < c | H_0) = \mathbb{P}(\sqrt{n}D < \sqrt{n}c | H_0) \approx F_K(\sqrt{n}c). \tag{10.71}$$

Therefore, the critical value $c \approx F_K^{-1}(1-\alpha)/\sqrt{n}$, and

$$\text{The size } \alpha \text{ KS test rejects } H_0 \quad \Longleftrightarrow \quad D \geq \frac{F_K^{-1}(1-\alpha)}{\sqrt{n}}. \tag{10.72}$$

Note that the size of this test is *approximately* α, but it becomes exactly α in the limit as $n \to \infty$. In other words, the test is *asymptotically exact.*

The p-value $\mathfrak{p}(x)$ of the actually observed data $x = (x_1, \ldots, x_n)$ is the probability, under H_0, to observe the value of the KS statistic $D = D(X)$ which is at least as large as the value $D_{\text{obs}} = D(x)$ that has been actually observed:

$$\mathfrak{p}(x) = \mathbb{P}(D \geq D_{\text{obs}} \,|\, H_0) = \mathbb{P}(\underbrace{\sqrt{n}D}_{\dot{\sim}\mathscr{K}} \geq \sqrt{n}D_{\text{obs}} \,|\, H_0) \approx 1 - F_K(\sqrt{n}D_{\text{obs}}). \quad (10.73)$$

Remark 10.5. Equivalently, the p-value can be obtained by finding the size α such that the observed data $x = (x_1, \ldots, x_n)$ lies exactly on the boundary of the rejection region[29] defined by inequality in (10.72). That is, by solving for α the following equation:

$$D_{\text{obs}} = \frac{F_K^{-1}(1-\alpha)}{\sqrt{n}}. \quad (10.74)$$

Remark 10.6. To both construct the KS test of a given size and to compute the p-value of the observed data, one needs to be able to compute the values of the CDF F_K of the Kolmogorov distribution. First approximations of $F_K(x)$ for $x = 0, 0.2, \ldots, 2.8$ were obtained by N. Kozhevnikov[30] [9]. A much more detailed table of approximate values of $F_K(x)$ for $x = 0.28, 0.29, \ldots, 3$ was derived by N. Smirnov [11, 12]. Based on Smirnov's approximations, the rejection regions (10.72) for some often-used size values are

$$\begin{aligned} \alpha = 0.01 &\iff D \geq 1.63/\sqrt{n}, \\ \alpha = 0.05 &\iff D \geq 1.36/\sqrt{n}, \\ \alpha = 0.10 &\iff D \geq 1.22/\sqrt{n}. \end{aligned} \quad (10.75)$$

To sum up the contributions, Kolmogorov defined the KS statistics (10.60) and derived its asymptotic distribution (10.67) under the null hypothesis, and Kozhevnikov and Smirnov calculated the tables of values of the asymptotic distribution[31].

The Kolmogorov-Smirnov test is widely used in many disciplines and various applications, and, unfortunately, sometimes it is misused. There are two common ways to apply the KS test incorrectly. Recall that the asymptotic result (10.67), which lies at the heart of the KS test, is obtained under the assumption that the reference distribution F_0 is *continuous*. If F_0 is discontinuous, then the Kolmogorov theorem (10.67) does not hold. Moreover, if F_0 is discontinuous, then the distribution of the KS statistic D under H_0 depends on F_0 (it does not depend in the continuous case). This means that the classical KS test (10.72), (10.73)[32] is not applicable for testing (10.57) if the reference distribution F_0 is discontinuous (discrete or mixed). More specifically, it can be shown that the use of the KS test for discontinuous reference distributions produces conservative results[33] [13, 14]. An extension and efficient implementation of the KS test for the case where the reference distribution is discontinuous was developed in [15].

[29] Recall the geometric interpretation of the p-value in Section 9.7.3.

[30] The author will be very grateful for any biographical information about Kozhevnikov. Was he Kolmogorov's student, collaborator?

[31] Another important contribution of Smirnov is an analog of the asymptotic result (10.67) for the two-sample test, which is discussed in the next section.

[32] Which is what is commonly implemented in statistical software packages.

[33] A statistical test is called *conservative* if its true size is less than its nominal size (i.e. the size it is supposed to have). For conservative tests it is harder to reject the null hypothesis, which leads to loss of power and overestimated p-values.

Another critical assumption of the Kolmogorov-Smirnov test is that the reference distribution F_0 must be completely specified *in advance*, before seeing the data that will be used for testing. In particular, the parameters of the reference distribution cannot be estimated from the data. Instead, they must be derived from some external considerations, such as the underlying theory about the data-generating mechanism, background information about the data, or from another completely independent data set. If this assumption is violated and the parameters are estimated from the data, then the resulting test will be too conservative: the corresponding p-value will be overestimated and the test will tend to accept the null hypothesis[34]. Specifying the reference distribution based on the data and then testing how well it fits the same data is an example of *post-hoc theorizing*, a practice that should be avoided.

[34] This bias towards the null hypothesis can be intuitively explained as follows: instead of specifying the reference distribution a priori, we specify a family of distributions, then estimate the parameters (i.e. find the distribution within the family that fits the data the best), and only after that we test whether the "tuned" reference distribution fits the data well. By doing this, we are essentially using data to help us accept the null hypothesis.

The requirement to specify the reference distribution in advance, without looking at the data, poses a serious obstruction for the use of the KS test in practical applications, where it is possible to specify a parametric model for the data, $X_1, \ldots, X_n \overset{\text{iid}}{\sim} F(x \mid \theta)$, but hard to specify the model parameter (or vector of parameters) $\theta \in \Theta$ a priori. To overcome this obstruction, a simulation-based modification of the test, which is also based on the KS statistic, was proposed in [16].

Let $\hat{\theta}_n$ be an estimate of θ obtained form the data, $F_0(x) = F(x \mid \hat{\theta}_n)$ be the reference distribution obtained from the parametric model by estimating θ from the data, and D_{obs} be the observed value of the KS statistic that measures the discrepancy between the eCDF of the data $\hat{F}_n$ and F_0,

$$D_{\text{obs}} = \sup_{x \in \mathbb{R}} \left| \hat{F}_n(x) - F(x \mid \hat{\theta}_n) \right|. \tag{10.76}$$

As discussed above, we cannot use the p-value given by (10.73), since the F_K-based formula overestimates the p-value[35]. By definition, the p-value is given by

$$\mathfrak{p}(x) = \mathbb{P}(D \geq D_{\text{obs}} \mid H_0), \tag{10.77}$$

where D is a random value of the KS statistic, whose realization D_{obs} we observe. We don't know the null distribution[36] of D, and, therefore, we cannot compute the p-value analytically. But we can generate its realizations by mimicking the process of how D_{obs} was generated under the assumption that H_0 is true:

[35] The mathematical reason behind the inapplicability of (10.73) is that the fact $\sqrt{n}D \overset{\cdot}{\sim} \mathscr{K}$ used in the derivation of the formula, does not hold any more.

[36] That is, the distribution under the null hypothesis.

1. Generate N independent samples of size n from the reference distribution F_0:

$$\begin{aligned} &X_1^{(1)}, \ldots, X_n^{(1)} \overset{\text{iid}}{\sim} F_0(x) = F(x \mid \hat{\theta}_n), \\ &\ldots \\ &X_1^{(N)}, \ldots, X_n^{(N)} \overset{\text{iid}}{\sim} F_0(x) = F(x \mid \hat{\theta}_n). \end{aligned} \tag{10.78}$$

2. Based on these samples, compute estimates $\hat{\theta}_n^{(1)}, \ldots, \hat{\theta}_n^{(N)}$ of θ using the same estimation method that was used for computing $\hat{\theta}_n$ from the original sample. In other words, if $\hat{\theta}_n = s(X_1, \ldots, X_n)$, then

$$\hat{\theta}_n^{(1)} = s(X_1^{(1)}, \ldots, X_n^{(1)}), \quad \ldots, \quad \hat{\theta}_n^{(N)} = s(X_1^{(N)}, \ldots, X_n^{(N)}). \tag{10.79}$$

3. For $i = 1, \dots, N$, compute[37] the realization D_i of D by computing the KS statistic that measures the discrepancy between the eCDF of sample $X_1^{(i)}, \dots, X_n^{(i)}$ and the corresponding "reference" distribution $F_0^{(i)}(x) = F(x \,|\, \hat{\theta}_n^{(i)})$,

$$D_i = \sup_{x \in \mathbb{R}} \left| \hat{F}_n^{(i)}(x) - F(x \,|\, \hat{\theta}_n^{(i)}) \right|, \quad i = 1, \dots, N. \tag{10.80}$$

[37] Compare (10.80) with how D_{obs} is computed in equation (10.76).

Having realizations $D_1, \dots, D_N$ of random statistic D it is now straightforward to estimate the probability in (10.77) by the fraction of realizations whose values are at least as large as the value D_{obs}[38],

$$\mathfrak{p}(x) \approx \frac{1}{N} \sum_{i=1}^{N} \mathbb{1}_{[D_{\text{obs}}, \infty)}(D_i), \tag{10.81}$$

where $\mathbb{1}_{[D_{\text{obs}}, \infty)}$ is the indicator function of the interval $[D_{\text{obs}}, \infty)$, that is, $\mathbb{1}_{[D_{\text{obs}}, \infty)}(D_i) = 1$ if $D_i \geq D_{\text{obs}}$ and $\mathbb{1}_{[D_{\text{obs}}, \infty)}(D_i) = 0$ if $D_i < D_{\text{obs}}$. As the number N of "synthetic" samples increases, the estimate (10.81) converges to the p-value (10.77).

[38] This approach to estimating the p-value is very similar in the spirit to the bootstrap method (Chapter 6), especially to the parametric bootstrap (Problem 6.2). It is also similar to how p-values are estimated in the permutation test, which is discussed in Section 10.4.

Remark 10.7. What value of N should be used in applications? In [16], where the focus is on fitting the power law distribution to the data, it is recommended to use $N \approx 2500$ to guarantee that the estimate of the p-value is accurate up to about 2 decimal digits. If sampling from $F_0(x) = F(x \,|\, \hat{\theta}_n)$ in (10.78) and computing estimates in (10.79) is not computationally intensive, then I would recommend to use at least $N = 10^4$ synthetic samples.

A couple of remarks describing alternatives to the KS test are in order.

Remark 10.8. The Kolmogorov-Smirnov test is based on the KS statistic (10.60), which is one way to measure the discrepancy between the eCDF $\hat{F}_n$ of the data and the hypothesized CDF F_0. There are also other ways which lead to other tests. For example,

$$\tilde{D} = \int_{-\infty}^{\infty} (\hat{F}_n(x) - F_0(x))^2 dx \quad \text{and} \quad \check{D} = \int_{-\infty}^{\infty} \frac{(\hat{F}_n(x) - F_0(x))^2}{F_0(x)(1 - F_0(x))} dx \tag{10.82}$$

lead, respectively, to the *Cramér-von Mises test* (CvM test) [17, 18] and the *Anderson-Darling test* (AD test) [19, 20]. The AD test tends to be more powerful than the KS and CvM tests when the discrepancy between the two CDFs occurs in the tails[39]. The KS test tends to be more powerful when the discrepancy occurs mostly in the center, and the CvM test tends to be somewhat between the KS and AD tests. None of the three tests, however, universally dominates the other.

[39] Notice that, compared with D and $\tilde{D}$, statistic $\check{D}$ puts more weight on observations in the tails of the reference distribution.

Remark 10.9. Thanks to the central limit theorem, the normal distribution plays a special, central role in statistical inference. Many samples are approximately normally distributed[40], and many statistical methods are based on the assumption that the sample itself or some statistic of the sample is normally distributed[41]. To visually check whether the sample is normality distributed, we can use the normal-quantile

[40] The reasons behind this empirical observation are discussed in Probability Refresher 1.8.

[41] For example, the Wald test assumes that $\hat{\theta}_n \sim \mathcal{N}$, and the t-test assumes that $X_1, \dots, X_n \overset{\text{iid}}{\sim} \mathcal{N}$.

plot[42]: the closer the plot resembles the line $y = ax + b$, the more accurately the normal distribution $\mathcal{N}(\mu, \sigma^2)$ with $\mu = b$ and $\sigma^2 = a^2$ describes the distribution of the sample. Given $X_1, \dots, X_n \overset{\text{iid}}{\sim} F$, to formally test

$$H_0 : F = \mathcal{N}(\mu, \sigma^2) \quad \text{vs} \quad H_1 : F \neq \mathcal{N}(\mu, \sigma^2), \tag{10.83}$$

[42] See Section 1.5.

we can use the Kolmogorov-Smirnov test (10.72), (10.73), since it is applicable for any continuous reference distribution F_0, in particular, for $F_0 = \mathcal{N}(\mu, \sigma^2)$. In addition to the normal-quantile plot and the KS test, a large number of other tests and diagnostics have been developed for testing for normality of the data [21]. Among all normality tests, the *Shapiro-Wilk test* (SW test) developed in [22] is often recommended as one of the best (if not the best) tests. For example, it has been shown in [23] that among four popular normality tests[43] of a given size, the SW test is the most powerful, i.e. has the smallest probability of type II error: failing to reject normality when the data is not normal. Another "structural" advantage of the SW test over the KS test is that the parameters μ and σ^2 of the normal distribution don't need to be specified in advance (as in the case of the KS test): the Shapiro-Wilk test tests (10.83) for some unknown μ and some unknown σ^2. This superiority of the SW test over the KS test comes at a price: the SW test is a *normality test*, and, unlike the KS test, it cannot be used for testing non-normal reference distributions[44].

[43] Kolmogorov-Smirnov, Shapiro-Wilk, Anderson-Darling, Lilliefors tests.

[44] This trade-off between efficiency and universality (wide applicability) is typical for statistical methods.

10.3.1 Two-Sample Kolmogorov-Smirnov Test

The *two-sample Kolmogorov-Smirnov test* (2KS test) is a variation of the KS test for the case where we have two samples

$$X_1, \dots, X_n \overset{\text{iid}}{\sim} F_X \quad \text{and} \quad Y_1, \dots, Y_m \overset{\text{iid}}{\sim} F_Y, \tag{10.84}$$

and we want to test whether the distributions F_X and F_Y the samples come from are the same or different[45]:

$$H_0 : F_X = F_Y \quad \text{vs} \quad H_1 : F_X \neq F_Y. \tag{10.85}$$

[45] This is one of the classic problems of non-parametric inference with many applications. The two samples may represent, for instance, the medical test results after administering a placebo and a treatment. Testing (10.85) helps to decide whether the treatment differs from the placebo.

By analogy with the one-sample test, the 2KS test is based of the KS statistic that measures the largest absolute difference between the empirical CDFs $\hat{F}_{X,n}$ and $\hat{F}_{Y,m}$ constructed from the samples $X_1, \dots, X_n$ and $Y_1, \dots, Y_m$,

$$D = \sup_{z \in \mathbb{R}} \left| \hat{F}_{X,n}(z) - \hat{F}_{Y,m}(z) \right|. \tag{10.86}$$

If the null hypothesis is true, then the eCDFs $\hat{F}_{X,n}$ and $\hat{F}_{Y,m}$ converge to the same CDF $F_X = F_Y$ as sample sizes $n, m \to \infty$, and, therefore, for large n and m, the value of statistic D is expected to be not far from zero. On the contrary, large values of D will contradict H_0. This leads to a general form of the two-sample KS test:

$$\text{The 2KS test rejects } H_0 \quad \Longleftrightarrow \quad D = \sup_{z\in\mathbb{R}} \left|\hat{F}_{X,n}(z) - \hat{F}_{Y,m}(z)\right| \geq c. \tag{10.87}$$

Just like for the one-sample KS test[46], to construct the 2KS test of size α, we need to find the critical value c such that

[46] See (10.63) and (10.64).

$$\mathbb{P}(D < c \,|\, H_0) = 1 - \alpha. \tag{10.88}$$

The exact asymptotic distribution of the KS statistic (10.86) under the null hypothesis H_0 was derived by Nikolai Smirnov in [11] and also mentioned in [12]. He proved that if the common distribution $F_X = F_Y$ is *continuous*, then the probability $\mathbb{P}(D < c \,|\, H_0)$ does not depend on it, and for any $x > 0$

$$\mathbb{P}\left(\sqrt{\frac{nm}{n+m}}D < x \,\middle|\, H_0\right) \to F_K(x), \quad \text{as } n, m \to \infty. \tag{10.89}$$

In other words, as the sample sizes increase[47], the distribution of continuous random variable $\sqrt{nm/(n+m)}D$ converges to the Kolmogorov distribution,

[47] Sample sizes n and m should increase to infinity such that the ratio n/m remains between two fixed numbers α and β, that is $0 < \alpha \leq n/m \leq \beta < \infty$.

$$\text{Under } H_0: \quad \sqrt{\frac{nm}{n+m}}D \xrightarrow{\mathbb{D}} K \sim \mathscr{K}, \quad \text{as } n, m \to \infty. \tag{10.90}$$

This asymptotic result allows to approximate the critical value c for the size α 2KS test. Since for large n and m, the null distribution of $\sqrt{nm/(n+m)}D$ is approximately the Kolmogorov distribution, we have:

$$\begin{aligned} 1 - \alpha = \mathbb{P}(D < c \,|\, H_0) &= \mathbb{P}\left(\sqrt{\frac{nm}{n+m}}D < \sqrt{\frac{nm}{n+m}}c \,\middle|\, H_0\right) \\ &\approx F_K\left(\sqrt{\frac{nm}{n+m}}c\right). \end{aligned} \tag{10.91}$$

Solving this equation for c, we obtain that

$$\text{The size } \alpha \text{ 2KS test rejects } H_0 \quad \Longleftrightarrow \quad D \geq \sqrt{\frac{n+m}{nm}} F_K^{-1}(1-\alpha). \tag{10.92}$$

As in the case of the one-sample test (10.72), this two-sample test is only *asymptotically exact*. For finite n and m, the size of the test is *approximately* α.

By analogy with computations in (10.73), the p-value of the actually observed data $x = (x_1, \ldots, x_n)$ and $y = (y_1, \ldots, y_m)$ is given by

$$\begin{aligned} \mathfrak{p}(x,y) = \mathbb{P}(D \geq D_{\text{obs}} \,|\, H_0) &= \mathbb{P}\left(\underbrace{\sqrt{\frac{nm}{n+m}}D}_{\dot\sim \mathscr{K}} \geq \sqrt{\frac{nm}{n+m}}D_{\text{obs}} \,\middle|\, H_0\right) \\ &\approx 1 - F_K\left(\sqrt{\frac{nm}{n+m}}D_{\text{obs}}\right), \end{aligned} \tag{10.93}$$

where $D_{\text{obs}} = D(x,y)$ is observed value of statistic (10.86). The smaller the p-value, the stronger the evidence against the null hypothesis that samples x and y were generated from the same probability distribution.

The two-sample Kolmogorov-Smirnov test has the same limitation as the original one-sample test: the distribution $F_X = F_Y$ considered under the null hypothesis must be *continuous*. If it is not continuous (discrete or mixed), then[48] the asymptotic result (10.90) does not hold any more, and the 2KS test becomes conservative, which leads to the loss of power and overestimated p-values. An extension of the 2KS test for the discontinuous case was developed in [24], section 2.2. An attractive alternative to the 2KS test in this case is the permutation test[49], which is conceptually simpler and computationally easier to implement.

[48] This follows directly from Smirnov's results obtained in [11].

[49] See Section 10.4.

Remark 10.10. It is worth mentioning a possible counter-intuitive behavior of the two-sample Kolmogorov-Smirnov test in a situation where one sample size is much larger then the other, $n \gg m$. It was shown in [25] that if the size of one sample grows while the size of the other sample remains constant, then, in some cases, instead of gaining power, the 2KS test can lose it[50]. This means, paradoxically, that the use of additional observations from the large sample weakens the ability of the test to reject the null hypothesis when it is false. Compare this result with the requirement on sample sizes formulated in Footnote 47.

[50] Loss of power is equivalent to increase of the probability of type II error (recall Fig. 9.5) and conservativity of the test.

Remark 10.11. The original Kolmogorov's proof of (10.67) for the one-sample test and Smirnov's proof of (10.90) for the two-sample test are based on completely different methods. A unified and simplified approach is given by Feller in [26].

Both the one- and two-sample Kolmogorov-Smirnov tests are among the most useful and general nonparametric tests, which are widely used in many disciplines.

10.4 Permutation Test

The *permutation test* belongs to a wide class of permutation methods whose use dates back to Ronald Fisher in 1935 [27]. It can be viewed as a more general and flexible alternative to the 2KS test for testing whether two distributions are the same.

As in the previous section, suppose that we have two independent samples from two populations[51] with CDFs F_X and F_Y,

$$X_1,\ldots,X_n \overset{\text{iid}}{\sim} F_X \quad \text{and} \quad Y_1,\ldots,Y_m \overset{\text{iid}}{\sim} F_Y, \tag{10.94}$$

and we want to test

$$H_0 : F_X = F_Y \quad \text{vs} \quad H_1 : F_X \neq F_Y. \tag{10.95}$$

[51] For example, the two samples can represent household incomes in two different states, cholesterol levels in men and women, mathematics test results in two different schools, or lengths of flight delays for two different airlines.

Let $s(X,Y) = s(X_1,\ldots,X_n,Y_1,\ldots,Y_m)$ be any test statistic that discriminates between H_0 and H_1, and whose large values contradict the null hypothesis. For example, $s(X,Y)$ can be the absolute difference of the sample means, $s(X,Y) = |\overline{X} - \overline{Y}|$, or it can be the KS statistic (10.86).

Remark 10.12. When constructing a statistical test, we usually try to define a test statistic s in such a way that it is possible to find[52] its *null distribution*, i.e. the distribution of s under the assumption that H_0 is true. This is because the null distribution of a test statistic is precisely what is needed for constructing a test of a given size (see (9.46)) and for computing the p-value (see (9.72)). The requirement of a findable null distribution limits the choice of test statistics, often severely. With the permutation test, however, any statistic whose large values contradict H_0 can be used. In this sense, the permutation test is more flexible[53] than the 2KS test, which is based on the Kolmogorov-Smirnov statistic (10.86).

[52] Ideally exactly or at least approximately.

[53] As we will see soon, there is a price to be paid for this flexibility.

If H_0 is true, then all $N = n+m$ of the random variables $X_1,\dots,X_n,Y_1,\dots,Y_m$ that constitute the data are IID from their common distribution,

$$\text{Under } H_0: \;\; X_1\dots,X_n,Y_1\dots,Y_m \overset{\text{iid}}{\sim} F = F_X = F_Y. \tag{10.96}$$

Let Z denote the sample of size N obtained by concatenating X and Y,

$$Z = (Z_1,\dots,Z_N) = (X_1,\dots,X_n,Y_1,\dots,Y_m), \tag{10.97}$$

and s_{obs} be the actually observed value of the test statistic.

The basic idea behind the permutation test is to generate the null distribution of s by calculating the values of s for all permutations of the data. Namely, conditional on the observed values $X_1,\dots,X_n,Y_1,\dots,Y_m$, each of the $N!$ permutations of the data has the same probability of being observed. The fact that we observed $X_1,\dots,X_n,Y_1,\dots,Y_m$ in that particular order happened merely by chance. Let $s_1,\dots,s_{N!}$ denote the values of s on all $N!$ permutations of the data[54]. Then, under H_0 and conditional on the observed values, each of $s_1,\dots,s_{N!}$ is equally likely. The discrete probability distribution $\mathbb{P}_\pi$ that puts probability mass $1/N!$ on each s_i, $i = 1,\dots,N!$, is called the *permutation distribution* of statistic s,

[54] There are at most $\binom{N}{n}$ distinct values among $s_1,\dots,s_{N!}$, and s_{obs} is one of them.

$$\mathbb{P}_\pi(s = s_i) = \frac{1}{N!}, \quad i = 1,\dots,N!. \tag{10.98}$$

Let us illustrate the concept of permutation distribution with a simple example.

Example 10.2. Suppose that $n = 2$, $m = 1$, and $Z = (X_1,X_2,Y_1) = (1,-1,0)$. Let $s(X,Y) = |\overline{X} - \overline{Y}|$ be the test statistic. Table 10.1 lists all $N! = (2+1)! = 6$ permutations of the data $(1,-1,0)$ and the corresponding values of the test statistic s. The permutation distribution of s is the discrete distribution defined as follows: $\mathbb{P}_\pi(s=0) = 1/6+1/6 = 1/3$ and $\mathbb{P}_\pi(s=1.5) = 1/6+1/6+1/6+1/6 = 2/3$.

The permutation distribution $\mathbb{P}_\pi$ of the test statistic can be viewed as its "empirical null distribution" derived from the observed data. In permutation testing, the permutation distribution is used instead of the unknown null distribution for computing the p-value:

$$\mathfrak{p} = \mathbb{P}(s \geq s_{\text{obs}} \,|\, H_0) \approx \mathbb{P}_\pi(s \geq s_{\text{obs}}) = \frac{1}{N!}\sum_{i=1}^{N!} \mathbb{1}_{[s_{\text{obs}},\infty)}(s_i). \tag{10.99}$$

Permutations of Z	Values of $s(Z)$
$(1,-1,0)$	$s_1 = \lvert\frac{1}{2}(1-1)-0\rvert = 0 = s_{\text{obs}}$
$(1,0,-1)$	$s_2 = \lvert\frac{1}{2}(1+0)-(-1)\rvert = 1.5$
$(-1,1,0)$	$s_3 = \lvert\frac{1}{2}(-1+1)-0\rvert = 0$
$(-1,0,1)$	$s_4 = \lvert\frac{1}{2}(-1+0)-1\rvert = 1.5$
$(0,1,-1)$	$s_5 = \lvert\frac{1}{2}(0+1)-(-1)\rvert = 1.5$
$(0,-1,1)$	$s_6 = \lvert\frac{1}{2}(0-1)-1\rvert = 1.5$

Table 10.1 Permutations of the data and the corresponding values of the test statistic.

A small value of $\mathfrak{p}$ implies that among all equiprobable under H_0 values $s_1,\ldots,s_{N!}$ of the test statistic the observed value s_{obs} is large (in the sense that it is larger than most other values). This indicates evidence against the null hypothesis.

Unfortunately, there is no theoretical guarantee that the approximation of the null distribution by the permutation distribution — and, therefore, the approximation of the p-value in (10.99) — is accurate. The permutation test tends to be conservative[55], especially if the sample sizes n and m are small. Nevertheless, despite some loss of statistical power[56] (weakened ability to reject the null hypothesis when it is false), the permutation test provides a credible method for testing hypotheses.

In practical applications, the total sample size $N = n+m$ is often sufficiently large so that summing over all $N!$ permutations in (10.99) is computationally prohibitive. Fortunately, we can estimate the p-value by rewriting $\mathbb{P}_\pi(s \geq s_{\text{obs}})$ as an expectation and using the law of larger numbers:

$$\mathfrak{p} \approx \mathbb{P}_\pi(s \geq s_{\text{obs}}) = \mathbb{E}_\pi[\mathbb{1}_{[s_{\text{obs}},\infty)}(s)] \approx \frac{1}{K}\sum_{k=1}^{K} \mathbb{1}_{[s_{\text{obs}},\infty)}(\tilde{s}_k), \tag{10.100}$$

where $\tilde{s}_1,\ldots,\tilde{s}_K$ is an IID sample from the permutation distribution[57],

$$\tilde{s}_1,\ldots,\tilde{s}_K \overset{\text{iid}}{\sim} \mathbb{P}_\pi. \tag{10.101}$$

To generate a sample unit $\tilde{s}_k \sim \mathbb{P}_\pi$:

1. Generate a random permutation τ from the uniform distribution on the set $\mathfrak{S}_N$ of all permutations of the set $\{1,\ldots,N\}$,

$$\tau \sim \mathscr{U}(\mathfrak{S}_N). \tag{10.102}$$

2. Permute the data $Z \rightsquigarrow Z_\tau$ according to the permutation τ,

$$Z_\tau = \left(Z_{\tau(1)},\ldots,Z_{\tau(N)}\right). \tag{10.103}$$

3. Compute the test statistic on the permuted data,

$$\tilde{s}_k = s(Z_\tau). \tag{10.104}$$

[55] See Footnote 33.

[56] This loss in power is the price for test's flexibility mentioned in Footnote 53.

[57] The idea of sampling from the permutation distribution was first proposed by M. Dwass [28].

Example 10.3. To illustrate this algorithm, suppose that $n = 2$, $m = 3$, the test statistic $s(X,Y) = |\overline{X} - \overline{Y}|$, and the generated permutation is $\tau = (3,5,2,1,4)$. Then the permuted data is

$$Z_\tau = \left(Z_{\tau(1)}, \ldots, Z_{\tau(N)}\right) = (Z_3, Z_5, Z_2, Z_1, Z_4) = (Y_1, Y_3, X_2, X_1, Y_2), \tag{10.105}$$

and the value of the test statistic on the permuted data is

$$s(Z_\tau) = \left| \frac{Y_1 + Y_3}{2} - \frac{X_2 + X_1 + Y_2}{3} \right|. \tag{10.106}$$

For most applications, setting the total number of generated random permutations K to a few thousand is usually sufficient for estimating the p-value in (10.100).

Remark 10.13. Theoretically, it is possible to construct the permutation test of a given size α, but from a practical point of view, it can rarely be used for the same computational reasons that make calculating the p-value in (10.99) prohibitive. To define the test of size α, let $s_{(1)} \leq \ldots \leq s_{(N!)}$ be the order statistics of $s_1, \ldots, s_{N!}$. Then the size α permutation test rejects H_0 if and only if $s(X,Y) \geq s_{(N_\alpha)}$, where $N_\alpha = N! - [\alpha N!]$ and $[x]$ denotes the integer part of x. For more details, see for instance [29].

Remark 10.14. The permutation test does not have any distributional assumptions, and, in this sense, is more general than the 2KS test, which assumes that the distribution $F = F_X = F_Y$ considered under H_0 is continuous. Moreover, the permutation test can be used in the case, where $X_1, \ldots, X_n, Y_1, \ldots, Y_m \in \mathbb{R}^d$ are vectors ($d \geq 2$) rather than scalars ($d = 1$). This is because many test statistics that discriminate between H_0 and H_1 have natural high-dimensional analogs and their null distributions can be approximated by the permutation distribution just like in the one-dimensional case. For example, in high dimensions, instead of $s(X,Y) = |\overline{X} - \overline{Y}|$ we can use $s(X,Y) = \|\overline{X} - \overline{Y}\|$ or a more sophisticated Hotelling's statistic[58].

[58] Specific examples are given, for instance, in [30].

Remark 10.15. Finally, the IID assumption, which is critical for all tests we discussed so far, can be "relaxed" and replaced with a weaker exchangeability assumption. Namely, instead of assuming that $X_1, \ldots, X_n$ and $Y_1, \ldots, Y_m$ are IID samples from F_X and F_Y, it is sufficient (for the validity of the permutation test) to assume that, under the null hypothesis, $X_1, \ldots, X_n, Y_1, \ldots, Y_m$ are exchangeable random variables. Random variables $Z_1, \ldots, Z_N$ are called *exchangeable* if their joint probability distribution is invariant with respect to any permutation of the random variables, i.e. if vectors $Z = (Z_1, \ldots, Z_N)$ and $Z_\tau = \left(Z_{\tau(1)}, \ldots, Z_{\tau(N)}\right)$ have the same joint distribution for any $\tau \in \mathfrak{S}_N$. If random variables are IID, then they are exchangeable, but the reverse is not true: exchangeable random variables are identically distributed, but not necessarily independent,

$$\text{IID} \Longrightarrow \text{Exchangeable} \Longrightarrow \text{Identically Distributed} \tag{10.107}$$

If $X_1, \ldots, X_n, Y_1, \ldots, Y_m$ are exchangeable under H_0, then[59] each of the $N!$ permutations of the data has the same probability of being observed, and (10.99) still holds.

[59] Just as under the stronger IID assumption.

Example 10.4. To better feel the difference between IID and exchangeability, let us consider a simple example. Let Z be a random 2-vector distributed according to a multivariate normal distribution,

$$Z = \begin{bmatrix} Z_1 \\ Z_2 \end{bmatrix} \sim \mathcal{N}\left(\begin{bmatrix} \mu \\ \mu \end{bmatrix}, \begin{bmatrix} \sigma^2 & a \\ a & \sigma^2 \end{bmatrix}\right). \tag{10.108}$$

If $a = 0$, then Z_1 and Z_2 are IID and exchangeable. If $a \neq 0$, then Z_1 and Z_2 are not IID (since they are correlated), but are still exchangeable, since vector (Z_2, Z_1) has the same multivariate distribution. Another example of a collection of exchangeable but not IID random variables is a sample obtained by simple random sampling[60].

[60] See Section 2.3.

Despite being known since 1935[61], the permutation test became practical and popular only with the availability of cheap computing power[62]. For instance, the permutation test has recently become one of the most commonly used statistical methods in biostatistical and genomic research. A very clear and concise introduction to the permutation methods, which includes permutation and randomization tests, confidence intervals, and illustrative examples, is given by M.D. Ernst in [30], and is highly recommended.

[61] "Actually, the statistician does not carry out this very tedious process..." R.A. Fisher [27].

[62] Computing the p-value (10.100) without a computer is indeed quite tedious.

10.5 Case Study: Cholesterol Levels in Men and Women

...I give you until forty, and who knows, maybe you'll survive the first heart attack. That won't stop you, though, nope. Not when the world is full of beer and Whoppers and chocolate ice cream.

– Stephen King, *Mr. Mercedes*

Blood cholesterol[63] is fat-like substance produced by the liver, which is essential for good health, particularly for the brain, nerves, and skin. It helps the body to build cell membranes, synthesize hormones and vitamin D that keep the bones and muscles healthy, and make acids that help to digest fat and absorb important nutrients. High level of blood cholesterol, however, can be harmful[64]. Excess cholesterol in the blood can build up to form a sticky artery-clogging plaque, which can lead to cardiovascular disease, such as angina (chest pain), stroke (brain attack), and myocardial infarction (heart attack). Although there is some disagreement on what is considered to be "too high," as a general rule, the higher the cholesterol level, the higher the risk of cardiovascular disease. But high cholesterol is not the only risk factor for cardiovascular disease. Other factors include age[65], sex[66], and smoking[67].

In most people, the cholesterol level depends on uncontrollable genetic factors and controllable lifestyle factors such as diet and exercise habits. The most common causes of high cholesterol are lack of physical activity, unhealthy diet, and smoking. There are two main types of cholesterol, HDL (high-density lipoprotein) and LDL (low-density lipoprotein), which have opposite effects on cardiovascular health:

[63] "All the bad things in English-speaking life end in -ol." – Padgett Powell, *Edisto Revisited.*

[64] "All things are poison, and nothing is without poison; the dosage alone makes it so a thing is not a poison." – Paracelsus.

[65] The risk of cardiovascular disease increases with age.

[66] Men are at higher risk than women. For example, the risk of a 50-year-old having a heart attack or stroke in the next ten years if they don't have any risk factors is 3% for men, and under 1% for women [31].

[67] Smoking increases the risk more than moderately high cholesterol levels.

- HDL is often referred to as "good" cholesterol because it helps to remove excess cholesterol from the bloodstream and reduce the risk of cardiovascular disease.
- LDL is often called "bad" cholesterol because high levels of LDL are associated with a higher risk of cardiovascular disease. We do need some amount of LDL, but an excess of it can lead to health problems.

The *total cholesterol* measures the overall amount of cholesterol in the blood[68], and its high levels tend to be unfavorable. People are diagnosed with *hypercholesterolemia* if their total cholesterol or LDL cholesterol is above the 'normal" level. The normal or desirable levels for the total cholesterol are often defined as levels below 200 mg/dL (milligrams per deciliter). More background information on blood cholesterol is provided in [31, 32].

[68] It includes HDL, LDL, and also triglycerides (a type of fat).

In this Case Study, we will compare the mean total cholesterol levels in the male and female populations across the world using the data freely available at the Gapminder[69] website [33]. Figure 10.5 shows how the men's and women's cholesterol levels in several selected countries evolved with time between 1980 and 2008.

[69] Gapminder is an independent Swedish foundation with no political, religious, or economic affiliations, whose mission is "to fight devastating ignorance with a fact-based worldview everyone can understand." `https://www.gapminder.org/`

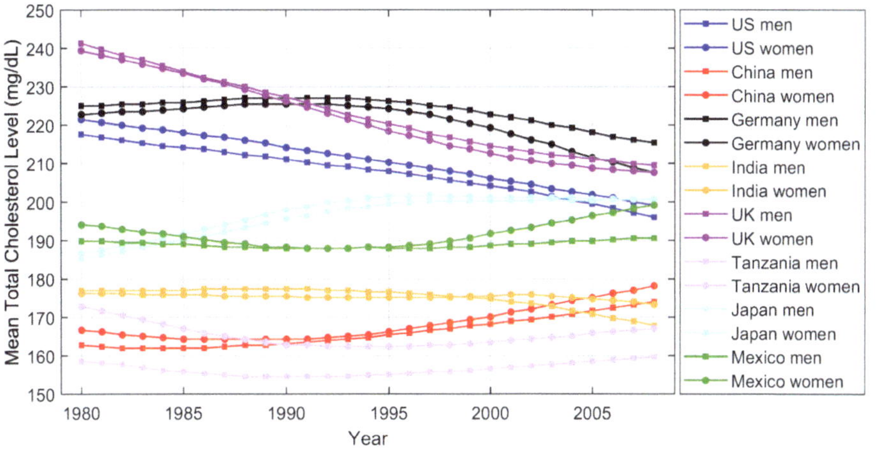

Fig. 10.5 Time-evolution of the men's and women's total cholesterol levels in selected countries.

The highest cholesterol levels in the world are found in the western European countries, such as Greenland, Iceland, and Germany. Although the UK's cholesterol level is also among the highest, it had one of the largest drops during the considered time period. On the other hand, the cholesterol levels in China and Japan increased, which perhaps could be explained by changes in diet, namely, increased consumption of animal products and fats. The lowest cholesterol levels are observed in African countries, including Niger, Congo, and Tanzania.

Among eight countries represented in Fig. 10.5, in 2008 in five countries women had higher mean total cholesterol than men. Let's look into this in more detail. Panel (a) of Fig. 10.6 shows the scatterplot of the women's $(Y_1, \ldots, Y_n)$ versus men's $(X_1, \ldots, X_n)$ cholesterol levels in 2008 for $n = 199$ different countries. The largest and the smallest differences between the men's and women's levels occurred in Greenland and Dominica, respectively. Panel (b) shows the boxplots for the corresponding cholesterol levels of male and female populations.

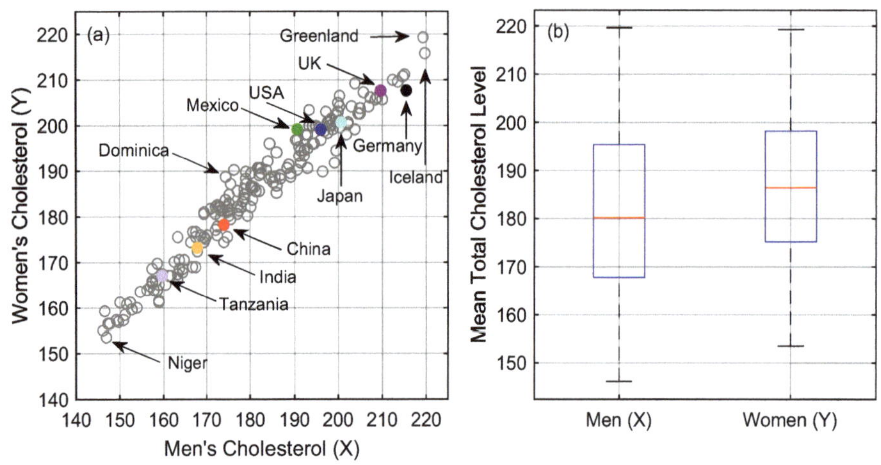

Fig. 10.6 The scatterplot (panel (a)) and boxplots (panel (b)) for the mean total cholesterol levels in men and women in 199 countries in 2008.

The boxplots suggest that the men and women cholesterol distributions are different, and, in particular, that on average women have a slightly higher cholesterol than men. Is this visual discrepancy due to statistical fluctuations, or is there a real difference between the two distributions? To formally address this question and test whether the mean total cholesterol levels in men $X_1,\ldots,X_n$ and in women $Y_1,\ldots,Y_n$ have the same distributions, we can use the two-sample Kolmogorov-Smirnov test or the permutation test. The p-value of the 2KS test (10.93) is

$$\mathfrak{p}_{2KS} \approx 0.0025, \tag{10.109}$$

which indicates very strong evidence against the null hypothesis that the men's and women's levels are identically distributed.

When using the permutation test, we have the freedom of choosing a test statistic $s(X,Y)$ that discriminates between the null and alternative hypotheses[70]. Let's consider two different statistics. Namely, let $s_1(X,Y) = |\overline{X} - \overline{Y}|$ and let $s_2(X,Y)$ be the KS statistic (10.86). The p-values of the permutation test corresponding to s_1 and s_2 are, respectively,

$$\mathfrak{p}_1 \approx 0.0067 \quad \text{and} \quad \mathfrak{p}_2 \approx 0.0026. \tag{10.110}$$

These p-values are computed via (10.100) using $K = 10^4$ samples from the permutation distributions of s_1 and s_2. Figure 10.7 shows the histograms of theses samples[71] together with the actually observed values of the test statistics. In both cases, the observed values are quite extreme and the p-values are very small. Note that in this example, the 2KS test and the permutation test with the KS statistic, despite be-

[70] Recall Remark 10.12.

[71] Using notation from the previous section, Fig 10.7 shows the histograms of $\tilde{s}_1,\ldots,\tilde{s}_K$ for the two chosen test statistics.

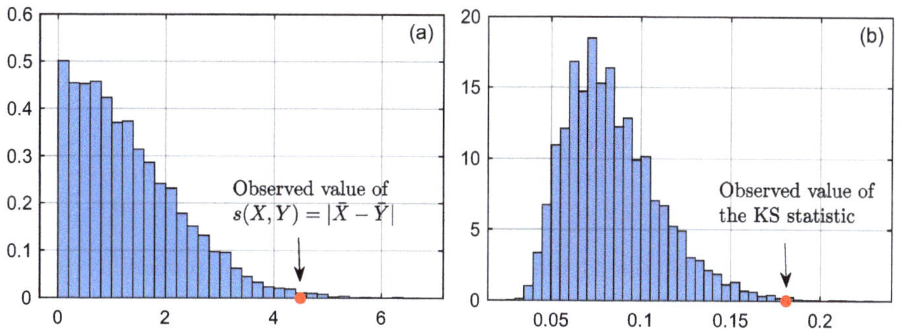

Fig. 10.7 Histograms of the values of $s_1(X,Y) = |\overline{X} - \overline{Y}|$ (panel (a)) and the KS statistic $s_2(X,Y)$ (panel (b)) obtained from $K = 10^4$ random permutations of the data.

ing different tests (derived using completely different arguments and assumptions), produce similar results, $\mathfrak{p}_{2KS} \approx \mathfrak{p}_2$, since they use the same test statistic.

To sum up, both the 2KS test and the permutation test assert that the Gapminder cholesterol data provides very strong evidence against the hypothesis that the men's and women's mean total cholesterol levels are identically distributed.

10.6 Likelihood Ratio Test

The *likelihood ratio test* (LRT) is one of the most popular and widely applicable methods for testing hypotheses in parametric settings. The LRT plays the same role in hypothesis testing as the maximum likelihood estimate (MLE) plays in point estimation. Both the MLE and the LRT are based on the concept of likelihood function[72], introduced by Ronald Fisher. The idea of using the likelihood function for testing can also be traced back to Fisher. However, the formal development and popularization of the likelihood ratio test were carried out primarily by Jerzy Neyman and Egon Pearson.

[72] See Section 8.2.

Let's start our discussion of the likelihood ratio test with the simplest scenario, where both the null and alternative hypotheses are simple[73]. In this case, the LRT is essentially an extension of the ABC example considered in Section 9.1.

[73] Recall that a hypothesis is called simple if it states that the parameter of interest θ equals to a specific value.

10.6.1 Likelihood Ratio Test for Simple Hypotheses

Let $X_1,\ldots,X_n$ be data modeled as an IID sample from $f(x\,|\,\theta)$,

$$X_1\ldots,X_n \overset{\text{iid}}{\sim} f(x\,|\,\theta), \quad \theta \in \Theta, \tag{10.111}$$

where $f(x\,|\,\theta)$ is a probability mass function or a probability density function[74], and

[74] For the LRT, it does not matter whether the stochastic model of the data is discrete or continuous.

the parameter space consists of just two points, $\Theta = \{\theta_0, \theta_1\}$. Suppose we want to test

$$H_0 : \theta = \theta_0 \quad \text{vs} \quad H_1 : \theta = \theta_1. \tag{10.112}$$

Remark 10.16. An important special case of (10.112) is testing $H_0 : f(x) = f_0(x)$ versus $H_1 : f(x) = f_1(x)$, where $f_0(x)$ and $f_1(x)$ are two rival distributions for modeling the data $X_1, \ldots, X_n \overset{\text{iid}}{\sim} f(x) \in \mathscr{F} = \{f(x) = \theta f_1(x) + (1-\theta) f_0(x), \ \ \theta \in \{0,1\}\}$.

Recall that the likelihood function is defined as the joint probability (in the discrete case) or as the joint density (in the continuous case) of the data $X = (X_1, \ldots, X_n)$, viewed as a function of parameter θ,

$$\mathscr{L}(\theta | X) = \prod_{i=1}^{n} f(X_i | \theta). \tag{10.113}$$

Remark 10.17. When we defined the likelihood function in Section 8.2, we denoted it by $\mathscr{L}_n(\theta)$ to emphasize that it is a function of θ only and that it is obtained based on a sample of size n. This notation was useful for studying the asymptotic properties of MLEs in Section 8.6. In this section, we will use a slightly different notation, $\mathscr{L}(\theta | X)$, which is more tailored to our current needs. This notation indicates that the likelihood function is a function of θ, but its values also depend on the data X.

The value of the likelihood function $\mathscr{L}(\theta | X)$ measures the consistency of θ with the data X: if $\mathscr{L}(\theta_1 | X) > \mathscr{L}(\theta_0 | X)$, then θ_1 is more consistent with the data X. In other words, if $\mathscr{L}(\theta_1 | X) > \mathscr{L}(\theta_0 | X)$, then it is more likely that X was sampled from $f(x | \theta_1)$ rather than from $f(x | \theta_0)$, and, therefore, it is more likely that H_0 is false and H_1 is true. This motivates the likelihood ratio test for simple hypotheses:

$$\text{The LRT rejects } H_0 \quad \Longleftrightarrow \quad \Lambda = \frac{\mathscr{L}(\theta_1 | X)}{\mathscr{L}(\theta_0 | X)} \geq c, \tag{10.114}$$

where Λ is called the *likelihood ratio statistic*. Note that the test (9.5) that Bob used in the ABC example is exactly the likelihood ratio test with $c = 1$. Let's consider another classical example of application of the LRT.

Example 10.5. Suppose that data is modeled with the normal model,

$$X_1, \ldots, X_n \overset{\text{iid}}{\sim} \mathscr{N}(\mu, \sigma^2), \tag{10.115}$$

where the variance σ^2 is known[75], $\mu \in \{\mu_0, \mu_1\}$ is the single model parameter, and we want to test

$$H_0 : \mu = \mu_0 \quad \text{vs} \quad H_1 : \mu = \mu_1. \tag{10.116}$$

Without loss of generality, let's assume that $\mu_0 < \mu_1$. The first step in constructing the LRT is computing the likelihood function:

[75] Note that if σ^2 is unknown, then hypotheses in (10.116) are not simple and the LRT (10.114) is not applicable.

$$\begin{aligned}\mathscr{L}(\mu|X) &= \prod_{i=1}^{n}\phi(X_i|\mu,\sigma^2) = \prod_{i=1}^{n}\frac{1}{\sqrt{2\pi}\sigma}\exp\left(-\frac{(X_i-\mu)^2}{2\sigma^2}\right) \\ &= \left(\frac{1}{\sqrt{2\pi}\sigma}\right)^n \exp\left(-\frac{\sum_{i=1}^n (X_i-\mu)^2}{2\sigma^2}\right).\end{aligned} \tag{10.117}$$

Next, the likelihood ratio statistic is

$$\Lambda = \frac{\mathscr{L}(\mu_1|X)}{\mathscr{L}(\mu_0|X)} = \exp\left(\frac{\sum_{i=1}^n (X_i-\mu_0)^2 - \sum_{i=1}^n (X_i-\mu_1)^2}{2\sigma^2}\right). \tag{10.118}$$

Simplifying this expression, we have:

$$\begin{aligned}\Lambda &= \exp\left(\frac{\sum_{i=1}^n (X_i-\mu_0)^2 - \sum_{i=1}^n (X_i-\mu_1)^2}{2\sigma^2}\right) \\ &= \exp\left(\frac{\sum_{i=1}^n (-2\mu_0 X_i + \mu_0^2 + 2\mu_1 X_i - \mu_1^2)}{2\sigma^2}\right) \\ &= \exp\left(\frac{n(\mu_0^2-\mu_1^2) + 2(\mu_1-\mu_0)n\overline{X}}{2\sigma^2}\right) = a^2\exp\left(b^2\overline{X}\right),\end{aligned} \tag{10.119}$$

where $a^2 = \exp\left(n(\mu_0^2-\mu_1^2)/2\sigma^2\right) > 0$ and $b^2 = (\mu_1-\mu_0)n/\sigma^2 > 0$ are positive constants[76], whose exact expressions, as we will see in a moment, are not important for constructing the test[77].

[76] Recall that we assumed that $\mu_1 > \mu_0$.

[77] That is why we simply denote them by a^2 and b^2.

The rejection region of the LRT has the following form:

$$\begin{aligned}\mathscr{R} &= \{X \in \mathbb{R}^n \mid \Lambda \geq c\} = \left\{X \in \mathbb{R}^n \mid a^2\exp\left(b^2\overline{X}\right) \geq c\right\} \\ &= \left\{X \in \mathbb{R}^n \mid \overline{X} \geq \tilde{c} = \frac{1}{b^2}\log\frac{c}{a^2}\right\}.\end{aligned} \tag{10.120}$$

This form of the rejection region for testing (10.116) with $\mu_1 > \mu_0$ makes sense intuitively: $H_0 : \mu = \mu_0$ is rejected if the sample mean $\overline{X}$, which is an estimate of μ, is sufficiently large. And the larger $\overline{X}$ is, the more evidence the data provides for rejecting the null hypothesis. Note that if $\mu_0 > \mu_1$, then $(\mu_1-\mu_0)n/\sigma^2 < 0$, and the rejection region would be $\mathscr{R} = \{X \in \mathbb{R}^n \mid \overline{X} \leq \tilde{c}\}$, which also agrees with intuition.

To construct the LRT of size α, we need to find the critical value c, or, equivalently, the critical value $\tilde{c}$ such that[78]

[78] Since H_0 is simple, the test size is simply the probability of type I error.

$$\begin{aligned}\alpha &= \mathbb{P}(\text{Type I error}) = \mathbb{P}(\text{Reject } H_0|H_0 \text{ is true}) = \mathbb{P}(\overline{X} \geq \tilde{c}|\mu = \mu_0) \\ &= \mathbb{P}\left(\underbrace{\frac{\overline{X}-\mu_0}{\sigma/\sqrt{n}}}_{\sim\mathscr{N}(0,1)} \geq \frac{\tilde{c}-\mu_0}{\sigma/\sqrt{n}} \,\middle|\, \mu = \mu_0\right) = 1 - \Phi\left(\frac{\tilde{c}-\mu_0}{\sigma/\sqrt{n}}\right)\end{aligned} \tag{10.121}$$

Solving this equation for $\tilde{c}$ gives

$$\tilde{c} = \mu_0 + \frac{\sigma}{\sqrt{n}}\Phi^{-1}(1-\alpha) = \mu_0 + \frac{\sigma}{\sqrt{n}}z_{1-\alpha}, \tag{10.122}$$

and, therefore,

$$\text{The size } \alpha \text{ LRT rejects } H_0 \quad \Longleftrightarrow \quad \overline{X} \geq \mu_0 + \frac{\sigma}{\sqrt{n}} z_{1-\alpha}. \tag{10.123}$$

Finally, using its geometric interpretation[79], the p-value $\mathfrak{p}(x)$ of the actually observed data $x = (x_1, \ldots, x_n)$ is the value of α such that x lies exactly on the boundary of the rejection region:

[79] Recall Fig. 9.9.

$$\overline{x} = \mu_0 + \frac{\sigma}{\sqrt{n}} z_{1-\mathfrak{p}(x)}, \tag{10.124}$$

which, after solving for $\mathfrak{p}(x)$, leads to

$$\mathfrak{p}(x) = 1 - \Phi\left(\frac{\overline{x} - \mu_0}{\sigma/\sqrt{n}}\right). \tag{10.125}$$

As expected, when the sample mean $\overline{x}$ increases, the p-value decreases, indicating increasing evidence against the null hypothesis.

Remark 10.18. Alternatively, we can compute the p-value $\mathfrak{p}(x)$ using its probabilistic interpretation: $\mathfrak{p}(x)$ is the probability of observing a test statistic value $\Lambda(X)$ that is at least as large as the value $\Lambda(x)$ that has been actually observed, under the assumption that H_0 is true:

$$\begin{aligned} \mathfrak{p}(x) &= \mathbb{P}(\Lambda(X) \geq \Lambda(x) \,|\, \mu = \mu_0) \\ &= \mathbb{P}\left(a^2 \exp\left(b^2 \overline{X}\right) \geq a^2 \exp\left(b^2 \overline{x}\right) \,|\, \mu = \mu_0\right) = \mathbb{P}\left(\overline{X} \geq \overline{x} \,|\, \mu = \mu_0\right) \\ &= \mathbb{P}\left(\underbrace{\frac{\overline{X} - \mu_0}{\sigma/\sqrt{n}}}_{\sim \mathcal{N}(0,1)} \geq \frac{\overline{x} - \mu_0}{\sigma/\sqrt{n}} \,\middle|\, \mu = \mu_0\right) = 1 - \Phi\left(\frac{\overline{x} - \mu_0}{\sigma/\sqrt{n}}\right). \end{aligned} \tag{10.126}$$

This computation is essentially the same as the computation in (10.121) with $\tilde{c} = \overline{x}$, which is the value of $\tilde{c}$ that defines the rejection region whose boundary contains the observed data x.

As mentioned in Section 9.6, in many cases, the most powerful tests do not exist, and when they do, it is often hard to find them. The case of simple hypotheses is a fortunate exception. It turns out that if both H_0 and H_1 are simple, then the likelihood ratio test is the most powerful test. This is exactly the statement of the *Neyman-Pearson Lemma* [34], a fundamental result in statistical inference.

Remark 10.19. Jerzy Neyman was not only one of the founders of modern statistics, but also a generous and compassionate man: he established at UC Berkeley a special scholarship program to help prepare talented underprivileged young people for a university education. Moreover, when he could not obtain federal grants, he supported his students out of his own pocket. Egon Pearson was the initiator of a long-term and fruitful collaboration with Neyman, who would most likely have left statistics for mathematics if not for a letter from Pearson.

Theorem 10.1 (Neyman-Pearson Lemma). *If* $H_0 : \theta = \theta_0$ *and* $H_1 : \theta = \theta_1$ *are simple hypotheses, then the size* α *LRT is the most powerful test at level* α. *That is, among all statistical tests with size* α *or less, the size* α *LRT has the largest power* $\beta(\theta_1) = 1 - \mathbb{P}($*Type II error*$)$, i.e. *the smallest probability of type II error.*

The statement of this lemma is not obvious or intuitive, and requires a proof. Moreover, the proof will provide us with some additional insights, and will reinforce and enhance our understanding of this lemma.

Proof. Let $c_\alpha > 0$ be the value of c such that the LRT (10.114) has size α, and let $\mathscr{R}^\dagger$ be the corresponding rejection region,

$$\text{The size } \alpha \text{ LRT rejects } H_0 \iff X \in \mathscr{R}^\dagger = \left\{ x \;\middle|\; \frac{\mathscr{L}(\theta_1|x)}{\mathscr{L}(\theta_0|x)} \geq c_\alpha \right\}. \quad (10.127)$$

Equivalently, the rejection region $\mathscr{R}^\dagger$ can be rewritten as follows:

$$\mathscr{R}^\dagger = \{x \mid \mathscr{L}(\theta_1|x) - c_\alpha \mathscr{L}(\theta_0|x) \geq 0\}. \quad (10.128)$$

Since H_0 is a simple hypothesis, the size of any test coincides with the probability of type I error[80]. Therefore, for the LRT of size α,

$$\mathbb{P}(X \in \mathscr{R}^\dagger | H_0) = \mathbb{P}(\text{Type I error}) = \alpha. \quad (10.129)$$

Consider now any test whose size is less than or equal to α. Let $\mathscr{R}$ be the rejection region this test. Then

$$\mathbb{P}(X \in \mathscr{R} | H_0) = \mathbb{P}(\text{Type I error}) \leq \alpha. \quad (10.130)$$

The Neyman-Pearson lemma states that the probability of type II error of the test with rejection region $\mathscr{R}^\dagger$ (the size α LRT) is less than or equal to the probability of type II error of the test with rejection region $\mathscr{R}$ (arbitrary test with size α or less)[81],

$$\mathbb{P}(X \notin \mathscr{R}^\dagger | H_1) \leq \mathbb{P}(X \notin \mathscr{R} | H_1). \quad (10.131)$$

To prove the last inequality, note that for any x,

$$\mathbb{1}_{\mathscr{R}^\dagger}(x)\,[\mathscr{L}(\theta_1|x) - c_\alpha \mathscr{L}(\theta_0|x)] \geq \mathbb{1}_{\mathscr{R}}(x)\,[\mathscr{L}(\theta_1|x) - c_\alpha \mathscr{L}(\theta_0|x)], \quad (10.132)$$

where $\mathbb{1}_{\mathscr{R}^\dagger}$ and $\mathbb{1}_{\mathscr{R}}$ are the indicator functions of regions $\mathscr{R}^\dagger$ and $\mathscr{R}$. Indeed, if $x \in \mathscr{R}^\dagger$, then $\mathbb{1}_{\mathscr{R}^\dagger}(x) = 1$ and $\mathscr{L}(\theta_1|x) - c_\alpha \mathscr{L}(\theta_0|x) \geq 0$ and

$$\begin{aligned} \mathbb{1}_{\mathscr{R}^\dagger}(x)\,[\mathscr{L}(\theta_1|x) - c_\alpha \mathscr{L}(\theta_0|x)] &= \underbrace{[\mathscr{L}(\theta_1|x) - c_\alpha \mathscr{L}(\theta_0|x)]}_{\geq 0} \\ &\geq \underbrace{[\mathscr{L}(\theta_1|x) - c_\alpha \mathscr{L}(\theta_0|x)]}_{\geq 0}\,\underbrace{\mathbb{1}_{\mathscr{R}}(x)}_{=0 \text{ or } 1} \end{aligned} \quad (10.133)$$

If $x \notin \mathscr{R}^\dagger$, then $\mathbb{1}_{\mathscr{R}^\dagger}(x) = 0$ and $\mathscr{L}(\theta_1|x) - c_\alpha \mathscr{L}(\theta_0|x) < 0$ and

Jerzy Neyman (1894-1981), a Polish-American mathematician and statistician, "whose work on hypothesis testing, confidence intervals, and survey sampling has revolutionized both theory and practice." [35]. Photo: Archives of the Mathematisches Forschungsinstitut Oberwolfach.

Egon Pearson (1895-1980), the son of Karl Pearson, was a leading British statistician, best known for the development of the likelihood ratio test, the Neyman-Pearson lemma, and major contributions to industrial statistics.

[80] Recall Definition 9.2.

[81] This is the same as saying that the size α LRT is the most powerful test at level α.

$$\mathbb{1}_{\mathscr{R}^\dagger}(x)\left[\mathscr{L}(\theta_1\,|\,x)-c_\alpha\mathscr{L}(\theta_0\,|\,x)\right]=0\geq\underbrace{\left[\mathscr{L}(\theta_1\,|\,x)-c_\alpha\mathscr{L}(\theta_0\,|\,x)\right]}_{<0}\underbrace{\mathbb{1}_{\mathscr{R}}(x)}_{=0\text{ or }1} \tag{10.134}$$

Using the definition of the likelihood function (10.113), we can rewrite (10.132) as follows:

$$\mathbb{1}_{\mathscr{R}^\dagger}(x)\left[\prod_{i=1}^{n}f(x_i\,|\,\theta_1)-c_\alpha\prod_{i=1}^{n}f(x_i\,|\,\theta_0)\right]\geq\mathbb{1}_{\mathscr{R}}(x)\left[\prod_{i=1}^{n}f(x_i\,|\,\theta_1)-c_\alpha\prod_{i=1}^{n}f(x_i\,|\,\theta_0)\right]. \tag{10.135}$$

Integrating with respect to x (if f is a PDF) or summing over all values of x (if f is a PMF) both sides of the last inequality yields[82]:

$$\mathbb{E}\left[\mathbb{1}_{\mathscr{R}^\dagger}(X)\,|\,H_1\right]-c_\alpha\mathbb{E}\left[\mathbb{1}_{\mathscr{R}^\dagger}(X)\,|\,H_0\right]\geq\mathbb{E}\left[\mathbb{1}_{\mathscr{R}}(X)\,|\,H_1\right]-c_\alpha\mathbb{E}\left[\mathbb{1}_{\mathscr{R}}(X)\,|\,H_0\right], \tag{10.136}$$

where $\mathbb{E}\left[\mathbb{1}_{\mathscr{R}^\dagger}(X)\,|\,H_i\right]$ and $\mathbb{E}\left[\mathbb{1}_{\mathscr{R}}(X)\,|\,H_i\right]$ denote the expected values of the indicator random variables $\mathbb{1}_{\mathscr{R}^\dagger}(X)$ and $\mathbb{1}_{\mathscr{R}}(X)$ given that H_i is true, $i=0,1$. But these expected values can be expressed in terms of probabilities[83]:

$$\begin{aligned}\mathbb{E}\left[\mathbb{1}_{\mathscr{R}^\dagger}(X)\,|\,H_1\right]&=0\times\mathbb{P}\big(\mathbb{1}_{\mathscr{R}^\dagger}(X)=0\,|\,H_1\big)+1\times\mathbb{P}\big(\mathbb{1}_{\mathscr{R}^\dagger}(X)=1\,|\,H_1\big)\\&=\mathbb{P}(X\in\mathscr{R}^\dagger\,|\,H_1)=1-\mathbb{P}(X\notin\mathscr{R}^\dagger\,|\,H_1).\end{aligned} \tag{10.137}$$

Similarly,

$$\begin{aligned}\mathbb{E}\left[\mathbb{1}_{\mathscr{R}}(X)\,|\,H_1\right]&=1-\mathbb{P}(X\notin\mathscr{R}\,|\,H_1),\\\mathbb{E}\left[\mathbb{1}_{\mathscr{R}^\dagger}(X)\,|\,H_0\right]&=\mathbb{P}(X\in\mathscr{R}^\dagger\,|\,H_0)=\alpha,\\\mathbb{E}\left[\mathbb{1}_{\mathscr{R}}(X)\,|\,H_0\right]&=\mathbb{P}(X\in\mathscr{R}\,|\,H_0)\leq\alpha.\end{aligned} \tag{10.138}$$

Finally, substituting these expressions back into (10.136), we have

$$1-\mathbb{P}(X\notin\mathscr{R}^\dagger\,|\,H_1)-c_\alpha\alpha\geq1-\mathbb{P}(X\notin\mathscr{R}\,|\,H_1)-c_\alpha\mathbb{P}(X\in\mathscr{R}\,|\,H_0). \tag{10.139}$$

Simplifying and rearranging the terms lead to the desired result:

$$\mathbb{P}(X\notin\mathscr{R}\,|\,H_1)-\mathbb{P}(X\notin\mathscr{R}^\dagger\,|\,H_1)\geq c_\alpha\big(\alpha-\underbrace{\mathbb{P}(X\in\mathscr{R}\,|\,H_0)}_{\leq\alpha}\big)\geq0, \tag{10.140}$$

which is equivalent to the statement (10.131) of the Neyman-Pearson lemma. □

In simple terms, the Neyman-Pearson lemma states that, in the case of simple hypotheses, the likelihood ratio test is the best available test. In particular, (10.123) is the best test for testing (10.116), and Bob did his best in the ABC example[84].

The Neyman-Pearson lemma was used in different testing applications in electrical engineering, signal processing, radar systems, and microeconomics. More recently, it has played a role in the discovery of the Higgs boson [36].

Remark 10.20. In reflections on his personal experiences in research [37], Neyman describes how he discovered a proof of the lemma. One evening he was working in

[82] Here we are using $\int g(x)f(x)dx=\mathbb{E}[g(X)]$, where $X\sim f(x)$, with $g(x)=\mathbb{1}_{\mathscr{R}^\dagger}(x)$ or $\mathbb{1}_{\mathscr{R}}(x)$ and $f(x)=\prod_{i=1}^n f(x_i|\theta_0)$ or $\prod_{i=1}^n f(x_i|\theta_1)$. Hope this footnote helps to digest the transition from (10.135) to (10.136).

[83] We are getting closer to proving (10.131).

[84] See Section 9.1.

his office trying to prove the lemma, when his wife and friends called him saying that it was time to go to a movie. Neyman was annoyed by this interruption, but as he was getting up from his desk to answer the call, the main idea of the proof came as a flash[85], and he suddenly understood how to prove the lemma. He could not recall the movie they saw that evening.

[85] Mystically, this experience of the main idea coming as a flash is not uncommon and has been described by many mathematicians. For more on this, see Poincaré's [38].

10.6.2 Generalized Likelihood Ratio Test

Let us now generalize the likelihood ratio test for the case, where the null and alternative hypotheses are not necessarily simple. Suppose that $X_1,\ldots,X_n$ is modeled as an IID sample form $f(x\,|\,\theta)$,

$$X_1\ldots,X_n \overset{\text{iid}}{\sim} f(x\,|\,\theta), \quad \theta\in\Theta, \tag{10.141}$$

where $f(x\,|\,\theta)$, as before, is a PMF or PDF, and the parameter space $\Theta=\Theta_0\sqcup\Theta_1$ is partitioned into two disjoint subsets. Suppose we want to test

$$H_0:\theta\in\Theta_0 \quad \text{vs} \quad H_1:\theta\in\Theta_1. \tag{10.142}$$

If subset Θ_i contains more than one parameter value, then the hypothesis $H_i:\theta\in\Theta_i$ is called *composite*. Suppose that H_0 or H_1, or both are composite hypotheses.

To generalize the LRT for the case of composite hypothesis, we need to modify the definition of the likelihood ratio statistic Λ, since its LRT definition in (10.114) does not make sense in the composite case[86]. A reasonable candidate definition is the following[87]:

$$\tilde{\Lambda}=\frac{\sup_{\theta\in\Theta_1}\mathscr{L}(\theta\,|\,X)}{\sup_{\theta\in\Theta_0}\mathscr{L}(\theta\,|\,X)}. \tag{10.143}$$

[86] In (10.114), θ_0 and θ_1 are specific values, and we cannot simply replace them with sets Θ_0 and Θ_1.

[87] If "sup" in (10.143) looks unfamiliar, see Remark 4.2.

Indeed, the quantity $\sup_{\theta\in\Theta_i}\mathscr{L}(\theta\,|\,X)$ measures the consistency of hypothesis H_i with the data X by finding the value of θ in Θ_i which is most consistent with the data. The larger $\tilde{\Lambda}$ is, the more consistent H_1 is with the data compared to H_0, and the more evidence we have to reject the null hypothesis. Notice also that if both H_0 and H_1 are simple, i.e. $\Theta_0=\{\theta_0\}$ and $\Theta_1=\{\theta_1\}$, then (10.143) reduces to (10.114).

Although definition (10.143) is conceptually sound, the likelihood ratio statistic for composite hypotheses is defined slightly differently. Namely,

$$\Lambda=\frac{\sup_{\theta\in\Theta}\mathscr{L}(\theta\,|\,X)}{\sup_{\theta\in\Theta_0}\mathscr{L}(\theta\,|\,X)}. \tag{10.144}$$

The difference between (10.143) and (10.144) is in the numerator: instead of maximizing[88] the likelihood function over Θ_1, it is maximized now over the full parameter space Θ. The main advantage of Λ over $\tilde{\Lambda}$ is that, as we will see, it is possible to derive its asymptotic null distribution, which is needed for computing the p-value[89].

[88] More precisely, finding the supremum.

[89] See Section 10.6.3.

Remark 10.21. There is a simple relationship between Λ and $\tilde{\Lambda}$, namely,

$$\Lambda = \max\{1, \tilde{\Lambda}\}. \tag{10.145}$$

It can be proved by considering two cases. If $\tilde{\Lambda} < 1$, then the value of θ that maximizes the likelihood function over Θ is in Θ_0, and therefore $\Lambda = 1$. If $\tilde{\Lambda} > 1$, then then the value of θ that maximizes the likelihood function over Θ is in Θ_1, and therefore $\Lambda = \tilde{\Lambda}$. The relationship (10.145) tells us that if Λ is large ($\Lambda > 1$), then it simply coincides with $\tilde{\Lambda}$, and large values of $\Lambda = \tilde{\Lambda}$ contradict the null hypothesis.

There is a close relationship between the likelihood ratio statistic Λ and maximum likelihood estimates of θ. Let

$$\hat{\theta}_{MLE} = \arg\max_{\theta \in \Theta} \mathscr{L}(\theta\,|\,X) \tag{10.146}$$

be the MLE of θ, and let $\hat{\theta}^0_{MLE}$ be the MLE of θ, where instead of Θ, its subset $\Theta_0 \subset \Theta$ is used as a parameter space,

$$\hat{\theta}^0_{MLE} = \arg\max_{\theta \in \Theta_0} \mathscr{L}(\theta\,|\,X). \tag{10.147}$$

If both $\hat{\theta}_{MLE}$ and $\hat{\theta}^0_{MLE}$ exist[90], then

$$\Lambda = \frac{\mathscr{L}(\hat{\theta}_{MLE}|X)}{\mathscr{L}(\hat{\theta}^0_{MLE}|X)}. \tag{10.148}$$

[90] If one of them does not exist, then the right-hand side of (10.148) is not defined. But the right-had side of (10.144) is always defined.

Now, after we tailored the definition of the likelihood ratio statistic to composite hypotheses, we can formulate the *generalized likelihood ratio test* (GLRT):

$$\text{The GLRT rejects } H_0 \quad \Longleftrightarrow \quad \Lambda = \frac{\sup_{\theta\in\Theta} \mathscr{L}(\theta\,|\,X)}{\sup_{\theta\in\Theta_0} \mathscr{L}(\theta\,|\,X)} \geq c. \tag{10.149}$$

The Neyman-Pearson Lemma states that the LRT is the most powerful test for simple hypotheses. However, if the hypotheses are composite, then the GLRT is not necessarily the most powerful test. Nevertheless, it often performs reasonably well[91]. Let's consider an example of application of the GLRT.

[91] Just like MLEs, on which the GLRT is essentially based, (10.148).

Example 10.6. Suppose that data is modeled with the normal model,

$$X_1, \ldots, X_n \overset{\text{iid}}{\sim} \mathcal{N}(\mu, \sigma^2), \tag{10.150}$$

where the variance σ^2 is known[92], $\mu \in \mathbb{R}$ is a model parameter, and we want to test

[92] Problem 10.7 will ask you to derive a GLRT for the case, where σ^2 is unknown.

$$H_0 : \mu = \mu_0 \quad \text{vs} \quad H_1 : \mu \neq \mu_0. \tag{10.151}$$

Here, $\Theta = \mathbb{R}$, $\Theta_0 = \{\mu_0\}$, and $\Theta_1 = (-\infty, \mu_0) \cup (\mu_0, \infty)$. While H_0 is simple, H_1 is a composite hypothesis. As in the previous Example 10.5, the likelihood function is given by

$$\mathscr{L}(\mu\,|X) = \left(\frac{1}{\sqrt{2\pi}\sigma}\right)^n \exp\left(-\frac{\sum_{i=1}^n (X_i-\mu)^2}{2\sigma^2}\right). \tag{10.152}$$

The likelihood ratio statistic is

$$\Lambda = \frac{\sup_{\mu\in\Theta}\mathscr{L}(\mu\,|X)}{\sup_{\mu\in\Theta_0}\mathscr{L}(\mu\,|X)} = \frac{\mathscr{L}(\hat{\mu}_{MLE}|X)}{\mathscr{L}(\mu_0\,|X)}, \tag{10.153}$$

where $\hat{\mu}_{MLE} = \overline{X}$ is the MLE of the mean[93]. Substituting the expressions for the likelihood function and the MLE into the previous equation, we have[94]:

$$\begin{aligned}\Lambda &= \exp\left(\frac{-\sum_{i=1}^n (X_i-\overline{X})^2 + \sum_{i=1}^n (X_i-\mu_0)^2}{2\sigma^2}\right)\\ &= \exp\left(\frac{\sum_{i=1}^n (-X_i^2 + 2X_i\overline{X} - \overline{X}^2 + X_i^2 - 2\mu_0 X_i + \mu_0^2)}{2\sigma^2}\right)\\ &= \exp\left(\frac{2n\overline{X}^2 - n\overline{X}^2 - 2\mu_0 n\overline{X} + n\mu_0^2}{2\sigma^2}\right) = \exp\left(\frac{(\overline{X}-\mu_0)^2}{2\sigma^2/n}\right).\end{aligned} \tag{10.154}$$

[93] Recall Example 8.2.

[94] In (10.154), the identity $\sum_{i=1}^n X_i = n\overline{X}$ is used to simplify the expression.

The rejection region of the GLRT has the following form:

$$\begin{aligned}\mathscr{R} &= \{X\in\mathbb{R}^n \,|\, \Lambda \geq c\} = \left\{X\in\mathbb{R}^n \,\middle|\, \exp\left(\frac{(\overline{X}-\mu_0)^2}{2\sigma^2/n}\right)\geq c\right\}\\ &= \left\{X\in\mathbb{R}^n \,\middle|\, \left|\frac{\overline{X}-\mu_0}{\sigma/\sqrt{n}}\right| \geq \tilde{c} = \sqrt{2\log c}\right\}.\end{aligned} \tag{10.155}$$

The form of this rejection region makes intuitive sense for testing (10.151): the null hypothesis $H_0 : \mu = \mu_0$ is rejected if the sample mean $\overline{X}$, which is an estimate of μ, is sufficiently different form the "null value" μ_0. And the larger the discrepancy between $\overline{X}$ and μ_0 is, the more evidence the data provides for rejecting H_0.

To construct the GLRT of size α, we need to find the critical value c, or, equivalently, the critical value $\tilde{c}$ such that[95]

$$\begin{aligned}\alpha &= \mathbb{P}(\text{Type I error}) = \mathbb{P}(\text{Reject } H_0\,|\,H_0 \text{ is true})\\ &= \mathbb{P}\left(\underbrace{\left|\frac{\overline{X}-\mu_0}{\sigma/\sqrt{n}}\right|}_{\sim\mathscr{N}(0,1)} \geq \tilde{c} \,\middle|\, \mu=\mu_0\right) = 2\Phi(-\tilde{c}).\end{aligned} \tag{10.156}$$

[95] Since H_0 is simple, the test size is simply the probability of type I error.

Solving this equation for $\tilde{c}$ gives

$$\tilde{c} = -\Phi^{-1}\left(\frac{\alpha}{2}\right) = -z_{\frac{\alpha}{2}} = z_{1-\frac{\alpha}{2}}, \tag{10.157}$$

and, therefore,

$$\text{The size } \alpha \text{ GLRT rejects } H_0 \quad \Longleftrightarrow \quad \left|\frac{\overline{X}-\mu_0}{\sigma/\sqrt{n}}\right| \geq z_{1-\frac{\alpha}{2}}. \tag{10.158}$$

Notice that in this example, the GLRT coincides with the Wald test (10.10).

Finally, the p-value $\mathfrak{p}(x)$ of the actually observed data $x = (x_1, \ldots, x_n)$ is such value of the size α that forces the boundary of the rejection region (10.158) to pass exactly through x:

$$\left|\frac{\overline{x}-\mu_0}{\sigma/\sqrt{n}}\right| = z_{1-\frac{\mathfrak{p}(x)}{2}}. \tag{10.159}$$

Solving this equation for $\mathfrak{p}(x)$ gives

$$\mathfrak{p}(x) = 2\Phi\left(-\left|\frac{\overline{x}-\mu_0}{\sigma/\sqrt{n}}\right|\right). \tag{10.160}$$

As expected, the p-value is a decreasing function of the distance $|\overline{x}-\mu_0|$ between the sample mean $\overline{x}$ and the null value μ_0: the larger the distance is, the smaller the p-value, and, thus, the more evidence against $H_0 : \mu = \mu_0$ the data provides.

10.6.3 The p-value of the GLRT

One of the reasons that contributes significantly to the popularity of the generalized likelihood ratio test is that the asymptotic behavior of the test statistic Λ allows to approximate the p-value for a wide range of testing problems (10.141), (10.142).

Recall[96] that, in the general case, the p-value is given by

[96] See (9.72).

$$\mathfrak{p}(x) = \sup_{\theta \in \Theta_0} \mathbb{P}(\Lambda(X) \geq \Lambda(x) \,|\, \theta). \tag{10.161}$$

This means that to compute the p-value, we need to know the *null distribution* of $\Lambda(X)$, that is, the distribution of $\Lambda(X)$ under the assumption that $H_0 : \theta \in \Theta_0$ is true. Let's return to the previous Example 10.6, and check whether it is possible to state something meaningful about the null distribution of $\Lambda(X)$ in that specific example. So, the likelihood ratio statistics is

$$\Lambda(X) = \exp\left(\frac{(\overline{X}-\mu_0)^2}{2\sigma^2/n}\right), \quad \text{where } X_1, \ldots, X_n \overset{\text{iid}}{\sim} \mathcal{N}(\mu, \sigma^2). \tag{10.162}$$

Under the null hypothesis, $H_0 : \mu = \mu_0$,

$$2\log\Lambda(X) = \frac{(\overline{X}-\mu_0)^2}{\sigma^2/n} = \Bigg(\underbrace{\frac{\overline{X}-\mu_0}{\sigma/\sqrt{n}}}_{\sim\mathcal{N}(0,1)}\Bigg)^2 \sim \chi_1^2, \tag{10.163}$$

where χ_1^2 is the chi-squared distribution with one degree of freedom.

Remark 10.22. Recall that if $Z_1,\dots,Z_q \overset{\text{iid}}{\sim} \mathcal{N}(0,1)$ are IID standard normal random variables, then, by definition, the squared Euclidean norm (squared length) of the Gaussian random vector $Z = (Z_1,\dots,Z_q) \in \mathbb{R}^q$ is distributed according to the *chi-squared distribution* with q degrees of freedom,

$$\|Z\|^2 = \sum_{i=1}^{q} Z_i^2 \sim \chi_q^2. \tag{10.164}$$

The chi-squared distribution was first discovered by the German geodesist Friedrich Helmert and later independently rediscovered by Karl Pearson.

It turns out that the appearance of the chi-squared distribution in (10.163) is not just a lucky coincidence, but a manifestation of a general result, known as the *Wilks theorem* [39]. Namely, it can be shown that if $\Theta \subset \mathbb{R}^d$ is a manifold[97] of dimension $\dim\Theta$ and $\Theta_0 \subset \Theta$ is a submanifold of dimension $\dim\Theta_0$, then, under certain regularity conditions[98], the null distribution of $2\log\Lambda(X)$ converges to the chi-squared distribution with $q = \dim\Theta - \dim\Theta_0$ degrees of freedom,

$$\text{Under } H_0: \quad 2\log\Lambda(X) \xrightarrow{\mathbb{D}} \chi_q^2, \quad \text{as } n \to \infty. \tag{10.165}$$

In Example 10.6, the full parameter space $\Theta = \mathbb{R}$, its subset $\Theta_0 = \{\mu_0\}$ consists of a single null value, their dimensions $\dim\Theta = 1$ and $\dim\Theta_0 = 0$, and the number of degrees of freedom $q = 1$, which is consistent with χ_1^2 in (10.163).

[97] If the differential-geometric term "manifold" is not familiar, think of $\dim\Theta$ and $\dim\Theta_0$ as the numbers of free parameters in Θ and Θ_0.

[98] Which, unfortunately, is hard to check in practice.

Remark 10.23. If $\Theta = \{\theta = (\theta_1,\dots,\theta_d)\} \subset \mathbb{R}^d$ and Θ_0 consists of all $\theta \in \Theta$ such that m components of θ are fixed to particular values, i.e.

$$\Theta_0 = \{\theta \in \Theta \mid \theta_1 = \theta_1^{(0)},\dots,\theta_m = \theta_m^{(0)}\}, \tag{10.166}$$

then $\dim\Theta = d$, $\dim\Theta_0 = d - m$, and $q = d - (d - m) = m$.

The proof of the Wilks theorem presented in [39] is rather nontrivial and falls outside the scope of this book. On the other hand, the result (10.165) is so counterintuitive and surprising[99] that the author owes the reader at least some explanation. An interested reader can find a sketch of proof of the Wilks theorem for a relatively simple special case in Appendix 10.9.

Using the asymptotic result (10.165), it is straightforward to approximate the p-value (10.161) of the GLRT:

$$\begin{aligned} \mathfrak{p}(x) &= \sup_{\theta\in\Theta_0} \mathbb{P}(\Lambda(X) \geq \Lambda(x)\,|\,\theta) = \sup_{\theta\in\Theta_0} \mathbb{P}(\underbrace{2\log\Lambda(X)}_{\dot\sim\chi_q^2} \geq 2\log\Lambda(x)\,|\,\theta) \\ &\approx \mathbb{P}(Q \geq 2\log\Lambda(x)) = 1 - F_{\chi_q^2}(2\log\Lambda(x)), \end{aligned} \tag{10.167}$$

where $Q \sim \chi_q^2$ and $F_{\chi_q^2}$ is the CDF of the χ_q^2 distribution.

[99] Indeed, how come that regardless of the statistical model for the data X (as long as it is regular in some sense), the null distribution of $2\log\Lambda(X)$ converges to the same χ^2 distribution?

The (generalized) likelihood ratio test that we discussed in this section is one of the most classical and popular statistical tests, widely used in various application domains. For example, it is used in high energy physics for the discovery of new phenomena [40], for multichannel radar detection in signal processing [41], and in empirical studies in psychology [42], to name but a few. It is also one of the most fundamental statistical tests since a number of other tests turns out to be asymptotically equivalent to the likelihood ratio test[100].

[100] For example, *Pearson's chi-squared test* is asymptotically equivalent to the GLRT as the sample size $n \to \infty$.

10.7 Case Study: Mendelian Inheritance

I had an inheritance from my father,
It was the moon and the sun.
And though I roam all over the world,
The spending of it's never done.
– Ernest Hemingway, *For Whom the Bell Tolls*

Genetics is an important branch of biology that studies heredity in organisms: how certain qualities or traits are passed down from parents to their offspring, from one generation to the next. As a scientific field, genetics has originated from experimental studies by Gregor Mendel, an Austrian-Czech biologist, mathematician, and a Christian monk, who explored the patterns of inheritance in pea plants and described them mathematically. After eight years of numerous tedious experiments with pea plant breeding, in 1865, Mendel formulated three fundamental laws of genetics [43]: the law of dominance, the law of segregation, and the law of independent assortment — collectively known as Mendel's laws of inheritance[101]. The ground-breaking nature of Mendel's discovery was understood only after his death, when other scientists rediscovered his results around 1900.

[101] A brief introduction to Mendel's laws of inheritance is provided in [44].

In this Case Study, we will statistically test Mendelian inheritance by constructing the generalized likelihood ratio test and computing the corresponding p-value. Suppose we breed two different pea plants, one with yellow round seeds and the other with green wrinkled seeds. Then, according to the law of dominance, all offspring plants in the first generation will have yellow round seeds, since the yellow trait is dominant to green and the round trait is dominant to wrinkled. If we now allow the plants in the first generation to self-fertilize, then the laws of segregation and independent assortment predict that progeny plants will have seeds of four possible types: yellow round, yellow wrinkled, green round, and green wrinkled. Moreover, the proportions of each type are, respectively, 9/16, 3/16, 3/16, and 1/16.

In his seminal paper [43], Mendel described the results of his experiments with pea plants, which served as the experimental basis for the proposed theory on inheritance. In one of the experiments, among $s = 556$ seeds yielded by 15 plants in the second generation, there were

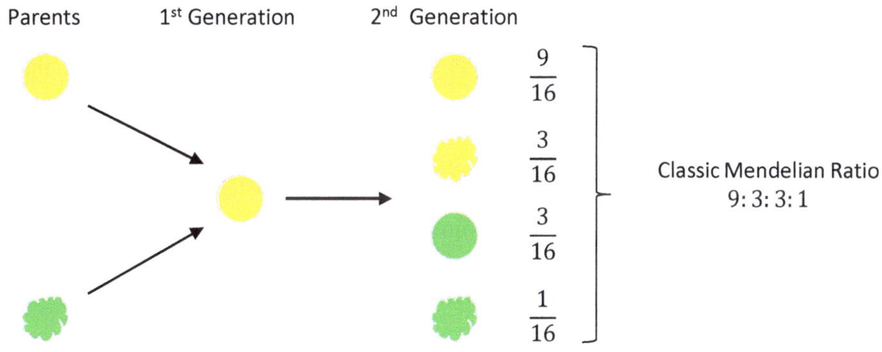

Fig. 10.8 The breeding process of pea plants and predictions of Mendel's laws of inheritance.

$$\begin{aligned} n_1 &= 315 \text{ yellow and round,} \\ n_2 &= 101 \text{ yellow and wrinkled,} \\ n_3 &= 108 \text{ green and round,} \\ n_4 &= 32 \text{ green and wrinkled.} \end{aligned} \tag{10.168}$$

Let us test whether Mendel's theoretical prediction that the vector of proportions of seed types

$$p_M = \left(\frac{9}{16}, \frac{3}{16}, \frac{3}{16}, \frac{1}{16}\right) \tag{10.169}$$

is consistent with the observed experimental data. Based on the data (10.168), should we reject or fail to reject the theoretical prediction (10.169)?

To answer this question and quantify the degree to which (10.168) and (10.169) are consistent, we need to assume a stochastic model for the Mendel data[102]. A natural model for the Mendel data is the *multinomial distribution*, which, as its name suggests, is a generalization of the binomial distribution for the case, where there are more than just two possible outcomes in an experiment.

[102] Without a stochastic model, the data (10.168) is just a collection of numbers. A stochastic model allows to interpret these numbers as a sample from a probability distribution and use the machinery of hypothesis testing.

Recall that the multinomial distribution is defined as follows. Let k be the number of possible outcomes in an experiment, and let $p = (p_1, \dots, p_k)$ be the vector of probabilities, where p_i is the probability of outcome i,

$$\sum_{i=1}^{k} p_i = 1 \quad \text{and} \quad p_i > 0, \;\; i = 1, \dots, k. \tag{10.170}$$

Let s be the total number of experiments, and N_i be the number of times outcome i has occurred,

$$\sum_{i=1}^{k} N_i = s \quad \text{and} \quad 0 \leq N_i \leq s, \;\; i = 1, \dots, k. \tag{10.171}$$

Then the random vector $N = (N_1, \dots, N_k)$ is said to be distributed according to the multinomial distribution with parameters s and p,

$$N = (N_1, \dots, N_k) \sim \mathscr{M}(s, p). \tag{10.172}$$

Let us model Mendel's data (10.168) by the multinomial distribution, where the total number of trials $s = 556$ and the number of possible outcomes $k = 4$,

$$\text{Mendel's Data:} \quad N = (N_1, \dots, N_{k=4}) \sim \mathscr{M}(s = 556, p), \tag{10.173}$$

where $p = (p_1, \dots, p_{k=4})$ is a vector of model parameters. In other words, we model the actually observed data $n = (n_1, n_2, n_3, n_4) = (315, 101, 108, 32)$ as a realization of random vector $N \sim \mathscr{M}(s, p)$. The parameter space Θ of this model is[103]

[103] Geometrically, Θ is a 3-simplex or tetrahedron.

$$\Theta = \left\{ p \;\middle|\; \sum_{i=1}^{k} p_i = 1, \;\; p_i > 0, \; i = 1, \dots, k = 4 \right\} \tag{10.174}$$

The question about the consistency between Mendel's theory and the experimental observations can then be formulated as a testing problem,

$$H_0: \underbrace{p = p_M}_{\text{Mendel's theory}} \quad \text{vs} \quad H_1: p \neq p_M. \tag{10.175}$$

Let's construct the generalized likelihood ratio test[104]. As usual, the first step is computing the likelihood function:

[104] Since H_1 is a composite hypothesis, we have to use the generalized version of the likelihood ratio test.

$$\mathscr{L}(p|N) = \mathbb{P}(N|s, p) = \frac{s!}{N_1! \dots N_k!} p_1^{N_1} \dots p_k^{N_k}, \tag{10.176}$$

where $\mathbb{P}(N|s,p)$ is the PMF of the multinomial distribution (viewed as a function of model parameters p). The likelihood ratio statistic is

$$\Lambda = \frac{\sup_{p \in \Theta} \mathscr{L}(p|N)}{\sup_{p \in \Theta_0 = \{p_M\}} \mathscr{L}(p|N)} = \frac{\mathscr{L}(\hat{p}_{MLE}|N)}{\mathscr{L}(p_M|N)}, \tag{10.177}$$

where $\hat{p}_{MLE}$ is the maximum likelihood estimate of vector p and p_M is Mendel's null value. It can be shown that the MLE is

$$\hat{p}_{MLE} = \left(\frac{N_1}{s}, \dots, \frac{N_k}{s} \right), \tag{10.178}$$

which is a very natural estimate: the probability p_i of outcome i is estimated by the proportion N_i/s of outcomes i in s experiments.

So, the likelihood ratio statistic is given by the following expression:

$$\Lambda = \frac{\mathscr{L}(\hat{p}_{MLE}|N)}{\mathscr{L}(p_M|N)} = \frac{\left(\frac{N_1}{s}\right)^{N_1} \dots \left(\frac{N_k}{s}\right)^{N_k}}{(p_{M,1})^{N_1} \dots (p_{M,k})^{N_k}} = \prod_{i=1}^{k} \left(\frac{N_i}{s p_{M,i}} \right)^{N_i}. \tag{10.179}$$

Now we can compute the p-value of the Mendelian data (10.168). Using the asymptotic approximation (10.167), we have:

$$\mathfrak{p}(n) \approx 1 - F_{\chi^2_q}(2\log\Lambda(n)) = 1 - F_{\chi^2_q}\left(2\sum_{i=1}^{k=4} n_i \log\left(\frac{n_i}{sp_{M,i}}\right)\right). \qquad (10.180)$$

What is q in this case? The dimension of the parameter space is $\dim\Theta = 3$, since it consists of 4-dimensional vectors (p_1, p_2, p_3, p_4) with positive coordinates that are not arbitrary[105], but satisfy one constraint: $p_1 + p_2 + p_3 + p_4 = 1$. The subset $\Theta_0 \subset \Theta$ consists of a single point, $\Theta_0 = \{p_M\}$, so it has zero dimension, $\dim\Theta_0 = 0$. Therefore, $q = \dim\Theta - \dim\Theta_0 = 3$, which yields

$$\mathfrak{p}(n) \approx 0.924. \qquad (10.181)$$

This p-value is very large, indicating that the observed experimental data does not provide any evidence against Mendel's theory.

In fact, the p-value 0.924 is so large that there is some controversy about whether Mendel's experimental data is "too good to be true." In 1936, Ronald Fisher analyzed Mendel's data and came to a conclusion that "the data of most, if not all, of the experiments have been falsified so as to agree closely with Mendel's expectations." In a more recent work [45], Pires and Branco provide a brief history of the controversy and offer a possible explanation[106], suggesting that, most likely, Mendel performed several experiments but reported only the results that best fit his theory.

105 If they were arbitrary, then we would have $\dim\Theta = 4$.

106 "Every explanation is after all an hypothesis." – Ludwig Wittgenstein, *Philosophical Occasions*.

10.8 Further Reading

This chapter focused on five popular statistical tests and their variations that are often used in applications. We discussed the Wald, Student's, Kolmogorov-Smirnov, permutation, and the likelihood ratio tests. The popularity of these tests stems from their wide applicability and relatively high power. As the main reference and a text for further reading on hypothesis testing, the same fundamental book as in Section 9.8 is recommended:

- Lehmann, E.L., Romano, J.P.: *Testing Statistical Hypotheses*. New York: Springer, 3rd Edition (2005). `https://doi.org/10.1007/0-387-27605-X`

Traditionally, introductory textbooks on statistical inference tend to focus more on parametric methods and rarely cover permutation tests. And yet permutation tests are among the primary methods for testing hypotheses in many application domains, especially in biostatistics, genetics, and social and life sciences. For further reading on permutation tests, their practical implementations, and other resampling techniques the following book is recommended:

- Good, P.: *Permutation Tests: A Practical Guide to Resampling Methods for Testing Hypotheses*. Springer New York, NY, 2nd Edition (2013). `https://doi.org/10.1007/978-1-4757-3235-1`

Finally, Section 7.3 of

- Davison, A.C.: *Statistical Models*. Cambridge University Press, 1st Edition (2003). `https://doi.org/10.1017/CBO9780511815850`

contains a good collection of examples of testing problems and their solutions.

10.9 Appendix: The Wilks Theorem

A rigorous proof of the Wilks theorem (10.165) is hard and lies outside our scope[107]. Yet, this asymptotic result is so surprising and unexpected that it requires at least some justification. This Appendix provides a sketch of the proof of the theorem for a special case, where the parameter $\theta \in \Theta = \mathbb{R}$ is one-dimensional, and $\Theta_0 = \{\theta_0\}$ consists of a single null value. So, the goal is to show that

$$\text{Under } H_0 : \theta = \theta_0 \qquad 2\log\Lambda(X) \xrightarrow{\mathbb{D}} \chi^2_{q=1}, \quad \text{as } n \to \infty. \tag{10.182}$$

Let $\hat{\theta}_{MLE}$ be the MLE of θ. Then[108]

$$2\log\Lambda(X) = 2\log\mathscr{L}(\hat{\theta}_{MLE}) - 2\log\mathscr{L}(\theta_0) = 2\ell(\hat{\theta}_{MLE}) - 2\ell(\theta_0), \tag{10.183}$$

where $\ell(\theta)$ is the log-likelihood function. Let's approximate the log-likelihood by the first three terms of its Taylor series at $\theta = \hat{\theta}_{MLE}$:

$$\begin{aligned}\ell(\theta) &= \sum_{k=0}^{\infty} \frac{\ell^{(k)}(\hat{\theta}_{MLE})}{k!}(\theta - \hat{\theta}_{MLE})^k \\ &\approx \ell(\hat{\theta}_{MLE}) + \ell'(\hat{\theta}_{MLE})(\theta - \hat{\theta}_{MLE}) + \frac{\ell''(\hat{\theta}_{MLE})}{2}(\theta - \hat{\theta}_{MLE})^2.\end{aligned} \tag{10.184}$$

Since the MLE maximizes the log-likelihood function, $\ell'(\hat{\theta}_{MLE}) = 0$. Therefore,

$$\ell(\theta) \approx \ell(\hat{\theta}_{MLE}) + \frac{\ell''(\hat{\theta}_{MLE})}{2}(\theta - \hat{\theta}_{MLE})^2. \tag{10.185}$$

Using this approximation for computing $\ell(\theta_0)$ in (10.183), we have:

$$\begin{aligned}2\log\Lambda(X) &\approx 2\ell(\hat{\theta}_{MLE}) - 2\left(\ell(\hat{\theta}_{MLE}) + \frac{\ell''(\hat{\theta}_{MLE})}{2}(\theta_0 - \hat{\theta}_{MLE})^2\right) \\ &= -\ell''(\hat{\theta}_{MLE})(\theta_0 - \hat{\theta}_{MLE})^2.\end{aligned} \tag{10.186}$$

Recall[109] that the maximum likelihood estimate $\hat{\theta}_{MLE}$ is asymptotically normal:

[107] The original Wilks's proof can be found in [39]. A more recent exposition is given by van der Vaart [46]. According to van der Vaart, Wilks's proof is not sufficiently rigorous and "would not be acceptable to present-day referees."

[108] Recall (10.148).

[109] See Section 8.6.2 and, in particular, (8.85).

$$\sqrt{n\mathscr{I}(\theta_\star)}(\hat{\theta}_{MLE} - \theta_\star) \xrightarrow{\mathbb{D}} \mathscr{N}(0,1), \qquad \text{as } n \to \infty, \tag{10.187}$$

where $\theta_\star$ is the true value of the parameter and $\mathscr{I}(\theta_\star)$ is the Fisher information. Under H_0, the true value of θ is θ_0, and, therefore,

$$\sqrt{n\mathscr{I}(\theta_0)}(\hat{\theta}_{MLE} - \theta_0) \xrightarrow{\mathbb{D}} \mathscr{N}(0,1), \qquad \text{as } n \to \infty. \tag{10.188}$$

Using this asymptotic result we have:

$$\begin{aligned} 2\log\Lambda(X) &\approx -\ell''(\hat{\theta}_{MLE})(\hat{\theta}_{MLE} - \theta_0)^2 \\ &= -\frac{\ell''(\hat{\theta}_{MLE})}{n\mathscr{I}(\theta_0)}\Big(\underbrace{\sqrt{n\mathscr{I}(\theta_0)}(\hat{\theta}_{MLE} - \theta_0)}_{Z_n}\Big)^2 = -\frac{\ell''(\hat{\theta}_{MLE})}{n\mathscr{I}(\theta_0)}Z_n^2, \end{aligned} \tag{10.189}$$

where random variable Z_n converges in distribution, as the sample size n increases, to the standard normal distribution,

$$Z_n = \sqrt{n\mathscr{I}(\theta_0)}(\hat{\theta}_{MLE} - \theta_0) \xrightarrow{\mathbb{D}} \mathscr{N}(0,1), \qquad \text{as } n \to \infty. \tag{10.190}$$

Next, recall[110] that the maximum likelihood estimate $\hat{\theta}_{MLE}$ is consistent, that is, it converges to the true value of the parameter, which is θ_0 under the null hypotheses,

[110] See (8.48).

$$\hat{\theta}_{MLE} \xrightarrow{\mathbb{P}} \theta_0, \qquad \text{as } n \to \infty. \tag{10.191}$$

Assuming the log-likelihood function satisfies certain smoothness conditions, this implies that

$$\ell''(\hat{\theta}_{MLE}) \xrightarrow{\mathbb{P}} \ell''(\theta_0), \qquad \text{as } n \to \infty. \tag{10.192}$$

Finally, while discussing the asymptotic normality of the MLE in Section 8.6.2, we derived[111] using the law of large numbers that

[111] See (8.83).

$$\frac{1}{n}\ell''(\theta_0) \xrightarrow{\mathbb{P}} -\mathscr{I}(\theta_0), \qquad \text{as } n \to \infty. \tag{10.193}$$

Combining now (10.189) with (10.192), (10.193), and then with (10.190), we have:

$$2\log\Lambda(X) \approx Z_n^2 \xrightarrow{\mathbb{D}} \chi_1^2, \qquad \text{as } n \to \infty. \tag{10.194}$$

Thus, in essence, the null distribution of the likelihood ratio statistic is a consequence of the nice asymptotic properties of the maximum likelihood estimate.

Problems

10.1. Wald Meets Poisson

Suppose that data is modeled as an IID sample from the Poisson distribution[112],

[112] For a definition and a brief description of the Poisson distribution, see Problem 8.2.

$$X_1,\dots,X_n \overset{\text{iid}}{\sim} \mathscr{P}(\lambda), \tag{10.195}$$

with parameter $\lambda > 0$, and suppose that we want to test

$$H_0 : \lambda = \lambda_0 \quad \text{vs} \quad H_1 : \lambda \neq \lambda_0, \tag{10.196}$$

where $\lambda_0 > 0$ is some constant (null value).

(a) Construct the Wald test of approximate size α.

(b) Let the null value $\lambda_0 = 1$, sample size $n = 20$, and the test size $\alpha = 0.05$. Generate the data $X_1,\dots,X_n \overset{\text{iid}}{\sim} \mathscr{P}(\lambda_0)$ and perform the Wald test constructed in part (a). Repeat this $m = 10^5$ times and compute the type I error rate, i.e. the proportion of times the Wald test rejects the true null hypothesis H_0.

(c) Let $x = (x_1,\dots,x_n)$ be the actually observed data. Compute the p-value $\mathfrak{p}(x)$ based on the Wald test constructed in part (a).

10.2. Wald Compares Drugs

Let A and B be two drugs that appear to have similar properties in terms of how they affect and cure patients. To test this formally, let's give drug A to n randomly chosen patients and drug B to m randomly chosen patients. Let N_A and M_B be the numbers of patients cured by drugs A and B, respectively. We can model N_A and M_B as realizations of binomial random variables,

$$N_A \sim \mathscr{B}(n, p_A) \quad \text{and} \quad M_B \sim \mathscr{B}(m, p_B), \tag{10.197}$$

where p_A and p_B are the probabilities of "success," i.e. the probabilities that the respective drugs cure a patient. So, our goal is to test

$$H_0 : p_A = p_B \quad \text{vs} \quad H_1 : p_A \neq p_B. \tag{10.198}$$

(a) Construct the Wald test of approximate size α.

(b) Suppose that drug A cured $n_A = 91$ out of $n = 100$ patients, and drug B cured $m_B = 191$ out of $m = 200$ patients. Find the p-value $\mathfrak{p}(n_A, n_B)$ of this data based on the Wald test constructed in part (a). What is your conclusion regarding the relative efficacy of the two drugs based on the obtained p-value?

Before solving this problem, it might be useful to revisit Example 10.1.

10.3. Understanding the Permutation Test

The goal of this problem is to test your basic understanding of the permutation test methodology. Suppose we have the following data:

$$X_1 = 2, X_2 = 5 \quad \text{and} \quad Y_1 = 1, Y_2 = 7, \tag{10.199}$$

and we want to test whether the distributions that generated these two samples have the same mean values:

$$H_0 : \mu_X = \mu_Y \quad \text{vs} \quad H_1 : \mu_X \neq \mu_Y. \tag{10.200}$$

To address this problem, let's use the permutation test with the absolute difference of the sample means $s(X,Y) = |\overline{X} - \overline{Y}|$ as the test statistic.

(a) Find the permutation distribution $\mathbb{P}_\pi$ of statistic s.

(b) Compute the p-value of the data based on the permutation test.

10.4. Mark Twain and the Quintus Curtius Snodgrass Letters

In early 1861, ten letters describing military adventures in the Civil War were published in the *New Orleans Daily Crescent*. Although these letters were signed "Quintus Curtius Snodgrass," Mark Twain was widely credited with the authorship of these letters until it was shown in [47] that Twain almost certainly did not write these ten letters[113]. In this problem, we will investigate this issue based on the proportions of the three-letter words in essays actually written by Mark Twain and in the Quintus Curtius Snodgrass letters.

In eight essays written by Twain, the proportions of the three-letter words are

$$T = (0.225, 0.262, 0.217, 0.24, 0.23, 0.229, 0.235, 0.217). \tag{10.201}$$

in the ten Snodgrass letters, the proportions are

$$S = (0.209, 0.205, 0.196, 0.21, 0.202, 0.207, 0.224, 0.223, 0.22, 0.201). \tag{10.202}$$

If (solely for the testing purposes!), we model Twain and Snodgrass as "text-generating mechanisms," then the larger the statistical discrepancy between samples T and S, the more evidence we will have against the hypothesis that the Snodgrass letters were written by Twain.

(a) Assuming that the samples T and S are normally distributed, perform the two-sample t-test (Welch t-test) for comparing the means of these samples and compute the corresponding p-value.

(b) Let's relax the normality assumption and, instead, just assume that samples T and S were generated by continuous distributions F_T and F_S. Ignoring the small sample sizes, perform the two-sample Kolmogorov-Smirnov test for testing whether the distributions F_T and F_S are the same and compute the corresponding p-value.

(c) Finally, perform the permutation test using two statistics, the absolute difference of the sample means and the KS statistic, and compute the corresponding p-values.

What is your conclusion regarding the authorship of the Quintus Curtius Snodgrass letters based on the performed statistical analysis?

10.5. Neyman-Pearson Lemma in Action

Suppose that data $X_1 \ldots, X_n$ is modeled as an IID sample,

$$X_1 \ldots, X_n \overset{\text{iid}}{\sim} f(x|\theta) = (1+\theta)x^\theta, \quad x \in (0,1), \ \theta \in \Theta, \tag{10.203}$$

where the parameter space $\Theta = \{\theta_0, \theta_1\}$ and $0 < \theta_0 < \theta_1$. Suppose we want to test

[113] "A lie can travel half way around the world while the truth is putting on its shoes." – Mark Twain.

$$H_0 : \theta = \theta_0 \quad \text{vs} \quad H_1 : \theta = \theta_1. \tag{10.204}$$

Let $x = (x_1, \dots, x_n)$ be the actually observed data. Find the p-value $\mathfrak{p}(x)$ using the most powerful test assuming that the sample size n is sufficiently large.

10.6. Generalized Likelihood Ratio Test
Suppose that data is modeled with the normal model,

$$X_1, \dots, X_n \overset{\text{iid}}{\sim} \mathcal{N}(\mu, \sigma^2), \quad \mu \in \mathbb{R}, \tag{10.205}$$

where σ^2 is known and μ is the single parameter. And suppose we want to test

$$H_0 : \mu \leq 0 \quad \text{vs} \quad H_1 : \mu > 0. \tag{10.206}$$

In Example 9.5, we constructed an ad hoc test with a rather natural rejection region $\mathscr{R} = \{X \in \mathbb{R}^n \mid \overline{X} \geq c\}$. The corresponding test of size α has the following form[114]:

[114] See (9.38).

$$\text{Reject } H_0 : \mu \leq 0 \quad \Longleftrightarrow \quad \overline{X} \geq \frac{\sigma}{\sqrt{n}} z_{1-\alpha}. \tag{10.207}$$

Construct the size α generalized likelihood ratio test and compare it with the ad hoc test (10.207).

10.7. Normal GLRT and Unknown Variance
In Example 10.6, we considered a normally modeled sample,

$$X_1, \dots, X_n \overset{\text{iid}}{\sim} \mathcal{N}(\mu, \sigma^2), \tag{10.208}$$

and constructed a generalized likelihood ratio test for testing

$$H_0 : \mu = \mu_0 \quad \text{vs} \quad H_1 : \mu \neq \mu_0. \tag{10.209}$$

We obtained that

$$\text{The size } \alpha \text{ GLRT rejects } H_0 \quad \Longleftrightarrow \quad \left| \frac{\overline{X} - \mu_0}{\sigma/\sqrt{n}} \right| \geq z_{1-\frac{\alpha}{2}}, \tag{10.210}$$

and the corresponding p-value of the observed data $x = (x_1, \dots, x_n)$ is given by

$$\mathfrak{p}(x) = 2\Phi\left(- \left| \frac{\overline{x} - \mu_0}{\sigma/\sqrt{n}} \right| \right). \tag{10.211}$$

These results were derived under the assumption that the variance σ^2 is known and the mean μ is the single model parameter. But what if σ^2 is unknown? In this case, we cannot use (10.210) and (10.211), since we don't know the value of σ. Construct the size α GLRT for the case where the variance σ^2 is unknown, and derive an expression for the corresponding p-value.

References

1. Morgenstern, O.: Abraham Wald, 1902-1950. *Econometrica* 19(4), 361-367 (1951). `https://doi.org/10.2307/1907462`
2. Student: The probable error of a mean. *Biometrika* 6(1), 1-25 (1908). `https://doi.org/10.2307/2331554`
3. Pearson, E.S.: *'Student': A Statistical Biography of William Sealy Gosset.* (R.L. Plackett and G.A. Barnard, eds.), Clarendon Press, Oxford (1990). `https://www.gwern.net/docs/statistics/decision/1990-pearson-studentastatisticalbiographyofwilliamsealygosset.pdf`
4. Pfanzagl, J., Sheynin, O.: Studies in the history of probability and statistics XLIV A forerunner of the t-distribution. *Biometrika* 83(4), 891-898 (1996). `https://doi.org/10.1093/biomet/83.4.891`
5. Welch, B.L.: The significance of the difference between two means when the population variances are unequal. *Biometrika* 29(3-4), 350-362 (1938). `https://doi.org/10.2307/2332010`
6. Satterthwaite, F.E.: Synthesis of variance. *Psychometrika* 6, 309-316 (1941). `https://doi.org/10.1007/BF02288586`
7. Starnes, D.S, Yates, D.S., Moore, D.S.: *The Practice of Statistics.* W.H. Freeman and Company, New York (2010).
8. Paul, S., You-Gan, W., Insha, U.: A review of the Behrens-Fisher problem and some of its analogs: does the same size fit all? *Revstat Statistical Journal* 17(4), 563-597 (2019). `https://www.ine.pt/revstat/pdf/REVSTAT_v17-n4-7.pdf`
9. Kolmogorov, A.N.: Sulla determinazione empirica di una legge di distribuzione. G. Ist. Ital. Attuari. 4: 83-91 (1933). English version: Kolmogorov, A.N.: On the empirical determination of a distribution law. In: A.N. Shiryayev, editor. *Selected Works of A.N. Kolmogorov.* Springer, Netherlands, 139-146 (1992). `https://doi.org/10.1007/978-94-011-2260-3_15`
10. Marsaglia, G., Tsang, W.W., Wang, J.: Evaluating Kolmogorov's distribution. *Journal of Statistical Software* 8 (18), 1-4 (2003). `https://doi.org/10.18637/jss.v008.i18`
11. Smirnov, N.V.: On the estimation of the discrepancy between empirical curves of distribution for two independent samples. *Bulletin Mathématique de l'Université de Moscou* 2, (1939).
12. Smirnov, N.V.: Table for estimating the goodness of fit of empirical distributions. *The Annals of Mathematical Statistics* 19(2), 279-281 (1948). `https://doi.org/10.1214/aoms/1177730256`
13. Kolmogorov, A.N.: Confidence limits for an unknown distribution function. *The Annals of Mathematical Statistics* 12(4), 461-463 (1941). `https://doi.org/10.1214/aoms/1177731684`
14. Noether, G.E.: Note on the Kolmogorov statistic in the discrete case. *Metrika* 7(1), 115-116 (1963). `https://doi.org/10.1007/bf02613966`
15. Dimitrova, D.S., Kaishev, V.K., Tan, S.: Computing the Kolmogorov-Smirnov distribution when the underlying CDF is purely discrete, mixed, or continuous. *Journal of Statistical Software* 95(10), 1-42 (2020). `https://doi.org/10.18637/jss.v095.i10`
16. Clauset, A., Shalizi, C.R., Newman, M.E.J.: Power-law distributions in empirical data. *SIAM Review* 51(4), 661-703 (2009). `http://www.jstor.org/stable/25662336`
17. Cramér, H.: On the composition of elementary errors. *Scandinavian Actuarial Journal* 1928(1), 13-74 (1928). `https://doi.org/10.1080/03461238.1928.10416862`
18. von Mises, R.: Wahrscheinlichkeitsrechnung. Leipzig-Wien, (1931).

19. Anderson, T.W., Darling, D.A.: Asymptotic theory of certain "goodness of fit" criteria based on stochastic processes. *The Annals of Mathematical Statistics* 23 (2), 193-212 (1952). https://doi.org/10.1214/aoms/1177729437
20. Anderson, T.W., Darling, D.A.: A test of goodness of fit. *Journal of the American Statistical Association* 49(268), 765-769 (1954). https://www.jstor.org/stable/2281537
21. Thode, H.J.: *Testing for Normality.* Marcel Dekker, New York (2002). https://doi.org/10.1201/9780203910894
22. Shapiro, S.S., Wilk, M.B.: An analysis of variance test for normality (complete samples). *Biometrika* 52(3-4), 591-611 (1965). https://doi.org/10.1093/biomet/52.3-4.591
23. Razali, N., Wah, Y.B.: Power comparisons of Shapiro-Wilk, Kolmogorov-Smirnov, Lilliefors and Anderson-Darling tests. *Journal of Statistical Modeling and Analytics* 2(1), 21-33 (2011). https://api.semanticscholar.org/CorpusID:18639594
24. Schröer, G., Trenkler, D.: Exact and randomization distributions of Kolmogorov-Smirnov tests two or three samples. *Computational Statistics & Data Analysis* 20(2), 185-202 (1995). https://doi.org/10.1016/0167-9473(94)00040-P
25. Gordon, A.Y., Klebanov, L.B.: On a paradoxical property of the Kolmogorov–Smirnov two-sample test. In *IMS (Institute of Mathematical Statistics) Collections* 7, 70-74 (2010) https://doi.org/10.1214/10-IMSCOLL707
26. Feller, W.: On the Kolmogorov-Smirnov limit theorems for empirical distributions. *The Annals of Mathematical Statistics*, 19(2), 177-189 (1948). https://doi.org/10.1214/aoms/1177730243
27. Fisher, R.A.: *The Design of Experiments.* Macmillan (9th ed., 1971), first published in 1935.
28. Dwass, M.: Modified randomization tests for nonparametric hypotheses. *The Annals of Mathematical Statistics*, 28(1), 181-187 (1957). https://doi.org/10.1214/aoms/1177707045
29. Chung, E., Romano, J.P.: Exact and asymptotically robust permutation tests. *The Annals of Statistics*, 41(2), 484-507 (2013). https://doi.org/10.1214/13-AOS1090
30. Ernst, M.D.: Permutation methods: a basis for exact inference. *Statistical Science* 19(4), 676-685 (2004). https://doi.org/10.1214/088342304000000396
31. Institute for Quality and Efficiency in Health Care: High cholesterol: Overview. https://www.ncbi.nlm.nih.gov/books/NBK279318/
32. CDC (Center for Disease Control and Prevention): About Cholesterol. https://www.cdc.gov/cholesterol/about.htm
33. Gapminder: Fat in blood, men. Fat in blood, women. https://www.gapminder.org/data/.
34. Neyman, J., Pearson, E.S.: On the problem of the most efficient tests of statistical hypotheses. *Philosophical Transactions of the Royal Society of London*, Ser. A, 231, 289-337 (1933). https://www.jstor.org/stable/91247
35. Lehmann, E. L.: Jerzy Neyman (1894-1981). In *Biographical Memoirs of the National Academy of Sciences*, 63, 395-420 (1994). http://www.nasonline.org/publications/biographical-memoirs/memoir-pdfs/neyman-jerzy.pdf
36. van Dyk, D.A.: The role of statistics in the discovery of a Higgs boson. *Annual Review of Statistics and Its Application*, 1(1), 41-59 (2014). https://doi.org/10.1146/annurev-statistics-062713-085841
37. Neyman, J.: A glance at some of my personal experiences in the process of research. In *Scientists at Work* (Eds. T. Dalenius et al.), 148-164 (1970).
38. Poincaré, H.: Mathematical creation. *Scientific American*, 179(2), 54-57 (1948). https://www.jstor.org/stable/24945870

39. Wilks, S.S.: The large-sample distribution of the likelihood ratio for testing composite hypotheses. *The Annals of Mathematical Statistics*, 9 (1), 60-62 (1938). https://doi.org/10.1214/aoms/1177732360
40. Cowan, G., Cranmer, K., Gross, E., Vitells, O.: Asymptotic formulae for likelihood-based tests of new physics. *The European Physical Journal C*, 71: 1554 (2011). https://doi.org/10.1140/epjc/s10052-011-1554-0
41. Liu, B., Chen, B., Michels, J.H.: A GLRT for multichannel radar detection in the presence of both compound Gaussian clutter and additive white Gaussian noise. *Digital Signal Processing*, 15(5), 437-454 (2005). https://doi.org/10.1016/j.dsp.2005.01.010
42. Glover, S., Dixon, P.: Likelihood ratios: A simple and flexible statistic for empirical psychologists. *Psychonomic Bulletin* & Review, 11(5), 791-806 (2004). https://doi.org/10.3758/BF03196706
43. Mendel, G.: Versuche über Pflanzen-hybriden. In: *Verhandlungen des Naturforschenden Vereins in Brünn, Bd. IV für das Jahr 1865*, Abhand-lungen, 3-47. English translation: http://www.esp.org/foundations/genetics/classical/gm-65.pdf
44. Miko, I.: Gregor Mendel and the principles of inheritance. *Nature Education*, 1(1), 134 (2008). https://www.nature.com/scitable/topicpage/gregor-mendel-and-the-principles-of-inheritance-593/
45. Pires, A.M., Branco, J.A.: A statistical model to explain the Mendel-Fisher controversy. *Statistical Science*, 25(4), 545-565 (2010). https://doi.org/10.1214/10-STS342
46. van der Vaart, A.W.: *Asymptotic Statistics*. Cambridge University Press, (1998). https://doi.org/10.1017/CBO9780511802256
47. Brinegar, C.S.: Mark Twain and the Quintus Curtius Snodgrass letters: a statistical test of authorship. *Journal of the American Statistical Association*, 58(301), 85-96 (1963). https://www.jstor.org/stable/2282956
48. Rohatgi, V.K.: *An Introduction to Probability Theory and Mathematical Statistics*. Wiley Series in Probability and Statistics. New York: Wiley (1976). https://doi.org/10.1002/9781118165676

11
Simple Linear Regression

Quite simple. Too simple for a learned loremaster in these suspicious days.
– J.R.R. Tolkien, *The Lord of the Rings*

Suppose we are interested in studying and learning about a certain phenomenon that can be schematically represented as a system that takes an "input" X and generates or associates with X a "response" Y,

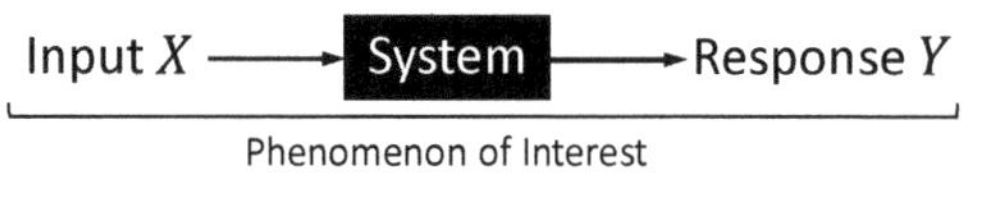

For example, X is the height of a parent and Y is the height of parent's child; X is the temperature of a star and Y is its luminosity; X and Y are the mean cholesterol levels, respectively, in men and women in a given country; X is the body size of an organism and Y is its metabolic rate; X is a current company performance measure and Y is the price of company's stock in a week from now. Examples are countless.

Let $D = \{(X_1, Y_1), \ldots, (X_n, Y_n)\}$ be the data, a collection of inputs and the corresponding responses, obtained by experiment or observation of the phenomenon. *Regression analysis* is the study of dependence between X and Y. The primary goal of regression analysis is to construct a *stochastic model* for predicting response Y from a new input X based on the observed data D. Since the question "how does Y depend on X?" is the single most important question in science and prediction is one of the most important tasks in science, regression analysis is the key "servant of all sciences" [1].

Regression analysis originated in the early XIX century in the works of Adrien-Marie Legendre and Carl Gauss, who used astronomical observations for studying the orbits of planets and comets around the Sun. The theory of regression and its methods were further developed, and applications were extended to other domains by Francis Galton, Karl Pearson, and Ronald Fisher. Regression analysis is still an active area of modern research and its methods play a central role in machine learning, information and data sciences. In the next two chapters, we will explore the key concepts and ideas of regression analysis with the main focus on the *linear*

Supplementary Information The online version contains supplementary material available at https://doi.org/10.1007/978-3-032-03848-7_11.

K. M. Zuev, *Fundamentals of Statistical Inference*,
International Series in Operations Research & Management Science 374,
https://doi.org/10.1007/978-3-032-03848-7_11

regression model — a mainstay of statistics in the last century, which still remains one of the most important and often used stochastic models. In this chapter, we will study the simplest version of the model, namely, the *simple linear regression model.*

11.1 Regression Function

Suppose that, given the data $(X_1,Y_1),\dots,(X_n,Y_n)$, we want to predict the response Y to a new, previously unseen input X. This prediction problem naturally leads to the concept of regression function, the key element of the regression methodology.

For the moment, just for simplicity, let's assume that responses do not depend on inputs[1] and let's try to predict Y based only on $Y_1,\dots,Y_n$. Let r denote our prediction. What is the best value for r? The answer depends on what exactly we mean by "best". One of the most popular and analytically convenient measures of prediction accuracy is the *mean squared error* (MSE), which is the expected squared difference between the true response and our prediction:

[1] Or that we simply don't have access to the input data $X_1,\dots,X_n$.

$$\text{MSE}[r] = \mathbb{E}\left[(Y-r)^2\right]. \tag{11.1}$$

If we accept it as a measure of prediction accuracy, then it is easy to find the best prediction, which minimizes it. The MSE is a quadratic function of r,

$$\text{MSE}[r] = r^2 - 2\mathbb{E}[Y]r + \mathbb{E}[Y^2], \tag{11.2}$$

and it is minimized when

$$r = \mathbb{E}[Y]. \tag{11.3}$$

So, the mean value $r = \mathbb{E}[Y]$ is the theoretically optimal (in the MSE sense) prediction of Y. Given the response data $Y_1,\dots,Y_n$, we can estimate it by the corresponding sample mean,

$$r \approx \hat{r} = \frac{1}{n}\sum_{i=1}^{n} Y_i. \tag{11.4}$$

This is a very natural and intuitive result: if responses do not depend on inputs, then the best prediction of the new response is simply the average of the previously observed responses.

Now let's return to the original problem: assuming that the inputs and responses are dependent, the goal is to predict the response Y to a future input X. Let $r(X)$ denote our prediction, which now explicitly depends on the input X. What function $r: X \to Y$ should we use? The best choice (again in the MSE sense) for $r(X)$ is the solution of the following minimization problem:

$$\text{MSE}[r] = \mathbb{E}\left[(Y-r(X))^2\right] \to \min. \tag{11.5}$$

Here, the expected value $\mathbb{E}$ is with respect to the joint distribution of X and Y. Since both the future input X and the corresponding response Y are uncertain, they are modeled as random quantities, and we want to find a prediction rule $r : X \to Y$ that predicts Y from X accurately on average across all X and Y.

Using the *law of total expectation* (LTE)[2],

$$\mathbb{E}\left[(Y-r(X))^2\right]=\mathbb{E}\left[\mathbb{E}\left[(Y-r(X))^2\,\middle|\,X\right]\right]. \tag{11.6}$$

[2] The LTE states that $\mathbb{E}[Y]=\mathbb{E}[\mathbb{E}[Y|X]]$ for any two random variables X and Y. It is briefly reviewed in Probability Refresher 11.13.

Here, the inner expected value is with respect to Y (given X) and the outer expected value is with respect to X. The inner expected value $\mathbb{E}\left[(Y-r(X))^2\,\middle|\,X\right]$ is a random variable whose value is $\mathbb{E}\left[(Y-r(X))^2\,\middle|\,X=x\right]$ when $X=x$. The conditional expectation

$$\begin{aligned}\mathbb{E}\left[(Y-r(X))^2\,\middle|\,X=x\right]&=\mathbb{E}\left[(Y-r(x))^2\,\middle|\,X=x\right]\\&=r(x)^2-2r(x)\mathbb{E}[Y\,|X=x]+\mathbb{E}[Y^2\,|X=x].\end{aligned} \tag{11.7}$$

Therefore,

$$\begin{aligned}\mathbb{E}\left[(Y-r(X))^2\,\middle|\,X\right]&=r(X)^2-2r(X)\mathbb{E}[Y\,|X]+\mathbb{E}[Y^2\,|X]\\&=(r(X)-\mathbb{E}[Y\,|X])^2+\underbrace{\mathbb{E}[Y^2\,|X]-\mathbb{E}[Y\,|X]^2}_{\mathbb{V}[Y|X]},\end{aligned} \tag{11.8}$$

where $\mathbb{V}[Y|X]$ is the conditional variance of Y given X (to learn more about this random variable, see Probability Refresher 11.13). So, the mean squared error is then given by

$$\begin{aligned}\mathrm{MSE}[r]&=\mathbb{E}\left[(r(X)-\mathbb{E}[Y\,|X])^2+\mathbb{V}[Y\,|X]\right]\\&=\mathbb{E}\left[(r(X)-\mathbb{E}[Y\,|X])^2\right]+\mathbb{E}\left[\mathbb{V}[Y\,|X]\right],\end{aligned} \tag{11.9}$$

and it is minimized when $r(X)=\mathbb{E}[Y\,|X]$. Indeed, the second term in (11.9) does not depend on $r(X)$, so we don't control it, and the first term is always nonnegative and equals zero precisely when $r(X)=\mathbb{E}[Y\,|X]$.

To summarize, the theoretically optimal — in the mean squared sense — prediction of response Y to the observed input $X=x$ is

$$r(x)=\mathbb{E}[Y\,|X=x]. \tag{11.10}$$

This function is called the *regression function*. It is a cornerstone of a general regression model, which is discussed in the next section. Intuitively, $r(x)$ provides a very natural prediction rule: after observing the new input $X=x$, predict the value of the corresponding response Y by its expected value, conditional on the fact that $X=x$. Note that if the response Y does not depend on the input X, then $\mathbb{E}[Y\,|X=x]=\mathbb{E}[Y]$ and (11.10) reduces to (11.3).

11.2 General Regression Model

In applications, the regression function $r(x) = \mathbb{E}[Y|X = x]$ is unknown[3], and it is what we need to infer from the data for making predictions. But it is important to realize that even if $r(x)$ is known, the true response Y to input $X = x$ will not be *exactly* equal to our prediction $r(x)$,

$$\text{True Response } \rightarrow Y \neq r(x) \leftarrow \text{ Our Prediction} \tag{11.11}$$

There are two reasons for that. The first, *technical* reason, is that almost always there are *measurement errors*. For example, if we hang weight $X = x$ (input) on a spring of length l_0 and stiffness k, then, according to Hooke's law, the length of the elongated spring Y (response) is given by $Y = l_0 + xg/k$, where $g \approx 9.81\ m/s^2$ is the gravitational acceleration. If, however, we perform this simple experiment in a laboratory n times[4] with the same weight x, we will obtain slightly different values $Y_1, \ldots, Y_n$ due to the measurement errors. The errors may be very small, if our measuring devices are very accurate. Nevertheless, they will be nonzero.

The second, more *conceptual* reason, is that in many applications, the response Y may have a range of values for a given input $X = x$. In other words, while input $X = x$ provides some information about Y, it does not determine it exactly. For example, according to genetics (and common sense), the height of a mother $X = x$ (input) does affect the height of her daughter Y (response)[5], but, of course, does not determine it exactly. So, instead of a deterministic relationship between the input and response, it is convenient to think about Y as a sample from a probability distribution, namely the conditional distribution of Y given $X = x$.

Nevertheless, we hope that our prediction will not be too bad and that $r(x)$ will be close to Y at least approximately, $Y \approx r(x)$. To account for the discrepancy between Y and $r(x)$, let's define the *prediction error*, also called the *statistical error*[6],

$$E = Y - r(x). \tag{11.12}$$

Rewriting this definition in terms of the response Y, we obtain that

$$Y = \underbrace{\mathbb{E}[Y|X = x]}_{r(x)} + E. \tag{11.13}$$

This equation is referred to as a *general regression model* (GRM). Note that it is not really a "model" in the exact sense of the word. While a model is a simplified version of reality, something that holds only approximately, the GRM holds exactly and is always true. Indeed, (11.13) is simply the definition of the prediction error: Y is the quantity we want to predict, $r(x)$ is our prediction, and the difference between them we call an error and denote by E. The GRM becomes a real model as soon as we start making assumptions about the regression function $r(x)$ and prediction error E.

[3] All that we know is the observed data $(X_1, Y_1), \ldots, (X_n, Y_n)$ and the value of the new input $X = x$. But the conditional distribution or expectation of the response Y given that $X = x$ is unknown.

[4] Making sure that the spring deformations are relatively small and the assumptions of Hooke's law are satisfied.

[5] The heights of mothers and their daughters are positively correlated.

[6] Traditionally, the statistical error is denoted by e or ε. Both notations, unfortunately, suggest (at least at a subconscious level) that the error is a small quantity, which may not be the case at all. To avoid this unnecessary association, we will use an uppercase E to denote the prediction error.

Since the value of the prediction error is uncertain, it is natural to model it as a random variable. Although the probability distribution of E may, in general, depend on the input[7], its conditional expectation, given that $X = x$, is always zero:

$$\mathbb{E}[E\,|\,X=x] = \mathbb{E}[Y\,|\,X=x] - r(x) = r(x) - r(x) = 0. \tag{11.14}$$

The response Y is, therefore, modeled as a sum of a deterministic prediction term $r(x)$ and a random zero-mean prediction error.

In the classical parametric approach to regression analysis, the regression function is assumed to have a certain functional form[8]. In other words, we assume that $r(x)$ belongs to a certain family of functions parameterized by a finite number of parameters,

$$r(x) \in \mathscr{F} = \{r(x\,|\,\beta),\ \ \beta \in B\}, \tag{11.15}$$

and then aim at estimating the vector of parameters β from the data[9].

Remark 11.1. In nonparametric approach, where the general philosophy is to make as few and weak assumptions as possible, the regression function $r(x)$ is estimated directly from the data, without assuming a particular parametric form (11.15). Methods of *nonparametric regression* are outside of our scope, and the reader is referred to [2] (Chapter 5) for a general overview of the field and to [3] for a more methodological exposition.

[7] For some inputs, our prediction could be more accurate, and for others, it could be less accurate.

[8] For example, based on the nature of the system that generated the data, it may be reasonable to assume that $r(x\,|\,\beta) = \beta_1 + \sin(\beta_2 x) + \exp(\beta_3 x^2)$, where $\beta_1, \beta_2, \beta_3 \in \mathbb{R}$.

[9] In regression analysis, parameters are traditionally denoted by β, not θ.

How to choose a suitable parametric form $\mathscr{F}$ for the regression function $r(x)$ and how to formulate a reasonable assumption about the prediction error E? If both the input and response are one-dimensional, $X, Y \in \mathbb{R}$, then a good starting point for addressing this question is to examine the scatterplot of the data.

11.3 Scatterplot Analysis

A *scatterplot* is a plot of the response values $Y_1, \ldots, Y_n$ versus the corresponding input values $X_1, \ldots, X_n$. It is an invaluable graphical tool for studying the relationship between two one-dimensional variables. Whenever possible, any regression analysis should start with the analysis of the corresponding scatterplot. Let's consider a typical example of scatterplot analysis, which will help us to formulate suitable assumptions about the regression function and prediction error.

Among many other things, Karl Pearson[10] explored the laws of inheritance by analyzing the transmission of different traits from generation to generation. Between 1893 and 1898, together with his assistant Alice Lee, he collected and analyzed data from over 1100 English families [4]. In particular, Pearson and Lee studied the relationship between heights of mothers and their adult daughters. To mitigate the effects of growth and height loss due to aging, all mothers were at most 65, and all daughters were at least 18.

[10] See page 109.

Figure 11.1(a) shows the scatterplot of the original Pearson-Lee data, where the mother's height X_i is considered as an input (predictor) and her daughter's height Y_i

as the corresponding response. The scatterplot contains $n = 1375$ data points (X_i, Y_i), but many of them have exactly the same location, since the data was rounded and each height was given to the nearest tenth of an inch. This rounding manifests itself in the highly structured and "non-random" zoomed portion of the scatterplot.

Alice Lee (1858-1939), an English mathematician and statistician, one of the first women to get a PhD from the University of London. She taught mathematics at Bedford College and simultaneously worked in Pearson's biometric laboratory. Her work was crucial in disproving the belief that skull size was related to intelligence. As a curious fact, Pearson volunteered to include the measurements of his head in the dataset analyzed by Alice Lee [5].

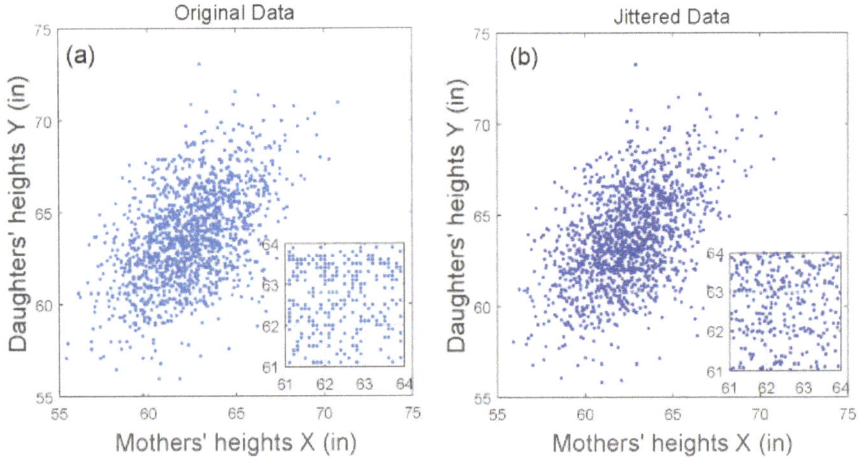

Fig. 11.1 Panel (a) displays the original Pearson-Lee data, with all heights rounded to the nearest tenth of an inch. Due to rounding, which is clearly visible in the zoomed portion of the plot, the scatterplot suffers from overplotting. Panel (b) presents the jittered data, where to each height a small random noise $\delta \sim \mathscr{U}(-0.05, 0.05)$ is added.

The effect of *overplotting* caused by rounding could be very misleading, since it is not clear how many real observations a single point on the scatterplot represents. The easiest way to resolve this issue is to use *jittering*: to add a small amount of noise to each X_i and Y_i,

$$X_i \to X_i + \delta_i, \quad Y_i \to Y_i + \delta_i', \quad \text{where } \delta_i, \delta_i' \overset{\text{iid}}{\sim} \mathscr{U}(-\varepsilon, \varepsilon), \tag{11.16}$$

so that each point representing a number of identical observation will become a tiny cloud of closely clustered points. Figure 11.1(b) shows the scatterplot of jittered data with $\varepsilon = 0.05$. The value $\varepsilon = 0.05$, which controls the strength of the jittering noise, is the largest value such that rounding the jittered values to the nearest tenth of an inch guarantees recovering the original nonjittered values. In what follows, we will work with the jittered data.

One important function of the scatterplot is to provide visual evidence on whether we can reasonably assume that the response variable Y indeed depends on the input variable X. This assumption is clearly reasonable for the Pearson-Lee data: when X increases, the scatter of Y-values shifts upwards[11]. This effect is illustrated in Fig. 11.2(a) with three vertical "slices" of the data.

[11] This is expected: higher mothers tend to have higher daughters.

It appears from examining the scatterplot that the mean value of Y increases as X increases. This suggests that the regression function $r(x) = \mathbb{E}[Y \mid X = x]$ is an in-

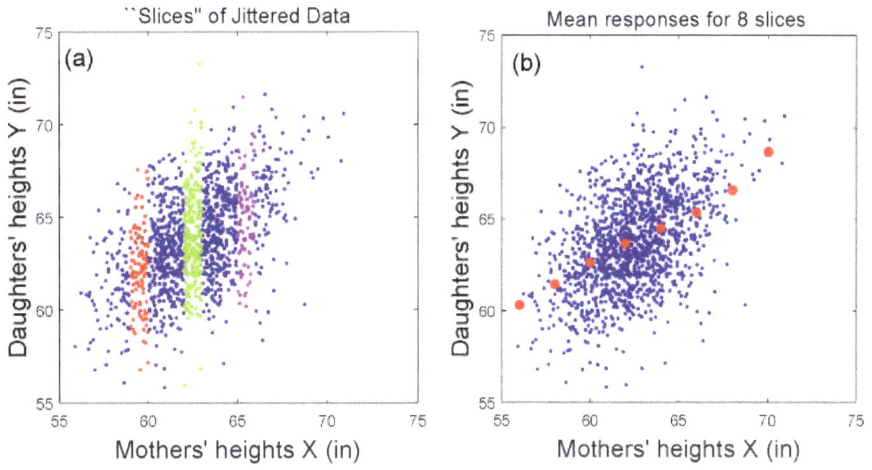

Fig. 11.2 Panel (a) highlights three vertical slices of the height data corresponding to $X \in (59,60)$, $X \in (62,63)$, and $X \in (65,66)$. Panel (b) shows eight estimated values of the regression function $r(x)$ for $x = 56, 58, \ldots, 70$ via nonparametric smoother (11.17) with $a = 1$.

creasing function of x. Let's investigate this in more detail by trying to estimate $r(x)$ nonparametrically directly from the data. For most x, the number of observations (X_i, Y_i) where $X_i = x$ is either zero (no such data points) or very small (just a few points). To alleviate this problem, let's relax the constraint $X = x$ by $X \in (x-a, x+a)$ for some small $a > 0$ end estimate $r(x)$ by the corresponding empirical average[12]:

$$r(x) = \mathbb{E}[Y \mid X = x] \approx \mathbb{E}[Y \mid X \in (x-a, x+a)] \approx \frac{1}{|S_a(x)|} \sum_{i: X_i \in (x-a, x+a)} Y_i, \quad (11.17)$$

[12] With the original constraint $X_i = x$, the sum in (11.17) would have zero or very few terms.

where $|S_a(x)|$ is the number of points in the vertical slice of the data centered at x and with width $2a$, that is, $S_a(x) = \{(X_i, Y_i) \mid X_i \in (x-a, x+a)\}$. Simply speaking, we estimate $r(x)$ by the average value of Y in the vertical slice $S_a(x)$[13].

[13] This method of estimating the regression function is called *nonparametric smoother*. It and its modifications lie at the core of many nonparametric regression methods.

Figure 11.2(b) shows the estimated values of regression function $r(x)$ for eight values $x = 56, 58, \ldots, 70$ with $a = 1$. The eight points almost perfectly lie on a straight line. This suggests that $r(x)$ can be accurately modeled by a linear function,

$$r(x) \in \mathscr{F} = \{\beta_0 + \beta_1 x, \ \ \beta_0, \beta_1 \in \mathbb{R}\}. \quad (11.18)$$

This model has two parameters, the intercept β_0 and the slope β_1, which, as we will see later, can be estimated from the data.

Let's return to Fig. 11.2(a) and look again at the vertical slices of the data. They appear to have approximately the same vertical spread[14]. This suggests that, while the expected value $\mathbb{E}[Y \mid X = x]$ increases with x, the variance $\mathbb{V}[Y \mid X = x]$ seems to be approximately constant. It means that it is reasonable to assume that

$$\mathbb{V}[Y \mid X = x] = \sigma^2, \quad (11.19)$$

[14] This seems not to be true for the leftmost and rightmost slices where $X \approx 56$ and $X \approx 70$, but those regions are likely to be undersampled: there are very few very short mothers and very few very tall mothers in the data.

where σ^2 is some positive constant. Using the general regression model (11.13), this assumption can be equivalently rewritten in therms of the prediction error:

$$\mathbb{V}[E\,|\,X=x]=\sigma^2. \tag{11.20}$$

The GRM (11.13), together with the linear model (11.18) for the regression function $r(x)$ and the properties (11.14) and (11.20) of the prediction error E, is called the simple linear regression model — one of the most popular and widely used stochastic models, which serves as the foundation for more complicated models.

11.4 Simple Linear Regression Model

Given data $(X_1,Y_1),\ldots,(X_n,Y_n)$, the *simple linear regression model* of Y_i on X_i is

$$Y_i=\beta_0+\beta_1X_i+E_i,\quad i=1,\ldots,n, \tag{11.21}$$

where β_0 and β_1 are model parameters, called the *regression parameters* or *regression coefficients*; X_i is the ith observation of input X, more traditionally called *predictor* or *feature*; Y_i is the corresponding i^{th} observation of response Y, also called *target* or *dependent variable*; and E_i is the *statistical* or *prediction error* in the ith case. The errors $E_1,\ldots,E_n$ are modeled as independent random variables[15], and it is assumed that, given X_i, error E_i has zero mean and constant variance, which is independent of X_i:

$$\mathbb{E}[E_i\,|\,X_i]=0 \quad\text{and}\quad \mathbb{V}[E_i\,|\,X_i]=\sigma^2. \tag{11.22}$$

[15] We assume that knowing the errors in some cases does not provide any information about the errors in other cases.

The last property of independence of the variance of the error on the value of input is called *homoscedasticity*, and the errors $E_1,\ldots,E_n$ satisfying this property are called *homoscedastic*.

It is important to emphasize that regression analysis is always done conditional on $X_1,\ldots,X_n$. We do not model these quantities. In general, they can be random or deterministic, but in both cases their values are assumed to be "observable" and known. We observe the value of X_i, and then model the corresponding response Y_i based on X_i with the idea to use the model constructed this way for predicting the response Y to a new input X after observing its value. In simple linear regression, $Y_1,\ldots,Y_n$ are modeled via (11.21) by modeling errors as independent random variables with properties (11.22). The conditional notation used in these properties explicitly indicates that the modeling is done conditional on the values of $X_1,\ldots,X_n$.

The laws of total expectation and total variance[16] imply that if (11.22) holds, then the unconditional mean and variance of E_i are also 0 and σ^2, respectively. Indeed:

[16] See Probability Refresher 11.13.

$$\begin{aligned}
\mathbb{E}[E_i]&=\mathbb{E}\big[\underbrace{\mathbb{E}[E_i\,|\,X_i]}_{0}\big]=0,\\
\mathbb{V}[E_i]&=\mathbb{E}\big[\underbrace{\mathbb{V}[E_i\,|\,X_i]}_{\sigma^2}\big]+\mathbb{V}\big[\underbrace{\mathbb{E}[E_i\,|\,X_i]}_{0}\big]=\sigma^2.
\end{aligned} \tag{11.23}$$

So, if $\mathbb{E}[E_i \mid X_i] = 0$ and $\mathbb{V}[E_i \mid X_i] = \sigma^2$, then $\mathbb{E}[E_i] = 0$ and $\mathbb{V}[E_i] = \sigma^2$. The reverse does not hold in general, unless X_i is deterministic, in which case the conditional distribution of E_i given X_i is the same as the marginal (unconditional) distribution of E_i and, therefore, $\mathbb{E}[E_i \mid X_i] = \mathbb{E}[E_i]$ and $\mathbb{V}[E_i \mid X_i] = \mathbb{V}[E_i]$.

Since in regression analysis the values of $X_1, \ldots, X_n$ are assumed to be known, it is mathematically convenient, regardless of their original nature, to treat them as fully *deterministic* fixed quantities (rather than as observable random variables). With this convention, the simple linear regression (SLR) model has the following form:

$$Y_i = \beta_0 + \beta_1 X_i + E_i, \quad i = 1, \ldots, n, \tag{11.24}$$

where X_i is a deterministic input, Y_i is the corresponding random response, and E_i is a random prediction error. The model assumptions are:

$$\begin{aligned} &1.\ E_1, \ldots, E_n \text{ are independent,} \\ &2.\ \mathbb{E}[E_i] = 0, \\ &3.\ \mathbb{V}[E_i] = \sigma^2. \end{aligned} \tag{11.25}$$

Remark 11.2. A calculus-based way to think about the SLR model (11.24), (11.25) is the following. The dependence of the response Y_i on the input X_i is governed by an unknown and possibly complicated function. Let's Taylor expand this function, keep just the constant and linear terms $\beta_0 + \beta_1 X_i$, and absorb all other non-linear in X_i terms as well as all other factors influencing the response (which are not captured by the input X_i) into the random error term E_i. The model is then obtained by assuming that errors $E_1, \ldots, E_n$ are independent and have zero mean and variance σ^2.

Why is the SLR model called "simple"? Because input X is a scalar quantity. In general, if X is a vector, then the model is called *multiple* linear regression. If response Y is a vector, then the model is referred to as *multivariate* linear regression[17].

Why is the SLR model called "linear"? Because the regression function $r(x) = \beta_0 + \beta_1 x$ is assumed to be linear in parameters, not because it is linear in input x. For example, while $r(x) = \beta_0 + \beta_1 \exp(x)$ is also a linear model[18], $r(x) = \beta_0 + \exp(\beta_1 x)$ is not. As we will see in the next section, the *linearity in parameters* allows us to easily estimate the model parameters β_0, β_1, and σ^2.

Why is the SLR model called "regression"? This term comes from genetics, and it was originally coined by Francis Galton [6] to describe situations in which extreme values of a trait measured in offspring are less extreme compared to the extreme values of the trait measured in their parents. In such cases, the trait values in the offspring *regress* away from the parent mean value toward the offspring mean value[19].

As an example, let us look at Fig. 11.3, where the eight estimated values of $r(x) = \mathbb{E}[Y \mid X = x]$ are plotted together with the line $y = x$. If all daughters had approximately the same heights as their mothers, then all data points would lie approximately on the line $y = x$, but they are not. Notice that the slope of the line $y = r(x)$, which is called the regression line[20], is positive. This is, of course, expected, since tall mothers tend to have tall daughters, and short mothers tend to have

[17] $\dim X = 1 \Rightarrow$ simple
$\dim X > 1 \Rightarrow$ multiple
$\dim Y > 1 \Rightarrow$ multivariate

[18] We can simply define $\tilde{x} = \exp(x)$ and consider $\tilde{x}$ as an input instead of x.

[19] To the "mediocre" point in Galton's terminology.

[20] It is represented by the eight big red points.

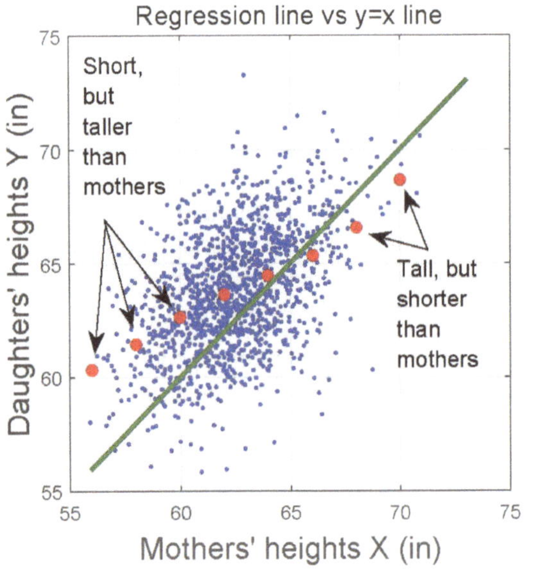

Fig. 11.3 Regression toward the mean: tall mothers have tall daughters, but shorter than themselves; short mothers have short daughters, but taller than themselves.

short daughters. The visual comparison of two lines, $y = x$ and $y = r(x)$, shows, however, that the slope of the regression line is less than one. This means that although tall mothers have tall daughters, their daughters are not as tall as their mothers. Likewise, although short mothers have short daughters, the daughters are higher than the mothers. This phenomenon is observed for various traits measured in the parent and offspring populations, and, since Galton's times, it is referred to as the *regression toward the mean*.

Why is the SLR model called a "model"? This question is left for contemplation[21].

[21] "If there were a little more silence, if we all kept quiet... maybe we could understand something." – Federico Fellini.

11.5 Ordinary Least Squares

The simple linear regression model (11.24), (11.25) has three parameters: the regression coefficients β_0 and β_1, and the variance σ^2. There are several methods for estimating β_0 and β_1, since there are several reasonable ways to fit a line to a cloud of points.The most common and popular method is *ordinary least squares* (OLS)[22].

Let $\hat{\beta}_0$ and $\hat{\beta}_1$ be some estimates of the regression coefficients. Then the corresponding estimate of the regression function

$$\hat{r}(x) = \hat{\beta}_0 + \hat{\beta}_1 x \tag{11.26}$$

[22] The OLS method was introduced by F. Galton in 1886, although earlier in the XIX century, related ideas were used by A. Legendre and C. Gauss for studying planetary and cometary orbits. For a comprehensive account of historical developments of linear regression (and statistics in general) see a wonderful book by S. Stigler [7].

is called the ***fitted line***, and the value of $\hat{r}(x)$ at $x = X_i$ is called the ***fitted value***,

$$\hat{Y}_i = \hat{r}(X_i) = \hat{\beta}_0 + \hat{\beta}_1 X_i. \tag{11.27}$$

The fitted value is simply our prediction of the response to input X_i based on the estimated regression function $\hat{r}(x)$. The difference between the true response value Y_i and the fitted value $\hat{Y}_i$ is called the *residual*,

$$\hat{E}_i = Y_i - \hat{Y}_i. \tag{11.28}$$

The residual $\hat{E}_i$ is the discrepancy between the true response Y_i to input X_i and the predicted response $\hat{Y}_i = \hat{r}(X_i)$. It can be viewed as an approximate realization of statistical error E_i. Geometrically, the residuals $\hat{E}_1, \dots, \hat{E}_n$ are simply the signed vertical distances between the responses $Y_1, \dots, Y_n$ and the fitted line, see Fig. 11.4.

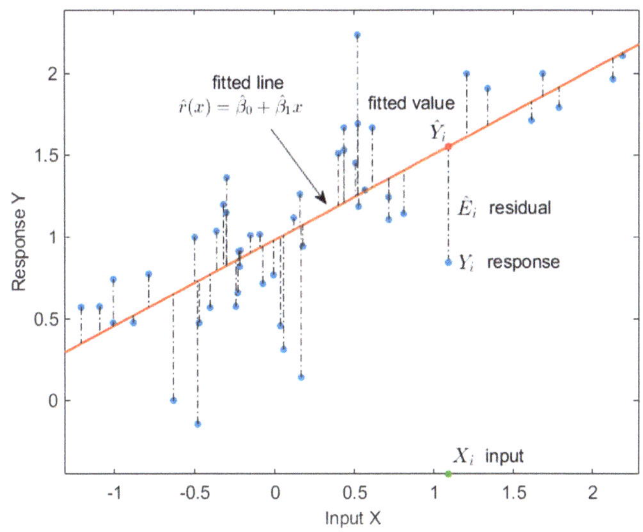

Fig. 11.4 The scatterplot of the data $(X_1, Y_1), \dots, (X_n, Y_n)$ and a fitted line $\hat{r}(x) = \hat{\beta}_0 + \hat{\beta}_1 x$. The residuals $\hat{E}_1, \dots, \hat{E}_n$ are represented by dash-dotted line segments. The residuals corresponding to the segments above/below the fitted line are positive/negative.

The OLS method finds the estimates $\hat{\beta}_0$ and $\hat{\beta}_1$ by minimizing the quantity called the *residual sum of squares* (RSS):

$$\mathrm{RSS}(\hat{\beta}_0, \hat{\beta}_1) = \sum_{i=1}^{n} \hat{E}_i^2 \to \min. \tag{11.29}$$

This minimization criterion is very natural: the squared residual $\hat{E}_i^2$ is a measure of goodness of fit of the regression line to the data point (X_i, Y_i), and $\mathrm{RSS}(\hat{\beta}_0, \hat{\beta}_1)$ is an overall measure of goodness of fit of the line to the cloud of points $(X_1, Y_1), \dots, (X_n, Y_n)$.

Remark 11.3. The absolute value of the residual $|\hat{E}_i|$ may appear as a more natural measure of fit than the squared residual $\hat{E}_i^2$. But the corresponding overall measure — the sum of absolute values of residuals — has a number of drawbacks compared to RSS: unlike $\sum_{i=1}^n \hat{E}_i^2 \to \min$, the optimization problem $\sum_{i=1}^n |\hat{E}_i| \to \min$ can't be solved analytically[23] and must be solved by iterative numerical methods. Moreover, the solution may not be unique resulting into an infinitely large family of fitted lines that all fit the data equally well.

[23] We usually solve optimization problems by differentiating functions and finding their critical points. But the absolute value is not differentiable.

Let's find the OLS estimates of the regression coefficients by solving (11.29). The residual sum of squares

$$\begin{aligned}\mathrm{RSS}(\hat{\beta}_0,\hat{\beta}_1) &= \sum_{i=1}^n (Y_i-\hat{Y}_i)^2 = \sum_{i=1}^n \left(Y_i-(\hat{\beta}_0+\hat{\beta}_1 X_i)\right)^2 \\ &= \sum_{i=1}^n \Big(\underbrace{Y_i-\hat{\beta}_1 X_i}_{\xi_i}-\hat{\beta}_0\Big)^2 = \sum_{i=1}^n (\xi_i-\hat{\beta}_0)^2.\end{aligned} \tag{11.30}$$

Recall that long time ago[24], in Problem 1.2, we established that the sample mean $\bar{\xi}$ of $\xi_1,\dots,\xi_n$ is the global minimizer of $\sum_{i=1}^n (\xi_i-\alpha)^2$ with respect to α, that is

[24] "When the world was young, and the woods were wide and wild..." – J.R.R. Tolkien, *The Lord of the Rings*.

$$\bar{\xi} = \arg\min_{\alpha\in\mathbb{R}} \sum_{i=1}^n (\xi_i-\alpha)^2. \tag{11.31}$$

This means that, whatever the value of $\hat{\beta}_1$ is, the optimal value of $\hat{\beta}_0$ (that minimizes the RSS) is the sample mean of quantities $\xi_i = Y_i - \hat{\beta}_1 X_i$:

$$\hat{\beta}_0 = \overline{Y_i - \hat{\beta}_1 X_i} = \overline{Y} - \hat{\beta}_1 \overline{X}. \tag{11.32}$$

Substituting this expression into the RSS, we have

$$\begin{aligned}\mathrm{RSS}(\overline{Y}-\hat{\beta}_1\overline{X},\hat{\beta}_1) &= \sum_{i=1}^n \left(Y_i-\hat{\beta}_1 X_i-\overline{Y}+\hat{\beta}_1\overline{X}\right)^2 \\ &= \sum_{i=1}^n \left(Y_i-\overline{Y}-\hat{\beta}_1(X_i-\overline{X})\right)^2 \\ &= \underbrace{\sum_{i=1}^n (Y_i-\overline{Y})^2}_{S_{YY}} - 2\hat{\beta}_1 \underbrace{\sum_{i=1}^n (X_i-\overline{X})(Y_i-\overline{Y})}_{S_{XY}} + \hat{\beta}_1^2 \underbrace{\sum_{i=1}^n (X_i-\overline{X})^2}_{S_{XX}} \\ &= S_{XX}\hat{\beta}_1^2 - 2\hat{\beta}_1 S_{XY} + S_{YY} = S_{XX}\left(\hat{\beta}_1-\frac{S_{XY}}{S_{XX}}\right)^2 - \frac{S_{XY}^2}{S_{XX}} + S_{YY}.\end{aligned} \tag{11.33}$$

Therefore, the *OLS estimates* of the regression coefficients that minimize RSS are:

$$\hat{\beta}_1 = \frac{S_{XY}}{S_{XX}} = \frac{\sum_{i=1}^n (X_i - \overline{X})(Y_i - \overline{Y})}{\sum_{i=1}^n (X_i - \overline{X})^2},$$
$$\hat{\beta}_0 = \overline{Y} - \hat{\beta}_1 \overline{X} = \frac{\sum_{i=1}^n (X_i - \overline{X})(\overline{Y} X_i - \overline{X} Y_i)}{\sum_{i=1}^n (X_i - \overline{X})^2}. \tag{11.34}$$

The quantities S_{XX} and S_{XY} are called the *sum of squares* and the *sum of cross-products*, respectively.

Remark 11.4. Notice that $S_{XX}/(n-1)$ is the sample variance of $X_1,\ldots,X_n$, and $S_{XY}/(n-1)$ is the sample covariance between two samples $X_1,\ldots,X_n$ and $Y_1,\ldots,Y_n$. The OLS estimate of the slope $\hat{\beta}_1$ of the regression line can, therefore, be viewed as the ratio of the sample covariance between the input X and response Y over the sample variance of the input X.

The OLS estimates $\hat{\beta}_0$ and $\hat{\beta}_1$ constructed from the data $(X_1,Y_1),\ldots,(X_n,Y_n)$ depend on the data only through fours statistics[25]: $\overline{X}$, $\overline{Y}$, S_{XX}, and S_{XY}. This means that any two data set $(X_1,Y_1),\ldots,(X_n,Y_n)$ and $(X_1',Y_1'),\ldots,(X_n',Y_n')$ for which these statistics are the same, $\overline{X} = \overline{X'}$, $\overline{Y} = \overline{Y'}$, $S_{XX} = S_{X'X'}$, and $S_{XY} = S_{X'Y'}$, will have identical fitted lines,

$$\hat{r}(x) = \overline{Y} - \frac{S_{XY}}{S_{XX}}\overline{X} + \frac{S_{XY}}{S_{XX}}x = \overline{Y} + (x - \overline{X})\frac{S_{XY}}{S_{XX}}. \tag{11.35}$$

The data sets can be substantially different, and yet, as long as the four statistics coincide, the fitted lines will coincide as well. This effect was beautifully demonstrated by Frank Anscombe [8]. He constructed four artificial data sets for which the OLS method fits the same regression line although their scatterplots are visually very different. Anscombe's "quartet" is shown in Fig. 11.5.

Among the four data sets, the SLR model seems to be suitable only for the first one, depicted in Fig. 11.5(a). For the data in Fig. 11.5(b), it seems more natural to fit a quadratic polynomial rather than a straight line. The scatterplot in Fig. 11.5(c) suggests that the SLR model would fit the data well if we remove the *outlier*, i.e. the data point which lies far from the rest of the data (and far away from the fitted line).

Finally, the data in Fig. 11.5(d) differs from the other data sets in that it contains only two distinct values of the input variable, $X = 8$ and $X = 19$. This means it provides information about the regression function $r(x) = \mathbb{E}[Y \mid X = x]$ only at two points. Moreover, there is only one response corresponding to the larger input $X = 19$. Without that point, we would not be able even to fit a line using OLS, since S_{XX} would be zero and estimates (11.34) would not be defined. Even if the SLR model is adequate for modeling the process that generated the data in Fig. 11.5(d), we can't make this judgment based on the given data; more data is needed. And we can't trust the fitted lined whose slope so heavily depends on a single observation, which, in fact, could be an outlier.

The Anscombe quartet demonstrates the importance of graphing data in regression analysis. Don't fit the regression line blindly, check the scatterplot first: do the (lonely) data points lie approximately on a line?[26]

25 Recall that in Statistics, a statistic is any function of the data.

26 "All the lonely people where do they all come from? All the lonely people where do they all belong?" – The Beatles, *Eleanor Rigby* [9].

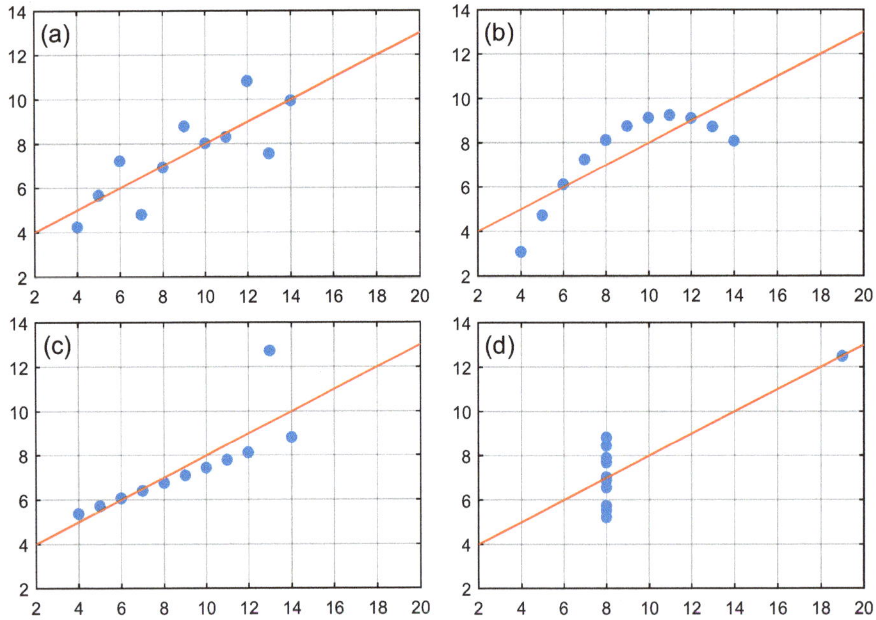

Fig. 11.5 Anscombe's "quartet": four very different data sets with the same fitted regression line.

11.6 OLS and MLE

The OLS method is not a method of statistical inference per se. It does not rely on any of the statistical assumptions (11.25); all it does is fit a line to a cloud of points $(X_1,Y_1),\ldots,(X_n,Y_n)$ using the optimization criterion (11.29). Interestingly, the OLS estimates (11.34), obtained this way, admit an important statistical interpretation.

Suppose, instead of (11.25), that prediction errors $E_1,\ldots,E_n$ are independent and normally distributed with zero mean and variance σ^2,

$$E_1,\ldots,E_n \overset{\text{iid}}{\sim} \mathcal{N}(0,\sigma^2). \tag{11.36}$$

Note that this assumption is stronger than assumptions (11.25): whereas (11.25) specify the mean and the variance of prediction errors, the normality assumption (11.36) specifies the full probability distribution of the errors. We will refer to the model $Y_i = \beta_0 + \beta_1 X_i + E_i$ with assumption (11.36) as the *normal simple linear regression model* (N-SLR).

If errors $E_1,\ldots,E_n$ are normally distributed, then so are the responses $Y_1,\ldots,Y_n$. Indeed, each response Y_i can be viewed as E_i shifted by a deterministic[27] quantity $\beta_0 + \beta_1 X_i$. Therefore, responses $Y_1,\ldots,Y_n$ are independent and

$$Y_i \sim \mathcal{N}(\beta_0 + \beta_1 X_i, \sigma^2), \quad i = 1,\ldots,n. \tag{11.37}$$

[27] Recall that $X_1,\ldots,X_n$ are deterministic quantities with known values.

The N-SLR model is a *parametric* model for $Y_1,\dots,Y_n$, since the set of all possible PDFs of $Y_1,\dots,Y_n$ under this model is fully parameterized by three parameters: β_0, β_1, and σ^2.[28] Let's find the MLEs of model parameters. The likelihood function, which is the joint PDF of $Y_1,\dots,Y_n$, viewed as a function of parameters, is

$$\begin{aligned}\mathscr{L}(\beta_0,\beta_1,\sigma^2) &= \prod_{i=1}^{n}\phi(Y_i|\beta_0+\beta_1X_i,\sigma^2)\\ &= \prod_{i=1}^{n}\frac{1}{\sqrt{2\pi\sigma^2}}\exp\left(-\frac{(Y_i-\beta_0-\beta_1X_i)^2}{2\sigma^2}\right)\\ &= (2\pi\sigma^2)^{-\frac{n}{2}}\exp\Bigg(-\frac{1}{2\sigma^2}\underbrace{\sum_{i=1}^{n}(Y_i-\beta_0-\beta_1X_i)^2}_{\mathrm{RSS}(\beta_0,\beta_1)}\Bigg)\\ &= (2\pi\sigma^2)^{-\frac{n}{2}}\exp\left(-\frac{1}{2\sigma^2}\mathrm{RSS}(\beta_0,\beta_1)\right).\end{aligned} \tag{11.38}$$

28 The original SLR model is nonparametric because β_0, β_1, and σ^2 do not fully define probability distributions of responses (only their means and variances).

Maximization of the likelihood function decouples: whatever the optimal value of σ^2 is, the optimal values of β_0 and β_1 are exactly those values that minimize the residual sum of squares $\mathrm{RSS}(\beta_0,\beta_1)$. This means that, under the normality assumption (11.36), the OLS estimates of regression coefficients are simply the MLEs,

$$\hat{\beta}_0^{OLS}=\hat{\beta}_0^{MLE} \quad\text{and}\quad \hat{\beta}_1^{OLS}=\hat{\beta}_1^{MLE}. \tag{11.39}$$

As we discussed in Section 8.6, MLEs have nice asymptotic properties: they are consistent, asymptotically normal, asymptotically unbiased, asymptotically efficient, and equivariant. The OLS estimates inherit all these properties if the normality assumption (11.36) holds[29].

To find the MLE of σ^2, we need to plug-in the OLS estimates $\hat{\beta}_0$ and $\hat{\beta}_1$[30] into (11.38), and then maximize the resulting likelihood, or, equivalently, the log-likelihood, with respect to σ^2,

$$\begin{aligned}\ell(\hat{\beta}_0,\hat{\beta}_1,\sigma^2) &= \log\mathscr{L}(\hat{\beta}_0,\hat{\beta}_1,\sigma^2)\\ &= -\frac{n}{2}\log 2\pi - n\log\sigma - \frac{1}{2\sigma^2}\mathrm{RSS}(\hat{\beta}_0,\hat{\beta}_1).\end{aligned} \tag{11.40}$$

29 And if (11.36) holds only approximately, then OLS estimates have these properties approximately, which is also not bad.

30 To keep the notation sweet and simple, we will be using $\hat{\beta}_0$ and $\hat{\beta}_1$ instead of $\hat{\beta}_0^{OLS}$ and $\hat{\beta}_1^{OLS}$ to denote the OLS estimates.

Differentiating with respect to σ, setting the derivative to zero, and solving for σ, we find the critical point $\hat{\sigma}$ of the log-likelihood:

$$\frac{d\ell}{d\sigma} = -\frac{n}{\sigma}+\frac{1}{\sigma^3}\mathrm{RSS}(\hat{\beta}_0,\hat{\beta}_1)=0 \quad\Longrightarrow\quad \hat{\sigma}^2=\frac{1}{n}\mathrm{RSS}(\hat{\beta}_0,\hat{\beta}_1). \tag{11.41}$$

Since the second derivative of l at $\hat{\sigma}$ is negative,

$$\left.\frac{d^2\ell}{d\sigma^2}\right|_{\sigma=\hat{\sigma}} = \left.\left(\frac{n}{\sigma^2}-\frac{3}{\sigma^4}\mathrm{RSS}(\hat{\beta}_0,\hat{\beta}_1)\right)\right|_{\sigma=\hat{\sigma}} = -\frac{2n^2}{\mathrm{RSS}(\hat{\beta}_0,\hat{\beta}_1)}<0, \tag{11.42}$$

the critical point is the global maximizer of ℓ and, therefore, the MLE,

$$\hat{\sigma}^2_{MLE} = \frac{1}{n}\mathrm{RSS}(\hat{\beta}_0, \hat{\beta}_1) = \frac{1}{n}\sum_{i=1}^{n}\hat{E}_i^2. \tag{11.43}$$

Remark 11.5. The MLE $\hat{\sigma}^2_{MLE}$ of the variance of prediction errors is biased[31]. An unbiased (even under the weaker nonparametric assumptions (11.25)) estimate is given by[32]

$$\hat{\sigma}^2_u = \frac{1}{n-2}\sum_{i=1}^{n}\hat{E}_i^2. \tag{11.44}$$

[31] But it is asymptotically unbiased.

[32] This result is proved for a more general case of multiple linear regression in Appendix 12.14.

Remark 11.6. Some — and for a good reason — may expect that instead of $1/n$ in MLE, the unbiased estimate should have the factor of $1/(n-1)$. This is, at least, exactly the difference between the biased MLE of variance in the normal model[33], $\hat{\sigma}^2_{MLE} = \frac{1}{n}\sum_{i=1}^{n}(X_i - \overline{X})^2$, and the unbiased sample variance $\hat{s}^2_n = \frac{1}{n-1}\sum_{i=1}^{n}(X_i - \overline{X})^2$. The unbiasedness of $\hat{s}^2_n$ follows from the independence and identical distribution of $X_1, \ldots, X_n$. The responses $Y_1, \ldots, Y_n$ in $\hat{\sigma}^2_u = \frac{1}{n-2}\sum_{i=1}^{n}(Y_i - \hat{Y}_i)^2$, however, are independent but not identically distributed. Moreover, from each Y_i we subtract not their mean $\overline{Y}$, but the corresponding fitted value $\hat{Y}_i$. That is why we should not expect the "expected" factor of $1/(n-1)$ in the expression for the unbiased estimate (11.44).

[33] See Example 8.2

11.7 Properties of OLS Estimates

In this section, we will discuss several important properties of the OLS estimates, $\hat{\beta}_0$ and $\hat{\beta}_1$, which justify the popularity of ordinary least squares.

11.7.1 OLS and the Data Centroid

The point $(\overline{X}, \overline{Y})$ is called the *centroid* of a set of points $(X_1, Y_1), \ldots, (X_n, Y_n)$ in $\mathbb{R}^2$. It can be shown that the centroid minimizes the sum of squared distances between a point (ξ, η) in the plane and each point (X_i, Y_i) in the set[34],

$$(\overline{X}, \overline{Y}) = \arg\min_{(\xi,\eta)\in\mathbb{R}^2}\sum_{i=1}^{n}\left((X_i - \xi)^2 + (Y_i - \eta)^2\right). \tag{11.45}$$

[34] Compare this result with Problem 1.2.

It immediately follows from (11.35) that the OLS fitted line always passes through the centroid of the data,

$$\hat{r}(\overline{X}) = \overline{Y}. \tag{11.46}$$

Conceptually, it means that the prediction $\hat{Y}$ of the response Y to the mean input $\overline{X}$ is the mean output, $\hat{Y} = \hat{r}(\overline{X}) = \overline{Y}$.

11.7.2 OLS and Residuals

The sum of all residuals is always zero. Indeed,

$$\sum_{i=1}^{n} \hat{E}_i = \sum_{i=1}^{n}(Y_i - \hat{Y}_i) = \sum_{i=1}^{n}(Y_i - \hat{\beta}_0 - \hat{\beta}_1 X_i)$$
$$= \sum_{i=1}^{n}(Y_i - (\overline{Y} - \hat{\beta}_1\overline{X}) - \hat{\beta}_1 X_i) = \underbrace{\sum_{i=1}^{n}(Y_i - \overline{Y})}_{0} + \hat{\beta}_1 \underbrace{\sum_{i=1}^{n}(\overline{X} - X_i)}_{0} = 0. \tag{11.47}$$

Conceptually, $\frac{1}{n}\sum_{i=1}^{n}(Y_i - \hat{Y}_i) = 0$ means that, on average, the fitted value $\hat{Y}_i$, which is the (OLS fitted) model prediction of the response to input X_i, neither overestimates nor underestimates the true response Y_i. In other words, the model prediction is unbiased on the data that was used for fitting the model[35].

[35] In machine learning and data science, the process of "fitting" a model is called "training", and the data $(X_1, Y_1), \ldots, (X_n, Y_n)$ used for training is called the *training data.*

11.7.3 OLS Estimates are Linear

There are different ways to estimate regression coefficients, as there are different ways to fit a line to a cloud of points[36]. But any estimate $\hat{\beta}$ of a regression coefficient is always some function of the data, $\hat{\beta} = \hat{\beta}(X_1, \ldots, X_n; Y_1, \ldots, Y_n)$. The dependence of $\hat{\beta}$ on the data may take different functional forms. An important class of estimates is a class of linear estimates, where the dependence on responses $Y_1, \ldots, Y_n$ (the quantities that we are modeling) is very simple — linear. Namely, an estimate $\hat{\beta}$ of a regression coefficient β is called *linear* if it is a linear function of responses,

[36] For instance, instead of minimizing the RSS, we can minimize the sum of the squared Euclidean distances from data points to the line.

$$\hat{\beta} = \sum_{i=1}^{n} \alpha_i Y_i, \tag{11.48}$$

where $\alpha_1, \ldots, \alpha_n$ may depend on the inputs $X_1, \ldots, X_n$ (but not on responses). Linear estimates are important because their simple dependence on responses allows to study their statistical properties analytically.

The OLS estimates (11.34) are linear. Indeed, the estimate of the slope is

$$\hat{\beta}_1 = \frac{S_{XY}}{S_{XX}} = \frac{\sum_{i=1}^{n}(X_i - \overline{X})(Y_i - \overline{Y})}{S_{XX}}$$
$$= \sum_{i=1}^{n} \frac{X_i - \overline{X}}{S_{XX}} Y_i - \frac{\overline{Y}}{S_{XX}} \underbrace{\sum_{i=1}^{n}(X_i - \overline{X})}_{0} = \sum_{i=1}^{n} \underbrace{\frac{X_i - \overline{X}}{S_{XX}}}_{\alpha_i} Y_i. \tag{11.49}$$

Similarly, the estimate of the intercept is

$$\hat{\beta}_0 = \overline{Y} - \hat{\beta}_1 \overline{X} = \frac{1}{n}\sum_{i=1}^{n} Y_i - \overline{X}\sum_{i=1}^{n}\frac{X_i - \overline{X}}{S_{XX}} Y_i = \sum_{i=1}^{n} \underbrace{\left[\frac{1}{n} - \frac{\overline{X}(X_i - \overline{X})}{S_{XX}}\right]}_{\alpha_i} Y_i. \tag{11.50}$$

Linearity of the OLS estimates is useful in establishing the next property.

11.7.4 OLS Estimates are Unbiased

It is straightforward to check that both OLS estimates are unbiased estimates of the regression coefficients by computing their expected values[37]:

[37] Recall that, according to the SLR model, $Y_i = \beta_0 + \beta_1 X_i + E_i$, where X_i is deterministic and $\mathbb{E}[E_i] = 0$.

$$\begin{aligned}
\mathbb{E}[\hat{\beta}_1] &= \mathbb{E}\left[\sum_{i=1}^{n}\frac{X_i - \overline{X}}{S_{XX}} Y_i\right] = \sum_{i=1}^{n}\frac{X_i - \overline{X}}{S_{XX}}\mathbb{E}[Y_i] \\
&= \sum_{i=1}^{n}\frac{X_i - \overline{X}}{S_{XX}}\mathbb{E}[\beta_0 + \beta_1 X_i + E_i] = \sum_{i=1}^{n}\frac{X_i - \overline{X}}{S_{XX}}(\beta_0 + \beta_1 X_i) \\
&= \frac{\beta_0}{S_{XX}}\underbrace{\sum_{i=1}^{n}(X_i - \overline{X})}_{0} + \frac{\beta_1}{S_{XX}}\sum_{i=1}^{n}(X_i - \overline{X})X_i \\
&= \frac{\beta_1}{S_{XX}}\sum_{i=1}^{n}(X_i - \overline{X})(X_i - \overline{X} + \overline{X}) \\
&= \frac{\beta_1}{S_{XX}}\left(\underbrace{\sum_{i=1}^{n}(X_i - \overline{X})^2}_{S_{XX}} + \overline{X}\underbrace{\sum_{i=1}^{n}(X_i - \overline{X})}_{0}\right) = \beta_1.
\end{aligned} \tag{11.51}$$

Using the unbiasedness of $\hat{\beta}_1$, we have:

$$\begin{aligned}
\mathbb{E}[\hat{\beta}_0] &= \mathbb{E}[\overline{Y} - \hat{\beta}_1 \overline{X}] = \frac{1}{n}\sum_{i=1}^{n}\mathbb{E}[Y_i] - \mathbb{E}[\hat{\beta}_1]\overline{X} = \\
&= \frac{1}{n}\sum_{i=1}^{n}(\beta_0 + \beta_1 X_i) - \beta_1 \overline{X} = \beta_0 + \beta_1 \overline{X} - \beta_1 \overline{X} = \beta_0.
\end{aligned} \tag{11.52}$$

In Section 11.7.2, we saw that the OLS fitted SLR model makes unbiased predictions on the data that was used for fitting the model. The unbiasedness of the OLS estimates of regression coefficients makes the model prediction on the new, previously unseen inputs also unbiased. Indeed, let $\hat{Y} = \hat{\beta}_0 + \hat{\beta}_1 X$ be the model prediction of the response to a new input X and $Y = \beta_0 + \beta_1 X + E$ be the true response. Then

$$\mathbb{E}[\hat{Y} - Y] = \mathbb{E}[\hat{\beta}_0 + \hat{\beta}_1 X - \beta_0 + \beta_1 X - E] = \beta_0 + \beta_1 X - \beta_0 - \beta_1 X = 0. \tag{11.53}$$

11.7.5 Variance of OLS Estimates

To quantify the variability of OLS estimates, let's compute their variances[38]. Using the linear representation (11.49) and model assumptions (11.25), we have

$$\begin{aligned}\mathbb{V}[\hat{\beta}_1] &= \mathbb{V}\left[\sum_{i=1}^{n}\frac{X_i-\overline{X}}{S_{XX}}Y_i\right] = \sum_{i=1}^{n}\left(\frac{X_i-\overline{X}}{S_{XX}}\right)^2\underbrace{\mathbb{V}[Y_i]}_{\sigma^2}\\ &= \frac{\sigma^2}{S_{XX}^2}\underbrace{\sum_{i=1}^{n}(X_i-\overline{X})^2}_{S_{XX}} = \frac{\sigma^2}{S_{XX}}.\end{aligned} \tag{11.54}$$

Here we used independence of $Y_1,\ldots,Y_n$ (induced by independence of $E_1,\ldots,E_n$) and the property that $\mathbb{V}[Y_i] = \mathbb{V}[\beta_0+\beta_1X_i+E_i] = \mathbb{V}[E_i] = \sigma^2$.

Similarly, using the linear representation (11.50), we have

$$\begin{aligned}\mathbb{V}[\hat{\beta}_0] &= \mathbb{V}\left[\sum_{i=1}^{n}\left(\frac{1}{n}-\frac{\overline{X}(X_i-\overline{X})}{S_{XX}}\right)Y_i\right] = \sum_{i=1}^{n}\left(\frac{1}{n}-\frac{\overline{X}(X_i-\overline{X})}{S_{XX}}\right)^2\underbrace{\mathbb{V}[Y_i]}_{\sigma^2}\\ &= \sigma^2\left(\underbrace{\sum_{i=1}^{n}\frac{1}{n^2}}_{1/n}-\frac{2\overline{X}}{nS_{XX}}\underbrace{\sum_{i=1}^{n}(X_i-\overline{X})}_{0}+\frac{\overline{X}^2}{S_{XX}^2}\underbrace{\sum_{i=1}^{n}(X_i-\overline{X})^2}_{S_{XX}}\right)\\ &= \sigma^2\left(\frac{1}{n}+\frac{\overline{X}^2}{S_{XX}}\right).\end{aligned} \tag{11.55}$$

Notice that both variances decrease as the sum of squares $S_{XX} = \sum_{i=1}^{n}(X_i-\overline{X})^2$ increases. Therefore, all other factors being equal, the larger the sample variance of $X_1,\ldots,X_n$, the more accurately the regression coefficients can be estimated.

[38] These results will be especially useful in the next section, where we will discuss how to test hypotheses about regression coefficients.

11.7.6 OLS Estimates are BLUE

In the previous sections, we have shown that the OLS estimates $\hat{\beta}_0$ and $\hat{\beta}_1$ are linear, unbiased, and we computed their variances. It turns out that $\hat{\beta}_0$ and $\hat{\beta}_1$ are the *best linear unbiased estimates* (BLUE) of β_0 and β_1. Here, "best" means that the OLS estimate has the smallest possible variance among all linear unbiased estimates, that is, the OLS estimate is the most accurate estimate in the class of linear unbiased estimates.

Remark 11.7. A standard measure of accuracy of an estimate $\hat{\beta}$ is its mean squared error. But since we consider only unbiased estimates, thanks to the bias-variance decomposition, $\text{MSE}[\hat{\beta}] = \mathbb{B}[\hat{\beta}]^2 + \mathbb{V}[\hat{\beta}] = \mathbb{V}[\hat{\beta}]$. So, for an unbiased estimate, its accuracy is quantified by its variance.

The result that OLS estimates are BLUE is called the *Gauss-Markov theorem*, and it holds not only for the SLR model, but also for a more general multiple regression model. The Gauss-Markov theorem, which guarantees that the OLS estimates are the most accurate estimates in a certain large class of estimates, contributes significantly to the popularity of OLS.

Let's prove the Gauss-Markov theorem for the OLS estimate of the slope $\hat{\beta}_1$[39]. To this end, let's find the BLUE estimate of β_1 and show that it coincides with the OLS estimate $\hat{\beta}_1$. Let $\check{\beta}_1$ be any linear unbiased estimate of β_1. Since $\check{\beta}_1$ is linear, it must have the following form[40]:

$$\check{\beta}_1 = \sum_{i=1}^{n} \alpha_i Y_i. \tag{11.56}$$

Since $\check{\beta}_1$ is unbiased, it must satisfy $\mathbb{E}[\check{\beta}_1] = \beta_1$. This condition implies that the coefficients $\alpha_1, \ldots, \alpha_n$ in (11.56) are not simply arbitrary coefficients but must satisfy the following equation:

$$\begin{aligned} \beta_1 = \mathbb{E}[\check{\beta}_1] = \mathbb{E}\left[\sum_{i=1}^{n} \alpha_i Y_i\right] &= \sum_{i=1}^{n} \alpha_i \mathbb{E}\left[Y_i\right] = \sum_{i=1}^{n} \alpha_i (\beta_0 + \beta_1 X_i) \\ &= \beta_0 \sum_{i=1}^{n} \alpha_i + \beta_1 \sum_{i=1}^{n} \alpha_i X_i. \end{aligned} \tag{11.57}$$

For this equation to always hold true for any β_0 and β_1, the coefficient in front of β_0 must be zero, and the coefficient in front of β_1 (in the right-hand side) must be one,

$$\sum_{i=1}^{n} \alpha_i = 0 \quad \text{and} \quad \sum_{i=1}^{n} \alpha_i X_i = 1. \tag{11.58}$$

So, any linear unbiased estimate $\check{\beta}_1$ of β_1 has form (11.56), where $\alpha_1, \ldots, \alpha_n$ satisfy (11.58). The BLUE estimate of β_1 is the linear unbiased estimate with the smallest possible variance. The variance of $\check{\beta}_1$ is

$$\mathbb{V}[\check{\beta}_1] = \mathbb{V}\left[\sum_{i=1}^{n} \alpha_i Y_i\right] = \sum_{i=1}^{n} \alpha_i^2 \underbrace{\mathbb{V}\left[Y_i\right]}_{\sigma^2} = \sigma^2 \sum_{i=1}^{n} \alpha_i^2. \tag{11.59}$$

Therefore, the BLUE estimate $\check{\beta}_1^{BLUE}$ of β_1 is the estimate of the form (11.56), where the vector of coefficients $\alpha = (\alpha_1, \ldots, \alpha_n)$ is the solution of the following constrained optimization problem:

[39] We will prove the Gauss-Markov theorem in full generality in Section 12.4.1.

[40] Recall the definition of a linear estimate in Section 11.7.3.

$$\sum_{i=1}^{n} \alpha_i^2 \to \min,$$
$$\text{s.t. } \sum_{i=1}^{n} \alpha_i = 0 \text{ and } \sum_{i=1}^{n} \alpha_i X_i = 1. \tag{11.60}$$

Before solving this problem analytically, let's first understand it and the solution that we are looking for geometrically. Figure 11.6 schematically visualizes the search space and the optimal solution of problem (11.60).

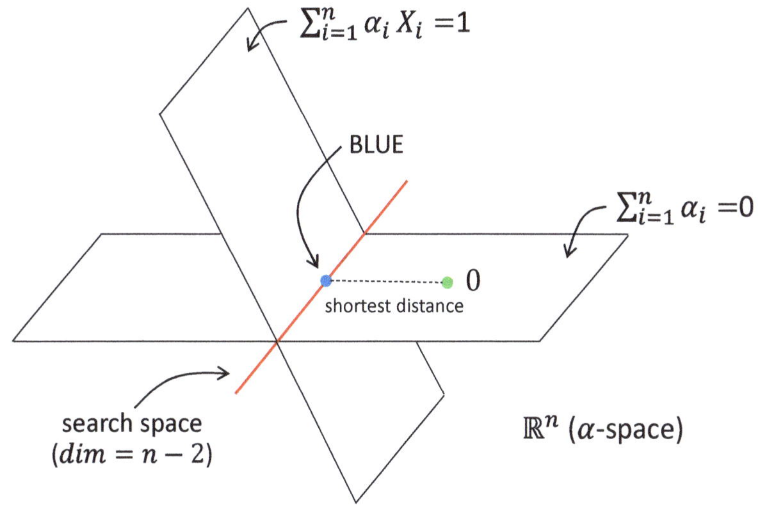

Fig. 11.6 The coefficients $(\alpha_1, \dots, \alpha_n)$ of the BLUE estimate are the coordinates of a point in the intersection of the two hyperplanes defined by the constraints, which is the closest to the origin.

The figure shows two hyperplanes of dimension $n-1$ in $\mathbb{R}^n$ (α-space) defined by the two constraints. The intersection of these hyperplanes, represented by the red line, forms the search space, which is a hyperplane of dimension $n-2$[41]. The objective function $\sum_{i=1}^{n} \alpha_i^2$ is the squared Euclidean distance from the origin (depicted as a green dot) to point $\alpha = (\alpha_1, \dots, \alpha_n)$. Therefore, the optimization problem (11.60) is equivalent to finding a point α (depicted as a blue dot) in the search space, which is closest to the origin.

This geometric interpretation and visualization show that a solution of (11.60) exists and is unique[42]. Let's find it using the *method of Lagrange multipliers*[43]. The Lagrangian function associated with the optimization problem is

$$L(\alpha, \lambda_1, \lambda_2) = \sum_{i=1}^{n} \alpha_i^2 + \lambda_1 \sum_{i=1}^{n} \alpha_i + \lambda_2 \left(\sum_{i=1}^{n} \alpha_i X_i - 1 \right), \tag{11.61}$$

where λ_1 and λ_2 are the Lagrange multipliers associated with the constraints. To find critical points of the Lagrangian L, i.e. points $(\alpha, \lambda_1, \lambda_2)$ where its gradient ∇L

[41] Or, as mathematicians say, codimension 2.

[42] This is because there exists a unique point $A \in \pi$ in a hyperplane π, which is the closest to another point $B \notin \pi$. The point A is simply the orthogonal projection of B onto π.

[43] A brief introduction to this widely used optimization method and useful illustrative examples are given in [10].

vanishes, we need to solve the following system of $n+2$ equations

$$\begin{cases} \dfrac{\partial L}{\partial \alpha_i} = 2\alpha_i + \lambda_1 + \lambda_2 X_i = 0, \quad i = 1,\ldots,n, \\ \dfrac{\partial L}{\partial \lambda_1} = \sum_{i=1}^{n} \alpha_i = 0, \\ \dfrac{\partial L}{\partial \lambda_2} = \sum_{i=1}^{n} \alpha_i X_i - 1 = 0. \end{cases} \tag{11.62}$$

for $\alpha_1,\ldots,\alpha_n,\lambda_1,\lambda_2$. Summing the first n equations and using that $\sum_{i=1}^{n}\alpha_i = 0$, we have:

$$n\lambda_1 + \lambda_2 \sum_{i=1}^{n} X_i = 0. \tag{11.63}$$

Next, multiplying the ith equation by X_i for $i = 1,\ldots,n$, and then summing the first n equations and using that $\sum_{i=1}^{n}\alpha_i X_i = 1$, we have:

$$2 + \lambda_1 \sum_{i=1}^{n} X_i + \lambda_2 \sum_{i=1}^{n} X_i^2 = 0. \tag{11.64}$$

It is straightforward now to solve the last two equation for λ_1 and λ_2:

$$\lambda_1 = \frac{2\overline{X}}{S_{XX}} \quad \text{and} \quad \lambda_2 = -\frac{2}{S_{XX}}. \tag{11.65}$$

Finally, substituting these expressions back into the first equation of system (11.62), we find that

$$\alpha_i = -\frac{1}{2}(\lambda_1 + \lambda_2 X_i) = -\frac{1}{2}\left(\frac{2\overline{X}}{S_{XX}} - \frac{2X_i}{S_{XX}}\right) = \frac{X_i - \overline{X}}{S_{XX}}. \tag{11.66}$$

So, the Lagrangian L has the single critical point

$$(\alpha_1,\ldots,\alpha_n,\lambda_1,\lambda_2) = \left(\frac{X_1 - \overline{X}}{S_{XX}},\ldots,\frac{X_n - \overline{X}}{S_{XX}}, \frac{2\overline{X}}{S_{XX}}, -\frac{2}{S_{XX}}\right). \tag{11.67}$$

Since the optimization problem (11.60) has a unique solution[44] and $\nabla L = 0$ is a necessary condition for optimality, it follows that

$$\alpha = \left(\frac{X_1 - \overline{X}}{S_{XX}},\ldots,\frac{X_n - \overline{X}}{S_{XX}}\right) \tag{11.68}$$

is precisely the solution of (11.60). Therefore, the BLUE estimate $\check{\beta}_1^{BLUE}$ of β_1 is

$$\check{\beta}_1^{BLUE} = \sum_{i=1}^{n} \alpha_i Y_i = \sum_{i=1}^{n} \frac{X_i - \overline{X}}{S_{XX}} Y_i, \tag{11.69}$$

[44] We established this above using the geometric interpretation of the problem, see Fig. 11.6.

and it coincides with the OLS estimate $\hat{\beta}_1$, see (11.49). Similarly, it is possible to show that the OLS estimate $\hat{\beta}_0$ is the BLUE estimate of the intercept β_0.

To sum up, if we are interested in finding linear ("simple") unbiased estimates of the regression coefficients of the simple linear regression model, then the OLS estimates are the best (most accurate) available estimates.

Remark 11.8. If, however, we don't care about simplicity/linearity and unbiasedness, then there exist more accurate (in the mean squared sense) estimates of the regression coefficients. For more details, see the references in Section 11.12.

11.8 Testing Hypotheses about Regression Coefficients

The simple linear regression model (our main object of study in this chapter)

$$Y_i = \beta_0 + \beta_1 X_i + E_i, \quad i = 1, \dots, n, \tag{11.70}$$

is accompanied by one of the two types of model assumptions about prediction errors $E_1, \dots, E_n$: (weaker) nonparametric assumptions

$$\begin{aligned} &1.\ E_1, \dots, E_n \text{ are independent,} \\ &2.\ \mathbb{E}[E_i] = 0, \\ &3.\ \mathbb{V}[E_i] = \sigma^2. \end{aligned} \tag{11.71}$$

or (stronger) parametric assumption

$$E_1, \dots, E_n \overset{\text{iid}}{\sim} \mathcal{N}(0, \sigma^2). \tag{11.72}$$

As we saw in Section 11.6, the parametric normality assumption (11.72) allows to interpret the OLS estimates $\hat{\beta}_0$ and $\hat{\beta}_1$ of regression coefficients β_0 and β_1 as the maximum likelihood estimates. As we will see here, the normality assumption also allows to test hypotheses about β_0 and β_1.

Suppose that we want to test

$$H_0 : \beta_1 = \beta_1^0 \quad \text{vs} \quad H_1 : \beta_1 \neq \beta_1^0, \tag{11.73}$$

where β_1^0 is a hypothesized value[45]. Testing problems of this type, where the null hypothesis is simple, fall into the framework of the Wald test[46]. To use the Wald test, we need two things. First, we need a point estimate of the parameter of interest, which we do have: the OLS estimate for the slope,

$$\hat{\beta}_1 = \sum_{i=1}^{n} \frac{X_i - \overline{X}}{S_{XX}} Y_i. \tag{11.74}$$

Second, we need this point estimate to be normally distributed[47]. While nonpara-

[45] The value β_1^0 could, for example, represent the "current theory" regarding the relationship between X and Y.

[46] See Section 10.1.

[47] See assumption (10.3).

metric assumptions (11.71) do not guarantee that $\hat{\beta}_1 \sim \mathcal{N}$ (even if the sample size n is large), the parametric assumption (11.72) does. Indeed, since errors $E_1, \dots, E_n$ are normally distributed, so are the responses $Y_1, \dots, Y_n$, and, therefore, so is $\hat{\beta}_1$, as a linear combination of normally distributed responses.

In Section 11.7.4, we established that the OLS estimate $\hat{\beta}_1$ is unbiased, $\mathbb{E}[\hat{\beta}_1] = \beta_1$, and in Section 11.7.5, we found its variance, $\mathbb{V}[\hat{\beta}_1] = \sigma^2/S_{XX}$. Therefore, thanks to assumption (11.72),

$$\hat{\beta}_1 \sim \mathcal{N}\left(\beta_1, \frac{\sigma^2}{S_{XX}}\right), \tag{11.75}$$

or, equivalently,

$$\frac{\hat{\beta}_1 - \beta_1}{\sigma/\sqrt{S_{XX}}} \sim \mathcal{N}(0,1). \tag{11.76}$$

Although the distribution of the random variable on the left-hand side is known exactly, the specific values of the random variable (its realizations) cannot be computed, since σ is unknown. An unbiased estimate $\hat{\sigma}_u^2$ of σ^2 is given by (11.44). If we replace σ with $\hat{\sigma}_u$, then the corresponding random variable will be "computable" and it will be distributed approximately according to the standard normal distribution,

$$\frac{\hat{\beta}_1 - \beta_1}{\hat{\sigma}_u/\sqrt{S_{XX}}} \mathrel{\dot{\sim}} \mathcal{N}(0,1). \tag{11.77}$$

Therefore, under the null hypothesis $H_0 : \beta_1 = \beta_1^0$, we have

$$\frac{\hat{\beta}_1 - \beta_1^0}{\hat{\sigma}_u/\sqrt{S_{XX}}} \mathrel{\dot{\sim}} \mathcal{N}(0,1), \tag{11.78}$$

and the (approximate) p-value obtained from the Wald test is[48]

[48] Compare with (10.20).

$$\mathfrak{p}_1^W \approx 2\Phi\left(-\left|\frac{\hat{\beta}_1 - \beta_1^0}{\hat{\sigma}_u/\sqrt{S_{XX}}}\right|\right). \tag{11.79}$$

The smaller the p-value $\mathfrak{p}_1^W$, the stronger the evidence against the null hypothesis provided by the regression data.

It can be shown that the exact null distribution[49] of the random variable on the left-hand side of (11.78) is the t-distribution with $(n-2)$ degrees of freedom [11][50]:

[49] That is, the distribution under the null hypothesis.

[50] A proof for a more general case of multiple linear regression is provided in Appendix 12.15.

$$\frac{\hat{\beta}_1 - \beta_1^0}{\hat{\sigma}_u/\sqrt{S_{XX}}} \sim \mathcal{T}(n-2). \tag{11.80}$$

This means that, instead of the approximate p-value provided by the Wald test, we can use the exact p-value obtained from the t-test[51]:

[51] Compare with (10.47).

$$\mathfrak{p}_1^T = 2F_{T,n-2}\left(-\left|\frac{\hat{\beta}_1 - \beta_1^0}{\hat{\sigma}_u/\sqrt{S_{XX}}}\right|\right), \tag{11.81}$$

where $F_{T,n-2}$ is the CDF of the t-distribution with $(n-2)$ degrees of freedom.

Remark 11.9. Since the t-distribution converges to the standard normal distribution[52], $\mathscr{T}(n) \to \mathscr{N}(0,1)$, as the number of degrees of freedom $n \to \infty$, we have that in the large-sample-size limit the two p-values are the same,

[52] See Fig. 10.2.

$$\lim_{n\to\infty} p_1^W = \lim_{n\to\infty} p_1^T. \tag{11.82}$$

Moreover, since the difference between $\mathscr{T}(n)$ and $\mathscr{N}(0,1)$ is negligible for $n \geq 30$, the two p-values are approximately the same $p_1^W \approx p_1^T$ for practical purposes. It becomes essential to use p_1^T instead of p_1^W only if the sample size is small, $n < 32$.

A very important special case of the testing problem (11.73), which is in fact the most frequent case that occurs in applications, is when the null value $\beta_1^0 = 0$,

$$H_0 : \beta_1 = 0 \quad \text{vs} \quad H_1 : \beta_1 \neq 0. \tag{11.83}$$

Here, we test whether there is a linear relationship between response Y and input X. Testing (11.83) is often referred to as testing the ***significance*** of the linear regression. If H_0 is true, then either Y does not depend on X at all, $Y = \beta_0 + E$ or the dependence on X is nonlinear and without the linear term, $Y = \beta_0 + \beta_2 X^2 + \beta_3 X^3 + \ldots + E$. If H_1 is true, then either Y depends on X precisely linearly and the SLR model $Y = \beta_0 + \beta_1 X + E$ is an excellent model for the data, or Y may also depend on higher order term, $Y = \beta_0 + \beta_1 X + \beta_2 X^2 + \beta_3 X^3 + \ldots + E$.

The corresponding p-values provided by the Wald and t-test are:

$$\mathrm{p}_1^W \approx 2\Phi\left(-\left|\frac{\hat{\beta}_1\sqrt{S_{XX}}}{\hat{\sigma}_u}\right|\right) \quad \text{and} \quad \mathrm{p}_1^T = 2F_{T,n-2}\left(-\left|\frac{\hat{\beta}_1\sqrt{S_{XX}}}{\hat{\sigma}_u}\right|\right). \tag{11.84}$$

Remark 11.10. If the 0.05 confidence level is used for rejecting the null hypothesis, then $\mathrm{p}_1^W < 0.05$ is equivalent to

$$|\mathfrak{z}_1| = \left|\frac{\hat{\beta}_1\sqrt{S_{XX}}}{\hat{\sigma}_u}\right| > -\Phi^{-1}(0.025) \approx 2. \tag{11.85}$$

The quantity $\mathfrak{z}_1$ is called a *z-score*[53] associated with the coefficient β_1. So, instead of the p-value, we can use the z-score directly: H_0 is rejected precisely when $|\mathfrak{z}_1| > 2$.

[53] "W is for Winnie embedded in ice. X is for Xerxes devoured by mice. Y is for Yorick whose head was knocked in. Z is for Zillah who drank too much gin." – Edward Gorey, *The Gashlycrumb Tinies*.

Similarly we can test a hypothesis about the intercept,

$$H_0 : \beta_0 = \beta_0^0 \quad \text{vs} \quad H_1 : \beta_0 \neq \beta_0^0, \tag{11.86}$$

where β_0^0 is some hypothesized value. Since the OLS estimate

$$\hat{\beta}_0 = \sum_{i=1}^{n}\left[\frac{1}{n} - \frac{\overline{X}(X_i - \overline{X})}{S_{XX}}\right] Y_i \tag{11.87}$$

is a linear combination of normally distributed[54] $Y_i \sim \mathcal{N}(\beta_0 + \beta_1 X_i, \sigma^2)$, it is also normally distributed, $\hat{\beta}_0 \sim \mathcal{N}$. The OLS estimate is unbiased and its variance is[55]

$$\mathbb{V}[\hat{\beta}_0] = \sigma^2 \left(\frac{1}{n} + \frac{\overline{X}^2}{S_{XX}} \right) \approx \hat{\sigma}_u^2 \left(\frac{1}{n} + \frac{\overline{X}^2}{S_{XX}} \right). \tag{11.88}$$

[54] Recall that we are working under the normality assumption (11.72).

[55] See Sections 11.7.4 and 11.7.5.

Therefore, under the null hypothesis $H_0 : \beta_0 = \beta_0^0$, we have

$$\frac{\hat{\beta}_0 - \beta_0^0}{\hat{\sigma}_u \sqrt{\frac{1}{n} + \frac{\overline{X}^2}{S_{XX}}}} \stackrel{\cdot}{\sim} \mathcal{N}(0,1), \tag{11.89}$$

and the (approximate) p-value obtained from the Wald test is

$$\mathfrak{p}_0^W \approx 2\Phi\left(- \left| \frac{\hat{\beta}_0 - \beta_0^0}{\hat{\sigma}_u \sqrt{\frac{1}{n} + \frac{\overline{X}^2}{S_{XX}}}} \right| \right). \tag{11.90}$$

It can be shown that the exact null distribution of the random variable on the left-hand side of (11.89) is the t-distribution with $(n-2)$ degrees of freedom [11]. The exact p-value obtained from the t-test is, therefore,

$$\mathfrak{p}_0^T = 2F_{T,n-2}\left(- \left| \frac{\hat{\beta}_0 - \beta_0^0}{\hat{\sigma}_u \sqrt{\frac{1}{n} + \frac{\overline{X}^2}{S_{XX}}}} \right| \right). \tag{11.91}$$

Thanks to the fast convergence $\mathcal{T}(n) \to \mathcal{N}(0,1)$, these two p-values are essentially identical (for all practical purposes) if the sample size $n \geq 32$.

11.9 Confidence Intervals for Regression Coefficients

In addition to testing hypotheses, the normality assumption $E_1, \ldots, E_n \stackrel{\text{iid}}{\sim} \mathcal{N}(0, \sigma^2)$ also allows to construct confidence intervals for the regression coefficients.

Let's start with the slope β_1. Our goal is to find a random interval $I_1 = (a_1, b_1)$ that contains β_1 with probability $1 - \alpha$,

$$\mathbb{P}(a_1 < \beta_1 < b_1) = 1 - \alpha, \qquad \alpha \in (0,1). \tag{11.92}$$

Since, as we established in the previous section[56], $\frac{\hat{\beta}_1 - \beta_1}{\hat{\sigma}_u / \sqrt{S_{XX}}}$ is approximately standard normal,

$$\mathbb{P}\left(z_{\frac{\alpha}{2}} < \frac{\hat{\beta}_1 - \beta_1}{\hat{\sigma}_u / \sqrt{S_{XX}}} < -z_{\frac{\alpha}{2}} \right) \approx 1 - \alpha, \tag{11.93}$$

[56] Namely, in (11.77).

where $z_{\frac{\alpha}{2}}$ is the standard normal quantile, $\Phi(z_{\frac{\alpha}{2}}) = \frac{\alpha}{2}$. Rewriting the inequalities explicitly in terms of β_1, we have:

$$\mathbb{P}\left(\hat{\beta}_1 + z_{\frac{\alpha}{2}} \frac{\hat{\sigma}_u}{\sqrt{S_{XX}}} < \beta_1 < \hat{\beta}_1 - z_{\frac{\alpha}{2}} \frac{\hat{\sigma}_u}{\sqrt{S_{XX}}}\right) \approx 1 - \alpha. \tag{11.94}$$

Therefore,

$$I_\alpha^1 = \left(\hat{\beta}_1 + z_{\frac{\alpha}{2}} \frac{\hat{\sigma}_u}{\sqrt{S_{XX}}},\ \hat{\beta}_1 - z_{\frac{\alpha}{2}} \frac{\hat{\sigma}_u}{\sqrt{S_{XX}}}\right) \tag{11.95}$$

is an approximate $100(1-\alpha)\%$ confidence interval for β_1. To construct an exact interval, thanks to the exact result (11.80), we simply need to replace the standard normal quantile $z_{\frac{\alpha}{2}}$ with the corresponding quantile $t_{n-2,\frac{\alpha}{2}}$ of the t-distribution with $(n-2)$ degrees of freedom.

Similarly, since the OLS estimate $\hat{\beta}_0$ of the intercept is approximately normally distributed,

$$\frac{\hat{\beta}_0 - \beta_0}{\hat{\sigma}_u \sqrt{\frac{1}{n} + \frac{\overline{X}^2}{S_{XX}}}} \mathrel{\dot\sim} \mathscr{N}(0,1), \tag{11.96}$$

the random interval

$$I_\alpha^0 = \left(\hat{\beta}_0 + z_{\frac{\alpha}{2}} \hat{\sigma}_u \sqrt{\frac{1}{n} + \frac{\overline{X}^2}{S_{XX}}},\ \hat{\beta}_0 - z_{\frac{\alpha}{2}} \hat{\sigma}_u \sqrt{\frac{1}{n} + \frac{\overline{X}^2}{S_{XX}}}\right) \tag{11.97}$$

is an approximate $100(1-\alpha)\%$ confidence interval for β_0. As above for β_1, the exact confidence interval for β_0 is obtained by replacing $z_{\frac{\alpha}{2}}$ with $t_{n-2,\frac{\alpha}{2}}$.

The width of confidence intervals I_α^0 and I_α^1 can be viewed as a measure of uncertainty in the values of the regression coefficients.

11.10 Prediction using the SLR Model

Let X_* be a new, previously unseen input, and suppose that we want to predict the corresponding response Y_*. According to the SLR model,

$$Y_* = \beta_0 + \beta_1 X_* + E_*, \tag{11.98}$$

and the best (in the MSE sense) prediction of the response Y_* is

$$r(X_*) = \beta_0 + \beta_1 X_*. \tag{11.99}$$

Since the regression coefficients β_0 and β_1 are unknown, so is the optimal prediction $r(X_*)$. But we can estimate it by replacing β_0 and β_1 with their OLS estimates,

$$\hat{Y}_* = \hat{r}(X_*) = \hat{\beta}_0 + \hat{\beta}_1 X_*. \tag{11.100}$$

If we work under the weaker nonparametric assumptions (11.71), then all we can do is to report the fitted value $\hat{Y}_* = \hat{r}(X_*)$, which is a point estimate of the optimal prediction $r(X_*)$. The stronger parametric assumption (11.72) allows to do more. Namely, the normality assumption allows to construct

- A *confidence interval* for $r(X_*)$, which is a deterministic unknown parameter[57],
- A *prediction interval* for the response $Y_* = r(X_*) + E_*$, which is an unobservable random variable: since $E_* \sim \mathcal{N}(0, \sigma^2)$, we have that $Y_* \sim \mathcal{N}(r(X_*), \sigma^2)$.

Let's first construct a confidence interval for $r(X_*)$. Recall that confidence intervals often have the following form: $I =$ (point estimate $\pm$ something), where "something" is called the margin of error (MOE). So, let's find a random interval $I = (\hat{r}(X_*) \pm \text{MOE})$, such that $\mathbb{P}(r(X_*) \in I) = 1 - \alpha$.

What is the probability distribution of $\hat{r}(X_*)$? The fitted value $\hat{r}(X_*)$ is a linear estimate[58], since both OLS estimates $\hat{\beta}_0$ and $\hat{\beta}_0$ are linear[59]:

$$\begin{aligned} \hat{r}(X_*) = \hat{\beta}_0 + \hat{\beta}_1 X_* &= \sum_{i=1}^{n} \left[\frac{1}{n} - \frac{\overline{X}(X_i - \overline{X})}{S_{XX}} \right] Y_i + X_* \sum_{i=1}^{n} \frac{X_i - \overline{X}}{S_{XX}} Y_i \\ &= \sum_{i=1}^{n} \left[\frac{1}{n} + \frac{(X_* - \overline{X})(X_i - \overline{X})}{S_{XX}} \right] Y_i. \end{aligned} \tag{11.101}$$

Thanks to the normality assumption $E_1, \ldots, E_n \overset{\text{iid}}{\sim} \mathcal{N}(0, \sigma)$, all responses $Y_1, \ldots, Y_n$ are also normal, and, therefore, $\hat{r}(X_*)$ is also normally distributed (as a linear combination of normal random variables), $\hat{r}(X_*) \sim \mathcal{N}$.

Since the OLS estimates $\hat{\beta}_0$ and $\hat{\beta}_1$ are unbiased[60], so is the fitted value $\hat{r}(X_*)$:

$$\mathbb{E}[\hat{r}(X_*)] = \mathbb{E}[\hat{\beta}_0 + \hat{\beta}_1 X_*] = \mathbb{E}[\hat{\beta}_0] + \mathbb{E}[\hat{\beta}_1] X_* = \beta_0 + \beta_1 X_* = r(X_*). \tag{11.102}$$

Using (11.54), (11.55), and (11.32), the variance of $\hat{r}(X_*)$ is

$$\begin{aligned} \mathbb{V}[\hat{r}(X_*)] &= \mathbb{V}[\hat{\beta}_0 + \hat{\beta}_1 X_*] = \mathbb{V}[\hat{\beta}_0] + X_*^2 \mathbb{V}[\hat{\beta}_1] + 2X_* \mathrm{Cov}(\hat{\beta}_0, \hat{\beta}_1) \\ &= \sigma^2 \left(\frac{1}{n} + \frac{\overline{X}^2}{S_{XX}} \right) + X_*^2 \frac{\sigma^2}{S_{XX}} + 2X_* \mathrm{Cov}(\overline{Y} - \hat{\beta}_1 \overline{X}, \hat{\beta}_1) \\ &= \sigma^2 \left(\frac{1}{n} + \frac{\overline{X}^2 + X_*^2}{S_{XX}} \right) + 2X_* \mathrm{Cov}(\overline{Y}, \hat{\beta}_1) - 2X_* \overline{X} \mathbb{V}[\hat{\beta}_1]. \end{aligned} \tag{11.103}$$

Since the variance $\mathbb{V}[\hat{\beta}_1]$ is given by (11.54), it remains to find the covariance between the mean response $\overline{Y}$ and the OLS estimate of the slope $\hat{\beta}_1$. Intuitively, $\overline{Y}$ and $\hat{\beta}_1$ should be uncorrelated, since the slope of the fitted line does not depend on how high or how low the cloud of points lies. Indeed, using independence of $Y_1, \ldots, Y_n$, which implies that $\mathrm{Cov}(Y_i, Y_j) = 0$ for $i \neq j$, we have:

57 The quantity $r(X_*)$ is a linear combination of parameters β_0 and β_1 with coefficients 1 and X_*. A linear combination of parameters is also a parameter.

58 Recall that an estimate is called linear if it is a linear combination of responses $Y_1, \ldots, Y_n$.

59 See Section 11.7.3.

60 See Section 11.7.4.

$$\begin{aligned}\mathrm{Cov}(\overline{Y},\hat{\beta}_1) &= \mathrm{Cov}\left(\frac{1}{n}\sum_{i=1}^{n}Y_i,\sum_{j=1}^{n}\frac{X_j-\overline{X}}{S_{XX}}Y_j\right) = \frac{1}{n}\sum_{i,j=1}^{n}\frac{X_j-\overline{X}}{S_{XX}}\mathrm{Cov}(Y_i,Y_j)\\ &= \frac{1}{n}\sum_{i=1}^{n}\frac{X_i-\overline{X}}{S_{XX}}\underbrace{\mathbb{V}[Y_i]}_{\sigma^2} = \frac{\sigma^2}{nS_{XX}}\underbrace{\sum_{i=1}^{n}(X_i-\overline{X})}_{0} = 0.\end{aligned} \tag{11.104}$$

So, the variance of $\hat{r}(X_*)$ is

$$\mathbb{V}[\hat{r}(X_*)] = \sigma^2\left(\frac{1}{n}+\frac{\overline{X}^2+X_*^2}{S_{XX}}\right) - 2X_*\overline{X}\frac{\sigma^2}{S_{XX}} = \sigma^2\left(\frac{1}{n}+\frac{(X_*-\overline{X})^2}{S_{XX}}\right). \tag{11.105}$$

After we found the mean and the variance of the fitted value $\hat{r}(X_*)$, its probability distribution is fully determined: it is normally distributed with mean $r(X_*)$ and variance written above. Therefore,

$$\frac{\hat{r}(X_*)-r(X_*)}{\sigma\sqrt{\frac{1}{n}+\frac{(X_*-\overline{X})^2}{S_{XX}}}} \sim \mathcal{N}(0,1). \tag{11.106}$$

If the value of σ were known, then we could immediately construct a confidence interval for $r(X_*)$ using this normality result. But σ is unknown. If we replace σ with it estimate $\hat{\sigma}_u$, then the random variable on the left-hand side will be approximately $\mathcal{N}(0,1)$-distributed, and this can be used for constructing an approximate confidence interval. However — and after Sections 11.8 and 11.9 this should not come as a surprise — it can be shown that the exact distribution is the t-distribution with $(n-2)$ degrees of freedom,

$$\frac{\hat{r}(X_*)-r(X_*)}{\hat{\sigma}_u\sqrt{\frac{1}{n}+\frac{(X_*-\overline{X})^2}{S_{XX}}}} \sim \mathcal{T}(n-2). \tag{11.107}$$

Therefore, the random interval

$$I_\alpha = \left(\hat{r}(X_*) \pm t_{n-2,\frac{\alpha}{2}}\hat{\sigma}_u\sqrt{\frac{1}{n}+\frac{(X_*-\overline{X})^2}{S_{XX}}}\right) \tag{11.108}$$

is a $100(1-\alpha)\%$ confidence interval for the optimal prediction $r(X_*)$ of the response Y_* to input X_*. If the sample size $n \geq 32$, then, for all practical purposes, the t-quantile $t_{n-2,\frac{\alpha}{2}}$ can be replaced with the normal quantile $z_{\frac{\alpha}{2}}$.

Remark 11.11. Note that the width of the confidence interval I_α depends on the input X_*. The width is minimized when $X_* = \overline{X}$, and it increases as the distance between X_* and $\overline{X}$ increases. This behavior is intuitively expected: the prediction $r(X_*)$ of the response to input X_* is most accurate at the center of the data, $X_* = \overline{X}$, since there are relatively many "examples" of input-response pairs (X_i,Y_i) near $\overline{X}$. But as X_* goes away from the center, the number of examples close to X_* decreases, the uncertainty about the value of $r(X_*)$ increases, and, as a result, the confidence interval becomes wider. This is schematically illustrated in Fig. 11.7.

Now let's construct a prediction interval for the response $Y_* = r(X_*) + E_*$. From a mathematical point of view, a prediction interval is the same as a confidence interval[61]: it is a random interval constructed from the data that contains the quantity of interest (Y_* in this case) with a specified probability.

[61] Different terminology is used to highlight the difference between $r(X_*)$, which is unknown and deterministic, and Y_*, which is unobservable and random.

The uncertainty in the value of Y_* comes from two sources: the uncertainty in the value of $r(X_*) = \beta_0 + \beta_1 X_*$ (which stems from the uncertainty in the values of regression coefficients) and the uncertainty in the value of the prediction error E_*. This suggests that the prediction interval for Y_* should be wider than the confidence interval for $r(X_*)$. To find the prediction interval for Y_*, let's consider the corresponding residual $\hat{E}_*$, which is the difference between the response Y_* and the fitted value $\hat{Y}_*$,

$$\hat{E}_* = Y_* - \hat{Y}_* = r(X_*) + E_* - \hat{r}(X_*). \tag{11.109}$$

What is the probability distribution of $\hat{E}_*$? The first term $r(X_*)$ is a deterministic quantity, the prediction error $E_* \sim \mathscr{N}(0, \sigma^2)$, and the fitted value $\hat{r}(X_*)$ is also normally distributed as we established above. Therefore, the residual is also normally distributed, $\hat{E}_* \sim \mathscr{N}$. Its expected value is

$$\mathbb{E}[\hat{E}_*] = \mathbb{E}[r(X_*)] + \mathbb{E}[E_*] - \mathbb{E}[\hat{r}(X_*)] = r(X_*) - r(X_*) = 0, \tag{11.110}$$

and its variance is

$$\begin{aligned} \mathbb{V}[\hat{E}_*] &= \mathbb{V}[E_*] + \mathbb{V}[\hat{r}(X_*)] = \sigma^2 + \sigma^2 \left(\frac{1}{n} + \frac{(X_* - \overline{X})^2}{S_{XX}} \right) \\ &= \sigma^2 \left(1 + \frac{1}{n} + \frac{(X_* - \overline{X})^2}{S_{XX}} \right). \end{aligned} \tag{11.111}$$

Therefore, after standardization, we have:

$$\frac{\hat{E}_*}{\sigma \sqrt{1 + \frac{1}{n} + \frac{(X_* - \overline{X})^2}{S_{XX}}}} = \frac{Y_* - \hat{r}(X_*)}{\sigma \sqrt{1 + \frac{1}{n} + \frac{(X_* - \overline{X})^2}{S_{XX}}}} \sim \mathscr{N}(0,1). \tag{11.112}$$

As before, we have the same technical issue: σ is unknown. Fortunately, the remedy is also the same: if we replace σ with $\hat{\sigma}_u$, then the random variable on the left-hand side will be distributed according to the t-distribution with $(n-2)$ degrees of freedom,

$$\frac{Y_* - \hat{r}(X_*)}{\hat{\sigma}_u \sqrt{1 + \frac{1}{n} + \frac{(X_* - \overline{X})^2}{S_{XX}}}} \sim \mathscr{T}(n-2). \tag{11.113}$$

This result allows to construct a $100(1-\alpha)\%$ prediction interval for the response Y_* to input X_*,

$$I_\alpha = \left(\hat{r}(X_*) \pm t_{n-2,\frac{\alpha}{2}} \hat{\sigma}_u \sqrt{1 + \frac{1}{n} + \frac{(X_* - \overline{X})^2}{S_{XX}}} \right). \tag{11.114}$$

This random interval contains Y_* with probability $1-\alpha$. As always for $n \geq 32$, replacing $t_{n-2,\frac{\alpha}{2}}$ with, perhaps, the more familiar normal quantile $z_{\frac{\alpha}{2}}$ leads to an approximate prediction interval[62].

The confidence interval (11.108) for the optimal prediction $r(X_*)$ of the response Y_* and the prediction interval (11.114) for the response Y_* itself are schematically shown in Fig. 11.7.

[62] Recall that the approximation $\mathscr{T}(n) \approx \mathscr{N}(0,1)$ is considered accurate for practical purposes when $n \geq 30$. See Fig. 10.2.

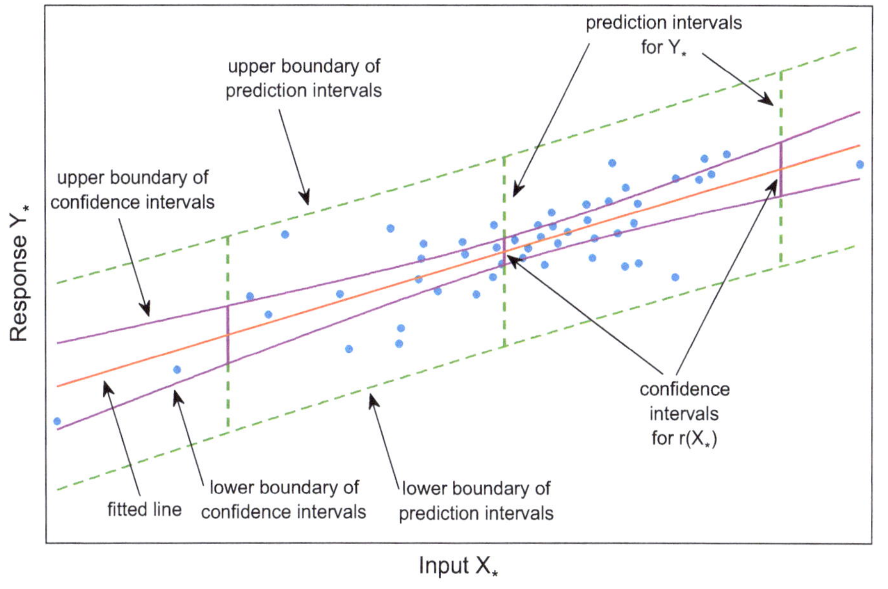

Fig. 11.7 Confidence intervals for $r(X_*) = \beta_0 + \beta_1 X_*$ and prediction intervals for $Y_* = r(X_*) + E_*$ under the normality assumption $E_1, \ldots, E_n, E_* \overset{\text{iid}}{\sim} \mathscr{N}(0, \sigma^2)$.

Both intervals are centered at the fitted value $\hat{r}(X_*)$, which is a point estimate of $r(X_*)$. However, as we expected, the prediction interval is wider, reflecting the greater uncertainty in the value of the response Y_* compared to its mean value $\mathbb{E}[Y_*] = r(X_*)$.

11.11 Case Study: Measuring the Hubble Constant

Dr. Flicker: "Why are you depressed, Alvy?"
Alvy: "The universe is expanding."
Dr. Flicker: "The universe is expanding?"

Alvy: "Well, the universe is everything, and if it's expanding, some day it will break apart and that will be the end of everything."
Mother: "Why is that your business? He's stopped doing his homework!"
Alvy: "What's the point?"
Mother: "What has the universe got to do with it! You're here in Brooklyn! Brooklyn is not expanding!"
Dr. Flicker: "It won't be expanding for billions of years, Alvy, and we've got to enjoy ourselves while we're here! Huh?... Huh?... A-ha-hah-ha..."
– Woody Allen, *Annie Hall* [12]

One of the greatest intellectual revelations of the twentieth century is the discovery that the universe is expanding. The universe is expanding not into something, but intrinsically. This process is hard to imagine based on our everyday experience. It is somewhat analogous to the expanding surface of an inflating balloon, although the inside and outside of the balloon do not exist: its surface is the entire universe. An empirical manifestation of the expanding universe is the observation that all distant galaxies are moving away from each other, and the distance between any two galaxies is increasing[63]. The present rate of expansion is estimated to be between 5% and 10% per billion years, and the current cosmological observations suggest that the expansion is accelerating due to a mysterious form of energy called "dark energy". To learn more about our wonderful universe and its expansion, the reader is recommended to check the classic Stephen Hawking's book [13].

[63] Just like the distance between any two points on the surface of an inflating balloon is increasing.

The expansion of the universe was theoretically predicted by Alexander Friedman in 1922 based on the Einstein field equations in general relativity[64]. In 1929, Friedman's prediction was confirmed[65] by Edwin Hubble, who analyzed the spectra of galaxies in his seminal paper [14] and found that all distant galaxies were moving away from Earth. Moreover, Hubble discovered that the more distant a galaxy is, the faster it is moving away. By analyzing the galaxy spectral data, he inferred the following linear relationship which is now known as *Hubble's law*,

[64] Unfortunately, Friedman's work was mostly unknown until about 1935.

[65] Unfortunately, Friedman never found out about this, for he died in 1925 from typhoid fever after eating an unwashed pear bought at a railway station.

$$v = H_0 d, \tag{11.115}$$

where v is the galaxy's recessional velocity (measured in km/s), i.e. the radial velocity with which the galaxy recedes from us, d is the distance to the galaxy (measured in Mpc[66]), and H_0 is the constant of proportionality, called the *Hubble constant*.

[66] One megaparsec (Mpc) is one million parsecs, which is approximately 3.26 million light years or 3.09×10^{19} km.

Remark 11.12. The Hubble constant is in fact a constant in space, but not in time. At any time, it is a constant throughout the universe, but its value changes over cosmic timescales. For that reason, H_0 is often referred to as the Hubble constant at present.

The Hubble constant is one of the most important parameters in cosmology, since it controls the rate of the universe's expansion. If we assume that galaxies have been always receding from us due to the expansion with constant speeds v governed by Hubble's law, then we can compute the time t_H it took the galaxies to move from us (from all "being in one place") to distances d,

$$t_H = \frac{d}{v} = \frac{d}{H_0 d} = \frac{1}{H_0}. \tag{11.116}$$

This time, called the *Hubble time*, is a rough estimate of the overall age of the universe, the time since the Big Bang. Estimating the Hubble constant is thus a very important cosmological problem.

The original Hubble's estimate of the constant was $H_0 \approx 500$ (km/s)/Mpc, which turned out to be a severe overestimation. The reason for this overestimation is twofold. First, because of the relatively poor techniques and telescopes in the 1920s, Hubble strongly underestimated the distances to individual galaxies[67]. Second, again because of the technological limitations, Hubble considered only "nearby" galaxies (within 2 Mpc from Earth), while Hubble law holds only for distant galaxies, whose gravitational interactions with the Milky Way are negligible. It is, in fact, a miracle that Edwin Hubble was able to discover the law given a very poor quality of his cosmological data, shown in Fig. 11.8.

[67] For example, Hubble underestimated the distance to the Andromeda galaxy by a factor of 2.5.

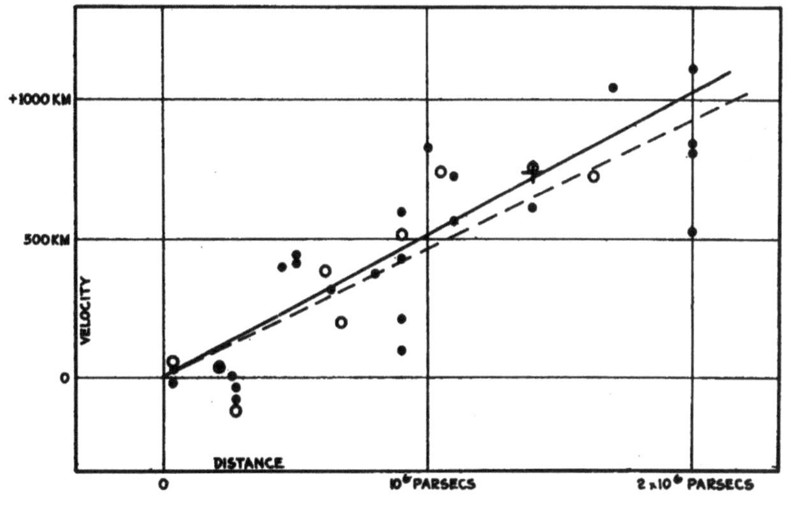

Fig. 11.8 The original Hubble diagram from [14] showing the velocity-distance relation.

The true value of the Hubble constant is currently the subject of a scientific debate, known as the "Hubble tension" [15]. Over the years, various methods have been developed and used by many research groups for measuring the Hubble constant. One set of methods, based on the observations from the "late universe," suggests an approximate value $H_0 \approx 73$ (km/s)/Mpc. However, another set of methods, based on the measurements of the "early universe," gives a different value $H_0 \approx 67$ (km/s)/Mpc. Initially, this discrepancy in the estimates was attributed to measurement errors. But with time, as technology has improved and techniques have become more precise, the errors have shrunk, and the discrepancy between the two estimates has become statistically significant. The reason for this discrepancy is currently unknown, and, perhaps, we will need new physics to explain it.

In this Case Study, we will estimate the Hubble constant H_0 using simple linear regression and galaxy data retrieved from the SIMBAD astronomical database [16]. Figure 11.9 shows a Hubble diagram, which is a plot of recessional velocity of a galaxy versus its distance from the Milky Way, for a sample of $n = 1000$ galaxies. The considerable scatter is due to measurement errors caused by observational uncertainties and galaxy motions.

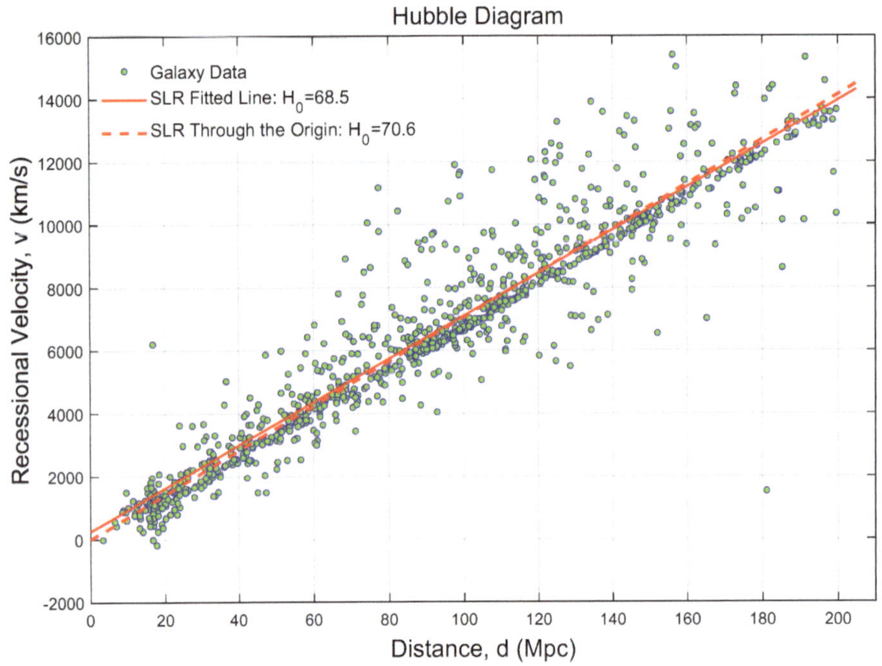

Fig. 11.9 Scatterplot of recessional velocity versus distance for a sample of $n = 1000$ galaxies. The solid line is the OLS fitted regression line with the slope $H_0 \approx 68.5$. The dashed line is the fitted line obtained via regression through the origin (see Problem 11.1) with the slope $H_0 \approx 70.6$. The galaxy data used in this Case Study is available at `https://link.springer.com/chapter/10.1007/978-3-032-03848-7_1`.

Consider the simple linear regression model for the galaxy data:

$$v_i = \beta_0 + \beta_1 d_i + E_i, \quad i = 1, \ldots, n. \tag{11.117}$$

The OLS estimates of the regression parameter are[68]

[68] See (11.34).

$$\begin{aligned} \hat{\beta}_1 &= \frac{\sum_{i=1}^{n}(d_i - \overline{d})(v_i - \overline{v})}{\sum_{i=1}^{n}(d_i - \overline{d})^2} = 68.5, \\ \hat{\beta}_0 &= \frac{\sum_{i=1}^{n}(d_i - \overline{d})(\overline{v} d_i - \overline{d} v_i)}{\sum_{i=1}^{n}(d_i - \overline{d})^2} = 251.1. \end{aligned} \tag{11.118}$$

The Hubble constant H_0 is the slope β_1 of the regression line. So, its estimate $\hat{H}_0$ is

$$\hat{H}_0 = \hat{\beta}_1 = 68.5 \text{ (km/s)/Mpc}, \tag{11.119}$$

which is close to the aforementioned estimate $H_0 \approx 67$ (km/s)/Mpc, based on the measurements of the early universe. The standard error of the OLS estimate $\hat{H}_0$ of the Hubble constant is[69]

$$\text{se}[\hat{H}_0] \approx \sqrt{\frac{\sum_{i=1}^{n}(v_i - \hat{\beta}_0 - \hat{\beta}_1 d_i)^2}{(n-2)\sum_{i=1}^{n}(d_i - \bar{d})^2}} = 0.78 \text{ (km/s)/Mpc}. \tag{11.120}$$

[69] It is obtained by combining formulas (11.54) and (11.44).

To construct a confidence interval for H_0 (Section 11.9) or to rigorously test the null hypothesis $H_0 = 67$ versus the alternative $H_0 \neq 67$ (Section 11.8), we need to assume that the prediction errors are normally distributed, $E_1, \ldots, E_n \overset{\text{iid}}{\sim} \mathcal{N}(0, \sigma^2)$.

Can we check this assumption, at least to some degree? The normality assumption is a statement about the random variables

$$E_i = v_i - \beta_0 - \beta_1 d_i, \tag{11.121}$$

which are, unfortunately, unobservable since β_0 and $\beta_1 = H_0$ are unknown. As long as $\beta_0 \approx \hat{\beta}_0$ and $\beta_1 \approx \hat{\beta}_1$, the corresponding residuals

$$\hat{E}_i = v_i - \hat{\beta}_0 - \hat{\beta}_1 d_i \tag{11.122}$$

can be viewed as observable approximate realizations of the random errors. Therefore, any deviation from normality in the errors should manifest itself in the residuals. Figure 11.10 shows the *residual plot* (panel (a)), which is a scatterplot of the residuals $\hat{E}_i$ versus the corresponding values of predictor d_i, and the normal-quantile plot[70] (panel (b)) for the sample of residuals $\hat{E}_1, \ldots, \hat{E}_n$.

[70] See Section 1.5.

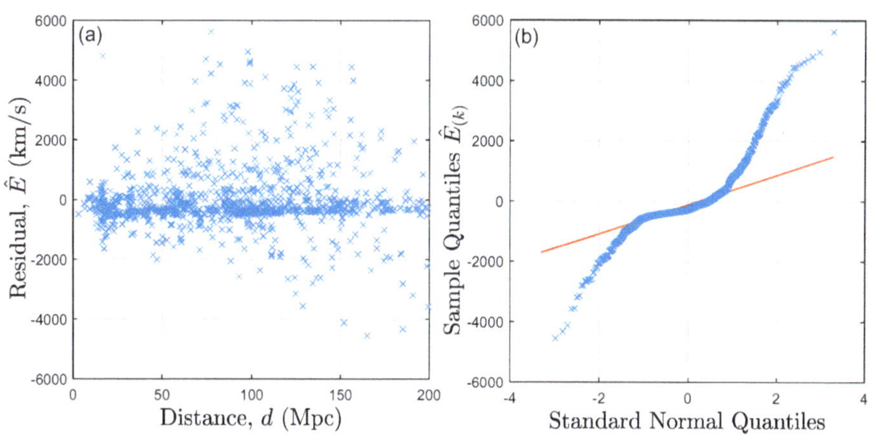

Fig. 11.10 Residual analysis: (a) the residual plot and (b) the normal-quantile plot for residuals. The outlier $(d, v) \approx (180, 2000)$ shown in Fig 11.9 is omitted here for visualization purposes.

While the homoscedasticity assumption $\mathbb{V}[E_i|d_i] = \sigma^2$ may be valid at least to some extent, the normality assumption $E_1, \ldots, E_n \overset{\text{iid}}{\sim} \mathcal{N}(0, \sigma^2)$ is very unlikely to hold. The residuals have long tails, which decay much slower than the normal tails. This means that the magnitude of some errors can be substantially larger than predicted by the normal model. Thus, the *residual analysis* indicates that the normality assumption most likely does not hold for the galaxy data shown in Fig. 11.9.

The Hubble time (11.116) corresponding to the estimate (11.119) is

$$\begin{aligned} t_H &= \frac{1}{H_0} \approx \frac{1\ \text{Mpc}}{68.5\ (\text{km/s})} = \frac{3.09 \times 10^{19}\ \text{km}}{68.5\ (\text{km/s})} \\ &= 4.5 \times 10^{17}\ \text{s} = 14.3\ \text{billion years}, \end{aligned} \tag{11.123}$$

which is close to the currently accepted estimate of the age of the universe 13.7 billion years.

Both the common sense and the Hubble law $v = H_0 d$ suggest that the regression line must go through the origin[71]. So, instead of the simple liner regression model (11.117), let's consider a regression model without the intercept β_0,

$$v_i = \beta_1 d_i + E_i, \quad i = 1, \ldots, n. \tag{11.124}$$

This model is called the *regression through the origin*. In can be shown[72] that in this case, the OLS estimate of the regression parameter β_1 is

$$\hat{\beta}_1 = \frac{\sum_{i=1}^n d_i v_i}{\sum_{i=1}^n d_i^2} = 70.6. \tag{11.125}$$

The fitted regression line $v = \hat{\beta}_1 d$ is represented by the dashed line in Fig. 11.9.

So, based on the regression through the origin model, the Hubble constant is

$$H_0 \approx \hat{\beta}_1 = 70.6\ (\text{km/s})/\text{Mpc}, \tag{11.126}$$

which is between the currently main "competing" estimates of $H_0 \approx 67\ (\text{km/s})/\text{Mpc}$ and $H_0 \approx 73\ (\text{km/s})/\text{Mpc}$, based on the late and early universe measurements.

The corresponding Hubble time is

$$t_H = \frac{1}{H_0} \approx \frac{1\ \text{Mpc}}{70.6\ (\text{km/s})} = 13.9\ \text{billion years}. \tag{11.127}$$

Remark 11.13. Originally, Einstein believed that the universe was static, and he (incorrectly) criticized Friedman's work, which theoretically predicted the expansion of the universe. In 1931, two years after Hubble's work [14] providing observational evidence for the expansion of the universe was published, Einstein visited the Mount Wilson Observatory near Pasadena to thank Hubble, who worked there on the 100-inch (2.5 meter) Hooker telescope[73], for his groundbreaking discovery, which laid the foundation for modern cosmology. As the story goes, Einstein called his static universe assumption the "biggest mistake" of his life.

[71] If the distance $d = 0$, then we must have the velocity $v = 0$.

[72] Problem 11.1 will ask you to prove this.

[73] The Hooker telescope was the largest telescope in the world from 1917 to 1949.

11.12 Further Reading

In this chapter, we discussed the simplest and arguably the most widely used regression model, the *simple linear regression model*, together with the most straightforward and popular method for estimating model parameters, *ordinary least squares*.

As a result of frequent use, the SLR model is sometimes misused and abused. One such example is the (mis)use of the regression model for *extrapolation*: predicting the response Y to the input X, which lies outside the range $[X_{(1)}, X_{(n)}]$ of "training" inputs $X_1, \ldots, X_n$ used for estimating the model parameters. The SLR model is, by construction, an *interpolation model* and should not be used for extrapolation without proper justification. The danger associated with using the model for extrapolation is schematically illustrated in Fig. 11.11.

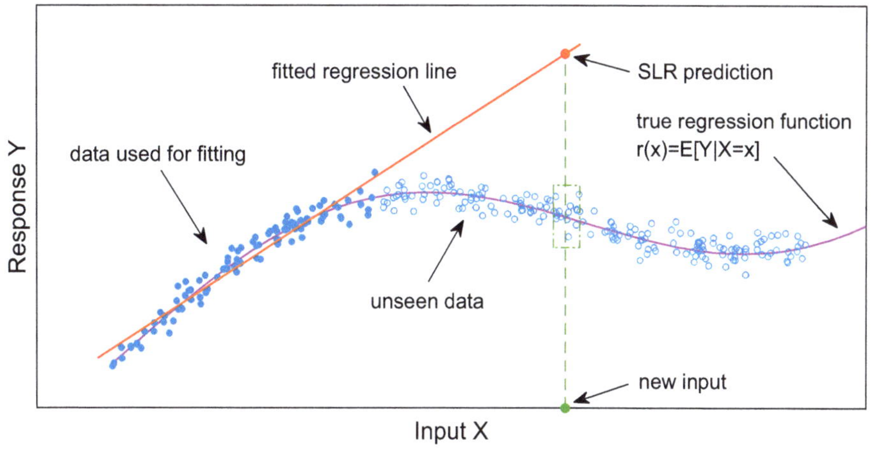

Fig. 11.11 Using the SLR model for extrapolation can lead to highly inaccurate predictions.

In spite of this danger, the unjustifiable use of the SLR model for extrapolation is widespread. Several examples from medical literature are provided in [17].

Another abuse of regression, which has to do with the presence of the so-called latent or lurking variables, is discussed in the following paper by George Box[74].

- Box, G.E.P: Use and abuse of regression. *Technometrics* 8(4), 625-629 (1966). `https://doi.org/10.2307/1266635`

The simple linear regression model studies the linear dependence between two variables, input X and response Y. Approximate linear (in fact, any) dependence between variables suggests but does not necessarily imply a *causal* relationship between them. As the famous but often forgotten mantra goes, "correlation does not imply causation"[75]. The dependence between X and Y can arise not only because X causes Y, but also because there may exist another variable Z, called the *confounding variable*, that causes both X and Y.

[74] George Box, who is widely considered one of the greatest statisticians of the 20th century, has famously stated that "all models are wrong, but some are useful". Using this opportunity and the space on the margin, I would like to highly recommend Box's essay on Science and Statistics [18].

[75] What is less widely appreciated is that causation does not imply correlation either. To convince yourself that this is the case, look at the first or fourth cloud of points in the last row of Fig. 5.2: the corresponding variables are strongly (nonlinearly) dependent, but uncorrelated. Causation implies some form of dependence between variables, but not necessarily linear.

As a simple illustrative example, the number of sold ice creams (X) is positively correlated with the number of drowning deaths (Y). But this, of course, does not imply that ice cream consumption causes drowning. The confounding variable Z in this example is temperature: during hot days, more people buy ice cream, more people participate in water-based activities, and, as result, more people drown.

As a more realistic example[76], several published studies have established a positive correlation between coffee consumption (X) and risk of lung cancer (Y), suggesting that coffee drinkers have a greater chance of developing lung cancer. A meta-analysis performed in [19] shows, however, that the seemingly present association between coffee consumption and lung cancer is confounded by tobacco smoking (Z), which is a known cause of lung cancer, and which in many countries and cultures is associated with coffee drinking[77]. For non-smokers, coffee drinking does not appear to be a risk factor for lung cancer.

[76] And an example which is dear to the author, who is an avid espresso lover.

[77] "Do you smoke? Only when I drink coffee." *Coffee and Cigarettes.*

A branch of statistics that studies causality and develops methods for establishing causal relationships from data is called *causal inference*. A comprehensive exposition of causal inference is provided in an award-winning book by Judea Pearl.

- Pearl, J: *Causality: Models, Reasoning and Inference*. Cambridge University Press, 2nd Edition (2009). `https://bayes.cs.ucla.edu/BOOK-2K/`

In this chapter, we considered only one (the most commonly used) method for estimating the regression coefficients: the method of ordinary least squares. But OLS is not the only way to fit a line to a cloud of points in the plane. For example, in some applications it might not be clear which of the two variables X and Y is the input (independent variable) and which is the response (dependent variable). In such problems, where variables require symmetrical treatment, should we use the OLS regression of Y on X, of X on Y, or maybe the bisector of these two regression lines? The following paper explores five different methods for obtaining linear regression fits to bivariate data and discusses their relative advantages and disadvantages:

- Isobe, T., Feigelson, E.D., Akritas, M.G., Babu, G.J.: Linear regression in astronomy. I. *Astrophysical Journal* 364, 104-113 (1990). `https://articles.adsabs.harvard.edu/full/1990ApJ...364..104I`

Appendix B in this paper provides readers with a short yet insightful review of linear regression techniques from a historical and multidisciplinary perspective.

11.13 Probability Refresher: The Law of Total Expectation

The law of total expectation is one of the most fundamental and useful formulas in probability theory. It allows us to compute expected values, which are hard to compute directly, *by conditioning*.

Let $Y \sim F_Y$ be a random variable with CDF F_Y. Then its expected value $\mathbb{E}[Y]$ is a *number*, a single-number summary of the probability distribution of Y[78],

[78] If the integral with respect to $dF_Y(y)$ looks confusing, see Remark 4.1.

$$\mathbb{E}[Y] = \int y dF_Y(y) = \begin{cases} \int y f_Y(y)dy, & \text{if } Y \text{ is continuous with PDF } f_Y, \\ \sum_i y_i p_Y(y_i), & \text{if } Y \text{ is discrete with PMF } p_Y. \end{cases} \tag{11.128}$$

Let $X \sim F_X$ be another random variable with CDF F_X. Then the conditional expectation of Y given that $X = x$ is, by definition, the expectation of Y with respect to the conditional distribution $F_{Y|X}(y|x) = \mathbb{P}(Y \leq y|X = x)$. If, for example, this distribution is continuous, then

$$\mathbb{E}[Y|X = x] = \int y f_{Y|X}(y|x)dy, \tag{11.129}$$

where $f_{Y|X}(y|x)$ is the conditional PDF of Y given $X = x$. As a mathematical object, the conditional expectation $\mathbb{E}[Y|X = x]$ is a *function* of x.

What is $\mathbb{E}[Y|X]$, as a mathematical object, when the value of X is random, not fixed to x? It is a *random variable*. By definition, $\mathbb{E}[Y|X]$ is a random variable whose value is $\mathbb{E}[Y|X = x]$ when $X = x$.

A bit more rigorously, if Ω is a sample space, and $X : \Omega \to \mathbb{R}$ is a random variable[79], then $\mathbb{E}[Y|X] : \Omega \to \mathbb{R}$ is a random variable defined as follows:

$$\mathbb{E}[Y|X](\omega) = \mathbb{E}[Y|X = X(\omega)]. \tag{11.130}$$

To find an expression for $\mathbb{E}[Y|X]$, we can first compute $\mathbb{E}[Y|X = x]$ and then "unfreeze" X by replacing x with X. This is illustrate in Example 11.1 below.

[79] Recall that a *random variable* is a deterministic function $X : \Omega \to \mathbb{R}$ that assigns a number $x = X(\omega) \in \mathbb{R}$ to a random outcome $\omega \in \Omega$.

Remark 11.14. The conditional expectation $\mathbb{E}[Y|X]$ is one of the central concepts in probability theory. It is useful not only for deriving the regression function, which is how we meet this concept, but also for computing unconditional expectations, studying compound random variables, defining martingales, to name but a few.

Since $\mathbb{E}[Y|X]$ is a random variable, we can compute its expectation. It turns out that its expected value is simply the original unconditional expected value of Y:

$$\mathbb{E}\left[\mathbb{E}[Y|X]\right] = \mathbb{E}[Y]. \tag{11.131}$$

This property is called the *law of total expectation* (LTE). The inner expected value $\mathbb{E}$ on the left-hand side is with respect to randomness in Y (for X is given) and the outer expected value is with respect to randomness in X. This equation is typically used from right to left for computing a hard-to-compute expectation $\mathbb{E}[Y]$ by introducing a new random variable X and then calculating the total expectation $\mathbb{E}\left[\mathbb{E}[Y|X]\right]$. The random variable X is chosen such that $\mathbb{E}\left[\mathbb{E}[Y|X]\right]$ is relatively easy to compute.

If X is discrete or continuous, then the form of the law of total expectation becomes more concrete and easier to digest. By writing the outer expectation (the one with respect to X) explicitly, we have:

$$\mathbb{E}[Y] = \begin{cases} \int \mathbb{E}[Y|X = x] f_X(x)dx, & \text{if } X \text{ is continuous with PDF } f_X, \\ \sum_x \mathbb{E}[Y|X = x] p_X(x), & \text{if } X \text{ is discrete with PMF } p_X. \end{cases} \tag{11.132}$$

To justify the LTE (11.131), let's prove it for a special case, where random variables X and Y are jointly continuous. In this case,

$$\begin{aligned}\mathbb{E}\left[\mathbb{E}[Y\,|\,X]\right] &= \int \mathbb{E}[Y\,|\,X=x]f_X(x)dx = \int\left(\int y f_{Y|X}(y|x)dy\right)f_X(x)dx \\ &= \int\int y\underbrace{f_{Y|X}(y|x)f_X(x)}_{\text{joint PDF } f_{X,Y}(x,y)}dxdy = \int\int y f_{X,Y}(x,y)dxdy \\ &= \int y\Bigg(\underbrace{\int f_{X,Y}(x,y)dx}_{\text{marginal PDF } f_Y(y)}\Bigg)dy = \int y f_Y(y)dy = \mathbb{E}[Y].\end{aligned} \tag{11.133}$$

Let's consider now a very simple example that illustrates a typical application of the law of total expectation.

Example 11.1. Let $X \sim \mathscr{U}[0,1]$ be a random variable uniformly distributed on the unit interval $[0,1]$. If $X = x$, let Y be a random variable uniformly distributed on the interval $[x,1]$, that is $(Y|X=x) \sim \mathscr{U}[x,1]$. What is the expected value of Y?

The most straightforward way to answer this question is the following. Since we know the PDF $f_X(x)$ of X and the conditional PDF $f_{Y,X}(y|x)$ of Y given $X = x$, we can find their joint PDF $f_{X,Y}(x,y) = f_X(x)f_{Y,X}(y|x)$. By integrating the joint PDF with respect to x, we can find the marginal PDF $f_Y(y)$ of Y[80]. Having found $f_Y(y)$, we can compute the expected value of Y by definition (11.128).

[80] By doing the outlined calculation, it can be shown that the PDF of Y is $f_Y(y) = -\log(1-y)$.

The law of total expectation provides a faster and computationally simpler way to the answer. So,

$$\mathbb{E}[Y] = \mathbb{E}\left[\mathbb{E}[Y\,|\,X]\right]. \tag{11.134}$$

To find the expression for $\mathbb{E}[Y\,|\,X]$, let's first compute $\mathbb{E}[Y\,|\,X=x]$. Since, given $X = x$, random variable Y is uniformly distributed on $[x,1]$, its conditional expectation is the middle point of the interval,

$$\mathbb{E}[Y\,|\,X=x] = \frac{1+x}{2}. \tag{11.135}$$

Therefore,

$$\mathbb{E}[Y\,|\,X] = \frac{1+X}{2} \quad \text{and} \quad \mathbb{E}[Y] = \mathbb{E}\left[\frac{1+X}{2}\right] = \frac{1+\mathbb{E}[X]}{2} = \frac{3}{4}. \tag{11.136}$$

Notice that in this approach, we don't need to find the full distribution $f_Y(y)$ of Y in order to find its expectation, which is just a single-number summary of $f_Y(y)$.

Finally, by analogy with $\mathbb{E}[Y\,|\,X]$, we can define the conditional variance $\mathbb{V}[Y\,|\,X]$. By definition, $\mathbb{V}[Y\,|\,X]$ is a random variable whose value is $\mathbb{V}[Y\,|\,X=x]$ when $X = x$. Here, $\mathbb{V}[Y\,|\,X=x]$ is the conditional variance of Y given that $X = x$. In other words, it is the variance of Y with respect to the conditional distribution of Y given $X = x$.

An analog of the law of total expectation for the variance is the *law of total variance* (LTV):

$$\mathbb{V}[Y] = \mathbb{E}[\mathbb{V}[Y \mid X]] + \mathbb{V}[\mathbb{E}[Y \mid X]]. \tag{11.137}$$

Here, the inner variance $\mathbb{V}$ and expectation $\mathbb{E}$ are with respect to randomness in Y and the outer expectation $\mathbb{E}$ and variance $\mathbb{V}$ are with respect to randomness in X.

Although the form of LTV is more complicated compared to that of LTE (11.131), its proof is rather simple and is based on the LTE. Using the variance property $\mathbb{V}[Z] = \mathbb{E}[Z^2] - \mathbb{E}[Z]^2$, the two terms on the right-hand side of (11.137) are

$$\begin{aligned} \mathbb{E}[\mathbb{V}[Y \mid X]] &= \mathbb{E}\left[\mathbb{E}[Y^2 \mid X] - \mathbb{E}[Y \mid X]^2\right] = \underbrace{\mathbb{E}\left[\mathbb{E}[Y^2 \mid X]\right]}_{\mathbb{E}[Y^2] \text{ (LTE)}} - \mathbb{E}\left[\mathbb{E}[Y \mid X]^2\right] \\ &= \mathbb{E}[Y^2] - \mathbb{E}\left[\mathbb{E}[Y \mid X]^2\right], \qquad (11.138) \\ \mathbb{V}[\mathbb{E}[Y \mid X]] &= \mathbb{E}\left[\mathbb{E}[Y \mid X]^2\right] - \big(\underbrace{\mathbb{E}[\mathbb{E}[Y \mid X]]}_{\mathbb{E}[Y] \text{ (LTE)}}\big)^2 = \mathbb{E}\left[\mathbb{E}[Y \mid X]^2\right] - \mathbb{E}[Y]^2 \end{aligned}$$

After adding these two terms, the two opposite-sign copies of the "ugly" term $\mathbb{E}\left[\mathbb{E}[Y \mid X]^2\right]$ get canceled, and $\mathbb{E}[Y^2] - \mathbb{E}[Y]^2$ results in the variance $\mathbb{V}[Y]$, as promised by the law of total variance (11.137).

Problems

11.1. Regression Through the Origin

In some applications the underlying theory (or simply the common sense) of the system under study suggests that if the input is zero, $X = 0$, then the corresponding response must also be zero, $Y = 0$. Given the bivariate data $(X_1, Y_1), \ldots, (X_n, Y_n)$, consider the following linear regression model of Y on X:

$$Y_i = \beta X_i + E_i, \qquad i = 1, \ldots, n, \tag{11.139}$$

where prediction errors $E_1, \ldots, E_n$ are assumed to be independent, have zero mean $\mathbb{E}[E_i] = 0$, and constant variance $\mathbb{V}[E_i] = \sigma^2$. This regression model is called the *regression through the origin*.

(a) Find the OLS estimate $\hat{\beta}$ of the regression parameter β.

(b) Is $\hat{\beta}$ an unbiased estimate of β?

(c) Find the standard error of $\hat{\beta}$.

(d) Find a condition on inputs that guarantees that $\hat{\beta}$ is a consistent estimate of β.

11.2. Controlling Inputs

Suppose we are studying the behavior of a certain system by perturbing it with inputs X_i and recording the corresponding outputs Y_i, $i = 1, \ldots, n$. We have full control of inputs and can choose them to be anywhere between -1 and 1. After collecting the data $(X_1, Y_1), \ldots, (X_n, Y_n)$, we plan to use the SLR model for studying the dependence between X and Y. How should $n = 100$ inputs $X_i \in [-1,1]$ be chosen in order to maximize the accuracy of the OLS estimate $\hat{\beta}_1$ of the slope of the fitted line?

11.3. Testing the Relationship Between Regression Coefficients
Consider the normal SLR model for the bivariate data $(X_1,Y_1),\dots,(X_n,Y_n)$:

$$Y_i = \beta_0 + \beta_1 X_i + E_i, \qquad E_1,\dots,E_n \overset{\text{iid}}{\sim} \mathcal{N}(0,\sigma^2). \tag{11.140}$$

Suppose that the theory (or previous experiments) of the system that generated the data suggests that $\beta_1 = \psi\beta_0$, where ψ is some constant, and we want to test this statement using the data. Construct a test of size α for testing

$$H_0 : \beta_1 = \psi\beta_0 \quad \text{vs} \quad H_1 : \beta_1 \neq \psi\beta_0. \tag{11.141}$$

11.4. Effect of Ignoring the Quadratic Term
Let $X_1,\dots,X_n$ be some (deterministic) inputs, and suppose that the corresponding responses $Y_1,\dots,Y_n$ are generated by the following model:

$$Y_i = a + bX_i + cX_i^2 + E_i, \qquad i = 1,\dots,n, \tag{11.142}$$

where $a, b,$ and c are some constants, and $E_1,\dots,E_n$ are independent random variables with zero mean. Let's ignore the quadratic term cX_i^2 (its presence could be hard to detect if c is relatively small), consider the SLR model, and fit the OLS regression line $y = \hat{r}(x) = \hat{\beta}_0 + \hat{\beta}_1 x$ to the bivariate data $(X_1,Y_1),\dots,(X_n,Y_n)$. Let $\hat{E}_i = Y_i - \hat{Y}_i$ be the ith residual, describing the discrepancy between the true response Y_i and the fitted value $\hat{Y}_i = \hat{r}(X_i)$.

(a) Find the expected value $\mathbb{E}[\hat{E}_i]$ of the ith residual
(b) Find the expected value of the average of all n residuals.

11.5. The SLR Model in Action
In this problem, to get a better empirical feel for the regression analysis, you will perform some of its elements that we discussed in Chapter 11. First, generate a set of inputs $X_1,\dots,X_n$ as a collection of $n = 100$ IID realizations of a uniform random variable $X \sim \mathcal{U}[0,10]$. Next, compute the corresponding responses $Y_1,\dots,Y_n$ using the following relationship[81]: $Y_i = \beta_0 + \beta_1 X_i + E_i$, where the coefficients $\beta_0 = 1$ and $\beta_1 = 0.1$, and $E_1,\dots,E_n \overset{\text{iid}}{\sim} \mathcal{N}(0,1)$. Finally, suppose that the relationship between X_i and Y_i is unknown and consider the normal SLR model of Y_i on X_i:

[81] The goal is to uncover this relationship from the data using the SLR model.

$$Y_i = \beta_0 + \beta_1 X_i + E_i, \qquad E_1,\dots,E_n \overset{\text{iid}}{\sim} \mathcal{N}(0,\sigma^2). \tag{11.143}$$

(a) Compute the OLS estimates $\hat{\beta}_0$ and $\hat{\beta}_1$ of the regression coefficients and plot the fitted line $y = \hat{\beta}_0 + \hat{\beta}_1 x$ together with the scatterplot of the data.
(b) It is hard to decide based solely on the scatterplot whether there is indeed a linear relationship between X and Y. Test the significance of the linear regression $H_0 : \beta_1 = 0$ versus $H_1 : \beta_1 \neq 0$ by using both the Wald and t-test. Compute the corresponding p-values of the data.

(c) Construct 95% confidence intervals for regression coefficients β_0 and β_1 based on both normal and t-distributions.

(d) Construct a 95% prediction interval for the response Y to input $X = 5$ and plot it on the figure from part (a).

References

1. Neyman, J.: Statistics – servant of all sciences. *Science* 122(3126), 401-406 (1955). https://doi.org/10.1126/science.122.3166.401
2. Wasserman, L.A.: All of *Nonparametric Statistics*. Springer Texts in Statistics (2006). https://www.springer.com/gp/book/9780387251455
3. Györfi, L., Kohler, M., Krzyżak, A., Walk, H.: *A Distribution-Free Theory of Nonparametric Regression*. Springer Texts in Statistics (2002). https://doi.org/10.1007/b97848
4. Pearson, K., Lee, A.: On the laws of inheritance in man. *Biometrika* 2, 357-463 (1903). https://doi.org/10.2307/2331507
5. Biography of Alice Elizabeth Lee https://mathshistory.st-andrews.ac.uk/Biographies/Lee/.
6. Galton, F.: Regression towards mediocrity in hereditary stature. *The Journal of the Anthropological Institute of Great Britain and Ireland* 15, 246-263 (1886). https://doi.org/doi:10.2307/2841583
7. Stigler, S.M.: *The History of Statistics: The Measurement of Uncertainty Before 1900.* Belknap Press of Harvard University Press (1986).
8. Anscombe, F.: Graphs in statistical analysis. *The American Statistician* 27(1), 17-21 (1973). https://doi.org/10.2307/2682899
9. The Beatles, *Eleanor Rigby*. https://youtu.be/HuS5NuXRb5Y?si=pQT945_Zwr526ZaU
10. Trench, W.F.: The method of Lagrange multipliers. A supplement to the author's book *Introduction to Real Analysis.* (2012). http://digitalcommons.trinity.edu/mono/7/
11. Montgomery, D.C., Peck, E.A., Vining, G.G.: *Introduction to Linear Regression Analysis, 6th Edition.* United Kingdom, Wiley (2021).
12. Allen, W.: *Annie Hall* (1977). https://youtu.be/5U1-OmAICpU?si=FcKch3hScFoxBXs9
13. Hawking, S.: *The Illustrated A Brief History of Time.* Bantam Books (1996).
14. Hubble, E.: A relation between distance and radial velocity among extra-galactic nebulae. *Proceedings of the National Academy of Sciences* 15(3), 168-173 (1929). https://doi.org/10.1073/pnas.15.3.168
15. NASA's Webb, Hubble Telescopes Affirm Universe's Expansion Rate, Puzzle Persists. https://science.nasa.gov/missions/webb/nasas-webb-hubble-telescopes-affirm-universes-expansion-rate-puzzle-persists/
16. SIMBAD Database: https://simbad.u-strasbg.fr/simbad/sim-tap/
17. Kuo, Y-H.: Extrapolation of correlation between 2 variables in 4 general medical journals. *The Journal of the American Medical Association* 287(21), 2815-2817 (2002). https://doi.org/10.1001/jama.287.21.2815
18. Box, G.E.P: Science and Statistics. *Journal of the American Statistical Association* 71(356), 791-799 (1976). https://doi.org/10.2307/2286841
19. Galarraga, V., Boffetta, P.: Coffee drinking and risk of lung cancer — a meta-analysis. *Cancer Epidemiology, Biomarkers and Prevention* 25(6), 951–957 (2016). https://doi.org/10.1158/1055-9965.EPI-15-0727

12
Multiple Linear Regression

The development of modern science is marked by a pronounced tendency toward indeterminism. A somewhat brutal description of this tendency may be states as follows. In relation to some phenomena, instead of trying to establish a (deterministic) functional relationship between a variable y, and some other variables $x_1, x_2, \ldots, x_n$, we try to build a (stochastic or probabilistic) model of these phenomena, predicting frequencies with which, in specified conditions, the same variable y will assume all of its possible values.

– Jerzy Neyman, *Statistics – Servant of All Sciences* [1]

The behavior of complex systems often depends on many different factors. For example, the level of blood cholesterol depends on age, diet, physical exercise, and various genetic factors. A stock price tomorrow may depend on its price over the last several days, the company's performance measures, and overall economic and financial factors. The number of points a chess player scores in a tournament depends on the ratings of the player's opponents, the player's physical and psychological conditions throughout the tournament, and who knows what else.

For most real systems, it is impossible to account for all factors that affect its behavior[1]. That is why we try to identify the most important factors, or simply factors that are available for measurements, and build a stochastic model of the system with the hope that it will mimic the behavior of the real system adequately, at least to some extent. In this final chapter, we will discuss the *multiple linear regression model*, which is an extension of the simple linear regression model for the case, where system's response Y depends not just on a single input X, but on a vector $X = (X^{(1)}, \ldots, X^{(p)})$ of p different inputs. Multiple linear regression is widely used in different areas of science and engineering, including biology and medicine, economics and finance, mechanical and electrical engineering, life and social sciences. Moreover, multiple linear regression serves as a foundation for many advanced methods and algorithms that have been developed in machine learning out of the attempts to generalize or improve on multiple linear regression. Let's start by formulating the model and introducing the corresponding notation and terminology.

[1] "There are a million things in this universe you can have and a million things you can't have. It's no fun facing that, but that's the way things are." – Captain Kirk, *Star Trek*.

Supplementary Information The online version contains supplementary material available at https://doi.org/10.1007/978-3-032-03848-7_12.

K. M. Zuev, *Fundamentals of Statistical Inference*,
International Series in Operations Research & Management Science 374,
https://doi.org/10.1007/978-3-032-03848-7_12

12.1 Multiple Linear Regression Model

Suppose that the response Y of a system we are interested in depends on p different inputs (also called predictors[2], features, and independent variables): $X^{(1)}, \dots, X^{(p)}$. Predictors can come from different sources. For example, they can be the original quantities that we measure during the observation and experimentation with the system. We can also include in the list of predictors some transformations of the original predictors, such as $\log X^{(j)}$ and $\exp X^{(j)}$, or interactions between them, such as $X^{(j)}X^{(k)}$. If one of the original predictors, $\tilde{X}$, is categorical[3] with k levels $c_1, \dots, c_k$, then we can represent it with k numerical predictors: $X^{(j)} = 1$ if $\tilde{X} = c_j$, and $X^{(j)} = 0$ otherwise, for $j = 1, \dots, k$.

[2] That is why we denote the total number of predictors by p. In view of this notation, in this chapter, we will use "predictor" more often than "input."

[3] This means that the possible values of $\tilde{X}$ are not numbers, as usual, but categories. For example, $\tilde{X} \in \{\text{hobbit}, \text{dwarf}, \text{elf}\}$ is categorical with $k = 3$ levels.

Remark 12.1. This representation of a categorical predictor $\tilde{X}$ is better than perhaps a more straightforward representation by a single numerical predictor $X \in \{1, \dots, k\}$, where $X = j$ if $\tilde{X} = c_j$, since the latter representation introduces an artificial linear order $c_1 < \dots < c_k$ into the unordered set of categories $\{c_1, \dots, c_k\}$.

Let X be a vector of all predictors $X^{(1)}, \dots, X^{(p)}$. In this chapter, for consistency, we will adopt the convention that *all vectors are column vectors*. So,

$$X = \begin{bmatrix} X^{(1)} \\ \vdots \\ X^{(p)} \end{bmatrix} = [X^{(1)}, \dots, X^{(p)}]^T \in \mathbb{R}^p, \tag{12.1}$$

where ^{T} denotes the operation of transposition (interchanging rows and columns).

Remark 12.2. Some may wonder why I use $X^{(1)}, \dots, X^{(p)}$ to denote predictors instead of simpler $X_1, \dots, X_p$. The reason is because I want to avoid the clash of notation with the previous chapter on the SLR model (where $X_1, \dots, X_n$ denoted the observed values of one single predictor). All formulas derived for multiple regression will reduce to the corresponding formulas for simple regression when $p = 1$.

Let $X_1, \dots, X_n \in \mathbb{R}^p$ be the input data, a collection of input vectors obtained by experiment or observation of the system. We can think of $X_1, \dots, X_n$ as n measurements or observations of the input vector X,

$$X_1 = \begin{bmatrix} X_1^{(1)} \\ \vdots \\ X_1^{(p)} \end{bmatrix}, \quad \dots, \quad X_n = \begin{bmatrix} X_n^{(1)} \\ \vdots \\ X_n^{(p)} \end{bmatrix}. \tag{12.2}$$

Let $Y_1, \dots, Y_n$ be the corresponding system responses. The data $(X_1, Y_1), \dots, (X_n, Y_n)$, for which a stochastic model is formulated and which later will be used for estimating the model parameters, is commonly referred to as the *training data*.

Given the training data $(X_1, Y_1), \dots, (X_n, Y_n)$, the *multiple linear regression model* of Y_i on X_i is

$$Y_i = \beta_0 + \beta_1 X_i^{(1)} + \ldots + \beta_p X_i^{(p)} + E_i, \quad i = 1, \ldots, n, \tag{12.3}$$

where $\beta_0, \ldots, \beta_p$ are model parameters, called the ***regression coefficients***, and $E_1, \ldots, E_n$ are the *statistical* or *prediction errors*. Note that if $p = 1$, then the multiple linear regression (MLR) model (12.3) reduces to the SLR model (11.21).

It is mathematically convenient to rewrite the MLR model in a more compact vector form. Let $\beta \in \mathbb{R}^{p+1}$ be the vector of regression coefficients, and, to account for the intercept β_0, let's add 1 as the first component of input vectors[4]

[4] We can think of 1 as the zeroth artificial predictor.

$$\beta = \begin{bmatrix} \beta_0 \\ \beta_1 \\ \vdots \\ \beta_p \end{bmatrix}, \quad X_1 = \begin{bmatrix} 1 \\ X_1^{(1)} \\ \vdots \\ X_1^{(p)} \end{bmatrix}, \quad \ldots, \quad X_n = \begin{bmatrix} 1 \\ X_n^{(1)} \\ \vdots \\ X_n^{(p)} \end{bmatrix}. \tag{12.4}$$

Using this notation, (12.3) can be equivalently written as follows[5]:

$$Y_i = X_i^T \beta + E_i, \quad i = 1, \ldots, n. \tag{12.5}$$

[5] As reminder, if $u, v \in \mathbb{R}^n$, then $u^T v = \sum_{i=1}^n u_i v_i$ is the *dot product* between them.

As in the case of the SLR model, we will consider two types of model assumptions about prediction errors $E_1, \ldots, E_n$: (weaker) nonparametric assumptions

1. $E_1, \ldots, E_n$ are independent,
2. $\mathbb{E}[E_i] = 0$, (12.6)
3. $\mathbb{V}[E_i] = \sigma^2$.

and (stronger) parametric assumption

$$E_1, \ldots, E_n \overset{\text{iid}}{\sim} \mathcal{N}(0, \sigma^2). \tag{12.7}$$

Our first task in the study of the MLR model is to learn how to estimate the regression coefficients $\beta_0, \ldots, \beta_p$ by extending the method of ordinary least squares introduced in Section 11.5 for the case of multiple predictors.

12.2 Ordinary Least Squares

As in the case of the SLR model, the OLS estimate of the vector β of regression coefficients is obtained by minimizing the *residual sum of squares* (RSS). Given β, the RSS is defined as the sum of squared differences between the true responses Y_i and the predicted responses $X_i^T \beta$ based on inputs X_i,

$$\text{RSS}(\beta) = \sum_{i=1}^{n} (Y_i - X_i^T \beta)^2. \tag{12.8}$$

The residual sum of squares $\mathrm{RSS}(\beta)$ can be viewed as a measure of goodness of fit of the hyperplane $y = r(x) = x^T\beta$ to the cloud of points $(X_1,Y_1),\ldots,(X_n,Y_n)$ in a high-dimensional space[6].

Let $\hat{\beta}$ denote the OLS estimate of β, that is, the value of β that minimizes the residual sum of squares[7],

$$\hat{\beta} = \arg\min_{\beta\in\mathbb{R}^{p+1}} \mathrm{RSS}(\beta). \tag{12.9}$$

It has an important linear-algebraic interpretation. To derive it, let's first rewrite the RSS in a matrix form. To this end, let's combine all training input vectors $X_1,\ldots,X_n$ into a single matrix:

$$\mathbb{X} = \begin{bmatrix} X_1^T \\ \vdots \\ X_n^T \end{bmatrix} = \underbrace{\begin{bmatrix} 1 & X_1^{(1)} & \ldots & X_1^{(p)} \\ \vdots & \vdots & \vdots & \vdots \\ 1 & X_n^{(1)} & \ldots & X_n^{(p)} \end{bmatrix}}_{n\times(p+1)\text{ matrix}} \tag{12.10}$$

The matrix $\mathbb{X}$ is called the *design matrix*[8]. Its (i,j) entry $\mathbb{X}_{ij} = X_i^{(j)}$ represents the value of the jth predictor in the ith observation, $i = 1,\ldots,n$ and $j = 0,\ldots,p$. The zeroth predictor ($j = 0$) is the artificial predictor, which is always equal to one.

The RSS can be written in terms of the design matrix $\mathbb{X}$ and the vector of responses $Y = (Y_1,\ldots,Y_n)^T$ as follows:

$$\mathrm{RSS}(\beta) = \sum_{i=1}^{n}(X_i^T\beta - Y_i)^2 = \|\mathbb{X}\beta - Y\|^2, \tag{12.11}$$

where $\|v\| = \sqrt{\sum_{i=1}^n v_i^2}$ is the Euclidean norm of vector $v\in\mathbb{R}^n$. Therefore[9],

$$\hat{\beta} = \arg\min_{\beta\in\mathbb{R}^{p+1}} \|\mathbb{X}\beta - Y\|. \tag{12.12}$$

This means that the OLS estimate $\hat{\beta}$ is a solution of the linear system of equations $\mathbb{X}\beta = Y$ in the *least squares sense*[10].

Remark 12.3. This interpretation of $\hat{\beta}$ allows a geometric visualization that shows that the optimization problem (12.9) always has a solution or, in other words, that the OLS estimate $\hat{\beta}$ of β always *exists*.

The design matrix $\mathbb{X}$ induces a linear map from $\mathbb{R}^{p+1}$ to $\mathbb{R}^n$, which maps vector $\beta\in\mathbb{R}^{p+1}$ to vector $\mathbb{X}\beta\in\mathbb{R}^n$,

$$\mathbb{X}:\mathbb{R}^{p+1}\to\mathbb{R}^n, \qquad \beta\mapsto\mathbb{X}\beta. \tag{12.13}$$

Let $\mathrm{Im}\,\mathbb{X}$ be the image of $\mathbb{X}$[11]. The image $\mathrm{Im}\,\mathbb{X}\subset\mathbb{R}^n$ is a subspace of $\mathbb{R}^n$, that is, a hyperplane that goes through zero[12]. For any $\beta\in\mathbb{R}^{p+1}$, by definition, $\mathbb{X}\beta\in\mathrm{Im}\,\mathbb{X}$. Therefore, solving optimization problem (12.12) is equivalent to finding a vector

[6] If $p = 1$, then (12.8) reduces to (11.30).

[7] Remark 1.11 reminds the notation used below.

[8] This name reflects the fact that we decide what quantities to use as predictors for the response. We design the MLR model by choosing $X^{(1)},\ldots,X^{(p)}$.

[9] Convince yourself that it does not really matter that we use the norm $\|\mathbb{X}\beta - Y\|$ below, not the squared norm.

[10] Recall that $\hat{x}$ is called a *least squares solution* of a linear system $Ax = b$ if $\hat{x}$ is a global minimizer of $\|Ax - b\|$, that is, if $\hat{x} = \arg\min_x \|Ax - b\|$.

[11] Recall that the image of an $n\times m$ matrix A is the set $\mathrm{Im}\,A = \{Ax\,|\,x\in\mathbb{R}^m\}$ of all vectors in $\mathbb{R}^n$ of the form Ax, where $x\in\mathbb{R}^m$.

[12] The fundamental theorem of linear algebra states that the dimension of this hyperplane is exactly the rank of the design matrix, $\dim\mathrm{Im}\,\mathbb{X} = \mathrm{rk}\,\mathbb{X}$.

$\mathbb{X}\hat{\beta}$ in the image $\operatorname{Im}\mathbb{X}$, which is the closet to the vector $Y \in \mathbb{R}^n$. Geometrically, it is obvious that such vector $\mathbb{X}\hat{\beta}$ exists and is unique: it is obtained by projecting Y orthogonally onto $\operatorname{Im}\mathbb{X}$,

$$\mathbb{X}\hat{\beta} = \operatorname{proj}_{\operatorname{Im}\mathbb{X}} Y. \tag{12.14}$$

This is schematically illustrated in Fig. 12.1. Note that although $\mathbb{X}\hat{\beta} \in \operatorname{Im}\mathbb{X}$ is unique, the OLS vector $\hat{\beta} \in \mathbb{R}^{p+1}$ may not be unique, since it may happen that more than one vector in $\mathbb{R}^{p+1}$ is mapped into vector $\mathbb{X}\hat{\beta}$. For example, if $\tilde{\beta} \in \operatorname{Ker}\mathbb{X}$ is in the kernel of $\mathbb{X}$[13], then $\mathbb{X}(\hat{\beta}+\tilde{\beta}) = \mathbb{X}\hat{\beta}$.

[13] Recall that the kernel of an $n \times m$ matrix A is the set $\operatorname{Ker}A = \{x \in \mathbb{R}^m \mid Ax = 0\}$ of all vectors in $\mathbb{R}^m$ that are solutions of the homogeneous system $Ax = 0$.

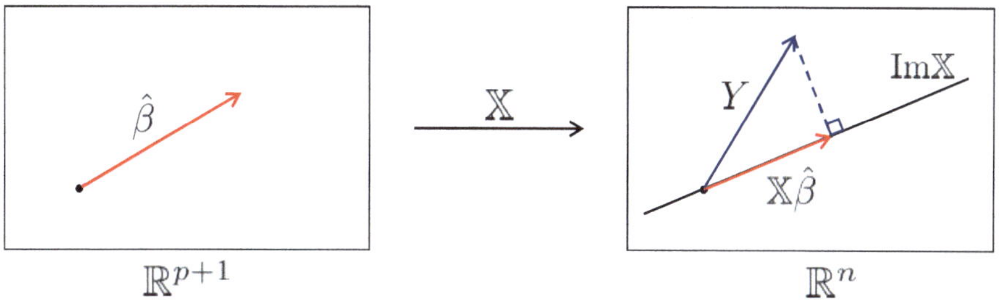

Fig. 12.1 Linear map $\mathbb{X} : \mathbb{R}^{p+1} \to \mathbb{R}^n$ induced by the design matrix. Vector $\hat{\beta}$ is the OLS estimate of β if and only if $\mathbb{X}\hat{\beta}$ is the closest to Y among all vectors in the image $\operatorname{Im}\mathbb{X}$.

Returning from geometry to analysis, to solve the optimization problem (12.9) and find $\hat{\beta}$, we first need to find the critical points of RSS by solving $\nabla_\beta \mathrm{RSS}(\beta) = 0$. The residual sum of squares is[14]

[14] Recall that if A and B are matrices (vectors are special cases of matrices), then $(AB)^T = B^T A^T$.

$$\begin{aligned}\mathrm{RSS}(\beta) &= \|\mathbb{X}\beta - Y\|^2 = (\mathbb{X}\beta - Y)^T(\mathbb{X}\beta - Y) = (\beta^T\mathbb{X}^T - Y^T)(\mathbb{X}\beta - Y) \\ &= \beta^T\mathbb{X}^T\mathbb{X}\beta - \beta^T\mathbb{X}^T Y - Y^T\mathbb{X}\beta + Y^T Y.\end{aligned} \tag{12.15}$$

The two middle terms, $\beta^T\mathbb{X}^T Y$ and $Y^T\mathbb{X}\beta$, are the same. Indeed, since $Y^T\mathbb{X}\beta \in \mathbb{R}$ is just a number, it does not change under transposition, and so[15]

[15] If A, B, and C are matrices, then, as it immediately follows from the above marginal note, $(ABC)^T = C^T B^T A^T$.

$$Y^T\mathbb{X}\beta = (Y^T\mathbb{X}\beta)^T = \beta^T\mathbb{X}^T Y. \tag{12.16}$$

Therefore, the RSS is

$$\mathrm{RSS}(\beta) = \beta^T\mathbb{X}^T\mathbb{X}\beta - 2\beta^T\mathbb{X}^T Y + Y^T Y. \tag{12.17}$$

Its gradient with respect to β is[16]

[16] Recall that the *gradient* of a function $f(x_1, \ldots, x_n)$ is the column vector of its partial derivatives, $\nabla_x f(x) = (\frac{\partial f}{\partial x_1}, \ldots, \frac{\partial f}{\partial x_n})^T$.

$$\nabla_\beta \mathrm{RSS}(\beta) = \nabla_\beta \left(\beta^T\mathbb{X}^T\mathbb{X}\beta\right) - 2\nabla_\beta \left(\beta^T\mathbb{X}^T Y\right). \tag{12.18}$$

To compute the last term, where the function under the gradient is linear in β, it is useful to establish the following simple property: if $a \in \mathbb{R}^{p+1}$ is a column vector, then the gradient of the dot product $\beta^T a$ is

$$\nabla_\beta \left(\beta^T a\right) = a. \tag{12.19}$$

Since this equation is an equality of two vectors, to prove it, it is sufficient to check that the components of the two vectors are the same. Indeed, the kth component of the gradient coincides with the kth component a_k of a:

$$\left(\nabla_\beta(\beta^T a)\right)_k = \frac{\partial}{\partial \beta_k}\left(\beta^T a\right) = \frac{\partial}{\partial \beta_k}\left(\sum_{j=0}^{p} \beta_j a_j\right) = a_k. \tag{12.20}$$

Now, applying this property, the gradient in last term in (12.18) is

$$\nabla_\beta \big(\beta^T \underbrace{\mathbb{X}^T Y}_{a}\big) = \mathbb{X}^T Y. \tag{12.21}$$

To compute the first term in (12.18), where the function under the gradient is quadratic in β, it is useful to derive the following property: if A is a $(p+1)\times(p+1)$ symmetric matrix[17], then the gradient of the quadratic form $\beta^T A \beta$ is

[17] Recall that a matrix A is *symmetric* if $A^T = A$ or, equivalently, $A_{ij} = A_{ji}$.

$$\nabla_\beta \left(\beta^T A \beta\right) = 2A\beta. \tag{12.22}$$

As above, let's prove this equation by checking that it holds at the level of vector components, namely, that $(\nabla_\beta(\beta^T A\beta))_k = (2A\beta)_k$ for all $k = 0,\ldots,p$. Indeed:

$$\begin{aligned}
\left(\nabla_\beta\left(\beta^T A\beta\right)\right)_k &= \frac{\partial}{\partial \beta_k}\left(\beta^T A \beta\right) = \frac{\partial}{\partial \beta_k}\left(\sum_{i,j=0}^{p} A_{ij}\beta_i\beta_j\right) \\
&= \sum_{i,j=0}^{p} A_{ij}\underbrace{\frac{\partial \beta_i}{\partial \beta_k}}_{\delta_{ik}}\beta_j + \sum_{i,j=0}^{p} A_{ij}\beta_i\underbrace{\frac{\partial \beta_j}{\partial \beta_k}}_{\delta_{jk}} \\
&= \sum_{i,j=0}^{p} A_{ij}\delta_{ik}\beta_j + \sum_{i,j=0}^{p} A_{ij}\beta_i\delta_{jk}{}^{18} \\
&= \sum_{j=0}^{p} A_{kj}\beta_j + \sum_{i=0}^{p} \underbrace{A_{ik}}_{A_{ki}}\beta_i = 2\sum_{j=0}^{p} A_{kj}\beta_j = (2A\beta)_k.
\end{aligned} \tag{12.23}$$

[18] Symbol δ_{ij} denotes the *Kronecker delta*: $\delta_{ij} = 1$ if $i = j$, and $\delta_{ij} = 0$ if $i \neq j$.

Since $\mathbb{X}^T\mathbb{X}$ is symmetric, this property allows to compute the first term in (12.18):

$$\nabla_\beta \big(\beta^T \underbrace{\mathbb{X}^T\mathbb{X}}_{A}\beta\big) = 2\mathbb{X}^T\mathbb{X}\beta. \tag{12.24}$$

Finally, plugging (12.21) and (12.24) back into (12.18), we have

$$\nabla_\beta \mathrm{RSS}(\beta) = 2\mathbb{X}^T\mathbb{X}\beta - 2\mathbb{X}^T Y. \tag{12.25}$$

Setting the gradient of the RSS to zero, we obtain that the OLS estimate $\hat{\beta}$ of the vector of regression coefficients is a solution of the following vector equation:

$$\mathbb{X}^T\mathbb{X}\beta = \mathbb{X}^T Y, \tag{12.26}$$

which is called the *normal equation*[19]. The normal equation is a linear system of $(p+1)$ equations on $\beta_0,\ldots,\beta_p$. In Remark 12.3, we used the geometric intuition to argue that the OLS estimate always exists, which implies that the normal equation has a solution. In Appendix 12.12, we prove this fact more rigorously using the fundamental theorem of linear algebra.

How to solve the normal equation (12.26)? Does it have a unique solution or infinitely many solutions[20]? The answer depends on the properties of matrix $\mathbb{X}^T\mathbb{X}$.

Remark 12.4. The $(p+1)\times(p+1)$ symmetric matrix $\mathbb{X}^T\mathbb{X}$ is the *Gram matrix*[21] associated with the columns $X^{(0)} = \mathbb{1}, X^{(1)},\ldots,X^{(p)}$ of the design matrix $\mathbb{X}$,

$$X^{(0)} = \mathbb{1} = \begin{bmatrix} 1 \\ \vdots \\ 1 \end{bmatrix}, \quad X^{(1)} = \begin{bmatrix} X_1^{(1)} \\ \vdots \\ X_n^{(1)} \end{bmatrix}, \quad \ldots, \quad X^{(p)} = \begin{bmatrix} X_1^{(p)} \\ \vdots \\ X_n^{(p)} \end{bmatrix}. \tag{12.27}$$

The jth column is the vector of values of the jth predictor across all n observations. For any choice of predictors, the Gram matrix $\mathbb{X}^T\mathbb{X}$ is always ***positive semi-definite*** (PSD)[22]. Indeed, for any $a \in \mathbb{R}^{p+1}$, $a^T\mathbb{X}^T\mathbb{X}a = (\mathbb{X}a)^T\mathbb{X}a = \|\mathbb{X}a\|^2 \geq 0$. It is strictly ***positive definite*** (PD)[23] if and only if it is nonsingular (invertible), which happens if and only if one of the following equivalent conditions is satisfied: a) columns $X^{(0)}, X^{(1)},\ldots,X^{(p)}$ are linearly independent, b) $\operatorname{Ker}\mathbb{X} = \{0\}$, c) $\operatorname{rk}\mathbb{X} = p+1$.

Generally speaking, the Gram matrix $\mathbb{X}^T\mathbb{X}$ can be either singular or nonsingular. Let's consider these two cases separately.

If $\mathbb{X}^T\mathbb{X}$ is nonsingular, then the normal equation has a unique solution:

$$\hat{\beta} = (\mathbb{X}^T\mathbb{X})^{-1}\mathbb{X}^T Y. \tag{12.28}$$

To show that the critical point $\hat{\beta}$ of the RSS is its global minimizer and, therefore, the OLS estimate of β, we need to check that the Hessian[24] of RSS is positive definite[25]. The first-order partial derivative of RSS with respect to β_k is simply the kth component of the gradient (12.25):

$$\begin{aligned} \frac{\partial \text{RSS}}{\partial \beta_k} &= \left(\nabla_\beta \text{RSS}(\beta)\right)_k = \left(2\mathbb{X}^T\mathbb{X}\beta - 2\mathbb{X}^T Y\right)_k \\ &= 2\sum_{j=0}^{p} \left(\mathbb{X}^T\mathbb{X}\right)_{kj}\beta_j - \left(2\mathbb{X}^T Y\right)_k. \end{aligned} \tag{12.29}$$

The (k,s) entry of the Hessian of the RSS is, thus,

[19] It is called "normal" because if $\hat{\beta}$ is a solution, then $Y - \mathbb{X}\hat{\beta}$ is normal (orthogonal) to $\operatorname{Im}\mathbb{X}$, see Fig. 12.1.

[20] Recall that a linear system can have no solutions, exactly one solution, or infinitely many solutions. Since the normal equation has a solution, only the last two cases are possible.

[21] The Gram matrix associated with vectors $v_1,\ldots,v_k \in \mathbb{R}^n$ is the $k\times k$ matrix G, whose elements are given by the dot product $G_{ij} = v_i^T v_j$.

[22] An $n\times n$ symmetric matrix A is PSD if $a^T A a \geq 0$ for any $a \in \mathbb{R}^n$.

[23] An $n\times n$ symmetric matrix A is PD if $a^T A a > 0$ for any $a \in \mathbb{R}^n, a \neq 0$.

[24] Recall that the *Hessian* of a function $f(x_1,\ldots,x_n)$ is the matrix of its second-order partial derivatives, $H(f) = (H_{ij}) = \left(\frac{\partial^2 f}{\partial x_i \partial x_j}\right)$.

[25] As a matter of fact, we don't need to check this. Since we know that the OLS estimate exists and is a solution of the normal equation, which has a unique solution, that solution must be the global minimizer. But we will check this anyway.

$$H(\text{RSS})_{ks} = \frac{\partial^2 \text{RSS}}{\partial \beta_s \partial \beta_k} = \frac{\partial}{\partial \beta_s}\left(2\sum_{j=0}^{p} \left(\mathbb{X}^T\mathbb{X}\right)_{kj}\beta_j\right) = 2\sum_{j=0}^{p} \left(\mathbb{X}^T\mathbb{X}\right)_{kj}\underbrace{\frac{\partial \beta_j}{\partial \beta_s}}_{\delta_{js}} \tag{12.30}$$
$$= 2\sum_{j=0}^{p} \left(\mathbb{X}^T\mathbb{X}\right)_{kj}\delta_{js} = 2\left(\mathbb{X}^T\mathbb{X}\right)_{ks}.$$

So, the full Hessian is, therefore, twice the Gram matrix $\mathbb{X}^T\mathbb{X}$,

$$H(\text{RSS}) = 2\mathbb{X}^T\mathbb{X}. \tag{12.31}$$

Since $\mathbb{X}^T\mathbb{X}$ is positive definite[26], so is the Hessian. Therefore, $\hat{\beta}$ in (12.28) is indeed the global minimizer of the RSS and, thus, the OLS estimate of the vector β of regression coefficients.

[26] See Remark 12.4.

Remark 12.5. In Problem 12.1, you will show that if $p = 1$ and $\hat{\beta} = (\hat{\beta}_0, \hat{\beta}_1)^T$, then, as expected, the OLS estimates $\hat{\beta}_0$ and $\hat{\beta}_1$ given by (12.28) reduce to the OLS estimate (11.34) derived for the case of simple linear regression.

Suppose now that the Gram matrix $\mathbb{X}^T\mathbb{X}$ is singular (noninvertible). This happens, for example, if the design matrix $\mathbb{X}$ is "flat," that is, if the number of its rows is less than the number of its columns, $n < p+1$. Indeed, if $n < p+1$, then the columns $X^{(0)}, \ldots, X^{(p)} \in \mathbb{R}^n$ of $\mathbb{X}$ are linearly dependent, and the Gram matrix $\mathbb{X}^T\mathbb{X}$ associated with these columns is automatically singular[27]. The flatness condition $n < p+1$ is equivalent to $p \geq n$, which means that the number of predictors is larger than or equal to the number of observations (sample size). Such data sets are sometimes called high-dimensional[28].

[27] See Remark 12.4.

[28] High-dimensional data sets often occur in biomedical applications (where we can measure many things on a small number of patients). Ridge regression and LASSO effectively deal with high-dimensional data. These methods are out of our scope, but we will talk about them briefly in Section 12.11.

If $\mathbb{X}^T\mathbb{X}$ is singular, then the normal equation (12.26) has infinitely many solutions, all delivering the smallest possible value of RSS. Besides, the formula (12.28) for the unique solution in the nonsingular case is no longer valid. From linear algebra [2], we know that the general solution (set of all solutions) of a linear system $Ax = b$ is given by $x = x^\star + \text{Ker}A$, where $x^\star$ is a particular solution and $\text{Ker}A = \{x \mid Ax = 0\}$ is the set of solutions of the corresponding homogeneous system $Ax = 0$. Therefore, the general solution of the normal equation $\mathbb{X}^T\mathbb{X}\beta = \mathbb{X}^T Y$ is

$$\hat{\beta} = \hat{\beta}^\star + \text{Ker}\,\mathbb{X}^T\mathbb{X}, \tag{12.32}$$

where $\hat{\beta}^\star$ is a particular solution. Since $\text{Ker}\,\mathbb{X}^T\mathbb{X} = \text{Ker}\,\mathbb{X}$ (this is proved in Appendix 12.12), the general solution of the normal equation simplifies to

$$\hat{\beta} = \hat{\beta}^\star + \text{Ker}\,\mathbb{X}. \tag{12.33}$$

A particular solution $\hat{\beta}^\star$ can be found using the concept of the pseudoinverse matrix, which is based on one of the most useful — both theoretically and numerically — matrix factorizations: the singular value decomposition [3].

Let A be any $m \times n$ matrix of rank $r > 0$. Then it can be decomposed as follows:

$$A = P\Sigma Q^T, \tag{12.34}$$

where matrix P is $m \times r$ with orthonormal columns, $P^T P = \mathbb{I}_r$; matrix Q is $n \times r$ with orthonormal columns, $Q^T Q = \mathbb{I}_r$; and matrix Σ is $r \times r$ diagonal, $\Sigma = \operatorname{diag}(\sigma_1, \ldots, \sigma_r)$, where $\sigma_1 \geq \ldots \geq \sigma_r > 0$ are called singular values of matrix A. The decomposition (12.34) is called the *singular value decomposition* (SVD) of A.

Remark 12.6. Since the first column of the design matrix $\mathbb{X}$ is the column of ones $X^{(0)} = \mathbb{1}$, the rank $\operatorname{rk}\mathbb{X} \geq 1$, and, therefore, the SVD of $\mathbb{X}$ always exists.

The SVD decomposition allows to generalize the notion of inverse matrix for the case when it does not exist [4]. Let A be an $m \times n$ matrix with positive rank[29], and let $A = P\Sigma Q^T$ be its SVD. Then the *pseudoinverse* (also known as the *Moor-Penrose inverse* [5, 6]) of A is defined as follows:

$$A^+ = Q\Sigma^{-1}P^T. \tag{12.35}$$

It can be shown[30] that the pseudoinverse A^+ has the following two properties:

- If A is nonsingular and its inverse A^{-1} exists, then $A^+ = A^{-1}$. This means that the pseudoinverse A^+ is indeed a true generalization of the inverse A^{-1}, since A^+ coincides with A^{-1} when the latter exists.
- If $\operatorname{Ker}A = \{0\}$, then $A^+ = (A^T A)^{-1}A^T$. This property tells us that if the kernel of A is trivial (consists only of the zero vector), then we don't need to compute the SVD of A in order to compute the pseudoinverse A^+ using its definition (12.35). Instead, we can compute A^+ using the matrix A directly: $A^+ = (A^T A)^{-1}A^T$.

[29] If $\operatorname{rk}A = 0$, then A is the zero matrix, and it is impossible to define its inverse in any meaningful way. Even zero number does not have the inverse.

[30] See Appendix 12.13

Let $\mathbb{X} = P\Sigma Q^T$ be the SVD of the design matrix, $\mathbb{X}^+ = Q\Sigma^{-1}P^T$ be its pseudoinverse, and let's consider the following vector

$$\hat{\beta}^\star = \mathbb{X}^+ Y. \tag{12.36}$$

It is straightforward to check that $\hat{\beta}^\star$ is a particular solution of the normal equation $\mathbb{X}^T\mathbb{X}\beta = \mathbb{X}^T Y$. Indeed, let's compute the left- and the right-hand sides of the equation and check that they are identical:

$$\begin{aligned} \mathbb{X}^T\mathbb{X}\hat{\beta}^\star &= \left(P\Sigma Q^T\right)^T P\Sigma \underbrace{Q^T Q}_{\mathbb{I}_r}\Sigma^{-1}P^T Y = Q\Sigma \underbrace{P^T P}_{\mathbb{I}_r}\underbrace{\Sigma\Sigma^{-1}}_{\mathbb{I}_r}P^T Y = Q\Sigma P^T Y, \\ \mathbb{X}^T Y &= \left(P\Sigma Q^T\right)^T Y = Q\Sigma P^T Y. \end{aligned} \tag{12.37}$$

So, if the Gram matrix $\mathbb{X}^T\mathbb{X}$ is singular, then the general solution of the normal equation is given by

$$\hat{\beta} = \mathbb{X}^+ Y + \operatorname{Ker}\mathbb{X}. \tag{12.38}$$

Combining the two cases, nonsingular (12.28) and singular (12.38), we have:

$$\hat{\beta} = \begin{cases} (\mathbb{X}^T\mathbb{X})^{-1}\mathbb{X}^T Y, & \text{if } \mathbb{X}^T\mathbb{X} \text{ is nonsingular} \Leftrightarrow \operatorname{Ker}\mathbb{X} = \{0\}, \\ \mathbb{X}^+ Y + \operatorname{Ker}\mathbb{X}, & \text{if } \mathbb{X}^T\mathbb{X} \text{ is singular} \Leftrightarrow \operatorname{Ker}\mathbb{X} \neq \{0\}. \end{cases} \tag{12.39}$$

But using the second property of the pseudoinverse, if $\operatorname{Ker}\mathbb{X} = \{0\}$, then in the nonsingular case we have that $(\mathbb{X}^T\mathbb{X})^{-1}\mathbb{X}^T = \mathbb{X}^+$. Therefore, in the nonsingular case, $\hat{\beta}$ can be written in the same form as in the singular case

$$\hat{\beta} = (\mathbb{X}^T\mathbb{X})^{-1}\mathbb{X}^T Y = \mathbb{X}^+ Y = \mathbb{X}^+ Y + \underbrace{\operatorname{Ker}\mathbb{X}}_{\{0\}}. \tag{12.40}$$

This means that the formula (12.38) derived for the singular case is, in fact universal: it holds both in the singular and nonsingular cases.

To sum up, the OLS estimate $\hat{\beta}$ of the vector β of regression coefficients in the MLR model is given by

$$\hat{\beta} = \mathbb{X}^+ Y + \operatorname{Ker}\mathbb{X}, \tag{12.41}$$

where $\mathbb{X}^+$ is the pseudoinverse of the design matrix and $\operatorname{Ker}\mathbb{X}$ is its kernel. The OLS estimate is unique if and only if the kernel of the design matrix is trivial, $\operatorname{Ker}\mathbb{X} = \{0\}$, that is, if consists only of the zero vector.

Remark 12.7. A particular OLS estimate $\hat{\beta}^\star = \mathbb{X}^+ Y$ has important geometric interpretation: it is an OLS estimate with the *smallest Euclidean norm*. To prove this, first let's check that $\hat{\beta}^\star$ belongs to the coimage of the design matrix,

$$\hat{\beta}^\star \in \operatorname{coIm}\mathbb{X}. \tag{12.42}$$

By the definition of the coimage of a matrix[31], to show this, we need to find a vector $a \in \mathbb{R}^n$ such that $\hat{\beta}^\star = \mathbb{X}^T a$ or, equivalently,

[31] See Appendix 12.12

$$\underbrace{Q\Sigma^{-1}P^T}_{\mathbb{X}^+} Y = \underbrace{Q\Sigma P^T}_{\mathbb{X}^T} a. \tag{12.43}$$

It is straightforward to check that this equality hold if we set $a = P\Sigma^{-2}P^T Y$.

Now, to see that $\hat{\beta}^\star$ is the smallest OLS estimate, consider an arbitrary OLS estimate $\hat{\beta}$, which as we know from (12.41), can be written as follows:

$$\hat{\beta} = \hat{\beta}^\star + \hat{\beta}^0, \quad \text{for some } \hat{\beta}^0 \in \operatorname{Ker}\mathbb{X}. \tag{12.44}$$

Its (squared) Euclidean norm is

$$\begin{aligned} \|\hat{\beta}\|^2 &= \|\hat{\beta}^\star + \hat{\beta}^0\|^2 = (\hat{\beta}^\star + \hat{\beta}^0)^T(\hat{\beta}^\star + \hat{\beta}^0) \\ &= \|\hat{\beta}^\star\|^2 + 2(\hat{\beta}^\star)^T\hat{\beta}^0 + \|\hat{\beta}^0\|^2. \end{aligned} \tag{12.45}$$

But thanks to the fundamental theorem of linear algebra[32], $\operatorname{coIm}\mathbb{X}$ and $\operatorname{Ker}\mathbb{X}$ are orthogonal subspaces of $\mathbb{R}^{p+1}$, which means that $\hat{\beta}^\star \in \operatorname{coIm}\mathbb{X}$ and $\hat{\beta}^0 \in \operatorname{Ker}\mathbb{X}$ are orthogonal $\hat{\beta}^\star \perp \hat{\beta}^0$ and, therefore, $(\hat{\beta}^\star)^T\hat{\beta}^0 = 0$. So,

[32] See Appendix 12.12

$$\|\hat{\beta}\|^2 = \|\hat{\beta}^\star\|^2 + \|\hat{\beta}^0\|^2 \leq \|\hat{\beta}^\star\|^2, \tag{12.46}$$

and the equality holds if and only if $\hat{\beta}^0 = 0$. This means that $\hat{\beta}^\star = \mathbb{X}^+Y$ is the smallest (in the Euclidean sense) OLS estimate. This result allows to schematically visualize in Fig. 12.2 the set of all OLS estimates given by (12.41).

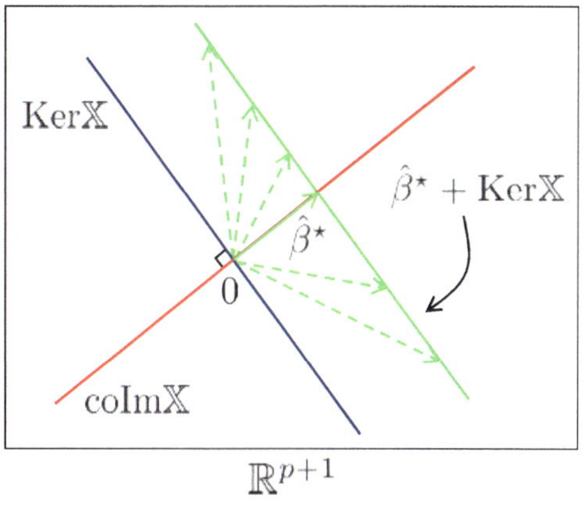

Fig. 12.2 Visualization of the set of all OLS estimates $\hat{\beta} = \mathbb{X}^+Y + \mathrm{Ker}\,\mathbb{X}$. The smallest OLS estimate, $\hat{\beta}^\star = \mathbb{X}^+Y \in \mathrm{coIm}\,\mathbb{X}$, is represented by the bold vector. A few other OLS estimates of the form $\hat{\beta}^\star + \hat{\beta}^0$, where $\hat{\beta}^0 \in \mathrm{Ker}\,\mathbb{X}$, are shown by dashed vectors. The set of all OLS estimates $\{\mathbb{X}^+Y + \mathrm{Ker}\,\mathbb{X}\}$ is obtained by shifting the kernel $\mathrm{Ker}\,\mathbb{X}$ by the vector $\hat{\beta}^\star$.

12.3 Geometry of OLS

The OLS estimate $\hat{\beta}$ is not unique when the Gram matrix $\mathbb{X}^T\mathbb{X}$ is singular, which occurs when the columns of the design matrix $\mathbb{X}$ are linearly dependent, that is, when at least one column is redundant by being a linear combination of the other columns. Let's drop all redundant columns in $\mathbb{X}$ (if there are any) and assume that $\mathbb{X}^T\mathbb{X}$ is noningular[33]. Then the OLS estimate is unique and given by

$$\hat{\beta} = (\mathbb{X}^T\mathbb{X})^{-1}\mathbb{X}^TY. \tag{12.47}$$

Let $X_* \in \mathbb{R}^{p+1}$ be a new previously unseen input vector. Then the prediction of the MLR model for the corresponding response Y_* is

$$\hat{Y}_* = X_*^T\hat{\beta} = X_*^T(\mathbb{X}^T\mathbb{X})^{-1}\mathbb{X}^TY. \tag{12.48}$$

The model predictions $\hat{Y}_1, \ldots, \hat{Y}_n$ for the training input vectors $X_1, \ldots, X_n$ are called the *fitted values*,

$$\hat{Y}_i = X_i^T\hat{\beta}, \tag{12.49}$$

and the p-dimensional hyperplane in $\mathbb{R}^{p+1}$ defined by

[33] All standard software packages have built-in functions for detecting these redundancies and removing the corresponding columns.

$$Y = X^T\hat{\beta} = \hat{\beta}_0 + \hat{\beta}_1 X^{(1)} + \ldots + \hat{\beta}_1 X^{(p)} \tag{12.50}$$

is called the *fitted plane*. It is schematically shown in Fig. 12.3[34].

[34] Compare Fig. 12.3 with Fig. 11.4 for the simple linear regression.

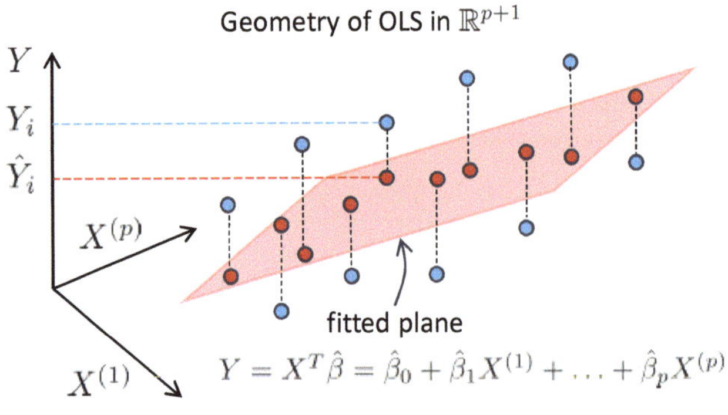

Fig. 12.3 Geometry of OLS in $\mathbb{R}^{p+1}$, where $p = 2$ and $n = 10$. The fitted plane plays the same role in multiple linear regression as the fitted line plays in simple linear regression.

Let $\hat{Y} \in \mathbb{R}^n$ be the vector of fitted values, then

$$\hat{Y} = \begin{bmatrix} \hat{Y}_1 \\ \vdots \\ \hat{Y}_n \end{bmatrix} = \begin{bmatrix} X_1^T\hat{\beta} \\ \vdots \\ X_n^T\hat{\beta} \end{bmatrix} = \begin{bmatrix} X_1^T \\ \vdots \\ X_n^T \end{bmatrix} \hat{\beta} = \mathbb{X}\hat{\beta} = \underbrace{\mathbb{X}(\mathbb{X}^T\mathbb{X})^{-1}\mathbb{X}^T}_{\mathbb{H}} Y = \mathbb{H}Y. \tag{12.51}$$

The matrix $\mathbb{H} = \mathbb{X}(\mathbb{X}^T\mathbb{X})^{-1}\mathbb{X}^T$ is called the *hat matrix*, since it puts the hat ^ on Y.

Since $\hat{\beta}$ is the solution of the normal equation, $\mathbb{X}^T\mathbb{X}\beta = \mathbb{X}^T Y$, we have:

$$0 = \mathbb{X}^T Y - \mathbb{X}^T \underbrace{\mathbb{X}\hat{\beta}}_{\hat{Y}} = \mathbb{X}^T(Y - \hat{Y}) = \left[X^{(0)}, \ldots, X^{(p)}\right]^T (Y - \hat{Y}). \tag{12.52}$$

This means that the *vector of residuals* $\hat{E} = Y - \hat{Y}$ is orthogonal to all columns $X^{(0)}, \ldots, X^{(p)}$ of the design matrix $\mathbb{X}$,

$$(Y - \hat{Y}) \perp X^{(j)}, \quad \text{for all } j = 0, \ldots, p. \tag{12.53}$$

Vectors $X^{(0)}, \ldots, X^{(p)} \in \mathbb{R}^n$ span a $(p+1)$-dimensional[35] subspace[36] in $\mathbb{R}^n$, which is exactly the image of $\mathbb{X}$,

[35] Recall that we assumed that the columns are linearly independent.

[36] Recall that a span of vectors $v_1, \ldots, v_k$ is a set (subspace) of all linear combinations of $v_1, \ldots, v_k$.

$$\begin{aligned} \operatorname{Im}\mathbb{X} &= \{\mathbb{X}\alpha \mid \alpha \in \mathbb{R}^{p+1}\} = \{\alpha_0 X^{(0)} + \ldots + \alpha_p X^{(p)} \mid \alpha_i \in \mathbb{R}\} \\ &= \operatorname{span}\{X^{(0)}, \ldots, X^{(p)}\}. \end{aligned} \tag{12.54}$$

The vector of residuals is, therefore, orthogonal to the image of the design matrix,

$$(Y - \hat{Y}) \perp \operatorname{Im}\mathbb{X}. \tag{12.55}$$

This property is schematically illustrated in Fig. 12.4[37].

[37] Figure 12.4 is a higher-dimensional version of the right side of Fig. 12.1.

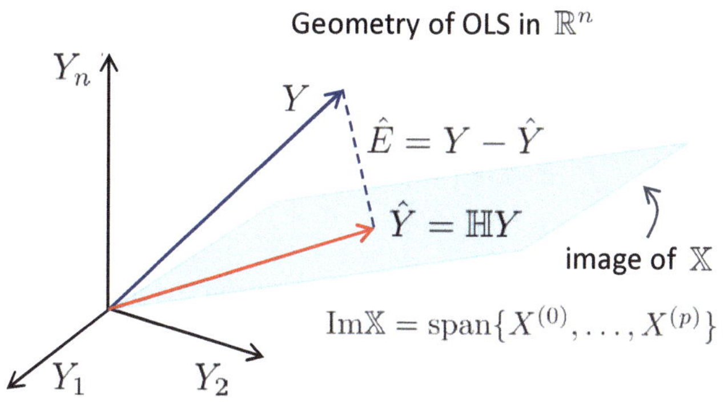

Fig. 12.4 Geometry of OLS in $\mathbb{R}^n$, where $p = 1$ and $n = 3$. The hat matrix $\mathbb{H}$ computes the orthogonal projection of vectors in $\mathbb{R}^n$ on its subspace $\operatorname{Im}\mathbb{X} \subset \mathbb{R}^n$.

The square $n \times n$ hat matrix $\mathbb{H} = \mathbb{X}(\mathbb{X}^T\mathbb{X})^{-1}\mathbb{X}^T$ induces a linear map $\mathbb{H} : \mathbb{R}^n \to \mathbb{R}^n$, which maps Y to $\hat{Y} = \mathbb{H}Y$. The property (12.55) means that the map defined by $\mathbb{H}$ is simply the orthogonal projection onto the image of the design matrix $\operatorname{Im}\mathbb{X}$, which is a subspace of $\mathbb{R}^n$.

When thinking about the OLS fitted multiple linear regression, it is useful to keep in mind Fig. 12.3 and Fig. 12.4, which describe the geometric features of this model.

12.4 Statistical Properties of OLS

As defined in Section 12.1, the multiple linear regression model for the training data $(X_1, Y_1), \ldots, (X_n, Y_n)$, where $X_i \in \mathbb{R}^{p+1}$ are the input vectors and Y_i are the corresponding responses, has the following form:

$$Y_i = X_i^T \beta + E_i, \qquad i = 1, \ldots, n. \tag{12.56}$$

Let's rewrite it in the matrix form:

$$Y = \begin{bmatrix} Y_1 \\ \vdots \\ Y_n \end{bmatrix} = \begin{bmatrix} X_1^T \beta \\ \vdots \\ X_n^T \beta \end{bmatrix} + \begin{bmatrix} E_1 \\ \vdots \\ E_n \end{bmatrix} = \mathbb{X}\beta + E, \tag{12.57}$$

where $E = (E_1, \ldots, E_n)^T$ is a vector of prediction errors.

Recall that the OLS estimate[38]

[38] As in the previous section, we assume that the training input vectors have been preprocessed so that $\mathbb{X}^T\mathbb{X}$ is nonsingular.

$$\hat{\beta} = (\mathbb{X}^T\mathbb{X})^{-1}\mathbb{X}^T Y \tag{12.58}$$

was derived without making any assumptions about the prediction errors. We simply used the geometric optimization criterion $\mathrm{RSS}(\beta) \to \min$ to fit a hyperplane, defined by β, to a cloud of points (see Fig. 12.3). But it turns out that $\hat{\beta}$ has very nice statistical properties under certain assumptions.

In this section, we will work under the weaker nonparametric assumptions:

$$\begin{aligned} &1.\ E_1,\dots,E_n \text{ are independent,}\\ &2.\ \mathbb{E}[E_i] = 0,\\ &3.\ \mathbb{V}[E_i] = \sigma^2. \end{aligned} \tag{12.59}$$

In terms of the vector of errors E, these assumptions can be reformulated as follows: the components of random vector E are independent, the mean vector of errors is zero, and the covariance matrix Σ_E of E is a scalar matrix[39],

[39] A scalar matrix is a multiple of the identity matrix. In statistical jargon, errors with a scalar covariance matrix are sometimes called *spherical errors*. In other words, errors are spherical if they are *homoscedastic* (same variance) and *uncorrelated*.

$$\mathbb{E}[E] = 0 \quad \text{and} \quad \Sigma_E = \mathbb{E}\left[EE^T\right] = \sigma^2\mathbb{I}_n. \tag{12.60}$$

It is straightforward to check that under these assumptions, as in the case of simple linear regression, $\hat{\beta}$ is an *unbiased* estimate of β:

$$\begin{aligned} \mathbb{E}[\hat{\beta}] &= \mathbb{E}\left[(\mathbb{X}^T\mathbb{X})^{-1}\mathbb{X}^T Y\right] = (\mathbb{X}^T\mathbb{X})^{-1}\mathbb{X}^T\mathbb{E}[Y]\\ &= (\mathbb{X}^T\mathbb{X})^{-1}\mathbb{X}^T\mathbb{E}[\mathbb{X}\beta + E] = (\mathbb{X}^T\mathbb{X})^{-1}\mathbb{X}^T\mathbb{X}\beta = \beta. \end{aligned} \tag{12.61}$$

Moreover, we can also find the covariance matrix of $\hat{\beta}$. By definition,

$$\Sigma_{\hat{\beta}} = \mathbb{E}\left[(\hat{\beta} - \mathbb{E}[\hat{\beta}])(\hat{\beta} - \mathbb{E}[\hat{\beta}])^T\right]. \tag{12.62}$$

The difference between the OLS estimate and its expected value is

$$\begin{aligned} \hat{\beta} - \mathbb{E}[\hat{\beta}] &= (\mathbb{X}^T\mathbb{X})^{-1}\mathbb{X}^T Y - \beta = (\mathbb{X}^T\mathbb{X})^{-1}\mathbb{X}^T(\mathbb{X}\beta + E) - \beta\\ &= (\mathbb{X}^T\mathbb{X})^{-1}\mathbb{X}^T E. \end{aligned} \tag{12.63}$$

The covariance matrix is then

$$\begin{aligned} \Sigma_{\hat{\beta}} &= \mathbb{E}\left[\left((\mathbb{X}^T\mathbb{X})^{-1}\mathbb{X}^T E\right)\left((\mathbb{X}^T\mathbb{X})^{-1}\mathbb{X}^T E\right)^T\right]\\ &= \mathbb{E}\left[(\mathbb{X}^T\mathbb{X})^{-1}\mathbb{X}^T EE^T\mathbb{X}(\mathbb{X}^T\mathbb{X})^{-1}\right]\\ &= (\mathbb{X}^T\mathbb{X})^{-1}\mathbb{X}^T\underbrace{\mathbb{E}\left[EE^T\right]}_{\sigma^2\mathbb{I}_n}\mathbb{X}(\mathbb{X}^T\mathbb{X})^{-1}\\ &= \sigma^2(\mathbb{X}^T\mathbb{X})^{-1}\mathbb{X}^T\mathbb{X}(\mathbb{X}^T\mathbb{X})^{-1} = \sigma^2(\mathbb{X}^T\mathbb{X})^{-1}. \end{aligned} \tag{12.64}$$

Although the design matrix $\mathbb{X}$ is known (it is a part of the training data), the common variance σ^2 of prediction errors is not: it is a parameter of the MLR model.

It can be shown – see Appendix 12.14 – that an unbiased estimate of σ^2 is given by

$$\hat{\sigma}_u^2 = \frac{1}{n-p-1}\sum_{i=1}^{n}(Y_i - \hat{Y}_i)^2. \tag{12.65}$$

Note that if $p = 1$ and we have the case of simple linear regression, then the above formula reduces to (11.44).

12.4.1 The Gauss-Markov Theorem

In this section, we will discuss the most important and famous result concerning the statistical properties of the OLS estimate: the *Gauss-Markov theorem*, which asserts that the OLS estimate is the best linear unbiased estimate (BLUE). This property generalizes the BLUE property of the OLS estimates in simple linear regression, which was discussed in Section 11.7.6, and contributes significantly to the popularity of OLS in applications.

Let $\beta \in \mathbb{R}^{p+1}$ be the vector of regression coefficients, and suppose that we are interested in estimating a parameter

$$\theta = \alpha^T\beta = \alpha_0\beta_0 + \ldots + \alpha_p\beta_p, \tag{12.66}$$

which is a linear combination of the regression coefficients. Here, $\alpha \in \mathbb{R}^{p+1}$ is a vector that we define and use to specify what exactly we are interested in estimating. To make it more concrete, if, for example, we want to estimate the jth regression coefficient β_j, then we set

$$\alpha = (0,\ldots,0,\underbrace{1}_{j},0,\ldots,0)^T \quad \Longrightarrow \quad \theta = \beta_j. \tag{12.67}$$

If we want to estimate the response Y_* to a new input $X_* \in \mathbb{R}^{p+1}$, then we set

$$\alpha = X_* \quad \Longrightarrow \quad \theta = X_*^T\beta. \tag{12.68}$$

In general, $\theta = \alpha^T\beta$ and its OLS estimate is

$$\hat{\theta} = \alpha^T\hat{\beta} = \alpha^T(\mathbb{X}^T\mathbb{X})^{-1}\mathbb{X}^TY. \tag{12.69}$$

The OLS estimate $\hat{\theta}$ is *unbiased* because $\hat{\beta}$ is unbiased:

$$\mathbb{E}[\hat{\theta}] = \mathbb{E}[\alpha^T\hat{\beta}] = \alpha^T\mathbb{E}[\hat{\beta}] = \alpha^T\beta = \theta. \tag{12.70}$$

Moreover, $\hat{\theta}$ is a *linear*[40] estimate:

[40] Recall that an estimate is called linear if it is a linear combination of responses $Y_1,\ldots,Y_n$.

$$\hat{\theta} = \underbrace{\alpha^T(\mathbb{X}^T\mathbb{X})^{-1}\mathbb{X}^T}_{1\times n} Y = \sum_{i=1}^{n} \left(\alpha^T(\mathbb{X}^T\mathbb{X})^{-1}\mathbb{X}^T\right)_i Y_i. \tag{12.71}$$

Let $\mathrm{LU}(\theta)$ be the set of all linear unbiased estimates of the parameter θ. Intuitively, $\mathrm{LU}(\theta)$ consists of relatively reasonable (unbiased) and simple (linear) estimates. The Gauss-Markov theorem states that $\hat{\theta} \in \mathrm{LU}(\theta)$ is the best estimate in $\mathrm{LU}(\theta)$, meaning it has the smallest possible variance (or, equivalently, the smallest possible mean squared error[41]). The precise statement is the following.

[41] Thanks to the bias-variance decomposition of MSE and unbiasedness, we have $\mathrm{MSE}[\tilde{\theta}] = \mathbb{V}[\tilde{\theta}]$ for any $\tilde{\theta} \in \mathrm{LU}(\theta)$.

Carl Friedrich Gauss (1777-1855) was a German mathematician, astronomer, and physicist, regarded by some as one of the top three mathematicians of all time, together with Archimedes and Newton.

Andrey Andreyevich Markov (1856-1922) was a Russian mathematician who made significant contributions to probability theory, number theory, and analysis.

Theorem 12.1 (Gauss-Markov Theorem). *Consider the MLR model,*

$$Y = \mathbb{X}\beta + E, \quad E = (E_1, \ldots, E_n)^T. \tag{12.72}$$

Suppose that the vector of errors has zero mean and scalar covariance matrix,

$$\mathbb{E}[E] = 0 \quad and \quad \Sigma_E = \mathbb{E}\left[EE^T\right] = \sigma^2\mathbb{I}_n. \tag{12.73}$$

Let $\hat{\theta}$ be the OLS estimate of $\theta = \alpha^T\beta$ and $\tilde{\theta} \in \mathrm{LU}(\theta)$ be any other linear unbiased estimate of θ. Then

$$\mathbb{V}[\hat{\theta}] \leq \mathbb{V}[\tilde{\theta}]. \tag{12.74}$$

Remark 12.8. Note that the assumption (12.73) is even weaker than the weak non-parametric assumption (12.59) because it does not assume that $E_1, \ldots, E_n$ are *independent*, only that they are *uncorrelated*. Originally, this theorem was proved by Carl Gauss in 1821 (in Latin!), but under a stronger assumption: he assumed that errors were both independent and normally distributed. In 1912, Andrey Markov noticed that independence and normality are not needed, and uncorrelatedness alone is sufficient for the conclusion of the theorem to hold. A short historical note on the contributions of Gauss and Markov (and Laplace and Aitken) is given in [7].

Proof. To prove that $\hat{\theta}$ has the smallest variance among all linear unbiased estimates, let's first find its variance so that we can compare it with the variance of an arbitrary estimate $\tilde{\theta} \in \mathrm{LU}(\theta)$. Using linearity of $\hat{\theta}$ (12.71) and the fact that Y_i and Y_j are uncorrelated (since E_i and E_j are uncorrelated), we have:

$$\begin{aligned}\mathbb{V}[\hat{\theta}] &= \mathbb{V}\left[\sum_{i=1}^{n} \left(\alpha^T(\mathbb{X}^T\mathbb{X})^{-1}\mathbb{X}^T\right)_i Y_i\right] = \sum_{i=1}^{n} \left(\alpha^T(\mathbb{X}^T\mathbb{X})^{-1}\mathbb{X}^T\right)_i^2 \underbrace{\mathbb{V}[Y_i]}_{\sigma^2} \\ &= \sigma^2\left(\alpha^T(\mathbb{X}^T\mathbb{X})^{-1}\mathbb{X}^T\right)\left(\alpha^T(\mathbb{X}^T\mathbb{X})^{-1}\mathbb{X}^T\right)^T \\ &= \sigma^2\alpha^T(\mathbb{X}^T\mathbb{X})^{-1}\mathbb{X}^T\mathbb{X}(\mathbb{X}^T\mathbb{X})^{-1}\alpha = \sigma^2\alpha^T(\mathbb{X}^T\mathbb{X})^{-1}\alpha.\end{aligned} \tag{12.75}$$

Here, in the third equality, we used the identity $\sum_{i=1}^{n} v_i^2 = vv^T$ applied to the $1\times n$ vector $v = \alpha^T(\mathbb{X}^T\mathbb{X})^{-1}\mathbb{X}^T$.

Now, let $\tilde{\theta} \in \mathrm{LU}(\theta)$ be any linear unbiased estimate of θ. Since $\tilde{\theta}$ is linear,

$$\tilde{\theta} = a^T Y, \tag{12.76}$$

where $a \in \mathbb{R}^n$ is a vector whose components do not depend on responses $Y_1, \ldots, Y_n$. To make the comparison between $\tilde{\theta}$ and $\hat{\theta}$, which is given by

$$\hat{\theta} = \underbrace{\alpha^T (\mathbb{X}^T \mathbb{X})^{-1} \mathbb{X}^T}_{a_0^T} Y = a_0^T Y, \tag{12.77}$$

easier, let's think of a^T as a perturbation of a_0^T. In other words, let's represent a^T as the sum of a_0^T and an arbitrary perturbation:

$$a^T = a_0^T + b^T, \tag{12.78}$$

where $b \in \mathbb{R}^n$ is an arbitrary vector (whose components do not depend on $Y_1, \ldots, Y_n$). This allows to represent $\tilde{\theta}$ as a perturbation of $\hat{\theta}$:

$$\tilde{\theta} = (a_0^T + b^T)Y = \hat{\theta} + b^T Y. \tag{12.79}$$

Since $\tilde{\theta}$ is an unbiased estimate of $\theta = \alpha^T \beta$, we must have that $\mathbb{E}[\tilde{\theta}] = \theta$ for any value of vector β. This leads to a condition on the vector b[42]:

$$\theta = \mathbb{E}[\tilde{\theta}] = \underbrace{\mathbb{E}[\hat{\theta}]}_{\theta} + b^T \mathbb{E}[Y] = \theta + b^T \mathbb{E}[\mathbb{X}\beta + E] = \theta + b^T \mathbb{X}\beta, \tag{12.80}$$

[42] Conceptually, since $\tilde{\theta}$ is unbiased, the perturbation term $b^T Y$ cannot be arbitrary. It must satisfy a certain condition, which we are about to derive.

which simplifies to

$$b^T \mathbb{X}\beta = 0. \tag{12.81}$$

This condition holds for any β if and only if

$$b^T \mathbb{X} = 0 \quad \text{or, equivalently,} \quad \mathbb{X}^T b = 0. \tag{12.82}$$

So, the set of all linear unbiased estimates of θ consists precisely of estimates of the form (12.79), where b satisfies (12.82),

$$\mathrm{LU}(\theta) = \{\tilde{\theta} = \hat{\theta} + b^T Y \mid b^T \mathbb{X} = 0\}. \tag{12.83}$$

The variance of $\tilde{\theta}$ is

$$
\begin{aligned}
\mathbb{V}[\tilde{\theta}] &= \mathbb{V}[\hat{\theta} + b^T Y] = \mathbb{V}[a_0^T Y + b^T Y] = \mathbb{V}[(a_0 + b)^T Y] \\
&= \mathbb{V}\left[\sum_{i=1}^{n} (a_0 + b)_i Y_i\right] = \sum_{i=1}^{n} (a_0 + b)_i^2 \mathbb{V}[Y_i] = \sigma^2 (a_0 + b)^T (a_0 + b) \\
&= \sigma^2 \left(\alpha^T (\mathbb{X}^T\mathbb{X})^{-1}\mathbb{X}^T + b^T\right) \left(\mathbb{X}(\mathbb{X}^T\mathbb{X})^{-1}\alpha + b\right) \\
&= \sigma^2 \Big(\alpha^T (\mathbb{X}^T\mathbb{X})^{-1}\alpha + \underbrace{b^T\mathbb{X}}_{0}(\mathbb{X}^T\mathbb{X})^{-1}\alpha + \alpha^T (\mathbb{X}^T\mathbb{X})^{-1}\underbrace{\mathbb{X}^T b}_{0} + b^T b\Big) \\
&= \underbrace{\sigma^2 \alpha^T (\mathbb{X}^T\mathbb{X})^{-1}\alpha}_{\mathbb{V}[\hat{\theta}]} + \sigma^2 b^T b = \mathbb{V}[\hat{\theta}] + \sigma^2 \|b\|^2 \geq \mathbb{V}[\hat{\theta}].
\end{aligned}
\tag{12.84}
$$

So, for any $\tilde{\theta} \in \mathrm{LU}(\theta)$ we have $\mathbb{V}[\tilde{\theta}] \geq \mathbb{V}[\hat{\theta}]$. Moreover, the equality occurs if and only if $b = 0$, that is, if $\tilde{\theta} = \hat{\theta}$. This precisely means that the OLS estimate $\hat{\theta}$ has the smallest possible variance among all linear unbiased estimates, or equivalently, that the OLS estimate $\hat{\theta}$ is BLUE: the best linear unbiased estimate. □

So, even under relatively weak nonparametric assumptions, the OLS estimate $\hat{\beta}$ of the vector of regression coefficients has nice statistical properties: it is unbiased (12.61), its covariance matrix can be explicitly computed (12.64), the variance of prediction errors can be estimated (12.65), and, thanks to the Gauss-Markov theorem, $\hat{\beta}$ is the most accurate estimate within a class of linear unbiased estimates.

12.5 Testing and Ranking Predictors by Significance

Once again, let's consider the multiple linear regression model for the training data $(X_1, Y_1), \ldots, (X_n, Y_n)$:

$$
Y = \mathbb{X}\beta + E, \tag{12.85}
$$

where $Y = (Y_1, \ldots, Y_n)^T$ is the vector of training responses, $\mathbb{X}$ is the $n \times (p+1)$ design matrix consisting of the training input vectors[43], $\beta = (\beta_0, \ldots, \beta_p)^T$ is the vector of regression coefficients, and $E = (E_1, \ldots, E_n)^T$ is the vector of prediction errors. If $X_* = (1, X_*^{(1)}, \ldots, X_*^{(p)})^T$ is a new input vector, then the prediction of the MLR model fitted by the OLS method for the corresponding response Y_* is

[43] To recall its precise definition, see (12.10).

$$
\hat{Y}_* = X_*^T \hat{\beta} = \hat{\beta}_0 + \hat{\beta}_1 X_*^{(1)} + \ldots + \hat{\beta}_p X_*^{(p)}. \tag{12.86}
$$

Why might we not be satisfied with the MLR model fitted by OLS, even when the true relationship between the system response Y_* and predictors $X_*^{(1)}, \ldots, X_*^{(p)}$ is approximately linear ($Y_* \approx X_*^T \beta$), and despite the nice statistical properties of OLS discussed in the previous section? There are two reasons related to *prediction accuracy* and *model interpretability*.

First, it is important to highlight that the mere fact that we measured and collected the data on predictors $X^{(1)}, \ldots, X^{(p)}$ does not imply that all these predictors indeed affect the system response and are relevant for prediction. Some of the predictors in the training data might not be associated with the response and might not be relevant for prediction. Moreover, in general, the larger the number of predictors p, the larger the number of irrelevant predictors among $X^{(1)}, \ldots, X^{(p)}$.

Remark 12.9. For example, consider a medical application where we want to predict the length (or severity or the long-term impact) of a certain not-well-understood disease (such as AIDS, COVID-19, or cancer) based on a number of biomarkers. We can measure and collect the data on hundreds, if not thousands, of biomarkers, but it does not mean that all those biomarkers are relevant for prediction. Similarly, not all measurable economic indicators are relevant for predicting stock prices.

If $X^{(j)}$ is an irrelevant predictor, not associated with the response, then we obviously want to exclude it from the model[44] and remove it from the list of predictors. However, it is extremely unlikely that the OLS estimate $\hat{\beta}_j$ corresponding to $X^{(j)}$ will be *exactly zero*.

[44] By deleting the corresponding column of the design matrix $\mathbb{X}$.

Remark 12.10. To see why this is the case, consider a special case with $p = 1$, which corresponds to simple linear regression. Let's generate the bivariate training data $(X_1, Y_1), \ldots, (X_n, Y_n)$, where X_i and Y_i are independent. For concreteness, suppose that X_i and Y_i are obtained by IID sampling from the standard normal distribution[45]. If we fit the regression line $Y = \beta_0 + \beta_1 X$ to the cloud of points $(X_1, Y_1), \ldots, (X_n, Y_n)$, then, despite the fact that X and Y are theoretically independent, the OLS estimate of the slope $\hat{\beta}_1$ will not be exactly zero. This is because the estimate depends on the *data*, which contains some, but not all, information about the independent *random variables* X and Y. As the sample size $n \to \infty$, the estimate $\hat{\beta}_1 \to 0$, but for a finite n, the probability $\mathbb{P}(\hat{\beta}_1 = 0) = 0$.

[45] The exact form of the distribution does matter; and X_i and Y_i may have different distributions.

As a result, the expression for model prediction $\hat{Y}_*$ will be "contaminated" by the irrelevant "noisy" term $\hat{\beta}_j X_*^{(j)}$. The more such irrelevant terms are present in (12.86), the higher the chance that the model prediction will not be accurate.

But even if all predictors $X^{(1)}, \ldots, X^{(p)}$ are associated with the response and are relevant for prediction, it can be difficult to *interpret the model* when the number of predictors p is large: to identify the main factors that determine the system response. To make the MLR model more interpretable, we may want to determine a smaller subset of predictors $\{X^{(j_1)}, \ldots, X^{(j_{\tilde{p}})}\} \subset \{X^{(1)}, \ldots, X^{(p)}\}$, where $\tilde{p} \ll p$, that are most strongly associated with the response and relevant for prediction. In other words, we may want to sacrifice some details — perhaps even accuracy — in order to get a "big picture" of the system under study and gain a better understanding of the main factors that govern its response.

This discussion suggests that we need a method for excluding irrelevant and, perhaps, weakly relevant predictors from the MLR model and, ideally, a way to rank the predictors with respect to their importance for the model. We can effectively exclude the predictor $X^{(j)}$ from the model by setting its regression coefficient β_j to zero in advance, and then estimating the remaining coefficients. But how to choose which β_j to set to zero? This question naturally calls for using hypothesis testing. Namely, let's test

$$H_0^{(j)} : \beta_j = 0 \quad \text{vs} \quad H_1^{(j)} : \beta_j \neq 0, \quad \text{for } j = 1, \ldots, p, \tag{12.87}$$

and use the obtained p-value[46] p_j as a measure of significance of the predictor $X^{(j)}$. Specifically: the smaller the p-value p_j, the stronger the evidence against $H_0^{(j)}$ provided by the training data $(\mathbb{X}, Y)$, the more significant $X^{(j)}$, and, therefore, the more important it is to keep $X^{(j)}$ in the model, that is, not to set $\beta_j = 0$.

To formally test the hypotheses in (12.87), the nonparametric assumptions (12.59) used in the previous section are not sufficient. We need stronger parametric assumptions. So, let's assume that the prediction errors are normally distributed:

$$E_1, \ldots, E_n \overset{\text{iid}}{\sim} \mathcal{N}(0, \sigma^2) \quad \text{or, equivalently,} \quad E \sim \mathcal{N}(0, \sigma^2 \mathbb{I}_n). \tag{12.88}$$

Since (12.88) implies (12.59), all the statistical properties of the OLS estimate $\hat{\beta}$ summarized at the end of the previous section still hold here. Additionally, thanks to the stronger normality assumption, the vector $\hat{\beta}$ has a new property that will allow us to test (12.87).

Namely, since the vector of errors E is a *Gaussian vector*[47], the vector of training responses $Y = \mathbb{X}\beta + E$ is also Gaussian. Therefore, the vector $\hat{\beta} = (\mathbb{X}^T\mathbb{X})^{-1}\mathbb{X}^T Y$ is Gaussian as well. Its mean $\mathbb{E}[\hat{\beta}] = \beta$ and covariance matrix $\Sigma_{\hat{\beta}} = \sigma^2(\mathbb{X}^T\mathbb{X})^{-1}$. Thus,

$$\hat{\beta} \sim \mathcal{N}\left(\beta, \sigma^2(\mathbb{X}^T\mathbb{X})^{-1}\right). \tag{12.89}$$

Its jth component $\hat{\beta}_j$, which is the OLS estimate of the regression coefficient associated with the predictor $X^{(j)}$, is therefore distributed according to the normal distribution:

$$\hat{\beta}_j \sim \mathcal{N}\left(\beta_j, \sigma^2 \xi_j\right), \quad \text{where } \xi_j = \left((\mathbb{X}^T\mathbb{X})^{-1}\right)_{jj}. \tag{12.90}$$

Remark 12.11. The jth diagonal element ξ_j of the matrix $(\mathbb{X}^T\mathbb{X})^{-1}$ is strictly positive, and consequently, the variance $\mathbb{V}[\hat{\beta}_j] = \sigma^2 \xi_j > 0$ is also strictly positive[48]. Indeed, since the Gram matrix $\mathbb{X}^T\mathbb{X}$ is positive definite[49], its inverse $(\mathbb{X}^T\mathbb{X})^{-1}$ is also positive definite. Therefore,

$$\xi_j = \left((\mathbb{X}^T\mathbb{X})^{-1}\right)_{jj} = e_j^T (\mathbb{X}^T\mathbb{X})^{-1} e_j > 0, \tag{12.91}$$

where $e_j = (0, \ldots, 0, 1, 0, \ldots, 0)^T$ is the jth vector of the standard basis of $\mathbb{R}^{p+1}$.

Under the null hypothesis $H_0^{(j)}$, we have that $\beta_j = 0$ and, therefore,

$$\hat{\beta}_j \sim \mathcal{N}\left(0, \sigma^2 \xi_j\right), \tag{12.92}$$

or, equivalently,

$$\frac{\hat{\beta}_j}{\sigma\sqrt{\xi_j}} \sim \mathcal{N}(0, 1). \tag{12.93}$$

46 Recall that the p-value can be viewed as a measure of evidence against the null hypothesis provided by the data and chosen test statistic (Section 9.7.4).

47 A random n-vector V is called Gaussian if it can be written as $V = AZ + a$, where A is an $n \times m$ matrix, Z is a random m-vector with IID standard normal components, and a is an n-vector.

48 It is useful to establish this fact, as we will soon need to use $\sqrt{\xi_j}$ in a denominator.

49 See Remark 12.4.

If the null hypothesis is true, then the quantity $|\hat{\beta}_j/(\sigma\sqrt{\xi_j})|$ is expected to be close to zero, and its large values will provide evidence against $H_0^{(j)}$. It would be natural to use this quantity as a test statistic, but unfortunately, we cannot compute its value since σ is unknown. Fortunately, we can estimate σ by $\sigma \approx \hat{\sigma}_u$, where $\hat{\sigma}_u^2$ is the unbiased estimate of σ^2 given by (12.65)[50]. Replacing σ with $\hat{\sigma}_u$ in (12.93), we have:

$$\mathfrak{z}_j = \frac{\hat{\beta}_j}{\hat{\sigma}_u\sqrt{\xi_j}} \overset{\cdot}{\sim} \mathcal{N}(0,1). \tag{12.94}$$

The quantity $\mathfrak{z}_j$ is called a *z-score* associated with the coefficient β_j[51]. Let's use the absolute value of the z-score $|\mathfrak{z}_j|$ as a test statistic and reject the null hypothesis if $|\mathfrak{z}_j|$ is sufficiently large. The test based on the statistic $W = |Z_j|$, where $\mathfrak{z}_j$ is the observed value of $Z_j \overset{\cdot}{\sim} \mathcal{N}(0,1)$, is nothing else but the Wald test, which we discussed in detail in Section 10.1 and used in Section 11.8 in the context of simple linear regression. The Wald test p-value $\mathfrak{p}_j^W$ can be computed[52] using its probabilistic definition: $\mathfrak{p}_j^W$ is the probability, under $H_0^{(j)}$, of observing a value of $W = |Z_j|$ that is at least as large as the value $w = |\mathfrak{z}_j|$ that has been actually observed:

$$\mathfrak{p}_j^W = \mathbb{P}\left(|Z_j| \geq |\mathfrak{z}_j| \,\middle|\, H_0^{(j)}\right) \approx 2\Phi\left(-|\mathfrak{z}_j|\right) = 2\Phi\left(-\left|\frac{\hat{\beta}_j}{\hat{\sigma}_u\sqrt{\xi_j}}\right|\right). \tag{12.95}$$

The smaller the p-value $\mathfrak{p}_j^W$, the stronger the evidence against the null hypothesis that $\beta_j = 0$ provided by the training data, and, therefore, the more important it is to keep the predictor $X^{(j)}$ in the MLR model.

Remark 12.12. If the 0.05 significance level is used[53], that is, if $H_0^{(j)} : \beta_j = 0$ is rejected when $\mathfrak{p}_j^W < 0.05$, then the predictor $X^{(j)}$ is kept in the model when

$$|\mathfrak{z}_j| > -\Phi^{-1}(0.025) \approx 2. \tag{12.96}$$

It can be shown that the exact null distribution of the z-score, that is, distribution of $\mathfrak{z}_j$ under $H_0^{(j)}$, is the t-distribution with $(n-p-1)$ degrees of freedom[54]:

$$\mathfrak{z}_j = \frac{\hat{\beta}_j}{\hat{\sigma}_u\sqrt{\xi_j}} \sim \mathcal{T}(n-p-1). \tag{12.97}$$

This means that, instead of the approximate p-value obtained from the Wald test based on the normal distribution, we can compute the exact p-value by using Student's t-test based on the t-distribution:

$$\mathfrak{p}_j^T = 2F_{T,n-p-1}\left(-\left|\frac{\hat{\beta}_j}{\hat{\sigma}_u\sqrt{\xi_j}}\right|\right), \tag{12.98}$$

[50] Note that $\hat{\sigma}_u$ is not an unbiased estimate of σ, but it is okay.

[51] This is a direct extension of the z-score (11.85) associated with β_1 in the context of simple linear regression.

[52] Just like in (10.20).

[53] It is important to remind that there is nothing special in the cutoff value 0.05. See Marginal Note 30 in Section 9.7.5.

[54] A proof of this result is provided in Appendix 12.15. This result is a generalization of (11.80) where $p = 1$, to an arbitrary number of predictors $p \in \{1,\ldots,n-2\}$.

where $F_{T,n-p-1}$ is the CDF of the t-distribution with $(n-p-1)$ degrees of freedom. Note that if $p=1$, then the p-values in (12.95) and (12.98) reduce to the p-values in (11.84) used for testing the significance of simple linear regression.

Recall[55] that the t-distribution $\mathscr{T}(d)$ converges to the standard normal distribution $\mathscr{N}(0,1)$ as $d\to\infty$, and the difference between $\mathscr{T}(d)$ and $\mathscr{N}(0,1)$ is practically negligible for $d\geq 30$. Therefore, $\mathfrak{p}_j^W \approx \mathfrak{p}_j^T$ as long as $n-p-1\geq 30$. Since in many applications the size n of the training data is substantially larger than the number of predictors p, and in particular, $n\geq p+31$, it is customary to use the standard normal approximation $\mathfrak{z}_j \stackrel{\cdot}{\sim} \mathscr{N}(0,1)$ and the Wald p-value given by (12.95). And this is what we will do unless otherwise stated.

[55] See Remark 11.9.

We can use the p-values associated with the regression coefficients for ranking predictors with respect to their significance for the MLR model. Namely,

$$X^{(i)} \text{ is more significant than } X^{(j)} \quad \Longleftrightarrow \quad \mathfrak{p}_i^W < \mathfrak{p}_j^W. \tag{12.99}$$

This criterion makes sense since the smaller the p-value, the stronger the evidence from the training data that the corresponding predictor should be kept in the model. Since $\mathfrak{p}_j^W \approx 2\Phi(-|\mathfrak{z}_j|)$ and the CDF Φ is a monotonically increasing function, we can rewrite the above criterion directly in terms of the z-scores:

$$X^{(i)} \text{ is more significant than } X^{(j)} \quad \Longleftrightarrow \quad |\mathfrak{z}_i| > |\mathfrak{z}_j|. \tag{12.100}$$

How can we use this ranking? First, as discussed above, the ranking could provide us with a better understanding of the main predictors that drive the system response. It is not likely that all predictors equally affect the system behavior, and identifying the most important factors is crucial for understanding the system.

The ranking also allows us to construct several "competing" MLR models. Namely, if there are p predictors, then we can define $(p+1)$ models by starting with the full MLR model that contains all predictors, and then construct the remaining p models by removing one predictor at a time in the order of increasing significance[56]. But how to choose the "best" model from this set of models? This task is called *model selection*[57] and it admits several reasonable solutions. The conceptually simplest approach is to split the available data $\mathscr{D}=\{(X_1,Y_1),\ldots,(X_n,Y_n)\}$ into two disjoint subsets: the *training data* $\mathscr{D}_t$ and the *validation data* $\mathscr{D}_v$. First, all models are fitted to the training data: the regression coefficients of all models are estimated using only $\mathscr{D}_t$. The prediction accuracy of each fitted model is then assessed using the validation data $\mathscr{D}_v$. The prediction accuracy of a fitted model can be quantified, for example, by the mean squared error on the validation data:

[56] The last model without predictors $Y_i=\beta_0+E_i$ is called the *base model* or the *null model*.

[57] Model selection and references are briefly discussed in Section 12.11.

$$\mathrm{Err} = \frac{1}{|\mathscr{D}_v|}\sum_{(X_i,Y_i)\in\mathscr{D}_v}(Y_i-\hat{Y}_i)^2, \tag{12.101}$$

which is often called the *test error*. Here, $|\mathscr{D}_v|$ is the size of the validation set, that is, the number of points (X_i,Y_i) in $\mathscr{D}_v$, and $\hat{Y}_i = X_i^T\hat{\beta}$ is the fitted value, where $\hat{\beta}$ is the OLS estimate based on $\mathscr{D}_t$. The model with the smallest test error is then declared to be the best. Finally, to obtain the final fitted model, the identified best model is fitted to the full data set $\mathscr{D}$. This method of model selection is called the *validation set approach*. It is illustrated in the next case study in Section 12.8.

12.6 Testing Groups of Predictors

In some applications, we may want to test the significance of a group of predictors simultaneously, instead of or in addition to testing the significance of each predictor individually, as in the previous section. For example, suppose that using the Wald test we failed to reject three hypotheses: $H_0^{(3)} : \beta_3 = 0$, $H_0^{(5)} : \beta_5 = 0$, and $H_0^{(7)} : \beta_7 = 0$. Before removing the corresponding predictors from the model, we may want to double-check this by testing a single hypothesis $H_0 : \beta_3 = \beta_5 = \beta_7 = 0$ versus the alternative that this is not true (at least one of the coefficients is not zero).

Another, somewhat more important example is when we have a categorical predictor[58] $\tilde{X}$ with, say, three levels c_1, c_2, and c_3. It can be represented with three numerical predictors:

[58] See Marginal Note 3.

$$X^{(k)} = \begin{cases} 1, & \text{if } \tilde{X} = c_k, \\ 0, & \text{if } \tilde{X} \neq c_k. \end{cases} \qquad k = 1, 2, 3. \tag{12.102}$$

Then to test the significance of predictor $\tilde{X}$, that is, to test whether it is associated with the response and relevant for prediction, we need to test $H_0 : \beta_1 = \beta_2 = \beta_3 = 0$, where β_1, β_2, and β_3 are the coefficients associated with $X^{(1)}, X^{(2)}$, and $X^{(3)}$, versus the alternative that at least one of β_1, β_2, and β_3 is not zero. In this case, testing the individual significance of $X^{(1)}, X^{(2)}$, and $X^{(3)}$ does not make sense, because they participate in the model as a whole.

So, suppose that we want to test

$$H_0 : (\beta_{j_1}, \ldots, \beta_{j_k}) = (0, \ldots, 0) \quad \text{vs} \quad H_1 : (\beta_{j_1}, \ldots, \beta_{j_k}) \neq (0, \ldots, 0). \tag{12.103}$$

Here, $j_1, \ldots, j_k \in \{1, \ldots, p\}$ are arbitrary k indices, and $\beta_{j_1}, \ldots, \beta_{j_k}$ are the regression coefficients corresponding to predictors $X^{(j_1)}, \ldots, X^{(j_k)}$ whose simultaneous significance we want to explore.

Let $\tilde{\beta} \in \mathbb{R}^{p+1-k}$ be the vector of regression coefficients of the *reduced model*, that is, a model which is obtained from the full MLR model by excluding predictors $X^{(j_1)}, \ldots, X^{(j_k)}$. In other words, $\tilde{\beta}$ is obtained from the original vector β by removing components $\beta_{j_1}, \ldots, \beta_{j_k}$. Let $\tilde{p} = p - k$ be the number of predictors in the reduced model. And let $\hat{\beta}$ and $\hat{\tilde{\beta}}$ be the OLS estimates of $\beta \in \mathbb{R}^{p+1}$ and $\tilde{\beta} \in \mathbb{R}^{\tilde{p}+1}$. It can be shown that, under the parametric normality assumption (12.88) and under the null hypothesis H_0, the quantity

$$\mathfrak{f} = \frac{\left(\text{RSS}(\hat{\tilde{\beta}}) - \text{RSS}(\hat{\beta})\right) / (p - \tilde{p})}{\text{RSS}(\hat{\beta}) / (n - p - 1)} \sim \mathscr{F}(p - \tilde{p}, n - p - 1) \tag{12.104}$$

is distributed according to the F-distribution with parameters $(p-\tilde{p})$ and $(n-p-1)$.

Remark 12.13. Recall that a non-negative random variable $\mathfrak{F}$ is said to be distributed according to the *F-distribution* with parameters d_1 and d_2, denoted $\mathfrak{F} \sim \mathscr{F}(d_1, d_2)$, if it has the following form:

$$\mathfrak{F} = \frac{Q_1/d_1}{Q_2/d_2}, \tag{12.105}$$

where $Q_1 \sim \chi^2_{d_1}$, $Q_2 \sim \chi^2_{d_2}$, and Q_1 and Q_2 are independent. The F-distribution is also called the Fisher-Snedecor distribution after Ronald Fisher and George Snedecor[59].

Remark 12.14. A full proof of (12.104) can be found, for instance, in [8]. The idea of the proof is straightforward: we need to prove that $Q_1 = \text{RSS}(\hat{\hat{\beta}}) - \text{RSS}(\hat{\beta})$ and $Q_2 = \text{RSS}(\hat{\beta})$ are independent and distributed according to the χ^2-distribution with $d_1 = (p-\tilde{p})$ and $d_2 = (n-p-1)$ degrees of freedom respectively. But given the derivations in Appendix 12.15, the aforementioned properties of Q_1 and Q_2 are somewhat expected[60].

The quantity $\mathfrak{f}$, which is called the *F-statistic*, can be written more explicitly in terms of the data:

$$\mathfrak{f} = \frac{n-p-1}{p-\tilde{p}} \cdot \frac{\|\tilde{\mathbb{X}}\hat{\hat{\beta}} - Y\|^2 - \|\mathbb{X}\hat{\beta} - Y\|^2}{\|\mathbb{X}\hat{\beta} - Y\|^2}, \tag{12.106}$$

where $\tilde{\mathbb{X}}$ is the design matrix of the reduced model, obtained from $\mathbb{X}$ by removing columns $j_1, \ldots, j_k$ corresponding to predictors $X^{(j_1)}, \ldots, X^{(j_k)}$.

A moment of thought shows that

$$\|\tilde{\mathbb{X}}\hat{\hat{\beta}} - Y\|^2 = \text{RSS}(\hat{\hat{\beta}}) \geq \text{RSS}(\hat{\beta}) = \|\mathbb{X}\hat{\beta} - Y\|^2. \tag{12.107}$$

Indeed, while $\text{RSS}(\hat{\beta})$ is the smallest possible value of the residual sum of squares, the value $\text{RSS}(\hat{\hat{\beta}})$ is obtained by first setting $\beta_{j_1} = \ldots = \beta_{j_k} = 0$ and only then minimizing the residual sum of squares over the remaining regression coefficients[61]. In other words, $\text{RSS}(\hat{\beta}) \leq \text{RSS}(\hat{\hat{\beta}})$ because $\hat{\beta}$ is the global minimizer of the residual sum of squares on $\mathbb{R}^{p+1}$ and $\hat{\hat{\beta}}$ is the minimizer on the subspace $\mathbb{R}^{\tilde{p}+1} \subset \mathbb{R}^{p+1}$.

Moreover, the larger the difference $\text{RSS}(\hat{\hat{\beta}}) - \text{RSS}(\hat{\beta})$ is, the more important are the coefficients $\beta_{j_1}, \ldots, \beta_{j_k}$ for minimizing the RSS, and, consequently, the more significant the corresponding predictors $X^{(j_1)}, \ldots, X^{(j_k)}$ are for the model. This means that large values of the F-statistic $\mathfrak{f}$ provide evidence against the null hypothesis H_0. This observation allows us to use the F-statistic as a test statistic and reject H_0 if its observed value is sufficiently large. The test based on $\mathfrak{F} \sim \mathscr{F}(p-\tilde{p}, n-p-1)$, where $\mathfrak{f}$ is the observed value of $\mathfrak{F}$, is called the *F-test*. The p-value provided by the F-test is

$$\mathfrak{p} = \mathbb{P}(\mathfrak{F} \geq \mathfrak{f} \mid H_0) = 1 - F_{p-\tilde{p}, n-p-1}(\mathfrak{f}), \tag{12.108}$$

[59] Who, among many other things, created the first academic department of statistics in the United States, at Iowa State University in 1947.

[60] Compare, for instance, Q_2 and random variable Q in the appendix.

[61] Note that the inequality (12.107) is consistent with the fact that an F-distributed random variable is non-negative.

where F_{d_1,d_2} is the CDF of the F-distribution with parameters d_1 and d_2. The smaller the p-value $\mathfrak{p}$, the stronger the evidence provided by the training data against the null hypothesis that $(\beta_{j_1},\dots,\beta_{j_k}) = (0,\dots,0)$, and, therefore, the more significant the group of predictors $X^{(j_1)},\dots,X^{(j_k)}$ is for the MLR model.

Remark 12.15. There exists an asymptotic relationship between the F-distribution and the more familiar χ^2-distribution. Namely, if $\mathfrak{F} \sim \mathscr{F}(d_1,d_2)$, then

$$d_1\mathfrak{F} \xrightarrow{\mathbb{D}} \chi^2_{d_1}, \quad \text{as } d_2 \to \infty. \tag{12.109}$$

Conceptually, this follows from $d_1\mathfrak{F} = Q_1 \cdot (d_2/Q_2)$, where $Q_1 \sim \chi^2_{d_1}$, $Q_2 \sim \chi^2_{d_2}$, and the strong concentration of the $\chi^2_{d_2}$-distribution around its mean d_2, as $d_2 \to \infty$[62]. Applying this result to our case, we have:

[62] Intuitively, this means that if d_2 is large, then $\mathbb{P}(Q_2 \approx d_2) \approx 1$.

$$(p-\tilde{p})\mathfrak{F} \xrightarrow{\mathbb{D}} \chi^2_{p-\tilde{p}}, \quad \text{as } n \to \infty. \tag{12.110}$$

Therefore, for the finite but large size n of the data,

$$(p-\tilde{p})\mathfrak{F} \mathrel{\dot\sim} \chi^2_{p-\tilde{p}}, \quad \text{if } n \gg 1. \tag{12.111}$$

This allows to approximate the p-value (12.108) using the χ^2-distribution:

$$\mathfrak{p} = \mathbb{P}(\mathfrak{F} \geq \mathfrak{f} \mid H_0) = \mathbb{P}((p-\tilde{p})\mathfrak{F} \geq (p-\tilde{p})\mathfrak{f} \mid H_0) \approx 1 - F_{\chi^2_{p-\tilde{p}}}((p-\tilde{p})\mathfrak{f}), \tag{12.112}$$

where $F_{\chi^2_{p-\tilde{p}}}$ is the CDF of the $\chi^2_{p-\tilde{p}}$-distribution. The test based on this p-value is called the *χ^2-test*. Although the χ^2-test may appear as more familiar and "user-friendly" than the F-test, it is safer to use the F-test in applications because — unlike $\mathscr{T}(d) \to \mathscr{N}(0,1)$ — the convergence of the F-distribution to the χ^2-distribution in (12.109) and (12.110) is slow. We don't want to wonder whether n is sufficiently large for (12.111) and (12.112) to be accurate.

Remark 12.16. What if $k = 1$ in (12.103)? In other words, what if the group of predictors that we want to test for significance consists of just one predictor $X^{(j)}$? Then, on the one hand, we can still use the F-test with the p-value given by[63]

[63] This is a special case of (12.108), where $p - \tilde{p} = 1$.

$$\mathfrak{p} = 1 - F_{1,n-p-1}(\mathfrak{f}). \tag{12.113}$$

On the other hand, we can use the t-test from the previous section with the p-value

$$\mathfrak{p}^T_j = 2F_{T,n-p-1}(-|\mathfrak{z}_j|). \tag{12.114}$$

Are these two tests and two p-values different? First, notice that there is a relationship between the t- and F-distributions. Namely, if $T \sim \mathscr{T}(d)$, then $T^2 \sim \mathfrak{F}(1,d)$. Indeed, using the definitions (12.196) and (12.105) of t- and F-distributions, we have[64]:

[64] Here, $Z \sim \mathscr{N}(0,1)$, $Q \sim \chi^2_d$, and Z and Q are independent.

$$T = Z\sqrt{\frac{d}{Q}} \sim \mathcal{T}(d) \quad \Longrightarrow \quad T^2 = \frac{Z^2/1}{Q/d} \sim \mathfrak{F}(1,d). \tag{12.115}$$

Second, it can be shown that the F-statistic equals the squared z-score[65]:

[65] This technical result is not trivial and requires both geometric perspective and linear algebraic arguments.

$$\mathfrak{f} = \mathfrak{z}_j^2. \tag{12.116}$$

Armed with these two results, let's find the relationship between the two p-values:

$$\begin{aligned} \mathfrak{p}_j^T &= 2F_{T,n-p-1}(-|\mathfrak{z}_j|) = \mathbb{P}(|\underbrace{T}_{\mathcal{T}(n-p-1)}| \geq |\mathfrak{z}_j|) = \mathbb{P}(\underbrace{T^2}_{\mathfrak{F}(1,n-p-1)} \geq \mathfrak{z}_j^2) \\ &= \mathbb{P}(T^2 \geq \mathfrak{f}) = 1 - F_{1,n-p-1}(\mathfrak{f}) = \mathfrak{p}. \end{aligned} \tag{12.117}$$

So, in this case, the t-test and F-test are identical and produce the same p-values.

12.7 Confidence Intervals for Regression Coefficients

In addition to testing hypotheses, the parametric normality assumption

$$E_1,\dots,E_n \overset{\text{iid}}{\sim} \mathcal{N}(0,\sigma^2) \quad \text{or, equivalently,} \quad E \sim \mathcal{N}(0,\sigma^2\mathbb{I}_n) \tag{12.118}$$

allows also to construct confidence intervals for regression coefficients.

A $100(1-\alpha)\%$ confidence interval for the regression coefficient β_j is a random interval computed from the data $(\mathbb{X},Y)$ that contains β_j with probability $(1-\alpha)$. So, to construct a confidence interval for β_j, we need to find an interval $I_j = (a_j,b_j)$, such that

$$\mathbb{P}(a_j < \beta_j < b_j) = 1-\alpha, \qquad \alpha \in (0,1). \tag{12.119}$$

Recall that if (12.118) holds, then β_j is normally distributed[66]:

[66] We established this in Section 12.5 in (12.90).

$$\hat{\beta}_j \sim \mathcal{N}\left(\beta_j, \sigma^2\xi_j\right), \qquad \text{where } \xi_j = \left((\mathbb{X}^T\mathbb{X})^{-1}\right)_{jj}. \tag{12.120}$$

Estimating the unknown variance $\sigma^2 \approx \hat{\sigma}_u^2$,

$$\hat{\beta}_j \mathrel{\dot{\sim}} \mathcal{N}\left(\beta_j, \hat{\sigma}_u^2\xi_j\right) \quad \text{and} \quad \frac{\hat{\beta}_j - \beta_j}{\hat{\sigma}_u\sqrt{\xi_j}} \mathrel{\dot{\sim}} \mathcal{N}(0,1). \tag{12.121}$$

Therefore,

$$\mathbb{P}\left(z_{\frac{\alpha}{2}} < \frac{\hat{\beta}_j - \beta_j}{\hat{\sigma}_u\sqrt{\xi_j}} < -z_{\frac{\alpha}{2}}\right) \approx 1-\alpha, \tag{12.122}$$

where $z_{\frac{\alpha}{2}} = \Phi^{-1}(\frac{\alpha}{2})$ is the standard normal quantile. Rewriting the inequalities explicitly in terms of β_j, we have:

$$\mathbb{P}\left(\hat{\beta}_j + z_{\frac{\alpha}{2}}\hat{\sigma}_u\sqrt{\xi_j} < \beta_j < \hat{\beta}_j - z_{\frac{\alpha}{2}}\hat{\sigma}_u\sqrt{\xi_j}\right) \approx 1-\alpha. \tag{12.123}$$

Therefore,

$$I_\alpha^j = \left(\hat{\beta}_j + z_{\frac{\alpha}{2}}\hat{\sigma}_u\sqrt{\xi_j},\ \hat{\beta}_j - z_{\frac{\alpha}{2}}\hat{\sigma}_u\sqrt{\xi_j}\right) \tag{12.124}$$

is an approximate $100(1-\alpha)\%$ confidence interval for β_j. The width of the confidence interval I_α^j can be viewed as a measure of uncertainty in the value of the regression coefficient β_j.

Now, after we have learned how to estimate the parameters of the MLR model, construct confidence intervals for the regression coefficients, and test and rank predictors for significance, let's consider an application of the model to the analysis of the prostate cancer data.

12.8 Case Study: Prostate Cancer

Some things are up to us and some are not up to us.
– Epictetus, *The Encheiridion*

The greatest of all contests – the struggle not to be overwhelmed by anything that happens.
– Marcus Aurelius, *Meditations*

We must take a less serious view of all things, tolerating them in a spirit of acceptance: it is more human to laugh at life than to weep tears over it.
– Seneca, *On the Tranquility of Mind*

The prostate is a small, walnut-shaped organ in males, located below the bladder, which helps to produce semen. Prostate cancer is one of the most common types of cancer affecting men. For example, in the United States, it is the second most common cancer in men after skin cancer and the second-leading cause of cancer death behind only lung cancer [9]. The risk of getting prostate cancer depends on many factors, including age: it rises rapidly after age 50; race: prostate cancer occurs more often in African men and less often in Asian men; family history: having a father or brother with prostate cancer more than doubles the risk of developing this cancer[67]; and it may depend on diet: men who eat high-fat diets and consume a lot of dairy products seem to have a higher chance of getting prostate cancer [10, 11]. Usually, prostate cancer grows slowly and, if diagnosed at an early stage, can be successfully treated. The catch is that at early stages, a person with cancer often does not have any symptoms, which often appear only at later stages of cancer development.

[67] Interestingly, the risk is higher for men who have a brother with prostate cancer compared to those who have a father with it.

In this Case Study, we will perform a regression analysis of the prostate cancer data, which was obtained and studied in [12] and also analyzed in [13]. The data contains clinical measurements of $p = 8$ predictors and one response variable, collected from $n = 97$ men who were scheduled to receive a radical prostatectomy (a surgical procedure to remove the prostate and some surrounding tissue after a prostate cancer diagnosis). Here is the list of predictors and the response variable:

- $X^{(1)} = lcavol$: log of the cancer volume,
- $X^{(2)} = lweight$: log of the prostate weight,
- $X^{(3)} = age$: patient age,
- $X^{(4)} = lbph$: log of the amount of benign prostatic hyperplasia[68],
- $X^{(5)} = svi$: seminal vesicle invasion[69],
- $X^{(6)} = lcp$: log of the capsular penetration[70],
- $X^{(7)} = gleason$: Gleason score[71],
- $X^{(8)} = pgg45$: percent of Gleason scores 4 or 5,
- $Y = lpsa$: log of the level of prostate-specific antigen.

The response Y is the log of the level of prostate-specific antigen (PSA), which is a protein produced by cells in the prostate. Cancer cells usually produce more PSA than normal cells. Therefore, high levels of PSA indicate a high chance of having prostate cancer. The level of PSA is measured by a blood test, called the PSA test, which often serves as an initial screening test for prostate cancer. However, the PSA test cannot conclude with certainty whether a man has prostate cancer since there is no cutoff value of the PSA level that can definitely indicate the presence of cancer. If the result of the PSA test is abnormal (PSA level is high), then a prostate biopsy is needed to determine with certainty if cancer is indeed present.

Figure 12.5 shows the scatterplots of the response Y versus all eight predictors[72]. Some predictors, such as *lcavol*, *lweight*, and *svi*, appear to be positively correlated with the response *lpsa*. The dependence between other predictors and the response looks more complicated. Some predictors, such as *svi* and *gleason*, have a very small range of values[73], and some are more widely spread.

Let's consider the multivariate linear regression model for this data:

$$Y_i = \beta_0 + \beta_1 X_i^{(1)} + \ldots + \beta_p X_i^{(p)} + E_i, \qquad i = 1, \ldots, n. \tag{12.125}$$

To estimate the model parameters and quantify its prediction accuracy, let's randomly split the set $\mathscr{D}$ of all $n = 97$ observations into two disjoint subsets: the training set $\mathscr{D}_t$ of size $n_t = 70$ and the validation set $\mathscr{D}_v$ of size $n_v = 27$. We will use the training data for fitting the model and testing hypotheses about its parameters, and the validation data for estimating the model prediction accuracy[74].

The OLS estimates of the regression coefficients based on the training data are given in Table 12.1.

$\hat{\beta}_0$	$\hat{\beta}_1$	$\hat{\beta}_2$	$\hat{\beta}_3$	$\hat{\beta}_4$	$\hat{\beta}_5$	$\hat{\beta}_6$	$\hat{\beta}_7$	$\hat{\beta}_8$
-0.098	0.521	0.499	-0.025	0.126	0.942	-0.051	0.215	0.003

Table 12.1 The OLS estimates of the regression coefficients of the MLR model (12.125), computed based on the randomly selected training data $\mathscr{D}_t \subset \mathscr{D}$ of size $n_t = 70$.

[68] BPH is a non-cancerous condition that occurs when the prostate grows larger than normal due to an overgrowth of cells.

[69] SVI is a condition when a tumor penetrates the muscular wall of the seminal vesicles. Typically caused by prostate cancer, but can also be caused by bladder or rectal cancer.

[70] Capsular penetration is a condition when prostate cancer reaches the outer wall of the prostate.

[71] The Gleason score describes how abnormal the cancer cells are and how quickly they are likely to grow and spread. Usually it ranges between 6 and 10, with 6 being the lowest grade cancer.

[72] The prostate cancer data used in this Case Study is available at `https://link.springer.com/chapter/10.1007/978-3-032-03848-7_1`.

[73] For instance, $svi = 1$ if prostate cancer reaches the outer wall of the prostate, and $svi = 0$ otherwise.

[74] See a short discussion on the validation set approach to model selection at the end of Section 12.5.

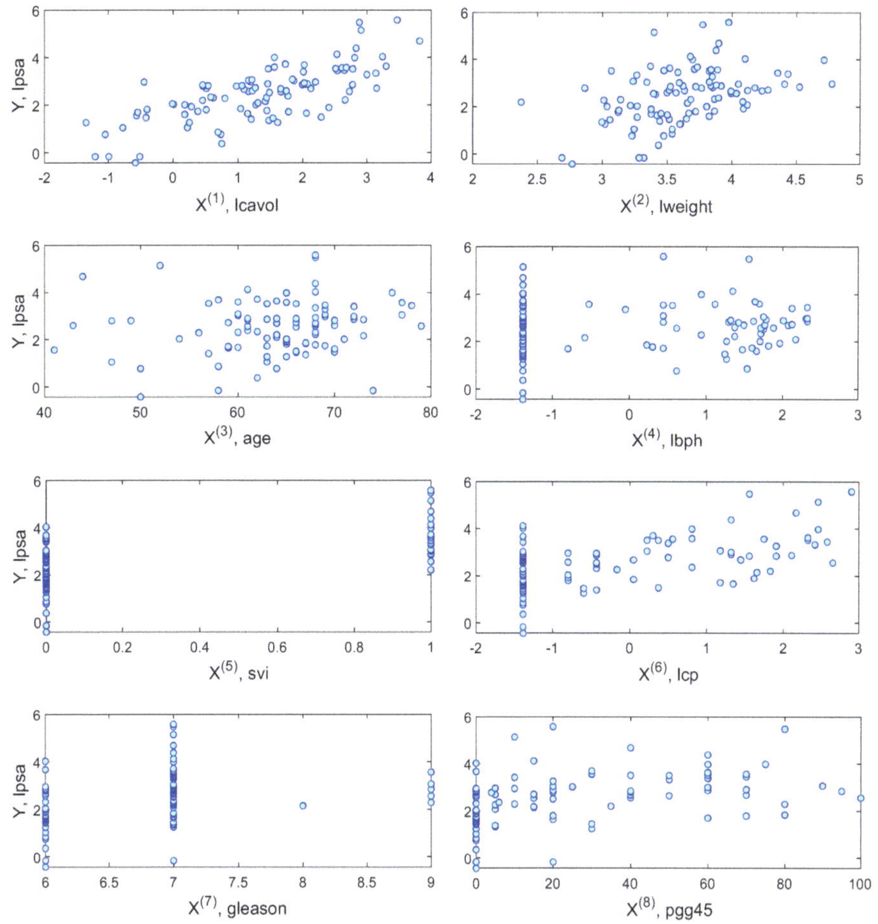

Fig. 12.5 Scatterplots of the response variable Y versus the eight predictors $X^{(1)},\dots,X^{(8)}$.

These estimates are computed via $\hat{\beta} = (\mathbb{X}_t^T\mathbb{X}_t)^{-1}\mathbb{X}_t^T Y_t$, where $\mathbb{X}_t$ is the $n_t \times (p+1)$ training design matrix, obtained from the full design matrix $\mathbb{X}$ by keeping only n_t rows that correspond to the training observations, and $Y_t \in \mathbb{R}^{n_t}$ is the vector of training responses, obtained from the full vector of responses Y by keeping only n_t entries that correspond to the training observations.

Note that the magnitudes of the OLS estimates cannot be considered directly as indicators of the importance or relevance of the corresponding predictors, since these magnitudes also depend on the scales of predictors. For instance, $\hat{\beta}_8$ has the smallest magnitude, but the values of the corresponding predictor $X^{(8)}$ =*pgg45* can be as large as 100, making the contribution term $\hat{\beta}_8 X^{(8)}$ potentially substantial. It can be shown[75] that the OLS estimates $\hat{\beta}_j$ are *equivariant* under the scaling of predictors: if $X^{(j)}$ is transformed as $X^{(j)} \to \alpha X^{(j)}$, then the corresponding OLS estimate $\hat{\beta}_j$ will be transformed as $\hat{\beta}_j \to \alpha^{-1}\hat{\beta}_j$, so that the contribution $\hat{\beta}_j X^{(j)}$ remains un-

[75] And Problem 12.2 will ask you to do that.

changed. This means that the magnitudes of the OLS estimates can be controlled by rescaling the corresponding predictors.

Remark 12.17. To eliminate the influence of different scales of predictors on the fitted model, we can *standardize* the predictors in the training portion of the data, that is, transform them so that they have zero mean and unit variance: for each $i \in \mathscr{D}_t$,

$$X_i^{(j)} \to \frac{X_i^{(j)} - \mu_j}{\sigma_j}, \quad \mu_j = \frac{1}{n_t} \sum_{i \in \mathscr{D}_t} X_i^{(j)}, \quad \sigma_j^2 = \frac{1}{n_t - 1} \sum_{i \in \mathscr{D}_t} (X_i^{(j)} - \mu_j)^2. \quad (12.126)$$

Standardization is strictly required for some statistical methods[76], but for the multivariate linear regression it is not mathematically necessary[77]. For the MLR model, standardization will affect the values of the OLS estimates and the confidence intervals for the regression coefficients, but it will not affect the z-scores of predictors, the corresponding p-values, and the model prediction accuracy.

[76] Such as ridge regression, LASSO, and support vector machines.

[77] Although it could be beneficial from a numerical point of view if the magnitudes of predictors are very different.

Assuming that the prediction errors $E_1, \ldots, E_n$ satisfy the parametric normality assumption (12.88), Table 12.2 provides the z-scores (12.93) associated with the predictors, the corresponding p-values (12.95) and (12.98) based on the Wald and t-tests, and the approximate 95% confidence intervals (12.124) for the regression parameters. All these quantities are computed based only on the training data.

Predictor	OLS $\hat{\beta}_j$	z-score $\mathfrak{z}_j$	p-value $\mathfrak{p}_j^W$	p-value $\mathfrak{p}_j^T$	95% CI $I_{0.05}^j$
lcavol	0.521	4.894	10^{-6}	7.6×10^{-6}	(0.312, 0.729)
lweight	0.499	2.130	0.0331	0.0372	(0.040, 0.958)
age	-0.025	-1.975	0.0483	0.0528	(-0.049, -0.0002)
lbph	0.126	1.763	0.0780	0.0830	(-0.014, 0.266)
svi	0.942	3.234	0.0012	0.0020	(0.371, 1.514)
lcp	-0.051	-0.474	0.6355	0.6372	(-0.264, 0.161)
gleason	0.215	0.945	0.3448	0.3485	(-0.231, 0.661)
pgg45	0.003	0.597	0.5508	0.5530	(-0.007, 0.013)

Table 12.2 The OLS estimates, z-scores, p-values, and confidence intervals for the regression coefficients, computed based on the randomly selected training data $\mathscr{D}_t \subset \mathscr{D}$ of size $n_t = 70$.

One interesting observation is that the p-values obtained from the t-test are consistently larger than the corresponding p-values obtained by the Wald test. This is a direct consequence of the fact that the t-distribution has longer tails than the standard normal distribution[78]. It implies that the exact t-test is always more conservative than the approximate Wald term with respect to rejecting the null hypothesis and declaring a predictor as insignificant. Another observation from the table is that a predictor $X^{(j)}$ is not significant at the 5% level ($p_j^W > 0.05$) if and only if the 95% confidence interval for β_j contains zero ($0 \in I_{0.05}^j$). Is this a mere coincidence or a general property? Problem 12.4 will ask you to investigate this question.

[78] In other words, the t-tails contain more probability mass than the normal tails. See Fig. 10.2.

There are four predictors that, according to both the Wald test and t-test, are not significant at the 5% level: *lbph*, *lcp*, *gleason*, and *pgg45*. The two test disagree on *age*: $\mathfrak{p}_3^W < 0.05$ but $\mathfrak{p}_3^T > 0.05$. The remaining three predictors, *lcavol*, *lweight*, and *svi*, are significant according to both tests.

Remark 12.18. One may notice a seeming discrepancy between the statement at the beginning of this section that age is a risk factor for prostate cancer and the empirical result that age is not significant, or at least, weakly significant. In fact, there is no discrepancy, since the empirical results are obtained not from a random sample of men, but from a sample of men who were already diagnosed with prostate cancer. The empirical results merely suggest that age is not important for predicting the level of PSA in men with prostate cancer.

Let's use the F-test to test the simultaneous significance of $k = 5$ predictors: *age*, *lbph*, *lcp*, *gleason*, and *pgg45*. That is, let's test

$$H_0 : (\beta_3, \beta_4, \beta_6, \beta_7, \beta_8) = 0 \quad \text{vs} \quad H_1 : (\beta_3, \beta_4, \beta_6, \beta_7, \beta_8) \neq 0 \tag{12.127}$$

using the training data $\mathscr{D}_t$. The value of the F-statistic (12.106) is

$$\mathfrak{f} = \frac{n_t - p - 1}{p - \tilde{p}} \cdot \frac{\|\tilde{\mathbb{X}}_t \hat{\tilde{\beta}} - Y_t\|^2 - \|\mathbb{X}_t \hat{\beta} - Y_t\|^2}{\|\mathbb{X}_t \hat{\beta} - Y_t\|^2} = 1.616. \tag{12.128}$$

Here, $n_t = 70$, $p = 8$, $\tilde{p} = 3$, $\tilde{\mathbb{X}}_t$ is the $n_t \times (\tilde{p} + 1)$ reduced training design matrix obtained from the training design matrix $\mathbb{X}_t$ by removing columns 3, 4, 6, 7, and 8, and $\hat{\tilde{\beta}} = (\tilde{\mathbb{X}}_t^T \tilde{\mathbb{X}}_t)^{-1} \tilde{\mathbb{X}}_t^T Y_t$ is the OLS estimate of the vector of regression coefficients of the reduced MLR model, obtained from the full model (12.125) by excluding predictors *age*, *lbph*, *lcp*, *gleason*, and *pgg45*. The p-value (12.108) produced by the F-test is

$$\mathfrak{p} = 1 - F_{p-\tilde{p}, n-p-1}(\mathfrak{f}) = 1 - F_{5,61}(\mathfrak{f}) = 0.169. \tag{12.129}$$

Therefore, the training data provides no evidence against the null hypothesis: the five predictors are not simultaneously significant even at the 15% level.

Now, in addition to the full MLR model (12.125), let's consider the following two regression models for the cancer data:

- The *base model*, which assumes that none of the predictors is relevant for predicting the response and has the following trivial form:

$$Y_i = \beta_0^b + E_i, \qquad i = 1, \ldots, n. \tag{12.130}$$

  The residual sum of squares in this case is $\text{RSS}(\beta_0^b) = \sum_{i=1}^n (Y_i - \beta_0^b)^2$, which is minimized at the mean response $\hat{\beta}_0^b = \overline{Y}$. Fitting the base model to the training data $\mathscr{D}_t$ yields

$$\hat{\beta}_0^b = \overline{Y}_t = 2.564. \tag{12.131}$$

- The *reduced model*, which is obtained from the full model by excluding *age*, *lbph*, *lcp*, *gleason*, and *pgg45* from the list of predictors:

$$Y_i = \tilde{\beta}_0 + \tilde{\beta}_1 X_i^{(1)} + \tilde{\beta}_2 X_i^{(2)} + \tilde{\beta}_5 X_i^{(5)} + E_i, \qquad i = 1, \ldots, n. \tag{12.132}$$

The vector of OLS estimates of regression coefficients is $\hat{\hat{\beta}} = (\tilde{\mathbb{X}}_t^T \tilde{\mathbb{X}}_t)^{-1} \tilde{\mathbb{X}}_t^T Y_t$, and its components are given in Table 12.3.

$\hat{\hat{\beta}}_0$	$\hat{\hat{\beta}}_1$	$\hat{\hat{\beta}}_2$	$\hat{\hat{\beta}}_5$
-0.349	0.540	0.555	0.844

Table 12.3 The OLS estimates of the regression coefficients of the reduced MLR model (12.132), computed based on the randomly selected training data $\mathscr{D}_t \subset \mathscr{D}$ of size $n_t = 70$.

Let's compare the base, full, and reduced regression models by estimating their prediction accuracy using the validation data $\mathscr{D}_v$. Namely, let's compute their test errors (12.101), that is, the mean squared errors on the validation data. For the base model, whose prediction for every input is the average training response, $\hat{\beta}_0^b = \overline{Y}_t = 2.564$, we have:

$$\mathrm{Err}_b = \frac{1}{n_v} \sum_{Y_i \in \mathscr{D}_v} (Y_i - \hat{\beta}_0^b)^2 = 0.949. \tag{12.133}$$

The test error of the base model is often referred to as the *base error rate*. For the full MLR model,

$$\mathrm{Err}_f = \frac{1}{n_v} \sum_{(X_i, Y_i) \in \mathscr{D}_v} (Y_i - X_i^T \hat{\beta})^2 = 0.583, \tag{12.134}$$

where $X_i^T = (1, X_i^{(1)}, \ldots, X_i^{(p)})$ and $\hat{\beta}$ is the OLS estimate whose components are given in Table 12.1. The full model clearly outperforms the base model on the validation set, suggesting that the eight predictors used in the full model are indeed important and relevant for predicting the response. Finally, for the reduced model,

$$\mathrm{Err}_r = \frac{1}{n_v} \sum_{(X_i, Y_i) \in \mathscr{D}_v} (Y_i - \tilde{X}_i^T \hat{\hat{\beta}})^2 = 0.479, \tag{12.135}$$

where $\tilde{X}_i^T = (1, X_i^{(1)}, X_i^{(2)}, X_i^{(5)})$ is obtained from X_i^T by removing the five excluded predictors, and $\hat{\hat{\beta}}$ is the OLS estimate whose components are given in Table 12.3. So, it appears beneficial to exclude the non-significant predictors from the model, since the model becomes simpler (easier to interpret) and slightly more accurate (at least on the given validation data). Of course, this conclusion requires further investigation involving larger data sets as well as the medical expertise.

Remark 12.19. We conclude this section by noting that if the standardization of predictors (12.126) was applied to the training part of the data, then absolutely the same transformation — with μ_j and σ_j computed on $\mathscr{D}_t$ — must be applied to the validation part of the data before computing the prediction errors in (12.134) and (12.135).

In other words, the transformation (12.126) must be applied to $X_i^{(j)}$ not only for all $i \in \mathscr{D}_t$, but also for all $i \in \mathscr{D}_v$. However, as stated previously, standardization of predictors is not necessary for fitting the MLR model.

12.9 Multivariate Multiple Linear Regression

We have started our study of linear regression in Chapter 11 with the simplest model, the simple linear regression (SLR) model, where both the input X and the response Y are scalar quantities. The multiple linear regression (MLR) model considered in this chapter extends the SLR model for the case, where the input is a vector consisting of p different predictors. Our last goal is to extend the MLR model for the case, where the response Y is a vector of q responses.

So, suppose that the overall system response Y we are interested in is a vector of q scalar responses $Y^{(1)},\dots,Y^{(q)}$,

$$Y = \begin{bmatrix} Y^{(1)} \\ \vdots \\ Y^{(q)} \end{bmatrix} = [Y^{(1)},\dots,Y^{(q)}]^T \in \mathbb{R}^q. \tag{12.136}$$

Let $X_1,\dots,X_n \in \mathbb{R}^{p+1}$ be the input vectors and $Y_1,\dots,Y_n \in \mathbb{R}^q$ be the corresponding system responses:

$$Y_1 = \begin{bmatrix} Y_1^{(1)} \\ \vdots \\ Y_1^{(q)} \end{bmatrix}, \quad \dots, \quad Y_n = \begin{bmatrix} Y_n^{(1)} \\ \vdots \\ Y_n^{(q)} \end{bmatrix} \tag{12.137}$$

Given the training data $(X_1,Y_1),\dots,(X_n,Y_n)$, let's consider the MLR model for each scalar response:

$$\begin{aligned} Y_i^{(k)} &= X_i^T\beta^{(k)} + E_i^{(k)}, \qquad k=1,\dots,q, \;\; i=1,\dots,n, \\ &= \beta_0^{(k)} + \beta_1^{(k)}X_i^{(1)} + \dots + \beta_p^{(k)}X_i^{(p)} + E_i^{(k)}, \end{aligned} \tag{12.138}$$

where $\beta^{(k)} = (\beta_0^{(k)},\dots,\beta_p^{(k)})^T \in \mathbb{R}^{p+1}$ is the vector of regression coefficients[79] and $E_i^{(k)}$ is the prediction error for the kth scalar response in the ith observation.

[79] Its own vector $\beta^{(k)}$ for each scalar response $Y^{(k)}$.

We can rewrite (12.138) in a more compact matrix form. To this end, let's combine all training response vectors $Y_1,\dots,Y_n$ into a single *response matrix*:

$$\mathbb{Y} = \begin{bmatrix} Y_1^T \\ \vdots \\ Y_n^T \end{bmatrix} = \underbrace{\begin{bmatrix} Y_1^{(1)} & \dots & Y_1^{(q)} \\ \vdots & \vdots & \vdots \\ Y_n^{(1)} & \dots & Y_n^{(q)} \end{bmatrix}}_{n\times q \text{ matrix}} \tag{12.139}$$

The (i,k) entry of the response matrix, $\mathbb{Y}_{ik} = Y_i^{(k)}$, is the value of the kth scalar response $Y^{(k)}$ in the ith observation. Next, let's combine all vectors of the regression coefficients $\beta^{(1)}, \ldots, \beta^{(q)}$ into a single ***coefficient matrix***:

$$B = \left[\beta^{(1)} \ldots \beta^{(q)}\right] = \underbrace{\begin{bmatrix} \beta_0^{(1)} & \ldots & \beta_0^{(q)} \\ \vdots & \vdots & \vdots \\ \beta_p^{(1)} & \ldots & \beta_p^{(q)} \end{bmatrix}}_{(p+1)\times q \text{ matrix}} \tag{12.140}$$

Finally, let E be the $n \times q$ matrix of prediction errors:

$$E = \underbrace{\begin{bmatrix} E_1^{(1)} & \ldots & E_1^{(q)} \\ \vdots & \vdots & \vdots \\ E_n^{(1)} & \ldots & E_n^{(q)} \end{bmatrix}}_{n\times q \text{ matrix}} \tag{12.141}$$

The collection of q MLR models (12.138) is then equivalent to

$$\mathbb{Y} = \mathbb{X}B + E, \tag{12.142}$$

where $\mathbb{X}$ is, as before, the $n \times (p+1)$ design matrix (12.10). This model is called the *multivariate multiple linear regression* (MMLR) model for the training data $(X_1, Y_1), \ldots, (X_n, Y_n)$. Note that if $q = 1$, then the MMLR model reduces to the MLR model for a scalar response. If, in addition, $p = 1$, then it becomes the SLR model.

We can estimate the regression coefficients in matrix B by the OLS method by minimizing the *residual sum of squares*. As in the cases of SLR and MLR models, the RSS is defined as the sum of squared differences between the true responses $Y_i^{(k)}$ and the corresponding predicted responses $X_i^T \beta^{(k)}$ to input vectors X_i,

$$\text{RSS}(B) = \sum_{k=1}^{q} \sum_{i=1}^{n} \left(Y_i^{(k)} - X_i^T \beta^{(k)}\right)^2. \tag{12.143}$$

To rewrite it in a coordinate-invariant matrix form, recall that in the univariate case of MLR model,

$$\text{RSS}(\beta) = \sum_{i=1}^{n} (Y_i - X_i^T \beta)^2 = (Y - \mathbb{X}\beta)^T (Y - \mathbb{X}\beta). \tag{12.144}$$

By analogy with the MLR case, let's consider the $q \times q$ matrix $(\mathbb{Y} - \mathbb{X}B)^T (\mathbb{Y} - \mathbb{X}B)$. Its (s,t) entry is

$$
\begin{aligned}
\left[(\mathbb{Y}-\mathbb{X}B)^T(\mathbb{Y}-\mathbb{X}B)\right]_{st} &= \sum_{i=1}^{n}[(\mathbb{Y}-\mathbb{X}B)^T]_{si}[\mathbb{Y}-\mathbb{X}B]_{it} \\
&= \sum_{i=1}^{n}[\mathbb{Y}-\mathbb{X}B]_{is}[\mathbb{Y}-\mathbb{X}B]_{it} = \sum_{i=1}^{n}\left(Y_i^{(s)}-X_i^T\beta^{(s)}\right)\left(Y_i^{(t)}-X_i^T\beta^{(t)}\right).
\end{aligned}
\tag{12.145}
$$

So, to get the RSS in (12.143), we need to set $s=t=k$ and sum all diagonal (k,k) entries of the matrix $(\mathbb{Y}-\mathbb{X}B)^T(\mathbb{Y}-\mathbb{X}B)$:

$$
\mathrm{RSS}(B) = \sum_{k=1}^{q}\left[(\mathbb{Y}-\mathbb{X}B)^T(\mathbb{Y}-\mathbb{X}B)\right]_{kk} = \mathrm{tr}\left[(\mathbb{Y}-\mathbb{X}B)^T(\mathbb{Y}-\mathbb{X}B)\right]. \tag{12.146}
$$

The OLS estimate $\hat{B}$ of the coefficient matrix B is the global minimizer of the RSS, that is, the value of B that makes the RSS as small as possible:

$$
\hat{B} = \arg\min_{B\in\mathbb{R}^{(p+1)\times q}} \mathrm{RSS}(B). \tag{12.147}
$$

The MMLR model is the collection of q MLR models, and the MMLR $\mathrm{RSS}(B)$ decomposes into the sum of MLR $\mathrm{RSS}(\beta^{(k)})$:

$$
\mathrm{RSS}(B) = \sum_{k=1}^{q}\underbrace{\sum_{i=1}^{n}\left(Y_i^{(k)}-X_i^T\beta^{(k)}\right)^2}_{\mathrm{RSS}(\beta^{(k)})} = \sum_{k=1}^{q}\mathrm{RSS}(\beta^{(k)}). \tag{12.148}
$$

This means that the minimization of $\mathrm{RSS}(B)$ with respect to the full coefficient matrix B is equivalent to the minimization of $\mathrm{RSS}(\beta^{(k)})$ with respect to $\beta^{(k)}$, the kth column of B, for all $k=1,\ldots,q$. But the global minimizer of $\mathrm{RSS}(\beta^{(k)})$ is simply the OLS estimate $\hat{\beta}^{(k)}$ of $\beta^{(k)}$ in the MLR model for the kth scalar response $Y^{(k)}$, which is given by[80]

$$
\hat{\beta}^{(k)} = \begin{bmatrix}\hat{\beta}_0^{(k)}\\ \vdots \\ \hat{\beta}_p^{(k)}\end{bmatrix} = (\mathbb{X}^T\mathbb{X})^{-1}\mathbb{X}^T\begin{bmatrix}Y_1^{(k)}\\ \vdots \\ Y_n^{(k)}\end{bmatrix} = (\mathbb{X}^T\mathbb{X})^{-1}\mathbb{X}^T Y^{(k)}. \tag{12.149}
$$

[80] Let's assume for simplicity that $\mathrm{Ker}\,\mathbb{X}=\{0\}$. Otherwise, the expression for the OLS estimate is $\hat{\beta}^{(k)}=\mathbb{X}^+Y^{(k)}+\mathrm{Ker}\,\mathbb{X}$.

The OLS estimates $\hat{\beta}^{(1)},\ldots,\hat{\beta}^{(q)}$ are the columns of the OLS estimate $\hat{B}$ of the coefficient matrix. Therefore,

$$
\hat{B} = (\mathbb{X}^T\mathbb{X})^{-1}\mathbb{X}^T\mathbb{Y}. \tag{12.150}
$$

This formula mimics the formula (12.28) for the OLS estimate in the case of the MLR model. The only difference is that instead of the vector of training responses Y on the right-hand-side, we now have the response matrix $\mathbb{Y}$.

12.10 Case Study: Mosaic Management Problem

> *Art is very different from science in that it creates worlds; it creates meaning and beauty and forms and structures and natures, while science discovers them.*
>
> – Peter Kreeft, *The Philosophy of Tolkien*

> *We are mosaics. Pieces of light, love, history, stars... Glued together with magic and music and words.*
>
> – Anita Krizzan

A *mosaic* is an art form — an abstract pattern, decorative design, or image — created by assembling small pieces of colored stone, glass, ceramic, or other materials. The first known mosaics were created in Mesopotamia[81] in the 3rd millennium BC, and they became particularly popular and widespread in Ancient Greece, Ancient Rome, and the Byzantine Empire. Mosaics were often used in Christian and early Islamic art. The individual pieces a mosaic is made of are called *tesserae*. While traditional tesserae are mostly natural pebbles and pieces of stone and glass, modern mosaics can be made of essentially any material including shells, coins, and bottle caps. Figure 12.6 shows an example of modern mosaic depicting the Nobel Prize-winning physicist Richard Feynman, located in Pasadena not far from the Caltech campus[82].

[81] A historic region where modern Iraq is located.

[82] More precisely, it is located on the back of a Ross Dress for Less clothing store on Shopper's Lane.

Fig. 12.6 A mosaic titled "The Motion of the Planet" was created by the artist Gifford Myers in 1997 in memory of Richard Feynman (1918-1988), a theoretical physicist who made seminal contributions to quantum electrodynamics and was a legendary professor at Caltech.

In this Case Study, we will study a statistical problem that originated from a consulting project[83]. A mosaic company assembles and sells $p = 100$ different types of mosaics. Each mosaic is made of different numbers of tesserae (mosaic pieces) of different types. For example, the mosaic in Fig. 12.6 consists of tesserae of three types: square tiles (336 pieces), blue irregular-shaped tiles (96 pieces), and spherical tiles (4 pieces). In total, the company uses $q = 1000$ distinct types of tesserae.

[83] The nature of the project was modified.

At the end of every week, the company management receives the data on how many mosaics of each type were assembled during that week and how many tesserae of each type were used to assemble those mosaics. Unfortunately, the numbers of tesserae are contaminated by errors, since some tesserae get broken during the assembling process, some are added to the inventory during the week, and some are simply hard to count. For example, it is not easy to count the number of blue irregular-shaped tiles in Fig. 12.6. Nevertheless, given $n = 200$ weekly observations, the company management wants to infer how many tesserae of each type are required for making a mosaic of each type.

Let's start by formalizing this problem. To describe the relationship between the mosaics and tesserae, let's consider the following function:

$$f : \mathbb{R}^p \to \mathbb{R}^q, \quad Y = f(X), \quad X = \begin{bmatrix} X^{(1)} \\ \vdots \\ X^{(p)} \end{bmatrix}, \quad Y = \begin{bmatrix} Y^{(1)} \\ \vdots \\ Y^{(q)} \end{bmatrix}, \tag{12.151}$$

where $X^{(j)}$ is the number of mosaics of type j and $Y^{(k)}$ is the number of tesserae of type k. This function describes how many tesserae of each type $Y^{(1)}, \ldots, Y^{(q)}$ are required to assemble given numbers of mosaics of each type $X^{(1)}, \ldots, X^{(p)}$. The quantities of interest are the values of f on the vectors of the standard basis of $\mathbb{R}^p$:

$$f_1 = f(1, 0, \ldots, 0), \ldots, f_p = f(0, \ldots, 0, 1), \tag{12.152}$$

since the components of the vector $f_j \in \mathbb{R}^q$ describes precisely how many tesserae of each type are needed for making one mosaic of type j.

Let $X_i = (X_i^{(1)}, \ldots, X_i^{(p)})^T$, where $X_i^{(j)}$ is the number of mosaics of type j made during week i, and let $Y_i = (Y_i^{(1)}, \ldots, Y_i^{(q)})^T$, where $Y_i^{(k)}$ is the number of tesserae of type k used during week i. We can think of $X_1, \ldots, X_n$ and $Y_1, \ldots, Y_n$ as observations of vectors X and $Y = f(X)$. Since the numbers of tesserae (components of Y_i) are contaminated by errors, we have

$$Y_i = f(X_i) + E_i, \quad i = 1, \ldots, n, \tag{12.153}$$

where $E_i \in \mathbb{R}^q$ is the vector of errors. The component $E_i^{(k)}$ of E_i is the error made in counting the number of tesserae of type k during week i. So, the problem is, given the data $(X_1, Y_1), \ldots, (X_n, Y_n)$, to infer the values of vectors $f_1, \ldots, f_p$.

A simple yet key observation that connects this problem to multivariate multiple linear regression is that f is a linear function[84]:

[84] If this is not obvious, convince yourself that this is true and obvious.

$$f(X+\tilde{X}) = f(X) + f(\tilde{X}) \quad \text{and} \quad f(\alpha X) = \alpha f(X). \tag{12.154}$$

Any linear function $f : \mathbb{R}^p \to \mathbb{R}^q$ can be written in a matrix form

$$f(X) = FX, \tag{12.155}$$

where F is a $q \times p$ matrix. Indeed, if $e_1, \dots, e_p$ are the vectors of the standard basis of $\mathbb{R}^p$, then, using linearity of f, we have:

$$\begin{aligned} f(X) &= f\left(\sum_{j=1}^{p} X^{(j)} e_j\right) = \sum_{j=1}^{p} X^{(j)} \underbrace{f(e_j)}_{f_j} = \sum_{j=1}^{p} X^{(j)} f_j \\ &= \underbrace{[f_1 \dots f_p]}_{F} X = FX \end{aligned} \tag{12.156}$$

Therefore, equations (12.153) can be rewritten as follows:

$$Y_i = FX_i + E_i, \quad i = 1, \dots, n, \tag{12.157}$$

and our goal is to infer the columns of matrix F from the data $(X_1, Y_1), \dots, (X_n, Y_n)$.

Let $\mathbb{X}$ be the design matrix and $\mathbb{Y}$ be the response matrix:

$$\mathbb{X} = \begin{bmatrix} X_1^T \\ \vdots \\ X_n^T \end{bmatrix} = \begin{bmatrix} X_1^{(1)} & \dots & X_1^{(p)} \\ \vdots & \vdots & \vdots \\ X_n^{(1)} & \dots & X_n^{(p)} \end{bmatrix} \quad \text{and} \quad \mathbb{Y} = \begin{bmatrix} Y_1^T \\ \vdots \\ Y_n^T \end{bmatrix} = \begin{bmatrix} Y_1^{(1)} & \dots & Y_1^{(q)} \\ \vdots & \vdots & \vdots \\ Y_n^{(1)} & \dots & Y_n^{(q)} \end{bmatrix} \tag{12.158}$$

Figure 12.7 visualizes these matrices[85]. Notice how darker rows of $\mathbb{X}$, representing weeks with fewer assembled mosaics, correspond to darker rows of $\mathbb{Y}$, representing weeks with fewer used tesserae.

The equations (12.157) can be written in the matrix form:

$$\mathbb{Y} = \mathbb{X} F^T + E, \tag{12.159}$$

where $E = [E^{(1)} \dots E^{(q)}]$ is the $n \times q$ matrix of errors. Notice that this is precisely the MMLR model (12.142), where the role of the coefficient matrix B is played by the matrix F^T, and the design matrix $\mathbb{X}$ does not have the column of ones[86].

In this application, we cannot assume that the prediction errors $E_i^{(k)}$, where $i = 1, \dots, n$ and $k = 1, \dots, q$, satisfy some of the "standard" assumptions that are often made for prediction errors in regression analysis. For example, it makes no sense to assume that $E_i^{(k)}$ is normally distributed, since the errors are integers. Furthermore, since some types of tesserae are easier to break and some types are harder to accurately count than others, the homoscedastic assumption of constant variance for the errors is also not justified.

Fortunately, we can still use the formula (12.150) for the OLS estimate of the coefficient matrix, since it was derived without any assumptions on the errors. And

[85] The (synthetically generated) data for this case Study is available at https://link.springer.com/chapter/10.1007/978-3-032-03848-7_1.

[86] Because in this case, it is not needed.

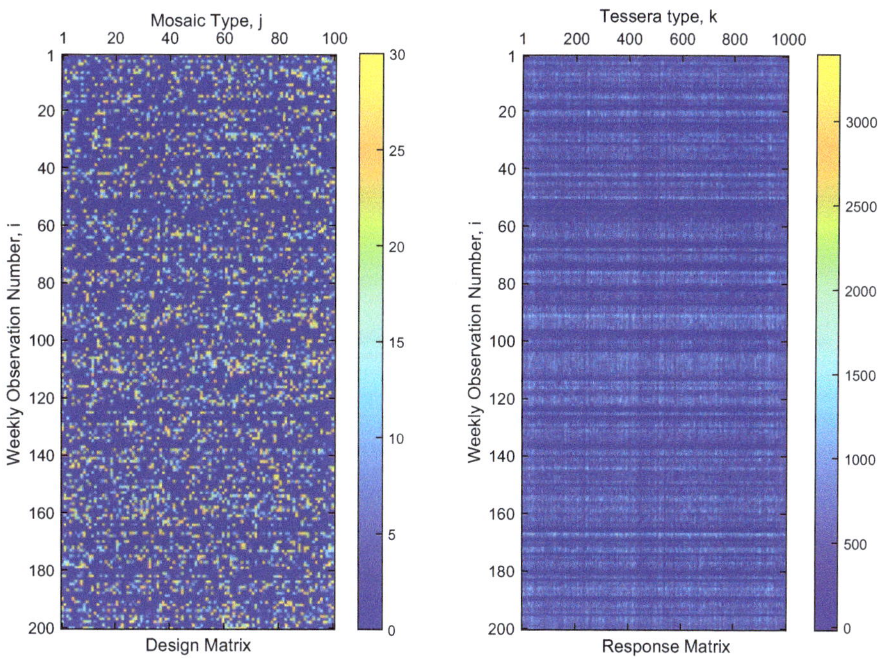

Fig. 12.7 The $n \times p$ design matrix $\mathbb{X}$ (left panel) and the $n \times q$ response matrix $\mathbb{Y}$ (right panel).

so,

$$\hat{F} = \hat{B}^T = \left((\mathbb{X}^T\mathbb{X})^{-1}\mathbb{X}^T\mathbb{Y}\right)^T = \mathbb{Y}^T\mathbb{X}(\mathbb{X}^T\mathbb{X})^{-1}. \quad (12.160)$$

The columns of matrix F can be estimated by the columns of matrix $\hat{F}$;

$$f_j \approx \hat{f}_j = \hat{F}e_j, \quad j = 1, \ldots, p. \quad (12.161)$$

The kth component $\hat{f}_j^{(k)}$ of vector $\hat{f}_j \in \mathbb{R}^q$ is the OLS estimate of the number of tesserae of type k required for assembling a mosaic of type j.

Figure 12.8 shows the results (components of $\hat{f}_j$) for the mosaic of type $j = 30$. According to the estimate, 23 different types of tesserae are needed to create this mosaic. However, if we look carefully, we will notice that the numbers of tesserae are not integers. For example, the number of tesserae of type 36 is 16.11, of type 114 is 8.99, of type 152 is 9.94, etc. Furthermore, the number of non-required tesserae is approximately, but not exactly, zero, fluctuating between -0.15 and 0.15. This observation should not be surprising, since the data in the response matrix $\mathbb{Y}$ was contaminated by errors, and also because we should not expect the matrix $\hat{F}$ computed numerically via (12.160) to have precisely integer values.

To enhance the OLS estimates (12.161), we can round all entries of $\hat{F}$ to the closest integer. After this enhancement, the OLS estimates become very accurate[87] even when the error magnitudes are relatively large. The reader is encouraged to

[87] Since the data for this Case Study was synthetically generated, the true values of tesserae per mosaic are known and available at `https://link.springer.com/chapter/10.1007/978-3-032-03848-7_1`.

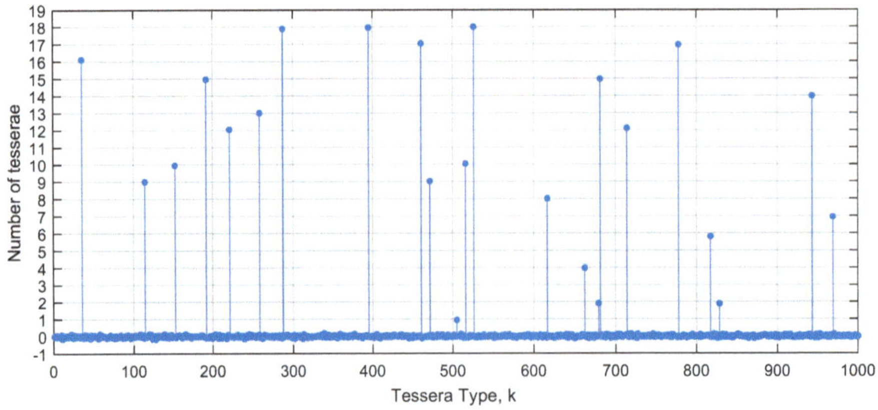

Fig. 12.8 Types and numbers of tesserae required for assembling one mosaic of type $j = 30$.

explore how the accuracy of this MMLR-based method depends on the number n of weekly observations. In particular, what happens if $n < p$?

12.11 Further Reading

This is not the end. It is not even the beginning of the end.
But it is, perhaps, the end of the beginning.
– Winston Churchill, *Speech at the Lord Mayor's Luncheon*

In this final chapter, we discussed two useful extensions of the simple linear regression model, which are often used in data analysis: the *multiple linear regression model*, for problems with multiple predictors, and the *multivariate multiple linear regression model*, where, in addition to multiple predictors, there are also several response variables.

A popular method for estimating the regression coefficients in all linear regression models is the *method of ordinary least squares*. Its popularity is based on the *Gauss-Markov theorem*[88] that states that, under relatively weak assumptions, the OLS estimate is *BLUE*: the best linear unbiased estimate. In other words, among all linear unbiased estimates, the OLS estimate has the smallest possible variance or, equivalently, the smallest possible mean squared error (MSE). But what if we do not care about linearity and unbiasedness, and simply want to find estimates of regression coefficients, possibly biased and nonlinear, which have small MSEs and lead to the fitted model with high prediction accuracy? Can we beat the OLS estimates and the OLS fitted model?

The answer is yes: there are several methods that produce biased nonlinear estimates and consistently outperform OLS. One of them is a "shrinkage method" called *ridge regression*, also known as *Tikhonov regularization*[89]. The main idea

[88] See Section 12.4.1.

[89] The method was originally developed in the context of integral equations by A.N. Tikhonov [14] in 1943, and then it was independently rediscovered in the statistical context by A.E. Hoerl and R. Kennard [15] in 1970.

behind ridge regression is to shrink the regression coefficients towards zero by penalizing their larger values. Specifically, the vector $\hat{\beta}_R$ of ridge estimates is obtained by solving the following optimization problem:

$$\hat{\beta}_R = \arg \min_{\beta \in \mathbb{R}^{p+1}} \left(\text{RSS}(\beta) + \lambda \sum_{j=1}^{p} \beta_j^2 \right), \tag{12.162}$$

where $\lambda \geq 0$ is a tuning parameter that controls how strongly the regression coefficients corresponding to predictors are "regularized" and shrunk towards zero. If $\lambda = 0$, then ridge regression reduces to OLS regression. If $\lambda \to \infty$, then the shrinkage penalty term becomes infinite, and ridge regression converges to the base model, where all $\beta_1 = \ldots = \beta_p = 0$ and there is only a constant term β_0.

Ridge regression often outperforms OLS regression in terms of *prediction accuracy*[90], especially in the case where the Gram matrix $\mathbb{X}^T\mathbb{X}$ is singular or ill-conditioned (close to being singular). This situation occurs when the predictors are linearly dependent or strongly correlated[91], and also in the case of high-dimensional data, where $p \gg n$[92]. Over the years, the concept of ridge regression has become one of the most essential and fruitful ideas in data analysis that has been used in many fields and inspired other statistical methods. An overview of ridge regression and its "magic" and "beauty" is provided by Trevor Hastie in

- Hastie, T: Ridge regularization: an essential concept in data science. *Technometrics* 62(4), 426-433 (2020).
`https://doi.org/10.1080/00401706.2020.1791959`

[90] After all, OLS is a special case of ridge regression with $\lambda = 0$.

[91] This situation is called *multicollinearity*.

[92] "High-dimensional" or "wide" data sets often occur in bio-medical applications.

Although ridge regression often makes more accurate predictions compared to OLS regression, it does not improve the OLS fitted model in terms of *model interpretability*: just like OLS regression, ridge regression includes all p predictors in the model, even if some of them are not relevant. Ridge regression shrinks the regression coefficients towards zero, but it does not shrink them all the way to zero[93]. As a result, ridge regression is not capable of excluding predictors from the model and improving the model interpretation by keeping only those predictors that are most strongly associated with the response and relevant for prediction.

[93] Unless $\lambda = \infty$, and the model degenerates to the base model.

In Section 12.5, we discussed a simple strategy for ranking and excluding less significant predictors from the model based on their z-scores (12.100). However, this strategy makes sense statistically only if the prediction errors are normally distributed (12.88), an assumption which may or may not hold and which is not easy to check. Moreover, after some predictors are excluded by setting their regression coefficients to zero, the remaining coefficients are still estimated via the plain-vanilla OLS, thereby leaving some room for improving the prediction accuracy.

A conceptually more sound and practically more popular method for improving both prediction accuracy and model interpretability is another shrinkage method called the *LASSO*, which stands for the Least Absolute Shrinkage and Selection Operator[94]. Inspired by ridge regression, the main idea behind the LASSO is also to shrink the regression coefficients towards zero, but using a different penalty term. Namely, the vector $\hat{\beta}_L$ of LASSO estimates is obtained by solving the following

[94] The method was originally introduced by F. Santosa and W.W. Symes in the context of seismology [16] in 1986, and then it was rediscovered and popularized in the statistical literature by R. Tibshirani [17] in 1996.

optimization problem:

$$\hat{\beta}_L = \arg\min_{\beta \in \mathbb{R}^{p+1}} \left(\text{RSS}(\beta) + \lambda \sum_{j=1}^{p} |\beta_j| \right), \tag{12.163}$$

where $\lambda \geq 0$ is a tuning regularization parameter. Like ridge regression, the LASSO interpolates between OLS regression and the base model as λ increases from zero to infinity. The key difference between the LASSO and ridge regression is that the l_1 penalty term $\lambda \sum_{j=1}^{p} |\beta_j|$ not only shrinks the regression coefficients towards zero, like the l_2 penalty term $\lambda \sum_{j=1}^{p} \beta_j^2$, but also forces some of the coefficients to be exactly zero when λ is sufficiently large yet finite[95]. The LASSO is thus capable of excluding predictors from the model and thereby improving the model interpretability. A great monograph treatment of the LASSO and its generalizations, such as the *elastic net* and *group lasso*, is provided in

[95] The reason behind this lies in the difference between the geometric properties of the l_1 and l_2 penalty terms.

- Hastie, T., Tibshirani, R., Wainwright, M: *Statistical Learning with Sparsity: The Lasso and Generalizations*. Chapman and Hall, CRC Press (2015). `https://hastie.su.domains/StatLearnSparsity/`

Both ridge regression and the LASSO define a family of regression models parameterized by the tuning parameter λ, which controls the strength of regularization. Different values of λ lead to different estimates of the regression coefficients and, therefore, to different fitted models that have different prediction accuracy. How to choose the value of λ? How to choose between ridge regression and the LASSO[96]? In other words, how to select the "best" model with the highest prediction accuracy out of (possibly infinitely) many candidate models obtained by different methods and different fitting techniques? This important task is called *model selection*.

[96] None of the two methods universally dominates the other in terms of prediction accuracy.

At the end of Section 12.5, we discussed the most straightforward approach to model selection, called the *validation set approach*, and illustrated it in the case study in Section 12.8. In this approach, the full data set is split into the training data, which is used for estimating the regression coefficients, and the validation data, which is used for estimating the prediction accuracy of the fitted models. But this simple method has several drawbacks. For instance, the best selected model depends on the partition of the data set into the training and validation subsets, and different partitions may and often do lead to different best models. Over the years, a number of methods for model selection has been proposed and used in various applications, such as the *Akaike information criterion* [18], the *Bayesian information criterion* [19], and *cross-validation* [20], to name but a few. An overview of model selection teachniques together with practical recommendations is given in

- Ding, J., Tarokh, V., Yang, Y: Model selection techniques: an overview. *IEEE Signal Processing Magazine* 35(6), 16-34 (2018). `https://doi.org/10.1109/MSP.2018.2867638`

12.12 Appendix: The Normal Equation Always Has a Solution

The goal of this Appendix is to prove that the normal equation always has a solution. Namely, for any matrix $\mathbb{X}$ and any vector Y there exists vector β satisfying[97]

$$\mathbb{X}^T\mathbb{X}\beta = \mathbb{X}^T Y. \tag{12.164}$$

[97] We obviously assume that $\mathbb{X}, Y$, and β have appropriate dimensions, so that all multiplications in (12.164) do exist.

The normal equation (12.164) has a solution if and only if the vector $\mathbb{X}^T Y$ belongs to the image[98] of matrix $\mathbb{X}^T\mathbb{X}$,

[98] See Marginal Note 11.

$$\mathbb{X}^T Y \in \operatorname{Im}\mathbb{X}^T\mathbb{X}. \tag{12.165}$$

Since $\mathbb{X}^T Y \in \operatorname{Im}\mathbb{X}^T$, it is, therefore, sufficient to prove that

$$\operatorname{Im}\mathbb{X}^T\mathbb{X} = \operatorname{Im}\mathbb{X}^T. \tag{12.166}$$

As a preliminary step, let's show that

$$\operatorname{Ker}\mathbb{X}^T\mathbb{X} = \operatorname{Ker}\mathbb{X}. \tag{12.167}$$

If $b \in \operatorname{Ker}\mathbb{X}$, then, by definition of the kernel[99], $\mathbb{X}b = 0$. Multiplying both sides by $\mathbb{X}^T$ from the left yields $\mathbb{X}^T\mathbb{X}b = 0$, which means that $b \in \operatorname{Ker}\mathbb{X}^T\mathbb{X}$. This proves that $\operatorname{Ker}\mathbb{X} \subset \operatorname{Ker}\mathbb{X}^T\mathbb{X}$. On the other hand, if $b \in \operatorname{Ker}\mathbb{X}^T\mathbb{X}$, then $\mathbb{X}^T\mathbb{X}b = 0$. Multiplying both sides by b^T from the left gives $b^T\mathbb{X}^T\mathbb{X}b = (\mathbb{X}b)^T\mathbb{X}b = \|\mathbb{X}b\|^2 = 0$. Since only zero vector has zero norm, we have that $\mathbb{X}b = 0$, which means that $b \in \operatorname{Ker}\mathbb{X}$. This proves that $\operatorname{Ker}\mathbb{X}^T\mathbb{X} \subset \operatorname{Ker}\mathbb{X}$. Here is a word-free summary of the above arguments[100]:

[99] See Marginal Note 13.

[100] "Mathematics is the esoteric language of civilization." – Yuri Manin.

$$\begin{aligned}
&\underline{\operatorname{Ker}\mathbb{X} \subset \operatorname{Ker}\mathbb{X}^T\mathbb{X}}:\\
&b \in \operatorname{Ker}\mathbb{X} \implies \mathbb{X}b = 0 \implies \mathbb{X}^T\mathbb{X}b = 0 \implies b \in \operatorname{Ker}\mathbb{X}^T\mathbb{X}\\
&\underline{\operatorname{Ker}\mathbb{X}^T\mathbb{X} \subset \operatorname{Ker}\mathbb{X}}:\\
&b \in \operatorname{Ker}\mathbb{X}^T\mathbb{X} \implies \mathbb{X}^T\mathbb{X}b = 0 \implies b^T\mathbb{X}^T\mathbb{X}b = (\mathbb{X}b)^T\mathbb{X}b = 0\\
&\qquad \implies \|\mathbb{X}b\| = 0 \implies \mathbb{X}b = 0 \implies b \in \operatorname{Ker}\mathbb{X}
\end{aligned} \tag{12.168}$$

So, $\operatorname{Ker}\mathbb{X}$ and $\operatorname{Ker}\mathbb{X}^T\mathbb{X}$ are subsets of each other and, therefore, are the same sets.

Now, to prove (12.166) using (12.167), we need to apply the fundamental theorem of linear algebra. With every $n \times m$ matrix A we can associate four *fundamental subspaces*: its kernel, image, cokernel, and coimage,

- *Kernel* (nullspace): $\operatorname{Ker}A = \{x \in \mathbb{R}^m \mid Ax = 0\}$ is a subspace of $\mathbb{R}^m$,
- *Image* (range or column space): $\operatorname{Im}A = \{Ax \mid x \in \mathbb{R}^m\}$ is a subspace of $\mathbb{R}^n$,
- *Cokernel*: $\operatorname{coKer}A = \operatorname{Ker}A^T = \{x \in \mathbb{R}^n \mid A^T x = 0\}$ is a subspace of $\mathbb{R}^n$,
- *Coimage* (row space): $\operatorname{coIm}A = \operatorname{Im}A^T = \{A^T x \mid x \in \mathbb{R}^n\}$ is a subspace of $\mathbb{R}^m$.

The *fundamental theorem of linear algebra* [21] states that $\operatorname{Ker}A$ and $\operatorname{coIm}A$ are not just arbitrary subspaces of $\mathbb{R}^m$, but they are *orthogonal complements*[101] of each

[101] The *orthogonal complement* of a subspace V of $\mathbb{R}^n$ is the subspace $V^\perp$ of all vectors in $\mathbb{R}^n$ that are orthogonal to all vectors in V, i.e. $V^\perp = \{u \in \mathbb{R}^n \mid u^T v = 0 \;\; \forall v \in V\}$. It can be shown that $(V^\perp)^\perp = V$.

other. Similarly, ImA and coKer$A$ are orthogonal complements in $\mathbb{R}^n$:

$$\text{coIm}A = (\text{Ker}A)^{\perp} \quad \text{and} \quad \text{Im}A = (\text{coKer}A)^{\perp}. \tag{12.169}$$

These orthogonal relationships between the four fundamental matrix subspaces are schematically illustrated in Fig. 12.9.

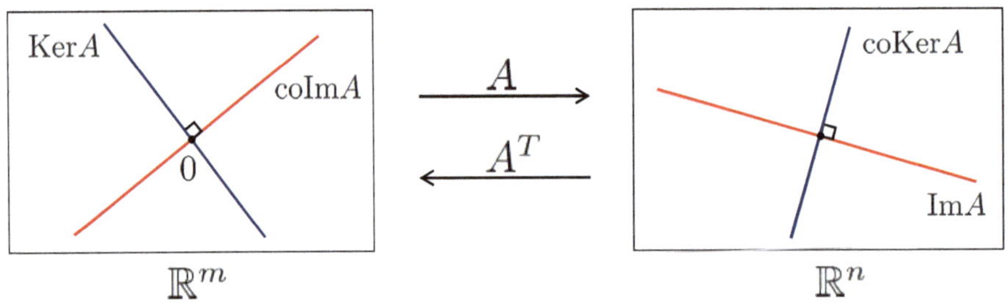

Fig. 12.9 Orthogonal complementarity of the fundamental subspaces associated with a matrix.

Now we are finally ready to prove (12.166). Using (12.167), (12.169), and the definitions of the cokernel and coimage, we have[102]:

[102] "Keep breathing, that's the key, breathe!" – Gimli, *The Lord of the Rings*.

$$\begin{aligned}\text{Im}\,\mathbb{X}^T\mathbb{X} &= \left(\text{coKer}\,\mathbb{X}^T\mathbb{X}\right)^{\perp} = \left(\text{Ker}\left(\mathbb{X}^T\mathbb{X}\right)^T\right)^{\perp} = \left(\text{Ker}\,\mathbb{X}^T\mathbb{X}\right)^{\perp} \\ &= (\text{Ker}\,\mathbb{X})^{\perp} = \text{coIm}\,\mathbb{X} = \text{Im}\,\mathbb{X}^T.\end{aligned} \tag{12.170}$$

Having established that, the rest (as it was pointed out at the beginning) is straightforward. For any matrix $\mathbb{X}$ and vector Y, the vector $\mathbb{X}^T Y \in \text{Im}\,\mathbb{X}^T = \text{Im}\,\mathbb{X}^T\mathbb{X}$. The latter, by the definition of the image, means that there exists a vector β such that $\mathbb{X}^T\mathbb{X}\beta = \mathbb{X}^T Y$, and, thus, the normal equation $\mathbb{X}^T\mathbb{X}\beta = \mathbb{X}^T Y$ has a solution.

12.13 Appendix: Properties of the Pseudoinverse

The goal of this Appendix is to prove two properties of the pseudoinverse matrix that were used in Section 12.2 for deriving the general formula for the OLS estimate

$$\hat{\beta} = \mathbb{X}^{+}Y + \text{Ker}\,\mathbb{X}. \tag{12.171}$$

If $A = P\Sigma Q^T$ is the SVD of an $m \times n$ matrix A of rank $r > 0$, then its pseudoinverse A^+ is defined in terms of the factors P, Σ and Q as follows[103]:

[103] "One of the great powers of the mathematical method is that I can define things how ever I like." – John Conway, *The Power of Mathematics*.

$$A^{+} = Q\Sigma^{-1}P^T. \tag{12.172}$$

The first property states that if A is nonsingular, then its pseudoinverse coincides with its ordinary inverse, $A^+ = A^{-1}$. To prove this, notice that if A is nonsingular, then it is a square matrix of full rank $r = m = n$, and, therefore, all matrices P, Σ and Q are square and nonsingular. Moreover, since P is $m \times m$ with orthonormal columns, $P^T P = \mathbb{I}_m$, it is an *orthogonal* matrix and its inverse is $P^{-1} = P^T$. Similarly, Q is an $n \times n$ orthogonal matrix, $Q^{-1} = Q^T$. Now we are ready to compute the inverse of A and check that it coincides with its pseudoinverse:

$$A^{-1} = \left(P\Sigma Q^T\right)^{-1} = (Q^T)^{-1}\Sigma^{-1}P^{-1} = Q\Sigma^{-1}P^T = A^+. \tag{12.173}$$

The second property claims that if A (not necessarily square) has a trivial kernel[104], $\mathrm{Ker}A = \{0\}$, then its pseudoinverse can be expressed in terms on A itself, without computing its SVD and then using (12.172). Namely,

[104] In other words, $x = 0$ is the unique solution of equation $Ax = 0$.

$$A^+ = (A^T A)^{-1} A^T. \tag{12.174}$$

To prove this, let's use *fundamental theorem of linear algebra* [21], which states that the dimension of kernel $\mathrm{Ker}A \subset \mathbb{R}^n$ is

$$\dim \mathrm{Ker}A = n - r. \tag{12.175}$$

Since the kernel is trivial, its dimension is zero, and, therefore, the rank of A is $r = n$. But this implies that matrix Q is $n \times n$ with orthonormal columns, $Q^T Q = \mathbb{I}_n$, and, thus, it is orthogonal, $Q^{-1} = Q^T$. Note that if A is not square, $m \neq n$, then P is not orthogonal (since it is not not square); it only has orthonormal columns, $P^T P = \mathbb{I}_n$. Now we are ready to compute $(A^T A)^{-1} A^T$ and check that it coincides with the pseudoinverse:

$$\begin{aligned}(A^T A)^{-1} A^T &= \left((P\Sigma Q^T)^T P\Sigma Q^T\right)^{-1} (P\Sigma Q^T)^T \\ &= \big(Q\Sigma \underbrace{P^T P}_{\mathbb{I}_n} \Sigma Q^T\big)^{-1} Q\Sigma P^T = \left(Q\Sigma^2 Q^T\right)^{-1} Q\Sigma P^T \\ &= (Q^T)^{-1}\Sigma^{-2}Q^{-1}Q\Sigma P^T = Q\Sigma^{-1}P^T = A^+.\end{aligned} \tag{12.176}$$

12.14 Appendix: Unbiased Estimate of σ^2

The goal of this Appendix is to prove that

$$\hat{\sigma}_u^2 = \frac{1}{n-p-1}\sum_{i=1}^{n}(Y_i - \hat{Y}_i)^2 \tag{12.177}$$

is an unbiased[105] estimate of the variance σ^2 of prediction errors $E_1, \ldots, E_n$, which is a parameter of the multiple linear regression model. The appearance of the sum of squared residuals $\sum_{i=1}^{n}(Y_i - \hat{Y}_i)^2$ in the expression for $\hat{\sigma}_u^2$ is somewhat expected, at least intuitively, based on Fig. 12.3. However, the factor $1/(n-p-1)$ looks strange[106]. To convince ourselves that this is the correct factor, let's compute the

[105] Even under the weak nonparametric assumptions on the prediction errors (12.59).

[106] At first glance, either $1/n$ or $1/(n-1)$ might appear more natural.

expected value of $\sum_{i=1}^{n}(Y_i-\hat{Y}_i)^2$ to verify that it equals $(n-p-1)\sigma^2$. This will imply that $\hat{\sigma}_u^2$ is unbiased.

First, let's rewrite the residual sum of squares $\sum_{i=1}^{n}(Y_i-\hat{Y}_i)^2$ in a matrix form. Since the vector of responses $Y=\mathbb{X}\beta+E$ and the vector of fitted values $\hat{Y}=\mathbb{X}\hat{\beta}$, we have:

$$\begin{aligned}\sum_{i=1}^{n}(Y_i-\hat{Y}_i)^2 &= (Y-\hat{Y})^T(Y-\hat{Y}) = (\mathbb{X}\beta+E-\mathbb{X}\hat{\beta})^T(\mathbb{X}\beta+E-\mathbb{X}\hat{\beta})\\ &= (E-\mathbb{X}(\hat{\beta}-\beta))^T(E-\mathbb{X}(\hat{\beta}-\beta))\\ &= E^TE-(\hat{\beta}-\beta)^T\mathbb{X}^TE-E^T\mathbb{X}(\hat{\beta}-\beta)+(\hat{\beta}-\beta)^T\mathbb{X}^T\mathbb{X}(\hat{\beta}-\beta).\end{aligned} \tag{12.178}$$

Note that the two middle terms are exactly the same. Indeed, since $(\hat{\beta}-\beta)\in\mathbb{R}^{p+1}$, $E\in\mathbb{R}^n$, and $\mathbb{X}$ is $n\times(p+1)$, both terms are scalars (1×1 matrices). But a scalar $a\in\mathbb{R}$ is invariant[107] with respect to transposition: $a^T=a$. Therefore,

[107] An object a is "invariant" with respect to some transformation f if it "does not change" under that transformation: $f(a)=a$.

$$\underbrace{(\hat{\beta}-\beta)^T\mathbb{X}^TE}_{\in\mathbb{R}} = \left((\hat{\beta}-\beta)^T\mathbb{X}^TE\right)^T = E^T\mathbb{X}(\hat{\beta}-\beta). \tag{12.179}$$

So, the residual sum of squares is

$$\sum_{i=1}^{n}(Y_i-\hat{Y}_i)^2 = E^TE-2E^T\mathbb{X}(\hat{\beta}-\beta)+(\hat{\beta}-\beta)^T\mathbb{X}^T\mathbb{X}(\hat{\beta}-\beta). \tag{12.180}$$

Let's find the expected value of each of the three terms. For the first one, it is easy:

$$\mathbb{E}[E^TE] = \mathbb{E}\left[\sum_{i=1}^{n}E_i^2\right] = \sum_{i=1}^{n}\mathbb{E}[E_i^2] = \sum_{i=1}^{n}\Big(\underbrace{\mathbb{V}[E_i]}_{\sigma^2}+\underbrace{\mathbb{E}[E_i]^2}_{0}\Big) = n\sigma^2. \tag{12.181}$$

The expected value of the (half of) second term is[108]

[108] In what follows, we try to express everything in terms of the vector of statistical errors E, since E is where the randomness comes from in the MLR model.

$$\begin{aligned}\mathbb{E}[E^T\mathbb{X}(\hat{\beta}-\beta)] &= \mathbb{E}\Big[E^T\mathbb{X}\Big(\underbrace{(\mathbb{X}^T\mathbb{X})^{-1}\mathbb{X}^TY}_{\hat{\beta}}-\beta\Big)\Big]\\ &= \mathbb{E}\Big[E^T\mathbb{X}\Big((\mathbb{X}^T\mathbb{X})^{-1}\mathbb{X}^T\underbrace{(\mathbb{X}\beta+E)}_{Y}-\beta\Big)\Big]\\ &= \mathbb{E}\Big[E^T\mathbb{X}\Big(\beta+(\mathbb{X}^T\mathbb{X})^{-1}\mathbb{X}^TE-\beta\Big)\Big] = \mathbb{E}\Big[E^T\mathbb{X}(\mathbb{X}^T\mathbb{X})^{-1}\mathbb{X}^TE\Big].\end{aligned} \tag{12.182}$$

How to proceed next? Imagine, for a moment, that the factor $E^T\mathbb{X}$, instead of being at the front, were at the end, and instead of $E^T\mathbb{X}(\mathbb{X}^T\mathbb{X})^{-1}\mathbb{X}^TE$, we had $(\mathbb{X}^T\mathbb{X})^{-1}\mathbb{X}^TEE^T\mathbb{X}$. Then it would be much easier to compute its expectation, since the random vectors E and E^T would be next to each other, and $\mathbb{E}[(\mathbb{X}^T\mathbb{X})^{-1}\mathbb{X}^TEE^T\mathbb{X}] = (\mathbb{X}^T\mathbb{X})^{-1}\mathbb{X}^T\mathbb{E}[EE^T]\mathbb{X}$, where $\mathbb{E}[EE^T]$ is simply the covariance matrix of E, which equals[109] $\sigma^2\mathbb{I}_n$. But, unfortunately, we can't move $E^T\mathbb{X}$ from the front to the end, since matrices don't commute,

[109] See (12.60).

$$\underbrace{E^T\mathbb{X}}(\mathbb{X}^T\mathbb{X})^{-1}\mathbb{X}^TE \neq (\mathbb{X}^T\mathbb{X})^{-1}\mathbb{X}^TE\underbrace{E^T\mathbb{X}}. \tag{12.183}$$

In general, matrices don't commute, $AB \neq BA$, but they always commute under the trace[110]! Namely, if AB is a square matrix, then $\mathrm{tr}(AB) = \mathrm{tr}(BA)$. This is straightforward to check using the definitions of trace and matrix multiplication. But how can this property be useful in our case? Note that $E^T\mathbb{X}(\mathbb{X}^T\mathbb{X})^{-1}\mathbb{X}^TE$ is a scalar:

[110] If A is a square $n \times n$ matrix, then its *trace* is the sum of its diagonal elements: $\mathrm{tr}A = \sum_{i=1}^n A_{ii}$.

$$\underbrace{E^T}_{1\times n}\ \underbrace{\mathbb{X}}_{n\times(p+1)}\ \underbrace{(\mathbb{X}^T\mathbb{X})^{-1}}_{(p+1)\times(p+1)}\ \underbrace{\mathbb{X}^T}_{(p+1)\times n}\ \underbrace{E}_{n\times 1} \in \mathbb{R}. \tag{12.184}$$

But any scalar $a \in \mathbb{R}$ can be viewed as a 1×1 matrix, and it coincides with its trace: $\mathrm{tr}\,a = a$. This observation allows us to replace $E^T\mathbb{X}(\mathbb{X}^T\mathbb{X})^{-1}\mathbb{X}^TE$ with its trace, and then, under the trace, we will be able to move $E^T\mathbb{X}$ from the front to the end (since, under the trace, matrices do commute). Let's execute this plan:

$$\begin{aligned}\mathbb{E}\left[E^T\mathbb{X}(\mathbb{X}^T\mathbb{X})^{-1}\mathbb{X}^TE\right] &= \mathbb{E}\left[\mathrm{tr}\left(\underbrace{E^T\mathbb{X}}(\mathbb{X}^T\mathbb{X})^{-1}\mathbb{X}^TE\right)\right]\\ &= \mathbb{E}\left[\mathrm{tr}\left((\mathbb{X}^T\mathbb{X})^{-1}\mathbb{X}^TE\underbrace{E^T\mathbb{X}}\right)\right].\end{aligned} \tag{12.185}$$

Finally, using the fact that the two operations — computing the expected value $\mathbb{E}$ and computing the trace tr — do commute[111], we have:

[111] Indeed: $\mathbb{E}[\mathrm{tr}A] = \mathbb{E}[\sum_{i=1}^n A_{ii}] = \sum_{i=1}^n \mathbb{E}[A_{ii}] = \mathrm{tr}(\mathbb{E}[A])$.

$$\begin{aligned}\mathbb{E}[E^T\mathbb{X}(\hat{\beta}-\beta)] &= \mathbb{E}\left[\mathrm{tr}\left((\mathbb{X}^T\mathbb{X})^{-1}\mathbb{X}^TEE^T\mathbb{X}\right)\right]\\ &= \mathrm{tr}\left(\mathbb{E}\left[(\mathbb{X}^T\mathbb{X})^{-1}\mathbb{X}^TEE^T\mathbb{X}\right]\right)\\ &= \mathrm{tr}\left((\mathbb{X}^T\mathbb{X})^{-1}\mathbb{X}^T\underbrace{\mathbb{E}\left[EE^T\right]}_{\sigma^2\mathbb{I}_n}\mathbb{X}\right)\\ &= \mathrm{tr}\left(\sigma^2(\mathbb{X}^T\mathbb{X})^{-1}\mathbb{X}^T\mathbb{X}\right) = \mathrm{tr}\left(\sigma^2\mathbb{I}_{p+1}\right) = \sigma^2(p+1).\end{aligned} \tag{12.186}$$

Remark 12.20. It is worth summarizing the method that we used for computing the expected value of the second term of (12.180), since it is often used in various settings. Suppose that we want to compute the expected value $\mathbb{E}[X]$ of a random variable X, which is hard to compute directly. Suppose that X can be represented as the dot product of two random k-vectors U and V, that is, $X = U^TV$. Then we can rewrite $\mathbb{E}[X]$ as follows:

$$\mathbb{E}[X] = \mathbb{E}[\mathrm{tr}X] = \mathbb{E}[\mathrm{tr}(U^TV)] = \mathbb{E}[\mathrm{tr}(VU^T)] = \mathrm{tr}(\mathbb{E}[VU^T]). \tag{12.187}$$

Note that, while U^TV is a random variable, VU^T is a random $k \times k$ matrix. If $\mathbb{E}[VU^T]$ is easy to compute[112], then (12.187) can be used for computing $\mathbb{E}[X]$. We will refer to this method as the *trace trick*.

[112] Or, at least, the expected values of the diagonal elements of VU^T are easy to compute.

Now let's compute the expected value of last term of (12.180). Since it is a scalar,

$$\underbrace{(\hat{\beta}-\beta)^T}_{1\times(p+1)} \underbrace{\mathbb{X}^T\mathbb{X}}_{(p+1)\times(p+1)} \underbrace{(\hat{\beta}-\beta)}_{(p+1)\times 1} \in \mathbb{R}, \tag{12.188}$$

we can use the trace trick again[113]:

$$\begin{aligned}
\mathbb{E}[(\hat{\beta}-\beta)^T\mathbb{X}^T\mathbb{X}(\hat{\beta}-\beta)] &= \mathbb{E}\Big[\mathrm{tr}\big(\underbrace{(\hat{\beta}-\beta)^T}\mathbb{X}^T\mathbb{X}(\hat{\beta}-\beta)\big)\Big] \\
&= \mathbb{E}\Big[\mathrm{tr}\big(\mathbb{X}^T\mathbb{X}(\hat{\beta}-\beta)\underbrace{(\hat{\beta}-\beta)^T}\big)\Big] \\
&= \mathrm{tr}\Big(\mathbb{E}\big[\mathbb{X}^T\mathbb{X}(\hat{\beta}-\beta)(\hat{\beta}-\beta)^T\big]\Big) \\
&= \mathrm{tr}\Big(\mathbb{X}^T\mathbb{X}\mathbb{E}\big[(\hat{\beta}-\beta)(\hat{\beta}-\beta)^T\big]\Big).
\end{aligned} \tag{12.189}$$

[113] "But above all, he loved sitting in the shade of a sprawling tree, poring over thick books with rough covers, filling white pages with elegant lines of mathematical symbols, and unconsciously listening to the rustle of leaves overhead as the dazzling sun hung motionless in the purest blue sky." – Arkady and Boris Strugatsky, *The Land of Crimson Clouds*.

Since the OLS estimate is unbiased, $\mathbb{E}[\hat{\beta}] = \beta$, the factor $\mathbb{E}\big[(\hat{\beta}-\beta)(\hat{\beta}-\beta)^T\big]$ is simply the covariance matrix $\Sigma_{\hat{\beta}}$ of $\hat{\beta}$, which is given by (12.64). Therefore,

$$\begin{aligned}
\mathbb{E}[(\hat{\beta}-\beta)^T\mathbb{X}^T\mathbb{X}(\hat{\beta}-\beta)] &= \mathrm{tr}\Big(\mathbb{X}^T\mathbb{X}\sigma^2(\mathbb{X}^T\mathbb{X})^{-1}\Big) \\
&= \mathrm{tr}\Big(\sigma^2\mathbb{I}_{p+1}\Big) = \sigma^2(p+1).
\end{aligned} \tag{12.190}$$

Finally, combining all three terms (12.181), (12.186), and (12.190) and plugging them into $\mathbb{E}[(12.180)]$, we have:

$$\mathbb{E}\left[\sum_{i=1}^{n}(Y_i-\hat{Y}_i)^2\right] = n\sigma^2 - 2\sigma^2(p+1) + \sigma^2(p+1) = \sigma^2(n-p-1). \tag{12.191}$$

This result implies that, as promised,

$$\hat{\sigma}_u^2 = \frac{1}{n-p-1}\sum_{i=1}^{n}(Y_i-\hat{Y}_i)^2 \tag{12.192}$$

is an unbiased estimate of σ^2, that is, $\mathbb{E}[\hat{\sigma}_u^2] = \sigma^2$.

12.15 Appendix: The Exact Null Distribution of Z-Score

The goal of this Appendix is to prove that the exact distribution of the z-score $\mathfrak{z}_j$ under the null hypothesis $H_0^{(j)}: \beta_j = 0$ is the t-distribution with $(n-p-1)$ degrees of freedom:

$$\mathfrak{z}_j = \frac{\hat{\beta}_j}{\hat{\sigma}_u\sqrt{\xi_j}} \sim \mathscr{T}(n-p-1). \tag{12.193}$$

The starting point of the proof is the observation that if we replace $\hat{\sigma}_u$ with σ in the denominator, then the null distribution of the resulting fraction will be known[114]:

$$Z = \frac{\hat{\beta}_j}{\sigma\sqrt{\xi_j}} \sim \mathcal{N}(0,1). \tag{12.194}$$

So, the z-score can be represented as follows:

$$\mathfrak{z}_j = \frac{\hat{\beta}_j}{\sigma\sqrt{\xi_j}} \cdot \frac{\sigma}{\hat{\sigma}_u} = Z\frac{\sigma}{\hat{\sigma}_u}, \quad \text{where } Z \sim \mathcal{N}(0,1). \tag{12.195}$$

On the other hand, one of the definitions of the t-distribution has a form similar to (12.195). Namely, a random variable T is said to have the t-distribution with d degrees of freedom, if it can be written as

$$T = Z\sqrt{\frac{d}{Q}}, \tag{12.196}$$

where $Z \sim \mathcal{N}(0,1)$, $Q \sim \chi^2_d$, and Z and Q are independent random variables.

So, let's "massage" the expression for the z-score and try to rewrite it as close to (12.196) as possible[115]:

$$\mathfrak{z}_j = Z\sqrt{\frac{\sigma^2}{\hat{\sigma}_u^2}} = Z\sqrt{\frac{\sigma^2}{\frac{1}{n-p-1}\sum_{i=1}^n (Y_i - \hat{Y}_i)^2}} = Z\sqrt{\frac{n-p-1}{\frac{1}{\sigma^2}\|Y-\hat{Y}\|^2}}. \tag{12.197}$$

What is the probability distribution of random variable $Q = \|Y - \hat{Y}\|^2/\sigma^2$?[116] Recall that the vector of residuals $\hat{E} = Y - \hat{Y}$ is orthogonal to the image of the design matrix $\operatorname{Im}\mathbb{X}$[117]. Therefore, $Y - \hat{Y}$ belongs to the orthogonal complement to the image:

$$Y - \hat{Y} \in (\operatorname{Im}\mathbb{X})^{\perp}. \tag{12.198}$$

Since the dimension of the image $\operatorname{Im}\mathbb{X} = \operatorname{span}\{X^0, \ldots, X^{(p)}\} \subset \mathbb{R}^n$ is $p+1$, the dimension of its orthogonal complement is

$$\dim(\operatorname{Im}\mathbb{X})^{\perp} = n - \dim\operatorname{Im}\mathbb{X} = n - p - 1. \tag{12.199}$$

This means that, although formally the vector $Y - \hat{Y}$ has n components, effectively it has only $(n-p-1)$ "degrees of freedom" since it belongs to the $(n-p-1)$-dimensional subspace of $\mathbb{R}^n$.

To find the distribution of $Q = \|Y - \hat{Y}\|^2/\sigma^2$, let's rewrite the vector of residuals $Y - \hat{Y}$ explicitly in terms of the vector of prediction errors E[118]. Recall that the vector of responses Y and the vector of fitted values $\hat{Y}$ are

$$Y = \mathbb{X}\beta + E \quad \text{and} \quad \hat{Y} = \mathbb{H}Y, \tag{12.200}$$

[114] See (12.93).

[115] In what follows, we use the expression (12.65) for $\hat{\sigma}_u^2$.

[116] The hope is obviously that $Q \sim \chi^2_d$, where the number of degrees of freedom is $d = n - p - 1$.

[117] See property (12.55), illustrated in Fig. 12.4.

[118] This will be useful since vector $E \sim \mathcal{N}(0, \sigma^2\mathbb{I}_n)$ is exactly where the randomness comes from in the MLR model.

where $\mathbb{H} = \mathbb{X}(\mathbb{X}^T\mathbb{X})^{-1}\mathbb{X}^T$ is the hat matrix, which orthogonally projects vectors in $\mathbb{R}^n$ onto the image $\operatorname{Im}\mathbb{X}$, see Fig. 12.4. Therefore,

$$\begin{aligned} Y - \hat{Y} &= Y - \mathbb{H}Y = (\mathbb{I}_n - \mathbb{H})Y = (\mathbb{I}_n - \mathbb{H})(\mathbb{X}\beta + E) \\ &= (\mathbb{X} - \mathbb{H}\mathbb{X})\beta + (\mathbb{I}_n - \mathbb{H})E \\ &= \underbrace{(\mathbb{X} - \mathbb{X}(\mathbb{X}^T\mathbb{X})^{-1}\mathbb{X}^T\mathbb{X})}_{0}\beta + (\mathbb{I}_n - \mathbb{H})E = (\mathbb{I}_n - \mathbb{H})E. \end{aligned} \tag{12.201}$$

The squared norm of the vector of residuals is then

$$\begin{aligned} \|Y - \hat{Y}\|^2 &= \|(\mathbb{I}_n - \mathbb{H})E\|^2 = ((\mathbb{I}_n - \mathbb{H})E)^T(\mathbb{I}_n - \mathbb{H})E \\ &= E^T(\mathbb{I}_n - \mathbb{H})^T(\mathbb{I}_n - \mathbb{H})E = E^T(\mathbb{I}_n - \mathbb{H})(\mathbb{I}_n - \mathbb{H})E \\ &= E^T(\mathbb{I}_n - 2\mathbb{H} + \mathbb{H}^2)E. \end{aligned} \tag{12.202}$$

The square of the hat matrix is just the hat matrix itself, $\mathbb{H}^2 = \mathbb{H}$. This is easy to check algebraically,

$$\mathbb{H}^2 = \mathbb{X}(\mathbb{X}^T\mathbb{X})^{-1}\mathbb{X}^T \cdot \mathbb{X}(\mathbb{X}^T\mathbb{X})^{-1}\mathbb{X}^T = \mathbb{X}(\mathbb{X}^T\mathbb{X})^{-1}\mathbb{X}^T = \mathbb{H}, \tag{12.203}$$

and this is obvious geometrically, since projecting twice is the same as projecting once. So, the squared norm is

$$\|Y - \hat{Y}\|^2 = E^T(\mathbb{I}_n - \mathbb{H})E = E^TE - E^T\mathbb{H}E. \tag{12.204}$$

The vector of prediction errors E is a Gaussian vector with normally distributed components[119]:

$$E_1, \ldots, E_n \overset{\text{iid}}{\sim} \mathcal{N}(0, \sigma^2) \quad \text{or, equivalently,} \quad E \sim \mathcal{N}(0, \sigma^2\mathbb{I}_n). \tag{12.205}$$

[119] Recall that we work under the parametric Gaussian assumption on errors.

Here, the components $E_1, \ldots, E_n$ are the coordinates of E in the *standard basis* $e_1, \ldots, e_n$ of $\mathbb{R}^n$, where $e_j = (0, \ldots, 0, 1, 0, \ldots, 0)^T$ has 1 in the jth position. Similarly, $\mathbb{H}$ is the matrix representation of the orthogonal projection $\operatorname{proj} : \mathbb{R}^n \to \operatorname{Im}\mathbb{X}$ in the standard basis. In view of the factor $\mathbb{H}E \in \operatorname{Im}\mathbb{X}$ in (12.204), it seems like a good idea to rewrite the expression for $\|Y - \hat{Y}\|^2$ in a more convenient basis that "respects" the decomposition $\mathbb{R}^n = \operatorname{Im}\mathbb{X} \oplus (\operatorname{Im}\mathbb{X})^{\perp}$ of $\mathbb{R}^n$ into the direct sum of its two orthogonal subspaces.

Namely, let $u_1, \ldots, u_{p+1}$ be an orthonormal basis[120] of $\operatorname{Im}\mathbb{X}$ and $v_1, \ldots, v_{n-p-1}$ be an orthonormal basis of $(\operatorname{Im}\mathbb{X})^{\perp}$. Then $u_1, \ldots, u_{p+1}, v_1, \ldots, v_{n-p-1}$ is an orthonormal basis of $\mathbb{R}^n$. Let W be the $n \times n$ matrix whose columns are the vectors of that basis,

[120] Recall that a basis is called orthonormal if all vectors in the basis have unit length and are orthogonal to each other.

$$W = \begin{bmatrix} u_1 \; u_2 \; \cdots \; u_{p+1} \; v_1 \; v_2 \; \cdots \; v_{n-p-1} \end{bmatrix} \tag{12.206}$$

Since the columns of W form an orthonormal basis, it is an orthogonal matrix:

$$WW^T = W^TW = \mathbb{I}_n. \tag{12.207}$$

Let $\tilde{E} = (\tilde{E}_1, \ldots, \tilde{E}_n)^T$ be the vector of coordinates of the vector E in the new basis, and $\tilde{\mathbb{H}}$ be the matrix representation of the orthogonal projection $\mathrm{proj} : \mathbb{R}^n \to \mathrm{Im}\,\mathbb{X}$ in the new basis. From linear algebra[121], we know that the relationships between the coordinates E and $\tilde{E}$ of the same vector in the two bases, and between the matrix representations $\mathbb{H}$ and $\tilde{\mathbb{H}}$ of the same linear map in the two bases, are given by

[121] See, for instance, [4] and check the "change of basis" item in the index.

$$E = W\tilde{E} \quad \text{and} \quad \mathbb{H} = W\tilde{\mathbb{H}}W^T. \tag{12.208}$$

Therefore, the squared norm (12.204) can be rewritten as follows:

$$\begin{aligned} \|Y - \hat{Y}\|^2 &= E^T E - E^T \mathbb{H} E = (W\tilde{E})^T W\tilde{E} - (W\tilde{E})^T W\tilde{\mathbb{H}}W^T W\tilde{E} \\ &= \tilde{E}^T W^T W\tilde{E} - \tilde{E}^T W^T W\tilde{\mathbb{H}}W^T W\tilde{E} = \tilde{E}^T\tilde{E} - \tilde{E}^T\tilde{\mathbb{H}}\tilde{E}. \end{aligned} \tag{12.209}$$

The advantage of the new basis and expression (12.209) over the standard basis and expression (12.204) is that the factor $\tilde{\mathbb{H}}\tilde{E}$ can now be computed explicitly. Indeed, by definition,

$$\tilde{E} = \tilde{E}_1 u_1 + \ldots \tilde{E}_{p+1} u_{p+1} + \tilde{E}_{p+2} v_1 + \ldots + \tilde{E}_n v_{n-p-1}. \tag{12.210}$$

Since $\tilde{\mathbb{H}}$ orthogonally projects vectors onto $\mathrm{Im}\,\mathbb{X}$, it leaves all vectors $u_i \in \mathrm{Im}\,\mathbb{X}$ unchanged and send all vectors $v_i \in (\mathrm{Im}\,\mathbb{X})^\perp$ to zero:

$$\tilde{\mathbb{H}}u_i = u_i \quad \text{and} \quad \tilde{\mathbb{H}}v_i = 0. \tag{12.211}$$

Therefore,

$$\tilde{\mathbb{H}}\tilde{E} = \tilde{E}_1 u_1 + \ldots \tilde{E}_{p+1} u_{p+1} \tag{12.212}$$

The square norm is then

$$\|Y - \hat{Y}\|^2 = \tilde{E}^T\tilde{E} - \tilde{E}^T\tilde{\mathbb{H}}\tilde{E} = \sum_{i=1}^{n} \tilde{E}_i^2 - \sum_{i=1}^{p+1} \tilde{E}_i^2 = \sum_{i=p+2}^{n} \tilde{E}_i^2. \tag{12.213}$$

What are the distributions of $\tilde{E}_1, \ldots, \tilde{E}_n$ provided that $E_1, \ldots, E_n$ are normally distributed (12.205)? Since $\tilde{E} = W^T E$ and $E \sim \mathcal{N}(0, \sigma^2 \mathbb{I}_n)$, the random vector $\tilde{E}$ is also Gaussian. Its mean is

$$\mathbb{E}[\tilde{E}] = \mathbb{E}[W^T E] = W^T \mathbb{E}[E] = 0. \tag{12.214}$$

Since W is orthogonal, the covariance matrix of $\tilde{E}$ is[122]

[122] Recall that if X and Y are random vectors and $Y = AX$, where A is a deterministic matrix, then covariance matrices of X and Y are related as follows: $\Sigma_Y = A\Sigma_X A^T$.

$$\Sigma_{\tilde{E}} = W^T \Sigma_E W = W^T \sigma^2 \mathbb{I}_n W = \sigma^2 \mathbb{I}_n. \tag{12.215}$$

So, $\tilde{E}_1, \ldots, \tilde{E}_n$ are also IID normally distributed with zero mean and variance σ^2,

$$\tilde{E}_1, \ldots, \tilde{E}_n \overset{\text{iid}}{\sim} \mathcal{N}(0, \sigma^2) \quad \text{or, equivalently,} \quad \tilde{E} \sim \mathcal{N}(0, \sigma^2 \mathbb{I}_n). \tag{12.216}$$

In other words, the multivariate normal distribution $\mathcal{N}(0, \sigma^2 \mathbb{I}_n)$ is invariant with respect to orthogonal transformations.

Finally, the random variable of interest Q that we were concerned about in the last two pages is

$$Q = \frac{1}{\sigma^2} \|Y - \hat{Y}\|^2 = \sum_{i=p+2}^{n} \left(\frac{\tilde{E}_i}{\sigma} \right)^2, \quad \text{where } \frac{\tilde{E}_{p+2}}{\sigma}, \ldots, \frac{\tilde{E}_n}{\sigma} \overset{\text{iid}}{\sim} \mathcal{N}(0,1). \quad (12.217)$$

Since Q is the sum of squares of $(n-p-1)$ IID standard normal random variance, it is, by definition[123], distributed according to the χ^2-distribution with $(n-p-1)$ degrees of freedom:

$$Q \sim \chi^2_{n-p-1}. \quad (12.218)$$

[123] Which is given in Remark 10.22.

The hope stated in the Marginal Note 116 has materialized!

To finish the proof that the z-score (12.197) has the form (12.196) and, therefore, is t-distributed, it remains to show that the random variables $Z = \hat{\beta}_j / (\sigma \sqrt{\xi_j})$ and Q are independent. Since $\sigma \sqrt{\xi_j}$ is deterministic, we need to show that the jth component $\hat{\beta}_j$ of vector $\hat{\beta}$ and Q are independent. How can we possibly do that? Notice that Q depends only on the last $(n-p-1)$ components $\tilde{E}_{p+2}, \ldots, \tilde{E}_n$ of vector $\tilde{E}$. If it turns out that $\hat{\beta}$ depends only on the first $(p+1)$ components $\tilde{E}_1, \ldots, \tilde{E}_{p+1}$, then this will imply the independence of the components of $\hat{\beta}$ and Q because $\tilde{E}_1, \ldots, \tilde{E}_n$ are IID. So, let's rewrite the vector of OLS estimates $\hat{\beta}$ in terms of $\tilde{E}$ and see what we get:

$$\hat{\beta} = (\mathbb{X}^T \mathbb{X})^{-1} \mathbb{X}^T Y = (\mathbb{X}^T \mathbb{X})^{-1} \mathbb{X}^T (\mathbb{X}\beta + E) = \beta + (\mathbb{X}^T \mathbb{X})^{-1} \mathbb{X}^T E. \quad (12.219)$$

It would be nice to have the hat matrix $\mathbb{H}$ in front of E instead of $(\mathbb{X}^T \mathbb{X})^{-1} \mathbb{X}^T$, since it would "kill" the $(\operatorname{Im} \mathbb{X})^{\perp}$-component of E by projecting it into $\operatorname{Im} \mathbb{X}$. Although we don't have $\mathbb{H}$ there, nothing can stop us from introducing it where we want it to be:

$$\hat{\beta} = \beta + (\mathbb{X}^T \mathbb{X})^{-1} \overbrace{\mathbb{X}^T \underbrace{\mathbb{X} (\mathbb{X}^T \mathbb{X})^{-1}}_{}}^{\mathbb{I}_{p+1}} \underbrace{\mathbb{X}^T}_{\mathbb{H}} E = \beta + (\mathbb{X}^T \mathbb{X})^{-1} \mathbb{X}^T \mathbb{H} E. \quad (12.220)$$

Using expressions (12.208), orthogonality of W, and (12.212), we finally obtain[124]:

[124] This time really finally!

$$\begin{aligned} \hat{\beta} &= \beta + (\mathbb{X}^T \mathbb{X})^{-1} \mathbb{X}^T W \tilde{\mathbb{H}} W^T W \tilde{E} = \beta + (\mathbb{X}^T \mathbb{X})^{-1} \mathbb{X}^T W \tilde{\mathbb{H}} \tilde{E} \\ &= \beta + (\mathbb{X}^T \mathbb{X})^{-1} \mathbb{X}^T W (\tilde{E}_1 u_1 + \ldots \tilde{E}_{p+1} u_{p+1}). \end{aligned} \quad (12.221)$$

So, the vector $\hat{\beta}$ indeed depends only on $\tilde{E}_1, \ldots, \tilde{E}_{p+1}$, which means that any of its components $\hat{\beta}_j$ and the random variable Q, which depends only on $\tilde{E}_{p+2}, \ldots, \tilde{E}_n$, are independent. This concludes the proof of (12.193): the exact null distribution of the z-score $\mathfrak{z}_j$ is the t-distribution with $(n-p-1)$ degrees of freedom[125].

[125] Congratulations on swallowing the largest appendix in the book. Now it is time for digesting it.

Problems

12.1. Coherence of the OLS Estimates for SLR and MLR Models
Let $(X_1,Y_1),\ldots,(X_n,Y_n)$, where $X_i \in \mathbb{R}^p$ and $Y_i \in \mathbb{R}$, be the multiple regression data. Recall that if the $n\times(p+1)$ design matrix $\mathbb{X}$ has trivial kernel, $\operatorname{Ker}\mathbb{X} = \{0\}$[126], then the vector of OLS estimates of the regression coefficients is given by

$$\hat{\beta} = (\mathbb{X}^T\mathbb{X})^{-1}\mathbb{X}^T Y. \tag{12.222}$$

[126] Or, equivalently, if its rank $\operatorname{rk}\mathbb{X} = p+1$.

Show that in the case of a single predictor, that is, when $p = 1$, the expressions for the components $\hat{\beta}_0$ and $\hat{\beta}_1$ of vector $\hat{\beta}$ coincide with the OLS estimates

$$\begin{aligned} \hat{\beta}_0 &= \overline{Y} - \hat{\beta}_1\overline{X}, \\ \hat{\beta}_1 &= \frac{S_{XY}}{S_{XX}} = \frac{\sum_{i=1}^n (X_i-\overline{X})(Y_i-\overline{Y})}{\sum_{i=1}^n (X_i-\overline{X})^2}. \end{aligned} \tag{12.223}$$

derived in Chapter 11 for the case of simple linear regression.

12.2. Equivariance of OLS
Equivariance is a useful "symmetry" property of the OLS estimate, which states that if predictors are rescaled or, more generally, linearly transformed, then the OLS estimate will be transformed "symmetrically."

(a) Let $X^{(1)},\ldots,X^{(p)}$ be the columns of the design matrix $\mathbb{X} = [X^{(0)}X^{(1)}\ldots X^{(p)}]$ corresponding to p predictors, where $X^{(0)} = \mathbb{1}$ is a column of ones. Suppose that the jth predictor is rescaled as follows:

$$X^{(j)} \to \tilde{X}^{(j)} = \alpha X^{(j)}, \tag{12.224}$$

where $\alpha \neq 0$ is some constant, and all other predictors remain the same. Show that the corresponding OLS estimate will be rescaled as

$$\hat{\beta}_j \to \hat{\hat{\beta}}_j = \alpha^{-1}\hat{\beta}_j, \tag{12.225}$$

leaving the contribution of the jth predictor unchanged, $\hat{\hat{\beta}}_j\tilde{X}^{(j)} = \hat{\beta}_j X^{(j)}$, and all other OLS estimates will remain the same.

(b) Let's generalize based on part (a). The transformation of the design matrix $\mathbb{X} \to \tilde{\mathbb{X}}$ induced by rescaling (12.224) can be written in a matrix form as follows:

$$\tilde{\mathbb{X}} = [X^{(0)}\ X^{(1)}\ \ldots\ X^{(j-1)}\ \alpha X^{(j)}\ X^{(j+1)}\ \ldots\ X^{(p)}] = \mathbb{X}D_\alpha^{(j)}, \tag{12.226}$$

where $D_\alpha^{(j)}$ is the $(p+1)\times(p+1)$ diagonal matrix with ones on the diagonal everywhere except for the jth element, which equals α. Let's consider a more general linear transformation of the design matrix:

$$\mathbb{X} \to \tilde{\mathbb{X}} = \mathbb{X}T, \tag{12.227}$$

where T is any $(p+1)\times(p+1)$ nonsingular matrix[127]. Find the relationship between the OLS estimate $\hat{\beta}$ based on the original design matrix and the OLS estimate $\hat{\hat{\beta}}$ based on the transformed design matrix. Assume that $\operatorname{Ker}\mathbb{X} = \{0\}$.

[127] Conceptually, (12.227) means that every column of $\mathbb{X}$ is replaced with an arbitrary linear combination of the columns of $\mathbb{X}$. In other words, instead of the original set of p predictors, we consider another set of p predictors, each of which is a linear combination of the old predictors.

12.3. Weighted Least Squares

Weighted least squares is an important generalization of ordinary least squares. Given the training data $(X_1,Y_1),\ldots,(X_n,Y_n)$, where $X_i \in \mathbb{R}^p$ and $Y_i \in \mathbb{R}$, consider the multiple linear regression model of Y_i on X_i,

$$Y_i = \beta_0 + \beta_1 X_i^{(1)} + \ldots + \beta_p X_i^{(p)} + E_i = X_i^T\beta + E_i, \qquad i = 1,\ldots,n. \tag{12.228}$$

Suppose that the observations $(X_1,Y_1),\ldots,(X_n,Y_n)$ have different reliabilities: some are more accurate and some are less accurate. This, for example, happens if the prediction errors $E_1,\ldots,E_n$ are not homoscedastic and have different variances $\mathbb{V}[E_i] = \sigma_i^2$: the larger the variance σ_i^2, the less accurate and reliable the corresponding observation (X_i,Y_i) is. To quantify the importance (reliability) of observations, let's assign to each observation (X_i,Y_i) a weight $w_i > 0$: the larger the weight w_i, the more reliable the observation (X_i,Y_i) and, therefore, the more important it is to accurately fit the model to (X_i,Y_i). For example, if the variances of errors are known, we can set $w_i = 1/\sigma_i^2$.

Given the weights $w_1,\ldots,w_n$, it is natural to quantify the overall goodness of fit of the MLR model to the data not by the RSS (12.8), where all observations are equally important, but by the *weighted residual sum of squares*, defined as follows:

$$\mathrm{wRSS}(\beta) = \sum_{i=1}^{n} w_i (Y_i - X_i^T\beta)^2. \tag{12.229}$$

The larger the weight w_i, the more important it is to make the term $(Y_i - X_i^T\beta)^2$ smaller and, thus, fit the model to the observation (X_i,Y_i) more accurately.

Let $\hat{\beta}_W$ denote the *weighted least squares* (WLS) estimate of β, that is, the value of β that minimizes the weighted residual sum of squares[128],

[128] Besides generalizing OLS, the WLS estimates are used for solving optimization problems. In particular, they are used in iterative numerical methods for finding MLEs and for fitting logistic regression.

$$\hat{\beta}_W = \arg\min_{\beta\in\mathbb{R}^{p+1}} \mathrm{wRSS}(\beta). \tag{12.230}$$

Derive the "weighted" analog of the normal equation (12.26) and find the expression for the WLS estimate $\hat{\beta}_W$. Compare it with the OLS estimate $\hat{\beta} = (\mathbb{X}^T\mathbb{X})^{-1}\mathbb{X}^T Y$. Assume that the design matrix has trivial kernel, $\operatorname{Ker}\mathbb{X} = \{0\}$.

12.4. Testing for Significance and Confidence Intervals

One interesting observation from Table 12.2 in Case Study 12.8 is that a predictor $X^{(j)}$ is not significant at the 5% level ($p_j^W > 0.05$) precisely when the 95% confidence interval $I_{0.05}^j$ for the regression coefficient β_j associated with $X^{(j)}$ contains zero. In this problem, you will show that this is not just a coincidence. Prove that the Wald test p-value $p_j^W > \alpha$ if and only if the confidence interval I_α^j contains zero.

12.5. The MLR Model in Action

In this problem, to gain practical experience with the multivariate regression model, you will perform the elements of the regression analysis discussed in Chapter 12.

First, generate a set of training input vectors $X_1, \ldots, X_n \in \mathbb{R}^p$, where the sample size $n = 100$, the number of predictors $p = 4$, and

$$
\begin{aligned}
X_1^{(1)}, \ldots, X_n^{(1)} &\overset{\text{iid}}{\sim} \mathcal{N}(0,1), \\
X_1^{(2)}, \ldots, X_n^{(2)} &\overset{\text{iid}}{\sim} \mathcal{P}(1), \\
X_1^{(3)}, \ldots, X_n^{(3)} &\overset{\text{iid}}{\sim} \mathcal{E}(1), \\
X_1^{(4)}, \ldots, X_n^{(4)} &\overset{\text{iid}}{\sim} \mathcal{U}[-1,1].
\end{aligned}
\tag{12.231}
$$

That is, the predictors are generated as IID samples from the normal, Poisson, exponential, and uniform distributions. Next, compute the corresponding training responses $Y_1, \ldots, Y_n$ using the following relationship[129]: $Y_i = \beta_0 + \beta_1 X_i^{(1)} + \beta_2 X_i^{(2)} + \beta_3 X_i^{(3)} + \beta_4 X_i^{(4)} + E_i$, where the vector of coefficients is $\beta = (1,2,3,0,0)^T$, and $E_1, \ldots, E_n \overset{\text{iid}}{\sim} \mathcal{N}(0,1)$. Note that the response Y does not depend on the third and fourth predictors, $X^{(3)}$ and $X^{(4)}$, since $\beta_3 = \beta_4 = 0$. Finally, suppose that the relationship between the response and the predictors is unknown and consider the MLR model of Y_i on X_i:

[129] The goal is to recover this relationship from the data using the MLR model.

$$
\begin{aligned}
&Y_i = \beta_0 + \beta_1 X_i^{(1)} + \beta_2 X_i^{(2)} + \beta_3 X_i^{(3)} + \beta_4 X_i^{(4)} + E_i, \\
&E_1, \ldots, E_n \overset{\text{iid}}{\sim} \mathcal{N}(0, \sigma^2).
\end{aligned}
\tag{12.232}
$$

(a) Visualize the training data by creating four scatterplots of the response Y versus each of the four predictors $X^{(1)}$, $X^{(2)}$, $X^{(3)}$, and $X^{(4)}$.

(b) Find the OLS estimate $\hat{\beta}$ of the vector $\beta \in \mathbb{R}^5$ of regression coefficients.

(c) To test the predictors $X^{(1)}$, $X^{(2)}$, $X^{(3)}$, and $X^{(4)}$ for significance, create a table, similar to Table 12.2, which contains the z-scores associated with the predictors, the corresponding p-values based on the Wald and t-test, the approximate 95% confidence intervals for the regression coefficients, and their OLS estimates computed in part (b).

(d) Consider a reduced model, which is obtained from the full model (12.232) by excluding $X^{(3)}$ and $X^{(4)}$ from the list of predictors:

$$
Y_i = \tilde{\beta}_0 + \tilde{\beta}_1 X_i^{(1)} + \tilde{\beta}_2 X_i^{(2)} + E_i, \quad E_1, \ldots, E_n \overset{\text{iid}}{\sim} \mathcal{N}(0, \sigma^2). \tag{12.233}
$$

Find the OLS estimate $\hat{\tilde{\beta}}$ of the vector $\tilde{\beta} \in \mathbb{R}^3$ of regression coefficients of the reduced model.

(e) Use the F-test to test the simultaneous significance of predictors $X^{(3)}$ and $X^{(4)}$. In other words, use the training data to test

$$
H_0 : (\beta_3, \beta_4) = 0 \quad \text{vs} \quad H_1 : (\beta_3, \beta_4) \neq 0. \tag{12.234}
$$

(f) Compare the full and reduced regression models in terms of their prediction accuracy by first generating the validation data of sample size $n = 100$ using the same process that was used to generate the training data, and then computing their test errors, that is, the mean squared errors on the validation data.

References

1. Neyman, J.: Statistics – servant of all sciences. *Science* 122(3126), 401-406 (1955). https://doi.org/10.1126/science.122.3166.401
2. Axler, S.: *Linear Algebra Done Right*. Forth Edition, Springer (2024). https://linear.axler.net/
3. Martin, C.D., Porter, M.A.: The extraordinary SVD. *The American Mathematical Monthly* 119(10), 838-851 (2012). https://doi.org/10.4169/amer.math.monthly.119.10.838
4. Strang, G: *Linear Algebra and its Applications*. Thomson Brooks/Cole, (2005).
5. Moore, E.H.: On the reciprocal of the general algebraic matrix. *Bulletin of the American Mathematical Society* 26, 394-395 (1920).
6. Penrose, R.: A generalized inverse for matrices. *Mathematical Proceedings of the Cambridge Philosophical Society* 51, 406-413 (1955). https://doi.org/10.1017/S0305004100030401
7. Plackett, R.L.: A historical note on the method of least squares. *Biometrika* 36(3/4), 458-460 (1949). https://doi.org/10.2307/2332682
8. Montgomery, D.C., Peck, E.A., Vining, G.G.: *Introduction to Linear Regression Analysis, 6th Edition.* United Kingdom, Wiley (2021).
9. American Cancer Society: Key Statistics for Prostate Cancer https://www.cancer.org/cancer/types/prostate-cancer/about/key-statistics.html
10. American Cancer Society: Prostate Cancer Risk Factors https://www.cancer.org/cancer/types/prostate-cancer/causes-risks-prevention/risk-factors.html
11. National Institute on Aging: Prostate Problems https://www.nia.nih.gov/health/prostate-health/prostate-problems
12. Stamey, T.A., Kabalin, J.N., Ferrari, M.: Prostate specific antigen in the diagnosis and treatment of adenocarcinoma of the prostate. II. Radiation treated patients. *Journal of Urology* 141(5), 1084-7 (1989). https://doi.org/10.1016/S0022-5347(17)41176-1
13. Hastie, T., Tibshirani, R., Friedman, J.: *The Elements of Statistical Learning, Second Edition*. Springer Series in Statistics (2009). https://link.springer.com/book/10.1007/978-0-387-84858-7
14. Tikhonov, A.N.: On the stability of inverse problems. *Doklady Akademii Nauk SSSR* 39(5), 195-198 (1943). https://cs.uwaterloo.ca/~y328yu/classics/Tikhonov43.pdf
15. Hoerl, A.E., Kennard, R.: Ridge regression: biased estimation for nonorthogonal problems. *Technometrics* 12(1), 55-67 (1970). https://doi.org/10.2307/1267351
16. Santosa, F., Symes, W.W.: Linear inversion of band-limited reflection seismograms. *SIAM Journal on Scientific and Statistical Computing* 7(4), 1307-1330 (1986). https://doi.org/10.1137/0907087

17. Tibshirani, R.: Regression shrinkage and selection via the Lasso. *Journal of the Royal Statistical Society. Series B (Methodological)* 58(1), 267-288 (1996).
`https://www.jstor.org/stable/2346178`
18. Akaike, H.: A new look at the statistical model identification. *IEEE Transactions on Automatic Control* 19(6), 716-723 (1974).
`https://doi.org/10.1109/TAC.1974.1100705`
19. Schwarz, G.: Estimating the dimension of a model. *The Annals of Statistics* 6(2), 461-464 (1978)
`https://doi.org/10.1214/aos/1176344136`
20. Allen, D.M.: The relationship between variable selection and data agumentation and a method for prediction. *Technometrics* 16(1), 125-127 (1974).
`https://doi.org/10.2307/1267500`
21. Strang, G.: The fundamental theorem of linear algebra. *The American Mathematical Monthly* 100(9), 848-855 (1993).
`https://www.jstor.org/stable/2324660`

Afterword

In conclusion, I would like to add... But no, that's enough.
– Janusz Korczak, *Playful Pedagogy*

This book was written not for instructors to teach, but for readers to learn from, although the latter does not of course exclude the former. That is why, in what follows, you will find solutions to all problems. To get the most out of these solutions, I strongly suggest reading them only after substantial and energy-consuming attempts to solve the corresponding problems. Learning is much more effective when there is active and prolonged engagement with the material.

This book originated from a set of lecture notes written for a course on statistical inference that I teach at the California Institute of Technology every spring. Converting them into the book turned out to be far more time-consuming than I originally expected. The notes were written for students whose mathematical and statistical background was familiar to me. Moreover, they were incomplete and lacked motivations, explanations, examples, and other important details, since all these gaps were filled and discussed with students during the lectures. While writing this book for a wider readership, I tried to make fewer assumptions about the technical background of my future readers, aimed at preserving the conversational tone of my lectures, and strived to ensure that the book was as self-contained, both motivationally and technically, as possible. Of course, I don't have the slightest clue whether I have achieved these goals.

I would, therefore, be very grateful if after reading the book, in addition to informing me about found errors and typos (as I kindly asked you in the Preface), you could also send me your general feedback on the book: what chapters, sections, and case studies are interesting and useful (if any), what needs to be changed and what should be preserved. Please send your feedback by emailing kostia@caltech.edu. Looking forward to hearing from you!

Konstantin M. Zuev
Pasadena, California
August 2025

K. M. Zuev, *Fundamentals of Statistical Inference*,
International Series in Operations Research & Management Science 374,
https://doi.org/10.1007/978-3-032-03848-7

Solutions

'The Answer to the Great Question... Of Life, the Universe and Everything... Is... Forty-two,' said Deep Thought, with infinite majesty and calm.
– Douglas Adams, *The Hitchhiker's Guide to the Galaxy*

Problems of Chapter 1

1.1 Data Summaries and Linear Transformations

By the definition of the sample mean,

$$\bar{y} = \frac{1}{n}\sum_{i=1}^{n} y_i = \frac{1}{n}\sum_{i=1}^{n}(\alpha + \beta x_i) = \alpha + \beta\frac{1}{n}\sum_{i=1}^{n} x_i = \alpha + \beta\bar{x}. \tag{1}$$

Now let's investigate how the sample median transforms. If $\beta > 0$, then linear transformation $y_i = \alpha + \beta x_i$ preserves the order in the sample. Therefore,

$$y_{(k)} = \alpha + \beta x_{(k)}, \quad k = 1, \ldots, n, \tag{2}$$

where $x_{(k)}$ and $y_{(k)}$ are the order statistics of the samples. Then, using the definition of the sample median,

$$\tilde{y} = \begin{cases} y_{(\frac{n+1}{2})} = \alpha + \beta x_{(\frac{n+1}{2})} = \alpha + \beta\tilde{x}, & n \text{ is odd,} \\ \frac{1}{2}\left(y_{(\frac{n}{2})} + y_{(\frac{n}{2}+1)}\right) = \frac{1}{2}\left(2\alpha + \beta x_{(\frac{n}{2})} + \beta x_{(\frac{n}{2}+1)}\right) = \alpha + \beta\tilde{x}, & n \text{ is even.} \end{cases} \tag{3}$$

If $\beta < 0$, then transformation $y_i = \alpha + \beta x_i$ reverses the order:

$$y_{(k)} = \alpha + \beta x_{(n-k+1)}, \quad k = 1, \ldots, n, \tag{4}$$

But still

K. M. Zuev, *Fundamentals of Statistical Inference*,
International Series in Operations Research & Management Science 374,
https://doi.org/10.1007/978-3-032-03848-7

$$\tilde{y} = \begin{cases} y_{(\frac{n+1}{2})} = \alpha + \beta x_{(n-\frac{n+1}{2}+1)} = \alpha + \beta x_{(\frac{n+1}{2})} = \alpha + \beta \tilde{x}, & n \text{ is odd}, \\ \frac{1}{2}\left(y_{(\frac{n}{2})} + y_{(\frac{n}{2}+1)}\right) = \frac{1}{2}\left(2\alpha + \beta x_{(\frac{n}{2}+1)} + \beta x_{(\frac{n}{2})}\right) = \alpha + \beta \tilde{x}, & n \text{ is even}. \end{cases} \tag{5}$$

So, in both cases

$$\tilde{y} = \alpha + \beta \tilde{x}. \tag{6}$$

Thus, both the sample mean and the sample median behave as any measure of center is expected to behave with respect to a linear transformation of the data: they are also transformed linearly.

By the definition of the interquartile range,

$$\text{IQR}_y = Q_3(y) - Q_1(y), \tag{7}$$

where $Q_1(y)$ and $Q_3(y)$ are the lower and upper quartiles of the sample y. Since $Q_1(y)$ and $Q_3(y)$ are the medians of the first half and the second half of the sample, they are related to $Q_1(x)$ and $Q_3(x)$ in the same way as $\tilde{y}$ is related to $\tilde{x}$, namely,

$$\begin{aligned} &\text{If } \beta > 0: \quad Q_1(y) = \alpha + \beta Q_1(x) \quad \text{and} \quad Q_3(y) = \alpha + \beta Q_3(x), \\ &\text{If } \beta < 0: \quad Q_1(y) = \alpha + \beta Q_3(x) \quad \text{and} \quad Q_3(y) = \alpha + \beta Q_1(x). \end{aligned} \tag{8}$$

Therefore,

$$\text{IQR}_y = |\beta| \text{IQR}_x. \tag{9}$$

Finally, the sample standard deviation is

$$s_y^2 = \frac{1}{n}\sum_{i=1}^{n}(y_i - \bar{y})^2 = \frac{1}{n}\sum_{i=1}^{n}(\alpha + \beta x_i - \alpha - \beta \bar{x})^2 = \beta^2 s_x^2, \tag{10}$$

and, therefore,

$$s_y = |\beta| s_x. \tag{11}$$

Thus, both the IQR and the sample standard deviation behave as any measure of spread is expected to behave with respect to a linear transformation of the data: they are invariant with respect to shifts (α does not affect them) and get appropriately rescaled.

1.2 Optimization interpretation of $\bar{x}$ and $\tilde{x}$

The optimization interpretation of the sample mean (1.38) can be readily established using the standard calculus:

$$\frac{d}{d\alpha}\sum_{i=1}^{n}(x_i - \alpha)^2 = 2\sum_{i=1}^{n}(\alpha - x_i) = 2\left(n\alpha - \sum_{i=1}^{n} x_i\right). \tag{12}$$

This derivative vanishes at

$$\alpha^* = \frac{1}{n}\sum_{i=1}^{n} x_i = \bar{x}. \tag{13}$$

This critical point $\alpha^* = \bar{x}$ is the global minimum of $f(\alpha) = \sum_{i=1}^{n}(x_i - \alpha)^2$ since

$$\frac{d^2}{d\alpha^2}\sum_{i=1}^{n}(x_i - \alpha)^2 = 2n > 0. \tag{14}$$

To derive the optimization interpretation of the sample median (1.39) is slightly more challenging. There are various ways how one can proceed. The most elegant way is perhaps to use induction.

Let's consider the case where n is odd, $n = 2k - 1$ (the case where n is even can be considered similarly). If $n = 1$, then (1.39) is trivial:

$$\arg\min_{\alpha}\sum_{i=1}^{1}|x_i - \alpha| = \arg\min_{\alpha}|x_1 - \alpha| = x_1 = \tilde{x}. \tag{15}$$

Suppose now that (1.39) is true for any sample $x_1 \ldots, x_{2k-1}$ of size $n = 2k - 1$. The goal is to demonstrate that in this case (1.39) is also true for any sample of size $n = 2k + 1$. The objective function[1] in (1.39) can be written as follows[2]:

$$\sum_{i=1}^{2k+1}|x_i - \alpha| = \sum_{i=1}^{2k+1}|x_{(i)} - \alpha| = |x_{(1)} - \alpha| + \sum_{i=2}^{2k}|x_{(i)} - \alpha| + |x_{(2k+1)} - \alpha|. \tag{16}$$

[1] That is the function that we want to minimize.

[2] All we are doing in (16) is rearranging the terms: instead of summing in the original order of $x_1, \ldots, x_n$, we sum the same terms, but in the order of the order statistics $x_{(1)}, \ldots, x_{(n)}$.

Since a point α^* where $\sum_{i=1}^{2k+1}|x_i - \alpha|$ achieves its minimum lies in the interval $[x_{(1)}, x_{(2k+1)}]$, we can assume that $x_{(1)} \leq \alpha \leq x_{(2k+1)}$. Therefore, $|x_{(1)} - \alpha| = \alpha - x_{(1)}$ and $|x_{(2k+1)} - \alpha| = x_{(2k+1)} - \alpha$. Thus,

$$\sum_{i=1}^{2k+1}|x_i - \alpha| = x_{(2k+1)} - x_{(1)} + \sum_{i=2}^{2k}|x_{(i)} - \alpha|. \tag{17}$$

This means that α^* does not really depend on the largest $x_{(2k+1)}$ and the smallest $x_{(1)}$ sample values. And so we have:

$$\begin{aligned}\arg\min_{\alpha}\sum_{i=1}^{2k+1}|x_i - \alpha| &= \arg\min_{\alpha}\sum_{i=2}^{2k}|x_{(i)} - \alpha| = \text{median}\{x_{(2)}, \ldots, x_{(2k)}\} \\ &= \text{median}\{x_{(1)}, \ldots, x_{(2k+1)}\} = \tilde{x},\end{aligned} \tag{18}$$

where the second equality holds by the induction hypothesis: we have assumed that (1.39) is true for any sample of size $n = 2k - 1$, in particular, for $x_{(2)}, \ldots, x_{(2k)}$; and the third equality holds since adding $x_{(1)}$ and $x_{(2k+1)}$ to the sample $\{x_{(2)}, \ldots, x_{(2k)}\}$ does not change the median.

1.3 Interpretation of Q-Q plots

Let's address the following more general question. Let x_q and y_q be the qth quantiles of the probability distributions with CDFs F and G, respectively:

$$F(x_q) = q \quad \text{and} \quad G(y_q) = q. \tag{19}$$

Suppose that for any $q \in (0,1)$ point (x_q, y_q) lies on the line $y = ax + b$, that is

$$y_q = ax_q + b. \tag{20}$$

What can we say about the relationship between F and G?

From (19) and (20),

$$F(x_q) = G(y_q) = G(ax_q + b). \tag{21}$$

Since this is true for *any* $q \in (0,1)$,

$$F(x) = G(ax+b) \quad \text{or, equivelantly,} \quad F\left(\frac{y-b}{a}\right) = G(y). \tag{22}$$

Let's now apply this general result to our problem. We know that approximately

$$x_{(k)} \approx az_{\frac{k}{n}} + b, \tag{23}$$

where $z_{\frac{k}{n}}$ and $x_{(k)}$ are the approximate $\left(\frac{k}{n}\right)^{\text{th}}$ quantiles of the standard normal distribution $\Phi = F$ and the eCDF of the sample $\widehat{F}_n = G$. Therefore,

$$\widehat{F}_n(y) \approx \Phi\left(\frac{y-b}{a}\right). \tag{24}$$

The CDF $\Phi\left(\frac{y-b}{a}\right)$ is the CDF of a normal random variable with mean b and variance a^2. Indeed, if $Y = aZ + b \sim \mathcal{N}(b, a^2)$, where $Z \sim \mathcal{N}(0,1)$, then its CDF is

$$F_Y(y) = \mathbb{P}(Y \le y) = \mathbb{P}(aZ + b \le y) = \mathbb{P}\left(Z \le \frac{y-b}{a}\right) = \Phi\left(\frac{y-b}{a}\right). \tag{25}$$

So, if the Q-Q plot $\left(z_{\frac{k}{n}}, x_{(k)}\right)$ falls roughly on the line $y = ax + b$, then the sample $x_1, \ldots, x_n$ has approximately normal distribution with parameters $\mu = b$ and $\sigma^2 = a^2$.

1.4 Readability of Q-Q plots

(a) Figure 1 shows the histograms and Q-Q plots for $m = 3$ samples of size $n = 20$ drawn from $\mathcal{N}(0,1)$. Neither the shape of histograms nor the Q-Q plots suggest that samples follow the standard normal distribution: the sample size is too small. For the second sample (middle Q-Q plot), a straight line can be fitted relatively accurately[3] to the Q-Q plot, but that line still differs substantially from the line $y = x$.

[3] If the smallest outlier in the sample is ignored.

(b) Figure 2 shows the histograms and Q-Q plots for samples of sizes $n = 50, 100$, and 1000. As expected, as the sample size increases, the histograms become more symmetric and bell-shaped, and the Q-Q plots start to follow the line$y = x$ more closely.

(c) My personal estimate is that about $n^* = 100$ points are necessary to produce a normal-quantile plot which is stable enough and falls closely on the line $y = x$.

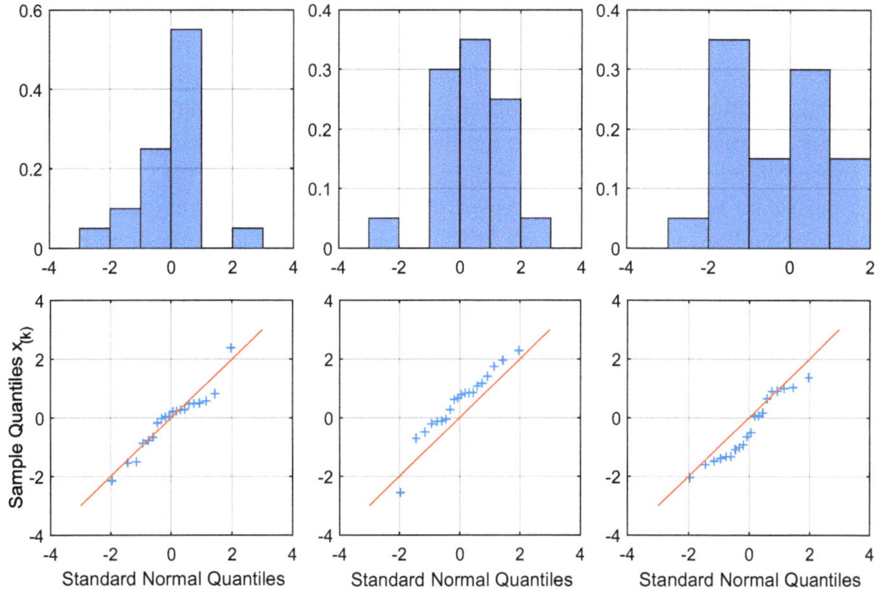

Fig. 1 Histograms and normal-qunatile plots for $m = 3$ samples $x_1, \ldots, x_n \sim \mathcal{N}(0,1)$, $n = 20$.

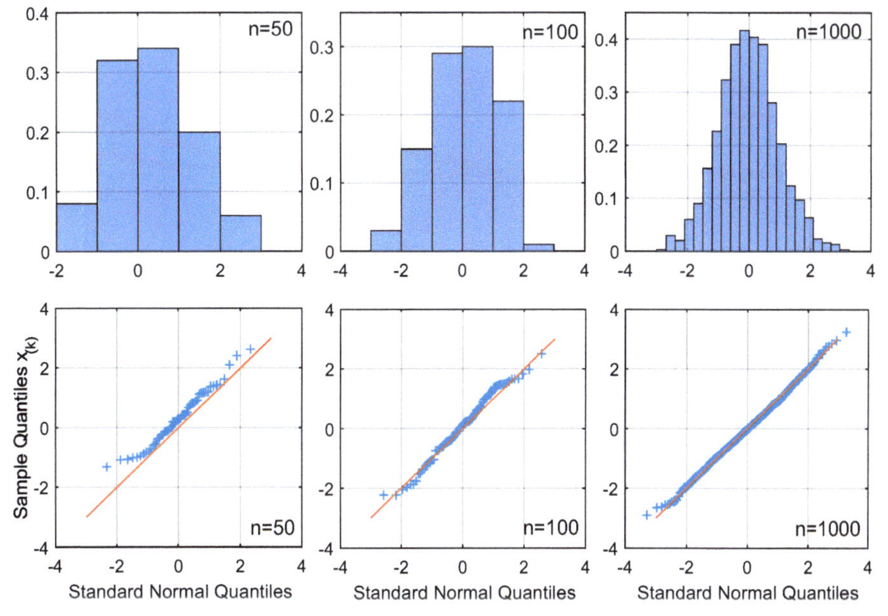

Fig. 2 Histograms and normal-quantile plots for $x_1, \ldots, x_n \sim \mathcal{N}(0,1)$, $n = 50, 100$, and 1000.

1.5 CO_2 Emissions Per Person

(a) Figure 3 shows the carbon dioxide emissions (tonnes per person) for seven countries: the United States, China, Russia, Germany, the United Kingdom, Japan, and India. While China is a clear leader in terms of the total pollution in 2014 (see Fig. 1.9), it emits less CO_2 per person than the United States, Russia, Japan, and Germany. Table 1 shows the ranking of countries with respect to the CO_2 pollution per-person in 2014. The top three absolute emitters, China, the United States, and India, are ranked only 34th, 9th, and 113th, respectively. The top three countries that emitted the most CO_2 per person are Qatar, Trinidad and Tobago, and Kuwait.

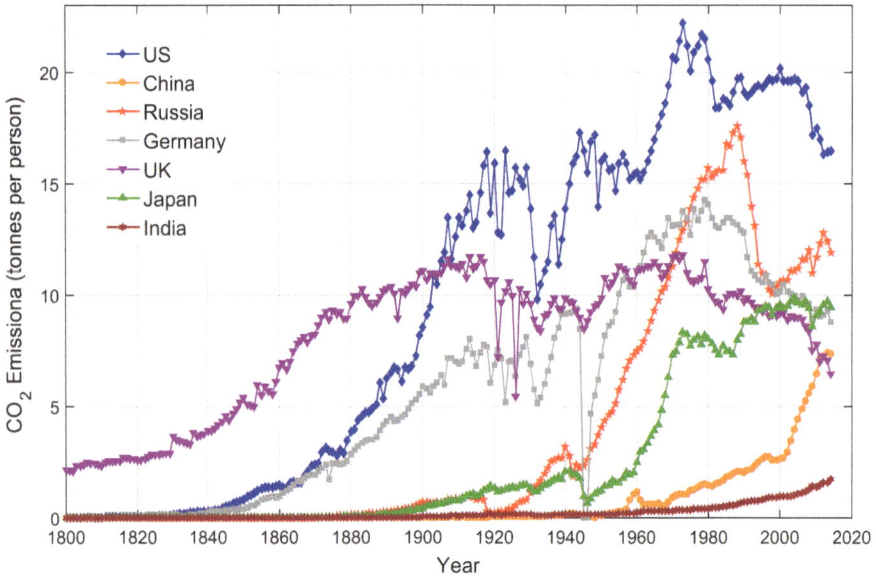

Fig. 3 Per-person CO_2 emissions from the burning of fossil fuels for the United States, China, Russia, Germany, the United Kingdom, Japan, and India.

Table 1 Ranking of countries in terms of the per-person CO_2 emissions in 2014.

Rank	Country	Tonnes of CO_2 per person
1	Qatar	45.4
2	Trinidad and Tobago	34.2
3	Kuwait	25.2
...	...	...
9	United States	16.5
...	...	...
34	China	7.4
...	...	...
113	India	1.73
...	...	...

(b) Table 2 shows the mean $\bar{x}$, range r_x, standard deviation s_x, IQR, and the five-number summary of the logarithms of per-person emissions of CO_2 in 2014.

Table 2 Numerical summaries of the log-per-person emissions of CO_2 during 2014.

$\bar{x}$	r_x	s_x	IQR	$x_{(1)}$	Q_1	$\tilde{x}$	Q_3	$x_{(n)}$
0.64	6.93	1.50	2.18	-3.11	-0.43	0.82	1.75	3.82

(c) Figure 4 summarizes the 2014 log-per-person emissions graphically. Interestingly, while, in terms of the total emissions, India was the third most polluting country in 2014 (after China and the United States), its pollution per person is even below the median.

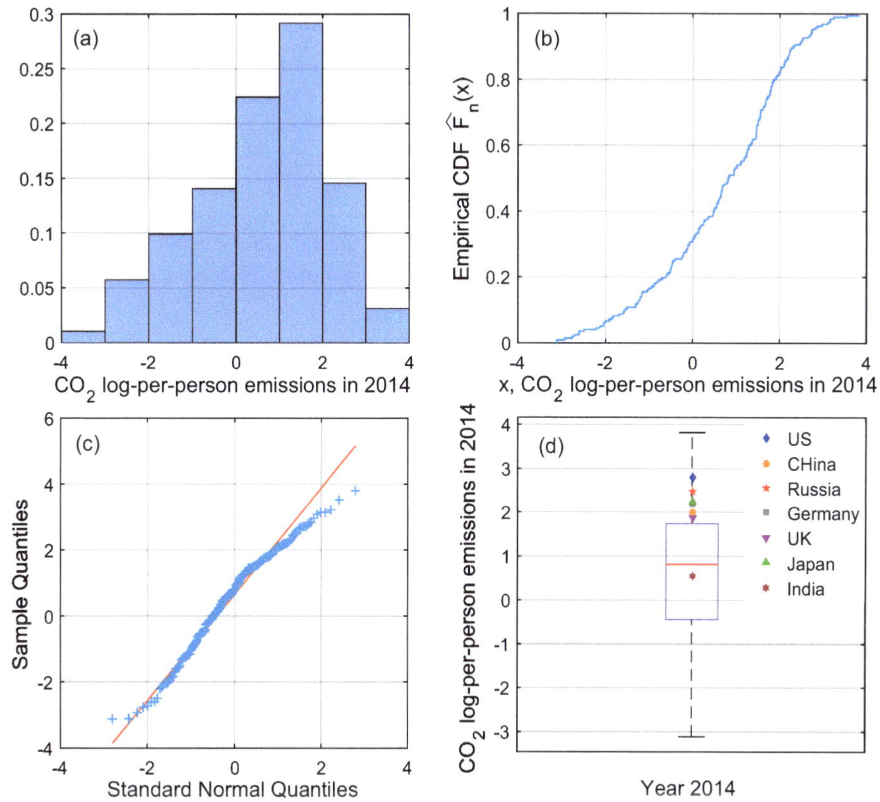

Fig. 4 Graphical summaries of the log-per-person CO_2 emissions in 2014: (a) histogram, (b) empirical CDF, (c) normal-quantile plot, and (d) boxplot.

The histogram and especially the normal-quantile plot suggest that the log-per-person emissions of CO_2 in 2014 are not normally distributed.

Problems of Chapter 2

2.1 Simple Random Sampling

(a) Since X_1 is drawn from $\mathscr{P} = \{x_1, \dots, x_N\}$ uniformly at random,

$$\mathbb{P}(X_1 = x_k) = \frac{1}{N}, \quad \text{for any } k = 1, \dots, N. \tag{26}$$

Next, since X_2 is drawn from $\mathscr{P} \setminus \{X_1\}$ uniformly at random, for any k:

$$\mathbb{P}(X_2 = x_k) = \mathbb{P}(X_1 \neq x_k)\mathbb{P}(X_2 = x_k | X_1 \neq x_k) = \frac{N-1}{N}\frac{1}{N-1} = \frac{1}{N}. \tag{27}$$

Similarly, for any $i = 1, \dots, n$,

$$\begin{aligned}
&\mathbb{P}(X_i = x_k) = \mathbb{P}(X_i = x_k | X_1 \neq x_k, \dots, X_{i-1} \neq x_k)\mathbb{P}(X_1 \neq x_k, \dots, X_{i-1} \neq x_k)\\
&= \frac{1}{N-(i-1)}\mathbb{P}(X_1 \neq x_k)\mathbb{P}(X_2 \neq x_k | X_1 \neq x_k)\dots\mathbb{P}(X_{i-1} \neq x_k | X_1 \neq x_k, \dots, X_{i-2} \neq x_k)\\
&\frac{1}{N-(i-1)}\frac{N-1}{N}\frac{N-2}{N-1}\cdots\frac{N-(i-1)}{N-(i-2)} = \frac{1}{N}.
\end{aligned} \tag{28}$$

(b) Here we use the result from part (a) and the fact that events $X_1 = x_k, X_2 = x_k, \dots, X_n = x_k$ are mutually exclusive, since the sampling is done without replacement:

$$\begin{aligned}
\mathbb{P}(x_k \in \mathscr{S}) &= \mathbb{P}(X_1 = x_k \ \text{ or } \ X_2 = x_k, \dots, \ \text{ or } \ X_n = x_k)\\
&= \sum_{i=1}^{n} \mathbb{P}(X_i = x_k) = \sum_{i=1}^{n} \frac{1}{N} = \frac{n}{N}.
\end{aligned} \tag{29}$$

(c) Using the notion of conditional probability,

$$\mathbb{P}(X_1 = x_k,\, X_2 = x_s) = \mathbb{P}(X_1 = x_k)\mathbb{P}(X_2 = x_s | X_1 = x_k) = \frac{1}{N}\frac{1}{N-1}. \tag{30}$$

(d) Similarly,

$$\begin{aligned}
&\mathbb{P}(X_i = x_i, \text{ for all } i = 1, \dots, n) = \mathbb{P}(X_1 = x_1)\, \mathbb{P}(X_i = x_i, \text{ for all } i = 2, \dots, n \mid X_1 = x_1)\\
&= \frac{1}{N}\, \mathbb{P}(X_2 = x_2 \mid X_1 = x_1)\, \mathbb{P}(X_i = x_i, \text{ for all } i = 3, \dots, n \mid X_1 = x_1,\, X_2 = x_2)\\
&= \frac{1}{N(N-1)}\, \mathbb{P}(X_3 = x_3 \mid X_1 = x_1,\, X_2 = x_2)\times\\
&\times \mathbb{P}(X_i = x_i, \text{ for all } i = 4, \dots, n \mid X_1 = x_1,\, X_2 = x_2,\, X_3 = x_3)\\
&= \dots = \frac{1}{N(N-1)\dots(N-n+1)}.
\end{aligned} \tag{31}$$

2.2 Weighted Sample Mean

(a) By definition, the estimate $\overline{X}_w$ is unbiased if $\mathbb{E}[\overline{X}_w] = \mu$. Using the fact that $\mathbb{E}[X_i] = \mu$ (Theorem 2.1), we have:

$$\mathbb{E}[\overline{X}_w] = \mathbb{E}\left[\sum_{i=1}^{n} w_i X_i\right] = \sum_{i=1}^{n} w_i \mathbb{E}[X_i] = \sum_{i=1}^{n} w_i \mu = \mu \sum_{i=1}^{n} w_i. \tag{32}$$

Therefore, the estimate $\overline{X}_w$ is unbiased if and only if

$$\sum_{i=1}^{n} w_i = 1. \tag{33}$$

(b) Our goal is to find the estimate $\overline{X}_w$, whose weights satisfy (33), and that has the smallest standard error, or, equivalently, the smallest possible variance,

$$\mathbb{V}\left[\overline{X}_w\right] = \mathbb{V}\left[\sum_{i=1}^{n} w_i X_i\right] = \sum_{i=1}^{n} w_i^2 \mathbb{V}[X_i] + 2\sum_{j>i} w_i w_j \mathrm{Cov}(X_i, X_j). \tag{34}$$

Recall (Theorems 2.1 and 2.2) that $\mathbb{V}[X_i] = \sigma^2$ and $\mathrm{Cov}(X_i, X_j) = -\frac{\sigma^2}{N-1}$. Therefore,

$$\mathbb{V}\left[\overline{X}_w\right] = \sigma^2 \sum_{i=1}^{n} w_i^2 - \left(2\sum_{j>i} w_i w_j\right)\frac{\sigma^2}{N-1}. \tag{35}$$

Since the weights satisfy (33),

$$1 = \left(\sum_{i=1}^{n} w_i\right)^2 = \sum_{i=1}^{n} w_i^2 + 2\sum_{j>i} w_i w_j. \tag{36}$$

From the last equation, we can express the last term in terms of the sum of squared weights and plug it into the expression for the variance (35):

$$\mathbb{V}\left[\overline{X}_w\right] = \sigma^2 \sum_{i=1}^{n} w_i^2 - \left(1 - \sum_{i=1}^{n} w_i^2\right)\frac{\sigma^2}{N-1} = \frac{N}{N-1}\sigma^2 \sum_{i=1}^{n} w_i^2 - \frac{\sigma^2}{N-1}. \tag{37}$$

To find the unbiased estimate $\overline{X}_w$ with the smallest variance, we thus need to solve the following constrained minimization problem:

$$\sum_{i=1}^{n} w_i^2 \longrightarrow \min, \quad \text{s.t.} \sum_{i=1}^{n} w_i = 1. \tag{38}$$

We can solve it using the method of *Lagrange multipliers*[4], or simply using the *Cauchy-Schwarz inequality*:

[4] This is a general method which is often used for solving constrained optimization problems.

$$\left(\sum_{i=1}^{n} x_i y_i\right)^2 \leq \left(\sum_{i=1}^{n} x_i^2\right)\left(\sum_{i=1}^{n} y_i^2\right), \quad \text{for any } x_1, \dots, x_n \text{ and } y_1, \dots, y_n. \tag{39}$$

The equality holds if and only if vectors $x = (x_1, \dots, x_n)$ and $y = (y_1, \dots, y_n)$ are parallel, i.e. one of them is a multiple of the other. Setting $x_i = w_i$ and $y_i = 1$, we have:

$$\sum_{i=1}^{n} w_i^2 \geq \frac{\left(\sum_{i=1}^{n} w_i\right)^2}{n} = \frac{1}{n}. \tag{40}$$

So, the smallest possible value of $\sum_{i=1}^{n} w_i^2$ under constraint $\sum_{i=1}^{n} w_i = 1$ is $1/n$. This value is achieved if and only if vectors $w = (w_1, \dots, w_n)$ and $(1, \dots, 1)$ are parallel, i.e. all $w_i = 1/n$. Thus, the sample mean $\overline{X}$ is the most efficient unbiased estimate. This result is intuitively expected: why should one give a larger weight $w_i > w_j$ to one sample unit than the other?

2.3 Estimating Proportions

(a) Using the definition of population variance,

$$\sigma^2 = \frac{1}{N}\sum_{i=1}^{N}(x_i - \mu)^2 = \frac{1}{N}\sum_{i=1}^{N} x_i^2 - \frac{2}{N}\mu\sum_{i=1}^{N} x_i + \mu^2 = \frac{1}{N}\sum_{i=1}^{N} x_i^2 - \mu^2. \tag{41}$$

The key observation is that, since x_i is either 0 or 1, $x_i^2 = x_i$. Therefore,

$$\sigma^2 = \frac{1}{N}\sum_{i=1}^{N} x_i - \mu^2 = \mu - \mu^2 = \mu(1-\mu). \tag{42}$$

(b) The expression for σ^2 in (42) suggests to use $\overline{X}(1-\overline{X})$ as its estimate. Let's check whether this estimate is unbiased or not. Using the fact that for any random variable Y, $\mathbb{E}[Y^2] = \mathbb{V}[Y] + \mathbb{E}[Y]^2$, and that $\mathbb{E}[\overline{X}] = \mu$ and $\mathbb{V}[\overline{X}]$ is given by (2.30), we have:

$$\begin{aligned}
\mathbb{E}[\overline{X}(1-\overline{X})] &= \mathbb{E}[\overline{X}] - \mathbb{E}[\overline{X}^2] = \mu - \mathbb{V}[\overline{X}] - \mathbb{E}[\overline{X}]^2 \\
&= \mu - \frac{\sigma^2}{n}\left(1 - \frac{n-1}{N-1}\right) - \mu^2 = \underbrace{\mu - \mu^2}_{\sigma^2} - \frac{\sigma^2}{n}\left(1 - \frac{n-1}{N-1}\right) \\
&= \sigma^2 - \frac{\sigma^2}{n}\left(1 - \frac{n-1}{N-1}\right) = \sigma^2 \frac{Nn-N}{Nn-n}.
\end{aligned} \tag{43}$$

Although $\overline{X}(1-\overline{X})$ is biased, we can immediately construct an unbiased estimate $\hat{\sigma}_u^2$ of σ^2 by properly normalizing $\overline{X}(1-\overline{X})$:

$$\hat{\sigma}_u^2 = \frac{Nn-n}{Nn-N}\,\overline{X}(1-\overline{X}). \tag{44}$$

Alternatively $\hat{\hat{\sigma}}_u^2$ can be obtained directly from the expression for $\hat{\sigma}_u^2$ using computations similar to (41) and (42).

(c) So, in this case, the population size is $N = 300$, the sample size is $n = 90$, the sample mean is $\overline{X} = 70/90$, and the required confidence level is given by $\alpha = 0.05$. The approximate $100(1-\alpha)\%$ confidence interval for μ is given by (2.69):

$$I_\alpha = \left(\overline{X} + z_{\frac{\alpha}{2}} \operatorname{se}[\overline{X}],\ \overline{X} - z_{\frac{\alpha}{2}} \operatorname{se}[\overline{X}]\right). \tag{45}$$

To compute it, we need to estimate the standard error

$$\operatorname{se}[\overline{X}] = \frac{\sigma}{\sqrt{n}}\sqrt{\left(1 - \frac{n-1}{N-1}\right)}. \tag{46}$$

We can't use the estimate derived in (2.40):

$$\widehat{\operatorname{se}}[\overline{X}] = \frac{\hat{\sigma}_u}{\sqrt{n}}\sqrt{\left(1 - \frac{n-1}{N-1}\right)}, \tag{47}$$

sine we can't compute $\hat{\sigma}_u$, since we don't know the individual responses of employees. But we can use $\hat{\hat{\sigma}}_u$ instead:

$$\operatorname{se}[\overline{X}] \approx \frac{\hat{\hat{\sigma}}_u}{\sqrt{n}}\sqrt{\left(1 - \frac{n-1}{N-1}\right)} = \sqrt{\frac{N-n}{Nn-N}}\sqrt{\overline{X}(1-\overline{X})}. \tag{48}$$

Plugging all numbers ($z_{0.025} \approx -2$) into (45), we have:

$$I_{\alpha=0.05} = (0.71, 0.85). \tag{49}$$

2.4 Finding the Sample Size

Our parameter of interest is the population mean

$$\mu = \frac{1}{N}\sum_{i=1} x_i, \tag{50}$$

where $x_i = 1$ if person i support the bill, and $x_i = 0$ otherwise. Since the approximate $100(1-\alpha)\%$ confidence interval for μ has the form $I_\alpha = \overline{X} \pm z_{\frac{\alpha}{2}} \operatorname{se}[\overline{X}]$, its length is $2|z_{\frac{\alpha}{2}}|\operatorname{se}[\overline{X}]$. Our goal is to find the minimum sample size n that guarantees that

$$2|z_{\frac{\alpha}{2}}|\operatorname{se}[\overline{X}] \leq 0.04, \quad \text{where } \alpha = 0.05. \tag{51}$$

Using approximate value $|z_{\frac{0.05}{2}}| \approx 2$, we have:

$$\operatorname{se}[\overline{X}] \leq 0.01. \tag{52}$$

The standard error of the sample mean is

$$\mathrm{se}[\overline{X}] = \frac{\sigma}{\sqrt{n}}\sqrt{1 - \frac{n-1}{N-1}}. \tag{53}$$

The population size N is unknown, but large enough so that the the finite population correction can be ignored (i.e. set to 1). This gives the following condition on n:

$$\mathrm{se}[\overline{X}] = \frac{\sigma}{\sqrt{n}} \leq 0.01, \tag{54}$$

or, equivalently,

$$n \geq 10^4\sigma^2. \tag{55}$$

The population variance is unknown, but from Problem 2.3 we know that

$$\sigma^2 = \mu - \mu^2. \tag{56}$$

The larger σ is, the larger the sample size n must be to achieve (55). The worst-case scenario occurs when σ^2 takes the largest possible value, which happens[5] when $\mu = 1/2$. Therefore, $\sigma^2_{\text{worst}} = \sigma^2_{\max} = 1/4$, and

[5] This is a simple calculus problem: $\mu - \mu^2 \to \max$. The maximizer is $\mu = 1/2$.

$$n \geq 2500. \tag{57}$$

The minimum size is then $n_{\min} = 2500$.

2.5 Variance of a Binary Population

(a) By definition, the population variance is

$$\sigma^2 = \frac{1}{N}\sum_{i=1}^{N}(x_i - \mu)^2 = \frac{1}{N}\sum_{i=1}^{N} x_i^2 - \mu^2. \tag{58}$$

Since each x_i is either α or β, the population mean μ is

$$\begin{aligned}\mu &= \frac{1}{N}\sum_{i=1}^{N} x_i = \frac{1}{N}\Bigg(\underbrace{\sum_{i:x_i=\alpha} x_i}_{\pi N \text{ terms}} + \underbrace{\sum_{i:x_i=\beta} x_i}_{(N-\pi N) \text{ terms}}\Bigg)\\ &= \frac{1}{N}(\alpha\pi N + \beta(N - \pi N)) = \alpha\pi + \beta(1-\pi).\end{aligned} \tag{59}$$

Similarly,

$$\begin{aligned}\frac{1}{N}\sum_{i=1}^{N} x_i^2 &= \frac{1}{N}\Bigg(\underbrace{\sum_{i:x_i=\alpha} x_i^2}_{\pi N \text{ terms}} + \underbrace{\sum_{i:x_i=\beta} x_i^2}_{(N-\pi N) \text{ terms}}\Bigg)\\ &= \frac{1}{N}\left(\alpha^2\pi N + \beta^2(N - \pi N)\right) = \alpha^2\pi + \beta^2(1-\pi).\end{aligned} \tag{60}$$

Finally,

$$\sigma^2 = \alpha^2\pi + \beta^2(1-\pi) - (\alpha\pi + \beta(1-\pi))^2, \tag{61}$$

and, after arithmetic simplification[6],

$$\sigma^2 = \pi(1-\pi)(\alpha-\beta)^2. \tag{62}$$

[6] "... which this margin is too narrow to contain." – Pierre de Fermat.

(b) Maximizing σ^2 with respect to π is equivalent to maximizing

$$\pi(1-\pi) = \pi - \pi^2 \to \max. \tag{63}$$

The maximum is achieved at $\pi = 1/2$ and the corresponding variance is

$$\sigma^2_{\max} = \frac{(\alpha-\beta)^2}{4}. \tag{64}$$

2.6 Matching the Accuracies of Surveys

(a) Recall (Corollary 2.2) that the standard error of the sample mean is given by

$$\text{se}[\overline{X}] = \frac{\sigma}{\sqrt{n}}\sqrt{1-\frac{n-1}{N-1}}, \tag{65}$$

where σ is the population standard deviation, N is the population size, and n is the sample size. To achieve the same accuracy in the surveys for Californian and Hawaiian doctors, we need the corresponding standard errors to coincide:

$$\frac{\sigma_C}{\sqrt{n_C}}\sqrt{1-\frac{n_C-1}{N_C-1}} = \frac{\sigma_H}{\sqrt{n_H}}\sqrt{1-\frac{n_H-1}{N_H-1}}. \tag{66}$$

Assuming that $\sigma_C = \sigma_H$ and solving for n_C gives

$$n_C = \frac{n_H N_C(N_H-1)}{N_C(N_H-n_H)+N_H(n_H-1)} \approx 1110. \tag{67}$$

(b) If $n_H = N_H$, then the standard error in the Hawaiian survey is zero, since in this case we survey the entire population of Hawaiian doctors. To achieve zero standard error in the Californian survey we thus also need to have $n_C = N_C$. The assumption $\sigma_H^2 = \sigma_C^2$ pays no role in this case.

2.7 The Bootstrap Method

(a) The obtained values of the population mean, the sample mean, and the standard error of the sample mean are[7]:

$$\mu = 0.024, \quad \overline{X} = -0.179, \quad \text{se}[\overline{X}] = 0.366. \tag{68}$$

[7] Since your Gaussian population and SRS sample are different, you will get slightly different values.

(b) The obtained value of the bootstrap estimate of the standard error is

$$\widehat{\text{se}}_B[\overline{X}] = 0.346. \tag{69}$$

(c) The obtained value of the modified bootstrap estimate of the standard error is

$$\widehat{\mathrm{se}}_B^*[\overline{X}] = 0.357. \tag{70}$$

(d) Figure 5 shows the required boxplots and the exact value of the standard error.

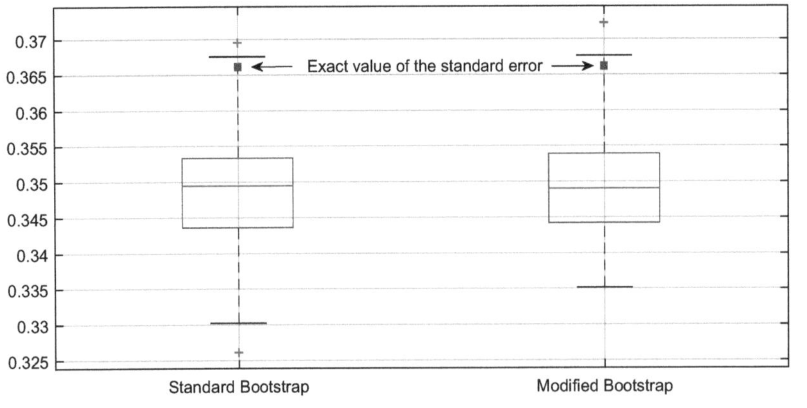

Fig. 5 Boxplots for 100 independent bootstrap estimates of the standard error $\mathrm{se}[\overline{X}]$ obtained by the standard (left boxplot) and modified (right boxplot) bootstrap methods.

2.8 The Body Mass Index of Male and Female Olympic Athletes

(a) The original data set [10] contains records for 11,538 Olympic athletes. After removing athletes with missing heights or masses, 10,858 athletes are left. After correcting masses and heights for the athletes in Table 3 and removing one athlete whose mass measurement is wrong, but the true value is unknown (see Table 3), $N = 10,857$ athletes are left for the purpose of the subsequent BMI analysis.

Table 3 Corrected data for the Rio 2016 Olympic athletes.

Athlete	Listed value	True value	Reference
Aslan Kakhidze	mass 39 kg	mass 86 kg	`https://g.co/kgs/YtGXp3`
Jeanelle Scheper	mass 39 kg	mass 60 kg	`https://g.co/kgs/BqWPyt`
Vitaly Fokeev	mass 45 kg	mass 86 kg	`https://g.co/kgs/n9Vf78`
Linet Arasa	height 2.03 m	height 1.73 m	`https://g.co/kgs/x1R83s`
Florian Temengil	height 1.53 m	height 2.11 m	`https://g.co/kgs/jQKKy2`
Florian Temengil	mass 125 kg	mass 112 kg	`https://g.co/kgs/jQKKy2`
Yvette Lewis	mass 150 kg	mass 75 kg	`https://g.co/kgs/Y1vrhY`
Darrell Wesh	mass 160 kg	mass 73 kg	`https://g.co/kgs/uyj4q6`
Maria Far[a]	mass 130 kg	NA	NA

[a] Maria Far is a Panamanian swimmer. In Rio de Janeiro, she competed in the women's 200 metre butterfly event. She obviously can't weight 130 kg, although even the official website `https://web.archive.org/web/20160825001159/https://www.rio2016.com/en/athlete/maria-far` claims that. Since I could not find the true value of her mass, her mass was considered as missing, and her record was removed from the data set.

(b) Figure 6(a) shows the boxplots for BMIs of male and female athletes. In general male athletes have higher BMI. There are $N_m = 5,862$ male and $N_f = 4,995$ female athletes. The mean values of the male and female BMIs are

$$\mu^{\mathrm{m}}_{\mathrm{BMI}} = 23.96\ \mathrm{kg/m^2} \quad \text{and} \quad \mu^{\mathrm{f}}_{\mathrm{BMI}} = 21.58\ \mathrm{kg/m^2}. \tag{71}$$

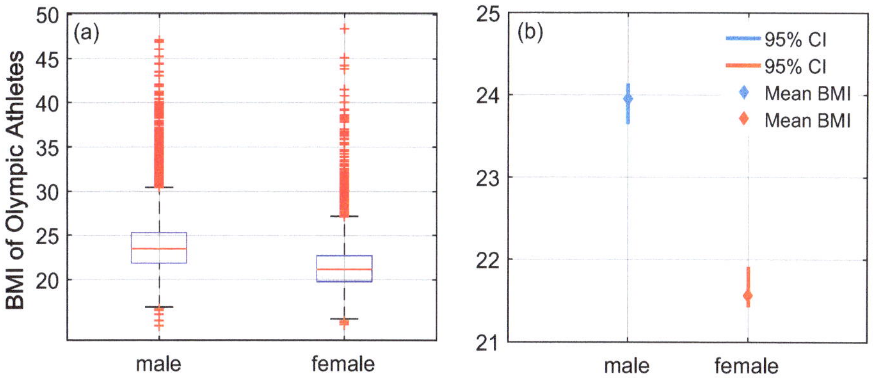

Fig. 6 Panel (a): Boxplots of male and female Olympic athletes participated in the Rio 2016 Olympic Games. Panel (b): 95% confidence intervals for $\mu^{\mathrm{m}}_{\mathrm{BMI}}$ and $\mu^{\mathrm{f}}_{\mathrm{BMI}}$ and their true values.

(c) The estimated values of $\mu^{\mathrm{m}}_{\mathrm{BMI}}$ and $\mu^{\mathrm{f}}_{\mathrm{BMI}}$ based on simple random samples $\mathscr{S}_Y$ and $\mathscr{S}_X$ of size $n = 500$ are[8]

$$\widehat{\mu}^{\mathrm{m}}_{\mathrm{BMI}} = \overline{Y} = 23.90\ \mathrm{kg/m^2} \quad \text{and} \quad \widehat{\mu}^{\mathrm{f}}_{\mathrm{BMI}} = \overline{X} = 21.67\ \mathrm{kg/m^2}. \tag{72}$$

The obtained analytical (2.40) and bootstrap (2.51) estimates of the standard errors of $\widehat{\mu}^{\mathrm{m}}_{\mathrm{BMI}}$ and $\widehat{\mu}^{\mathrm{f}}_{\mathrm{BMI}}$ are

$$\begin{aligned} \widehat{\mathrm{se}}_A[\widehat{\mu}^{\mathrm{m}}_{\mathrm{BMI}}] &= 0.124\ \mathrm{kg/m^2}, & \widehat{\mathrm{se}}_A[\widehat{\mu}^{\mathrm{f}}_{\mathrm{BMI}}] &= 0.123\ \mathrm{kg/m^2}, \\ \widehat{\mathrm{se}}_B[\widehat{\mu}^{\mathrm{m}}_{\mathrm{BMI}}] &= 0.132\ \mathrm{kg/m^2}, & \widehat{\mathrm{se}}_B[\widehat{\mu}^{\mathrm{f}}_{\mathrm{BMI}}] &= 0.124\ \mathrm{kg/m^2}. \end{aligned} \tag{73}$$

The 95% confidence intervals (2.69) for $\mu^{\mathrm{m}}_{\mathrm{BMI}}$ and $\mu^{\mathrm{f}}_{\mathrm{BMI}}$ are shown in Figure 6(b).

(d) My height is 1.78 m, my mass (as of December 2024) is 79 kg, my BMI is 24.9 kg/m^2. Five male athletes with the BMI closet to mine are: Novak Djokovic (SRB, tennis), Lucas Rossi (ARG, hockey), Angel Kodinov (BUL, canoe), Christos Afroudakis (GRE, aquatics), and Jack Beaumont (GBR, rowing).

[8] Once again, as in all problems involving random sampling, you will obtain slightly different values.

Problems of Chapter 3

3.1 Bias, Standard Error, and Mean Squared Error

(a) By definition (3.8), the bias of $\hat{\theta}_n^{(1)}$ is

$$\begin{aligned}\mathbb{B}[\hat{\theta}_n^{(1)}] &= \mathbb{E}[\hat{\theta}_n^{(1)}] - \theta = \mathbb{E}[2\overline{X}] - \theta = 2\mathbb{E}\left[\frac{1}{n}\sum_{i=1}^{n} X_i\right] - \theta \\ &= 2\frac{1}{n}\sum_{i=1}^{n}\underbrace{\mathbb{E}[X_i]}_{\theta/2} - \theta = 2\frac{1}{n}\sum_{i=1}^{n}\frac{\theta}{2} - \theta = \frac{n\theta}{n} - \theta = 0.\end{aligned} \tag{74}$$

The estimate $\hat{\theta}_n^{(1)}$ is, therefore, unbiased for any n.

By definition (3.7), the squared standard error[9] of estimate $\hat{\theta}_n^{(1)}$ is

[9] It is more convenient to work with the *squared* standard error, since it does not contain the annoying square root in its expression.

$$\text{se}[\hat{\theta}_n^{(1)}]^2 = \mathbb{V}[\hat{\theta}_n^{(1)}] = \mathbb{V}[2\overline{X}] = 4\mathbb{V}\left[\frac{1}{n}\sum_{i=1}^{n} X_i\right] = 4\frac{1}{n^2}\sum_{i=1}^{n}\mathbb{V}[X_i]. \tag{75}$$

Since $X_i \sim \mathscr{U}[0,\theta]$, its variance is

$$\mathbb{V}[X_i] = \mathbb{E}[X_i^2] - \mathbb{E}[X_i]^2 = \int_0^\theta x^2\frac{1}{\theta}dx - \left(\frac{\theta}{2}\right)^2 = \frac{\theta^2}{3} - \frac{\theta^2}{4} = \frac{\theta^2}{12}. \tag{76}$$

So, the standard error is then

$$\text{se}[\hat{\theta}_n^{(1)}] = \sqrt{4\frac{1}{n^2}\sum_{i=1}^{n}\frac{\theta^2}{12}} = \frac{\theta}{\sqrt{3n}}. \tag{77}$$

Finally, using the bias-variance decomposition of the mean squared error (3.11),

$$\text{MSE}[\hat{\theta}_n^{(1)}] = \mathbb{B}[\hat{\theta}_n^{(1)}]^2 + \text{se}[\hat{\theta}_n^{(1)}]^2 = \frac{\theta^2}{3n}. \tag{78}$$

(b) To compute the bias and standard error of $\hat{\theta}_n^{(2)} = X_{(n)}$, we first need to find the probability distribution of the order statistic $X_{(n)}$. By definition, its CDF is

$$\begin{aligned}F_{X_{(n)}}(x) &= \mathbb{P}(X_{(n)} \le x) = \mathbb{P}(\max\{X_1,\ldots,X_n\} \le x) \\ &= \mathbb{P}(X_i \le x \text{ for all } i = 1,\ldots,n) = \prod_{i=1}^{n}\mathbb{P}(X_i \le x) = \prod_{i=1}^{n}\frac{x}{\theta} = \left(\frac{x}{\theta}\right)^n.\end{aligned} \tag{79}$$

The PDF of $X_{(n)}$ is, therefore,

$$f_{X_{(n)}}(x) = \frac{d}{dx}F_{X_{(n)}}(x) = \frac{nx^{n-1}}{\theta^n}. \tag{80}$$

Now we can compute the bias of $\hat{\theta}_n^{(2)} = X_{(n)}$:

$$\begin{aligned}\mathbb{B}[\hat{\theta}_n^{(2)}] &= \mathbb{E}[\hat{\theta}_n^{(2)}] - \theta = \mathbb{E}[X_{(n)}] - \theta = \int_0^\theta x f_{X_{(n)}}(x)dx - \theta \\ &= \int_0^\theta \frac{nx^n}{\theta^n}dx - \theta = \frac{n}{\theta^n}\frac{\theta^{n+1}}{n+1} - \theta = \frac{n\theta}{n+1} - \theta = -\frac{\theta}{n+1}.\end{aligned} \tag{81}$$

So, the estimate $\hat{\theta}_n^{(2)}$ is biased, but the bias converges to zero as $n \to \infty$.

The squared standard error of $\hat{\theta}_n^{(2)}$ is

$$\begin{aligned}\text{se}[\hat{\theta}_n^{(2)}]^2 &= \mathbb{V}[\hat{\theta}_n^{(2)}] = \mathbb{E}[X_{(n)}^2] - \underbrace{\mathbb{E}[X_{(n)}]^2}_{\text{found in (81)}} = \int_0^\theta x^2 f_{X_{(n)}}(x)dx - \left(\frac{n\theta}{n+1}\right)^2 \\ &= \int_0^\theta \frac{nx^{n+1}}{\theta^n}dx - \left(\frac{n\theta}{n+1}\right)^2 = \frac{n}{\theta^n}\frac{\theta^{n+2}}{n+2} - \left(\frac{n\theta}{n+1}\right)^2 \\ &= \frac{n\theta^2}{n+2} - \left(\frac{n\theta}{n+1}\right)^2 = \frac{n(n+1)^2\theta^2 - n^2(n+2)\theta^2}{(n+2)(n+1)^2} \\ &= \frac{n\theta^2((n+1)^2 - n(n+2))}{(n+2)(n+1)^2} = \frac{n\theta^2}{(n+2)(n+1)^2}.\end{aligned} \tag{82}$$

The standard error is, thus,

$$\text{se}[\hat{\theta}_n^{(2)}] = \frac{\theta}{n+1}\sqrt{\frac{n}{n+2}}. \tag{83}$$

The mean-squared error of $\hat{\theta}_n^{(2)}$ can be also computed via the bias-variance decomposition as in part (a):

$$\begin{aligned}\text{MSE}[\hat{\theta}_n^{(2)}] &= \mathbb{B}[\hat{\theta}_n^{(2)}]^2 + \text{se}[\hat{\theta}_n^{(2)}]^2 = \frac{\theta^2}{(n+1)^2} + \frac{n\theta^2}{(n+2)(n+1)^2} \\ &= \frac{(n+2)\theta^2 + n\theta^2}{(n+2)(n+1)^2} = \frac{2\theta^2}{(n+1)(n+2)}.\end{aligned} \tag{84}$$

(c) To compare the accuracy of two estimates, we need to compare their mean squared errors[10]: the smaller the MSE, the more accurate the estimate is. When the sample size n increases, both MSEs decrease to zero. However, $\text{MSE}[\hat{\theta}_n^{(2)}]$ decays approximately as $1/n^2$, while $\text{MSE}[\hat{\theta}_n^{(1)}]$ decays slower, approximately as $1/n$. This means that for large n, the estimate $\hat{\theta}_n^{(2)} = X_{(n)}$ is more accurate than $\hat{\theta}_n^{(1)} = 2\overline{X}$.

[10] There are other measures of accuracy, but the MSE is one of the most popular and widely used measures.

A more careful analysis shows that $\text{MSE}[\hat{\theta}_n^{(1)}] = \text{MSE}[\hat{\theta}_n^{(2)}]$ if $n = 1, 2$, and $\text{MSE}[\hat{\theta}_n^{(1)}] > \text{MSE}[\hat{\theta}_n^{(2)}]$ if $n \geq 3$:

$$\text{MSE}[\hat{\theta}_n^{(1)}] > \text{MSE}[\hat{\theta}_n^{(2)}] \quad \Leftrightarrow \quad \frac{1}{3n} > \frac{2}{(n+1)(n+2)} \quad \Leftrightarrow \quad n^2 - 3n + 2 > 0 \tag{85}$$

For natural $n \in \mathbb{N}$, the last inequality holds if and only if $n = 3, 4, \ldots$.

3.2 Confidence Interval for the Right Bound of a Uniform Distribution

(a) Since $X_1, \ldots, X_n \overset{\text{iid}}{\sim} \mathscr{U}[0, \theta]$, all $X_i \in [0, \theta]$. Therefore, the maximum value $X_{(n)} = \max\{X_1, \ldots, X_n\} \in [0, \theta]$ and $Q = X_{(n)}/\theta \in [0, 1]$. To show that the distribution of Q does not depend on θ, let's find its CDF. For $x \in [0, 1]$, we have:

$$\begin{aligned} F_Q(x) &= \mathbb{P}(Q \leq x) = \mathbb{P}(X_{(n)} \leq x\theta) = \mathbb{P}(\max\{X_1, \ldots, X_n\} \leq x\theta) \\ &= \mathbb{P}(X_1 \leq x\theta, \ldots, X_n \leq x\theta) = \prod_{i=1}^{n} \mathbb{P}(X_i \leq x\theta) \\ &= \prod_{i=1}^{n} \mathbb{P}\Big(\underbrace{\frac{X_i}{\theta}}_{\sim \mathscr{U}[0,1]} \leq x\Big) = \prod_{i=1}^{n} x = x^n. \end{aligned} \tag{86}$$

So, Q is indeed a pivot: although its value depends on θ, its probability distribution does not.

(b) Our goal is to find an interval $I_\alpha = (a, b)$, where a and b are some functions of the data $X_1, \ldots, X_n$, such that

$$\mathbb{P}(a < \theta < b) = 1 - \alpha. \tag{87}$$

The problem gives a hint that a confidence interval can be constructed using the random variable Q. Using the derived formula for the CDF of Q, we have:

$$\mathbb{P}(c < Q < d) = F_Q(d) - F_Q(c) = d^n - c^n. \tag{88}$$

Using the definition of $Q = X_{(n)}/\theta$,

$$d^n - c^n = \mathbb{P}\left(c < \frac{X_{(n)}}{\theta} < d\right) = \mathbb{P}\left(\frac{X_{(n)}}{d} < \theta < \frac{X_{(n)}}{c}\right). \tag{89}$$

To achieve the required confidence level, let's set $d = 1$ and $c = \alpha^{1/n}$. Then we have

$$1 - \alpha = \mathbb{P}\left(X_{(n)} < \theta < \frac{X_{(n)}}{\alpha^{1/n}}\right). \tag{90}$$

This means that

$$I_\alpha = \left(X_{(n)}, \frac{X_{(n)}}{\alpha^{1/n}}\right) \tag{91}$$

is an $100(1-\alpha)\%$ confidence interval for θ.

It is important to highlight that the desired confidence interval is not unique. Note that, as it follows from (89), *any* choice of c and d such that $d^n - c^n = 1 - \alpha$ will give us an $100(1-\alpha)\%$ confidence interval for θ of the following form:

$$I_\alpha = \left(\frac{X_{(n)}}{d}, \frac{X_{(n)}}{c}\right), \quad d^n - c^n = 1 - \alpha. \tag{92}$$

For example, we could set $c = 0$ and $d = (1-\alpha)^{1/n}$. This would lead to the following $100(1-\alpha)\%$ confidence interval:

$$I_\alpha = \left(\frac{X_{(n)}}{(1-\alpha)^{1/n}}, \infty\right). \tag{93}$$

However, it can be shown that (91) is geometrically the shortest $100(1-\alpha)\%$ confidence interval for θ among all intervals of the from (92), and, therefore, the most useful in applications, since the shorter the interval, the more informative it is. The length l of the interval (92) is

$$l = X_{(n)}\left(\frac{1}{c} - \frac{1}{d}\right). \tag{94}$$

To find the shortest interval, we thus need to solve the following constrained optimization problem:

$$f(c,d) = \frac{1}{c} - \frac{1}{d} \to \min, \quad \text{subject to} \quad d^n - c^n = 1 - \alpha. \tag{95}$$

The solution to this problem is given by $c = \alpha^{1/n}$ and $d = 1$, which leads to the shortest $100(1-\alpha)\%$ confidence interval (91).

Problems of Chapter 4

4.1 Empirical CDF

The covariance between random variables $\hat{F}_n(x)$ and $\hat{F}_n(y)$ is

$$\mathrm{Cov}(\hat{F}_n(x), \hat{F}_n(y)) = \mathbb{E}\left[\hat{F}_n(x)\hat{F}_n(y)\right] - \mathbb{E}\left[\hat{F}_n(x)\right]\mathbb{E}\left[\hat{F}_n(y)\right]. \tag{96}$$

Since for any x, $\hat{F}_n(x)$ is an unbiased estimate of $F(x)$ (see Theorem 4.1),

$$\mathbb{E}\left[\hat{F}_n(x)\right] = F(x) \quad \text{and} \quad \mathbb{E}\left[\hat{F}_n(y)\right] = F(y). \tag{97}$$

The expected value of the product $\hat{F}_n(x)\hat{F}_n(y)$ is

$$
\begin{aligned}
\mathbb{E}\left[\hat{F}_n(x)\hat{F}_n(y)\right] &= \frac{1}{n^2}\mathbb{E}\left[\sum_{i=1}^{n} H(x-X_i)\sum_{j=1}^{n} H(y-X_j)\right] \\
&= \frac{1}{n^2}\mathbb{E}\left[\sum_{i=1}^{n} H(x-X_i)H(y-X_i) + \sum_{i\neq j} H(x-X_i)H(y-X_j)\right] \\
&= \frac{1}{n^2}\left(\sum_{i=1}^{n} \mathbb{E}\left[H(x-X_i)H(y-X_i)\right] + \sum_{i\neq j} \mathbb{E}\left[H(x-X_i)H(y-X_j)\right]\right).
\end{aligned}
\tag{98}
$$

Since random variables X_i and X_j are independent when $i \neq j$, so are random variables $H(x-X_i)$ and $H(y-X_j)$. Therefore[11],

$$
\mathbb{E}\left[H(x-X_i)H(y-X_j)\right] = \underbrace{\mathbb{E}[H(x-X_i)]}_{F(x)}\underbrace{\mathbb{E}[H(y-X_j)]}_{F(y)} = F(x)F(y). \tag{99}
$$

[11] Random variables $H(x-X_i)$ and $H(y-X_j)$ are the Bernoulli random variables with parameters $F(x)$ and $F(y)$, respectively. See (4.24), (4.25).

Let's look now at the random variable $H(x-X_i)H(y-X_i)$:

$$
H(x-X_i)H(y-X_i) = \begin{cases} 1, & \text{if } X_i \leq \min\{x,y\}, \\ 0, & \text{if } X_i > \min\{x,y\}. \end{cases} \tag{100}
$$

Therefore,

$$
\mathbb{E}\left[H(x-X_i)H(y-X_i)\right] = \mathbb{P}(X_i \leq \min\{x,y\}) = F(\min\{x,y\}). \tag{101}
$$

Combining all the above results, we have:

$$
\begin{aligned}
\operatorname{Cov}(\hat{F}_n(x),\hat{F}_n(y)) &= \frac{1}{n^2}\left(\sum_{i=1}^{n} F(\min\{x,y\}) + \sum_{i\neq j} F(x)F(y)\right) - F(x)F(y) \\
&= \frac{1}{n^2}\left(nF(\min\{x,y\}) + (n^2-n)F(x)F(y)\right) - F(x)F(y) \\
&= \frac{F(\min\{x,y\})}{n} + F(x)F(y)\left(\frac{n-1}{n} - 1\right) \\
&= \frac{F(\min\{x,y\}) - F(x)F(y)}{n}.
\end{aligned}
\tag{102}
$$

If $x = y$, then (102) boils down to

$$
\operatorname{Cov}(\hat{F}_n(x),\hat{F}_n(x)) = \mathbb{V}\left[\hat{F}_n(x)\right] = \operatorname{se}\left[\hat{F}_n(x)\right]^2 = \frac{F(x)(1-F(x))}{n}, \tag{103}
$$

which is equivalent to (4.21).

4.2 Nonparametric Confidence Band for the CDF

A general $(1-\alpha)$ nonparametric confidence band for the CDF F is given by (4.40). To construct a 95% confidence band, we need to set $\alpha = 0.05$. The corresponding confidence band will have then the folowing from:

$$\left\{x : \max\left\{\hat{F}_n(x) - 0.0407,\ 0\right\} \leq F(x) \leq \min\left\{\hat{F}_n(x) + 0.0407,\ 1\right\}\right\}, \tag{104}$$

where $\hat{F}_n(x)$ is the eCDF. Figure 7 shows both the eCDF (a point estimate of F) and the confidence band (104), a set estimate of F.

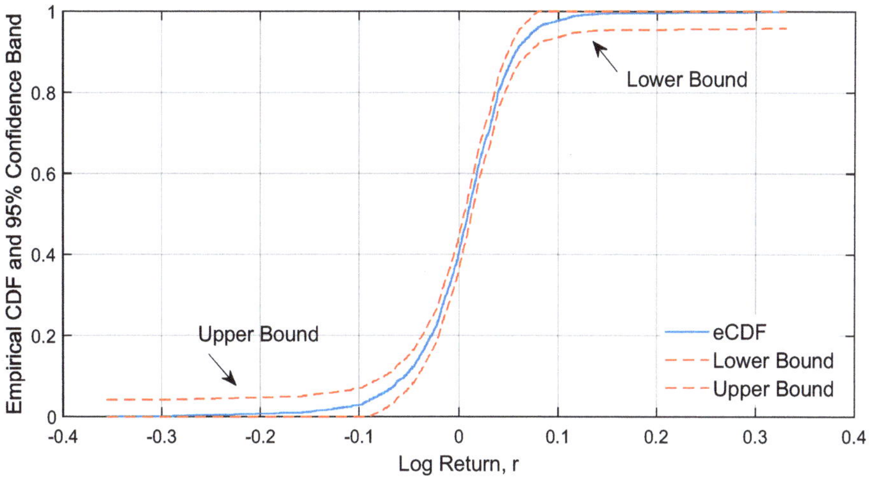

Fig. 7 The empirical CDF of the S&P 500 log returns and a 95% nonparametric confidence band for the CDF of the log returns.

4.3 Skewness and Kurtosis

By definition, the plug-in estimate of parameter $\theta = \mathfrak{f}(F)$ is $\hat{\theta}_n = \mathfrak{f}(\hat{F}_n)$, where $\hat{F}_n$ is the eCDF. Since the skewness is

$$\varkappa = \mathfrak{f}(F) = \frac{\int (x-\mu)^3 dF(x)}{\left(\int (x-\mu)^2 dF(x)\right)^{3/2}} = \frac{\int (x - \int x dF(x))^3 dF(x)}{\left(\int (x - \int x dF(x))^2 dF(x)\right)^{3/2}}, \tag{105}$$

its plug-in estimate is[12]

$$\begin{aligned}\hat{\varkappa}_n = \mathfrak{f}(\hat{F}_n) &= \frac{\int (x - \int x d\hat{F}_n(x))^3 d\hat{F}_n(x)}{\left(\int (x - \int x d\hat{F}_n(x))^2 d\hat{F}_n(x)\right)^{3/2}} \\ &= \frac{\int (x-\overline{X})^3 d\hat{F}_n(x)}{\left(\int (x-\overline{X})^2 d\hat{F}_n(x)\right)^{3/2}} = \frac{\frac{1}{n}\sum_{i=1}^n (X_i - \overline{X})^3}{\left(\frac{1}{n}\sum_{i=1}^n (X_i-\overline{X})^2\right)^{3/2}} = \frac{\frac{1}{n}\sum_{i=1}^n (X_i - \overline{X})^3}{\hat{\sigma}_n^3},\end{aligned} \tag{106}$$

where $\hat{\sigma}_n^2$ is the plug-in estimate of the variance (4.49).

Similarly, for the kurtosis,

$$\kappa = \mathfrak{f}(F) = \frac{\int (x-\mu)^4 dF(x)}{\left(\int (x-\mu)^2 dF(x)\right)^2} = \frac{\int (x - \int x dF(x))^4 dF(x)}{\left(\int (x - \int x dF(x))^2 dF(x)\right)^2} \tag{107}$$

[12] Computations in (106) and (108) are similar to those in Example 4.1.

we have:

$$\begin{aligned}\hat{\kappa}_n = \mathfrak{f}(\hat{F}_n) &= \frac{\int (x - \int x d\hat{F}_n(x))^4 d\hat{F}_n(x)}{\left(\int (x - \int x d\hat{F}_n(x))^2 d\hat{F}_n(x)\right)^2} \\ &= \frac{\int (x - \overline{X})^4 d\hat{F}_n(x)}{\left(\int (x - \overline{X})^2 d\hat{F}_n(x)\right)^2} = \frac{\frac{1}{n}\sum_{i=1}^n (X_i - \overline{X})^4}{\left(\frac{1}{n}\sum_{i=1}^n (X_i - \overline{X})^2\right)^2} = \frac{\frac{1}{n}\sum_{i=1}^n (X_i - \overline{X})^4}{\hat{\sigma}_n^4}.\end{aligned} \tag{108}$$

It is worth mentioning that both estimates, $\hat{\varkappa}_n$ and $\hat{\kappa}_n$, are strongly sensitive to outliers. So, before using them for statistical analysis of the data, it is important, if possible, to check that the data is "clean" and does not contain errors with large magnitudes.

4.4 Estimating the Effectiveness of a New Drug

The parameter of interest is $\theta = p_1 - p_2$, where p_1 and p_2 are the probabilities of recovery from the disease due to the new and old drugs respectively. The plug-in estimate of θ is then

$$\hat{\theta}_n = \hat{p}_{1,n} - \hat{p}_{2,n}, \tag{109}$$

where $\hat{p}_{1,n}$ and $\hat{p}_{2,n}$ are the plug-in estimates of p_1 and p_2. Let's find $\hat{p}_{1,n}$ first.

Let X be an "indicator" random variable that equals to 1 if a patient received the new drug recovers and equals to 0 if not. Then

$$p_1 = \mathbb{P}(X = 1). \tag{110}$$

Let F_X be the CDF of the binary random variable $X \in \{0, 1\}$. Then we can rewrite p_1 in terms of F_X as follows[13]:

$$p_1 = \mathbb{P}(X = 1) = 1 - \mathbb{P}(X = 0) = 1 - \mathbb{P}(X \leq 0) = 1 - F_X(0). \tag{111}$$

[13] Here we use the fact that the events $X = 0$ and $X \leq 0$ are equivalent.

The plug-in estimate of p_1 is then

$$\hat{p}_{1,n} = 1 - \hat{F}_{X,n}(0), \tag{112}$$

where $\hat{F}_{X,n}$ is the eCDF. We can find $\hat{F}_{X,n}$ from the data. Let

$$X_i = \begin{cases} 1, & \text{if patient } i \text{ in group 1 recovered,} \\ 0, & \text{if patient } i \text{ in group 1 did not recover.} \end{cases} \tag{113}$$

Then, by definition, the eCDF $\hat{F}_{X,n}$ is

$$\hat{F}_{X,n}(x) = \frac{1}{n}\sum_{i=1}^n H(x - X_i). \tag{114}$$

Therefore,

$$\hat{p}_{1,n} = 1 - \frac{1}{n}\sum_{i=1}^n H(-X_i) \tag{115}$$

Notice that if $X_i = 0$, then $H(-X_i) = H(0) = 1$; and if $X_i = 1$, then $H(-X_i) = H(-1) = 0$. This means that $H(-X_i) = 1 - X_i$. Finally,

$$\hat{p}_{1,n} = 1 - \frac{1}{n}\sum_{i=1}^{n}(1-X_i) = \frac{1}{n}\sum_{i=1}^{n} X_i = \frac{n_1}{n}, \tag{116}$$

which is a very natural and intuitive estimate[14] of p_1.

Similarly, the plug-in estimate of p_2 is

$$\hat{p}_{2,n} = \frac{n_2}{n}. \tag{117}$$

[14] The calculations above show that this natural estimate is, in fact, the plug-in estimate.

So, the plug-in estimate of θ is, therefore,

$$\hat{\theta}_n = \hat{p}_{1,n} - \hat{p}_{2,n} = \frac{n_1 - n_2}{n} = \frac{1}{18} \approx 0.056. \tag{118}$$

Problems of Chapter 5

5.1 Jackknife: Theoretical Analysis

(a) The plug-in estimate is

$$\hat{\theta}_n = \mathfrak{f}(\hat{F}_n) = \min\{x : \hat{F}_n(x) = 1\} = X_{(n)} = \max\{X_1, \ldots, X_n\}. \tag{119}$$

(b) The bias of the plug-in estimate is (see Problem 3.1(b)):

$$\mathbb{B}[\hat{\theta}_n] = \mathbb{B}[X_{(n)}] = -\frac{\theta}{n+1} = O\left(\frac{1}{n}\right). \tag{120}$$

It satisfies the assumption (5.12) with $a = -\theta$ and $b = \theta$:

$$\begin{aligned}\mathbb{B}[\hat{\theta}_n] &= -\frac{\theta}{n+1} = -\frac{\theta}{n} + \frac{\theta}{n^2} - \underbrace{\frac{\theta}{n^2(n+1)}}_{O\left(\frac{1}{n^3}\right)} \\ &= -\frac{\theta}{n} + \frac{\theta}{n^2} + O\left(\frac{1}{n^3}\right).\end{aligned} \tag{121}$$

(c) The bias-corrected estimate (also called the jackknifed estimate) is

$$\hat{\theta}_n^J = n\hat{\theta}_n - (n-1)\bar{\theta}_n^J, \tag{122}$$

where $\bar{\theta}_n^J$ is the mean of jackknife replications $\hat{\theta}_n^{(-1)}, \ldots, \hat{\theta}_n^{(-n)}$ of $\hat{\theta}_n$,

$$\bar{\theta}_n^J = \frac{1}{n}\sum_{i=1}^{n}\hat{\theta}_n^{(-i)}. \tag{123}$$

Let $i^* \in \{1,\dots,n\}$ be that number for which $X_{i^*} = X_{(n)}$. In other words, let X_{i^*} be the largest value in the sample $X_1,\dots,X_n$. Then the jackknife replication $\hat{\theta}_n^{(-i)}$ is

$$\hat{\theta}_n^{(-i)} = \max\{X_1,\dots,X_{i-1},X_{i+1},\dots,X_n\} = \begin{cases} X_{(n-1)}, & \text{if } i = i^*, \\ X_{(n)}, & \text{if } i \neq i^*. \end{cases} \tag{124}$$

Therefore,

$$\bar{\theta}_n^J = \frac{1}{n}\sum_{i=1}^{n}\hat{\theta}_n^{(-i)} = \frac{1}{n}\sum_{i\neq i^*}\underbrace{\hat{\theta}_n^{(-i)}}_{X_{(n)}} + \frac{1}{n}\underbrace{\hat{\theta}_n^{(-i^*)}}_{X_{(n-1)}} = \frac{n-1}{n}X_{(n)} + \frac{1}{n}X_{(n-1)}. \tag{125}$$

Plugging this back into (122), we have

$$\hat{\theta}_n^J = nX_{(n)} - (n-1)\left(\frac{n-1}{n}X_{(n)} + \frac{1}{n}X_{(n-1)}\right) = \frac{2n-1}{n}X_{(n)} - \frac{n-1}{n}X_{(n-1)}. \tag{126}$$

(d) Using the formula for the expected value of the order statistic $X_{(k)}$ sampled from the uniform distribution $\mathscr{U}[0,\theta]$,

$$\mathbb{E}[X_{(k)}] = \frac{k\theta}{n+1}, \tag{127}$$

we can find the bias of the jackknifed estimate:

$$\begin{aligned} \mathbb{B}[\hat{\theta}_n^J] = \mathbb{E}[\hat{\theta}_n^J] - \theta &= \frac{2n-1}{n}\frac{n\theta}{n+1} - \frac{n-1}{n}\frac{(n-1)\theta}{n+1} - \theta \\ &= -\frac{\theta}{n(n+1)} = O\left(\frac{1}{n^2}\right). \end{aligned} \tag{128}$$

So, the bias of the jackknifed estimate $\hat{\theta}_n^J$ is an order of magnitude smaller than then bias of the original plug-in estimate $\hat{\theta}_n$.

(e) Let's briefly recall how we derived the expression $\hat{\theta}_n^J = n\hat{\theta}_n - (n-1)\bar{\theta}_n^J$ for the jackknife estimate based on the assumption (5.12). First, we computed the expected value $\mathbb{E}[\bar{\theta}_n^J - \hat{\theta}_n]$ to see how different $\bar{\theta}_n^J - \hat{\theta}_n$ is from $\mathbb{B}[\hat{\theta}_n]$ on average. Based on this computation, we found that to estimate $\mathbb{B}[\hat{\theta}_n]$, we need to multiply $\bar{\theta}_n^J - \hat{\theta}_n$ by $(n-1)$, that is $\mathbb{B}[\hat{\theta}_n] \approx (n-1)(\bar{\theta}_n^J - \hat{\theta}_n) = \hat{\mathbb{B}}_J[\hat{\theta}_n]$. This inflation factor $(n-1)$ was determined by (5.12). Finally, to get the jackknifed estimate $\hat{\theta}_n^J$, we subtracted the estimated bias $\hat{\mathbb{B}}_J[\hat{\theta}_n]$ from the original estimate.

Let's perform the same steps, except that now, instead of (5.12), let's use the exact expression for the bias derived in part (b), $\mathbb{B}[\hat{\theta}_n] = -\theta/(n+1)$. First,

$$\bar{\theta}_n^J - \hat{\theta}_n = \underbrace{\frac{n-1}{n}X_{(n)} + \frac{1}{n}X_{(n-1)}}_{\bar{\theta}_n^J} - \underbrace{X_{(n)}}_{\hat{\theta}_n} = \frac{X_{(n-1)} - X_{(n)}}{n}. \tag{129}$$

Using (127), its expected value is

$$\mathbb{E}[\bar{\theta}_n^J - \hat{\theta}_n] = \frac{\frac{(n-1)\theta}{n+1} - \frac{n\theta}{n+1}}{n} = -\frac{\theta}{n(n+1)}. \tag{130}$$

By comparing this expected value with the exact bias $\mathbb{B}[\hat{\theta}_n] = -\theta/(n+1)$, we see that to obtain the bias on the left-hand-side of the above equation, we need to simply "inflate" $\bar{\theta}_n^J - \hat{\theta}_n$ by multiplying it by n. In other words,

$$\hat{\hat{\mathbb{B}}}_J[\hat{\theta}_n] = n(\bar{\theta}_n^J - \hat{\theta}_n) \tag{131}$$

is an unbiased estimate of $\mathbb{B}[\hat{\theta}_n]$. Note that the inflation factor here is n and not $(n-1)$ as in (5.17) because (131) is based on the exact expression for the bias and not on its asymptotic behavior (5.12).

Finally, to find the corresponding jackknifed estimate, we need to subtract the estimated bias $\hat{\hat{\mathbb{B}}}_J[\hat{\theta}_n]$ from the original estimate[15]:

[15] Compare with (5.19).

$$\hat{\hat{\theta}}_n^J = \hat{\theta}_n - \hat{\hat{\mathbb{B}}}_J[\hat{\theta}_n] = X_{(n)} - n\frac{X_{(n-1)} - X_{(n)}}{n} = 2X_{(n)} - X_{(n-1)}. \tag{132}$$

The original jackknifed estimate $\hat{\theta}_n^J$ derived in part (c) converges in probability to the new jackknifed estimate $\hat{\hat{\theta}}_n^J$ as the sample size increases:

$$\begin{aligned}\hat{\hat{\theta}}_n^J - \hat{\theta}_n^J &= (2X_{(n)} - X_{(n-1)}) - \left(\frac{2n-1}{n}X_{(n)} - \frac{n-1}{n}X_{(n-1)}\right) \\ &= \frac{X_{(n)} - X_{(n-1)}}{n} \xrightarrow{\mathbb{P}} 0, \quad \text{as } n \to \infty.\end{aligned} \tag{133}$$

(f) Once again using (127), the bias of $\hat{\hat{\theta}}_n^J$ is

$$\mathbb{B}[\hat{\hat{\theta}}_n^J] = \mathbb{E}[\hat{\hat{\theta}}_n^J] - \theta = 2\mathbb{E}[X_{(n)}] - \mathbb{E}[X_{(n-1)}] - \theta = \frac{2n\theta}{n+1} - \frac{(n-1)\theta}{n+1} - \theta = 0. \tag{134}$$

So, $\hat{\hat{\theta}}_n^J$ is an unbiased estimate of θ! The asymptotic assumption (5.12) allowed us to construct $\hat{\theta}_n^J$ and reduce the bias of $\hat{\theta}_n$ by an order of magnitude. But using more information — the exact formula for the bias — allows to construct $\hat{\hat{\theta}}_n^J$ and eliminate the bias completely[16].

[16] Estimate $\hat{\hat{\theta}}_n^J$ was derived by Sri Aditya Deevi, a Caltech electrical engineering graduate student.

5.2 Jackknife: Practical Implementation

(a) Since $X_1, \dots, X_n \overset{\text{iid}}{\sim} \mathcal{N}(\mu, 1)$, the sample mean $\overline{X} \sim \mathcal{N}\left(\mu, \frac{1}{n}\right)$. The plug-in estimate $\hat{\theta}_n = e^{\overline{X}}$ is then log-normally distributed and, using (5.53),

$$\mathbb{E}[\hat{\theta}_n] = e^{\mu+\frac{1}{2n}}. \tag{135}$$

Therefore, the bias of $\hat{\theta}_n$ is

$$\mathbb{B}[\hat{\theta}_n] = \mathbb{E}[\hat{\theta}_n] - \theta = e^{\mu+\frac{1}{2n}} - e^{\mu} = e^{\mu}(e^{\frac{1}{2n}} - 1). \tag{136}$$

Using the Taylor series expansion of the exponential function, $e^x = 1 + x + x^2/2! + O(x^3)$ as $x \to 0$, we have

$$\begin{aligned}\mathbb{B}[\hat{\theta}_n] &= e^{\mu}\left(1 + \frac{1}{2n} + \frac{1}{8n^2} - 1 + O\left(\frac{1}{n^3}\right)\right) \\ &= \frac{e^{\mu}}{2n} + \frac{e^{\mu}}{8n^2} + O\left(\frac{1}{n^3}\right), \quad \text{as } n \to \infty.\end{aligned} \tag{137}$$

The assumption (5.52) is thus indeed satisfied with $a = e^{\mu}/2$ and $b = e^{\mu}/8$.

(b) From (136), the exact value of the bias is

$$\mathbb{B}[\hat{\theta}_n] = e^{\mu}(e^{\frac{1}{2n}} - 1) = 0.744. \tag{138}$$

From (5.17), the obtained value (based on the generated sample) of the jackknife estimate of the bias is

$$\hat{\mathbb{B}}_J[\hat{\theta}_n] = (n-1)(\bar{\theta}_n^J - \hat{\theta}_n) = 0.662. \tag{139}$$

(c) The values of $\mathbb{B}_1$ and $\mathbb{B}_2$ obtained in simulation are:

$$\mathbb{B}_1 = 0.659 \quad \text{and} \quad \mathbb{B}_2 = -0.086. \tag{140}$$

As expected, $\mathbb{B}_1 \approx \mathbb{B}[\hat{\theta}_n]$ and $|\mathbb{B}_2| < |\mathbb{B}_1|$.

Problems of Chapter 6

6.1 Bootstrapping Fiji Earthquakes

(a) The scatterplot of the data, number of stations (Y) versus earthquake magnitude (X) is shown in Fig. 8. As expected, the two variables are strongly positively correlated: the larger the magnitude of a seismic event, the larger the number of stations that registered it.

(b) To find the plug-in estimate of the Pearson correlation coefficient, recall[17] that the plug-in estimate of the variance

$$\sigma_X^2 = \mathbb{V}[X] = \mathbb{E}[(X - \mathbb{E}[X])^2] \tag{141}$$

is

[17] See Example 4.1, equation (4.49).

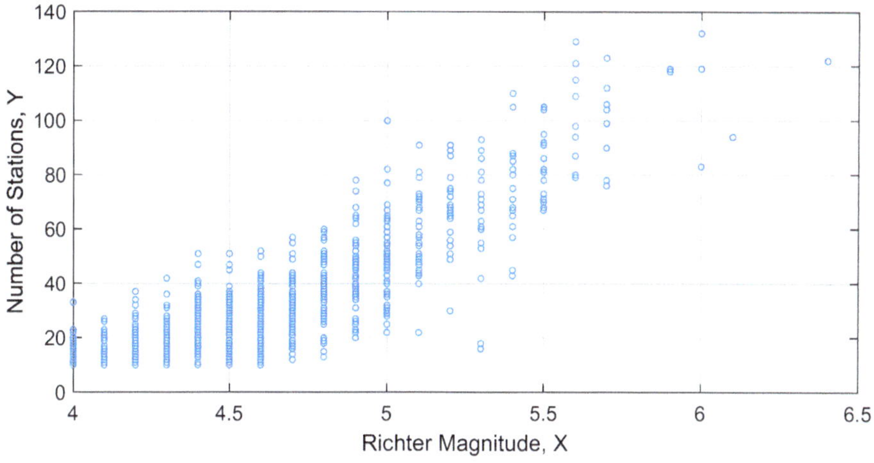

Fig. 8 Scatterplot $(X_1, Y_1), \ldots, (X_n, Y_n)$ of the number of stations (Y) vs earthquake magnitudes (X) for $n = 1000$ seismic event in the Fiji earthquakes dataset.

$$\hat{\sigma}_n^2 = \frac{1}{n}\sum_{i=1}^{n}(X_i - \overline{X})^2. \tag{142}$$

Similarly, the plug-in estimate of the covariance

$$\sigma_{XY}^2 = \mathrm{Cov}(\mathrm{X}, \mathrm{Y}) = \mathbb{E}[(X - \mathbb{E}[X])(Y - \mathbb{E}[Y])] \tag{143}$$

is

$$\hat{c}_n^2 = \frac{1}{n}\sum_{i=1}^{n}(X_i - \overline{X})(Y_i - \overline{Y}). \tag{144}$$

Therefore, the plug-in estimate of

$$\rho_{XY} = \frac{\mathrm{Cov}(\mathrm{X}, \mathrm{Y})}{\sqrt{\mathbb{V}[X]\mathbb{V}[Y]}} \tag{145}$$

is

$$\hat{\rho}_n = \frac{\sum_{i=1}^{n}(X_i - \overline{X})(Y_i - \overline{Y})}{\sqrt{\sum_{i=1}^{n}(X_i - \overline{X})^2 \sum_{i=1}^{n}(Y_i - \overline{Y})^2}}, \tag{146}$$

which is simply the sample correlation coefficient (5.41). For the Fiji earthquakes dataset, the value of the plug-in estimate is

$$\hat{\rho}_n = 0.85. \tag{147}$$

(c) The bootstrap estimate of the standard error of $\hat{\rho}_n$ based on $B = 10^3$ bootstrap samples is

$$\widehat{\mathrm{se}}_B[\hat{\rho}_n] = 0.01. \tag{148}$$

The histogram of the corresponding bootstrap replications $\hat{\rho}_n^{(1)}, \ldots, \hat{\rho}_n^{(B)}$ of $\hat{\rho}_n$ is shown in Fig. 9.

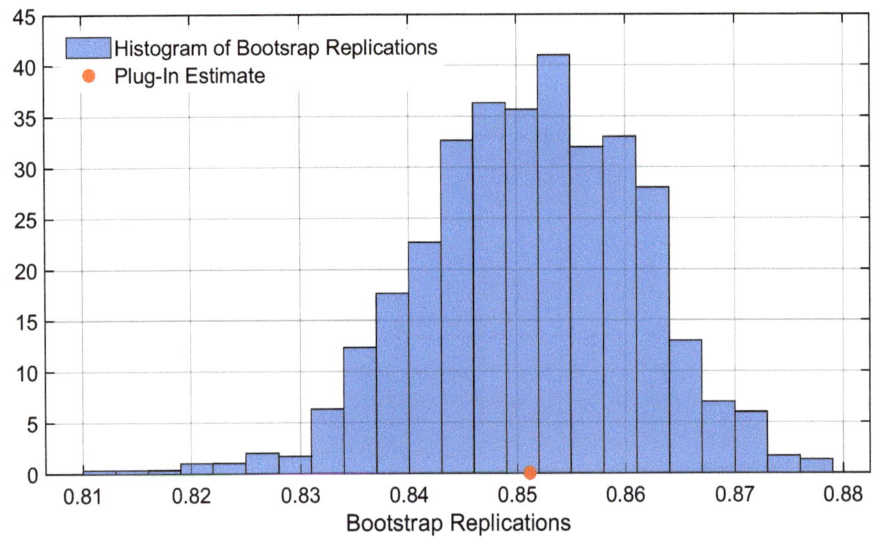

Fig. 9 Histogram of $B = 1000$ bootstrap replications $\hat{\rho}_n^{(1)}, \ldots, \hat{\rho}_n^{(B)}$ of the plug-in estimate $\hat{\rho}_n$ together with the original value (147).

(d) The normal and pivotal 95% ($\alpha = 0.05$) confidence interval for ρ_{XY} are approximately the same:

$$\begin{aligned}
&\text{Normal CI:}\ I_\alpha = \left(\hat{\rho}_n + z_{\frac{\alpha}{2}} \widehat{\mathrm{se}}_B[\hat{\rho}_n], \hat{\rho}_n - z_{\frac{\alpha}{2}} \widehat{\mathrm{se}}_B[\hat{\rho}_n]\right) = (0.832, 0.870),\\
&\text{Pivotal CI:}\ I_\alpha = (\hat{\rho}_n - \hat{\mathfrak{g}}_{1-\frac{\alpha}{2}}, \hat{\rho}_n - \hat{\mathfrak{g}}_{\frac{\alpha}{2}}) = (0.833, 0.871),
\end{aligned} \tag{149}$$

where z_α is the αth quantile of the standard normal distribution, $z_\alpha = \Phi^{-1}(\alpha)$, and $\hat{\mathfrak{g}}_\alpha$ is the αth sample quantile of $\hat{\rho}_n^{(1)} - \hat{\rho}_n, \hat{\rho}_n^{(2)} - \hat{\rho}_n, \ldots, \hat{\rho}_n^{(B)} - \hat{\rho}_n$.

6.2 Parametric vs Nonparametric Bootstrap

(a) If $X_1, \ldots, X_n \overset{\text{iid}}{\sim} \mathscr{U}[0, \theta]$, then the CDF of $\hat{\theta}_n = X_{(n)}$ is

$$F_{\hat{\theta}_n}(x) = \mathbb{P}(X_{(n)} \leq x) = \mathbb{P}(X_1 \leq x, \ldots, X_n \leq x) = \prod_{i=1}^{n} \mathbb{P}(X_i \leq x) = \left(\frac{x}{\theta}\right)^n. \tag{150}$$

Therefore, the PDF is

$$f_{\hat{\theta}_n}(x) = \frac{d}{dx} F_{\hat{\theta}_n}(x) = \frac{n x^{n-1}}{\theta^n}. \tag{151}$$

(b) Figure 10 shows the histograms of $B = 10^4$ parametric and nonparametric bootstrap replications of $\hat{\theta}_n = X_{(n)}$ together with the true distribution (151). A parametric replication is obtained by first generating an IID sample of size n from the uniform distribution $\mathscr{U}[0,\hat{\theta}_n]$, and then taking the maximum value on the sample. In this example, the nonparametric bootstrap performs very poorly compared to the parametric bootstrap.

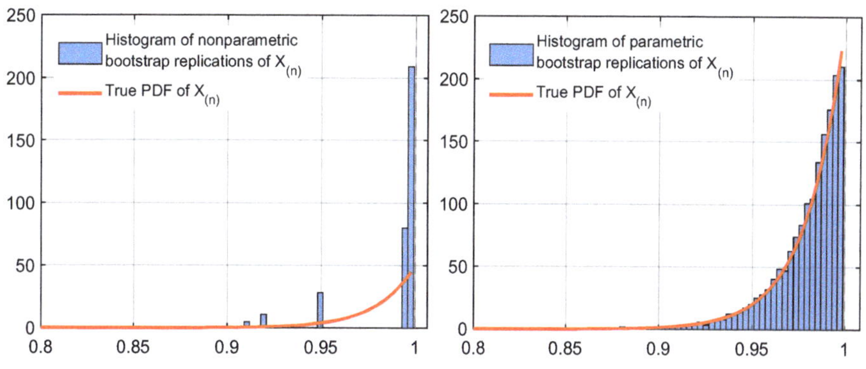

Fig. 10 Histograms of $B = 10^4$ parametric (right) and nonparametric (left) bootstrap replications of $\hat{\theta}_n = X_{(n)}$ together with the true PDF (151).

Note that in general, bootstrap methods often perform poorly when they try to approximate the sampling distribution of an extreme order statistic (e.g. minimum $X_{(1)}$, or maximum value $X_{(n)}$ like in this example). Depending on the original sample $X_1,\ldots,X_n$, the histogram of parametric bootstrap replications of $X_{(n)}$ could be worse than the one in the right panel of Fig. 10, and it could also differ substantially from the true distribution of $X_{(n)}$. Nevertheless, in this example the parametric bootstrap outperforms the nonparametric method because it uses the (correct) information that the data is uniformly distributed.

Problems of Chapter 7

7.1 Exponential Distribution

The MOM estimate of the single parameter $\lambda > 0$ of the exponential distribution is the solution of the following equation

$$m_1(\lambda) = \hat{m}_1, \tag{152}$$

where $\hat{m}_1 = \overline{X}$ is the sample moment and $m_1(\lambda)$ is the first theoretical moment (mean) of $\mathscr{E}(\lambda)$, which is

$$
\begin{aligned}
m_1(\lambda) = \mathbb{E}[X] &= \int_0^{\infty} x f(x|\lambda)dx = \lambda \int_0^{\infty} x e^{-\lambda x} dx \\
&= -\int_0^{\infty} x de^{-\lambda x} = \underbrace{-xe^{-\lambda x}\Big|_{x=0}^{x\to\infty}}_{0} + \int_0^{\infty} e^{-\lambda x} dx = -\frac{e^{-\lambda x}}{\lambda}\Big|_{x=0}^{x\to\infty} = \frac{1}{\lambda}.
\end{aligned} \tag{153}
$$

Therefore, the MOM estimate of λ is

$$
\hat{\lambda}_{MOM} = \frac{1}{\overline{X}}. \tag{154}
$$

It can be shown that although this estimate is biased, it is asymptotically unbiased.

7.2 Uniform Distribution with Unknown Support

In this problem, we have two unknown parameters: α and β. Therefore, to estimate them using the method of moments, we will have to match the first two theoretical moments with the corresponding sample moments. The first two moments of the uniform distribution $\mathscr{U}[\alpha, \beta]$ with PDF is $f(x|\alpha, \beta) = 1/(\beta - \alpha)$ for $x \in [\alpha, \beta]$ are

$$
\begin{aligned}
m_1(\alpha, \beta) = \mathbb{E}[X] &= \int_{\alpha}^{\beta} \frac{x}{\beta - \alpha} dx = \frac{\alpha + \beta}{2}, \\
m_2(\alpha, \beta) = \mathbb{E}[X^2] &= \int_{\alpha}^{\beta} \frac{x^2}{\beta - \alpha} dx = \frac{\alpha^2 + \alpha\beta + \beta^2}{3}.
\end{aligned} \tag{155}
$$

The MOM estimates $\hat{\alpha}_{MOM}$ and $\hat{\beta}_{MOM}$ are, therefore, the solution of the following system of two equations:

$$
\begin{aligned}
m_1(\alpha, \beta) = \hat{m}_1 \quad &\Rightarrow \quad \frac{\alpha + \beta}{2} = \frac{1}{n} \sum_{i=1}^{n} X_i = \overline{X}, \\
m_2(\alpha, \beta) = \hat{m}_2 \quad &\Rightarrow \quad \frac{\alpha^2 + \alpha\beta + \beta^2}{3} = \frac{1}{n} \sum_{i=1}^{n} X_i^2 = \overline{X^2}.
\end{aligned} \tag{156}
$$

The unique solution of this system, satisfying $\alpha < \beta$, is

$$
\begin{aligned}
\hat{\alpha}_{MOM} &= \overline{X} - \sqrt{3\left(\overline{X^2} - \overline{X}^2\right)}, \\
\hat{\beta}_{MOM} &= \overline{X} + \sqrt{3\left(\overline{X^2} - \overline{X}^2\right)}.
\end{aligned} \tag{157}
$$

Note that the expression under the square root is always non negative.

7.3 MOM for a Bernoulli-Like Model

In this problem, we have just one model parameter θ. The first theoretical moment of $p(x|\theta)$ is

$$m_1(\theta) = \mathbb{E}[X] = 1 \times \underbrace{\mathbb{P}(X = 1\,|\,\theta)}_{p(1\,|\,\theta)} + 2 \times \underbrace{\mathbb{P}(X = 2\,|\,\theta)}_{p(2\,|\,\theta)} = \theta + 2(1-\theta) = 2 - \theta. \quad (158)$$

The MOM estimate of θ is then the solution of

$$m_1(\theta) = \hat{m}_1 \quad \Rightarrow \quad 2 - \theta = \overline{X}, \quad (159)$$

which is

$$\hat{\theta}_{MOM} = 2 - \overline{X} = \frac{3}{5}. \quad (160)$$

7.4 Estimating the Effectiveness of a New Drug (Revisited)
In Example 7.2, we derived the MOM estimate for the Bernoulli model $\mathscr{B}(p)$,

$$\hat{p}_{MOM} = \overline{X}. \quad (161)$$

Therefore, the MOM estimate of θ is

$$\begin{aligned} \hat{\theta}_{MOM} &= \hat{p}_{1,MOM} - \hat{p}_{2,MOM} = \overline{X} - \overline{Y} \\ &= \frac{1}{n} \underbrace{\sum_{i=1}^{n} X_i}_{n_1} - \frac{1}{n} \underbrace{\sum_{i=1}^{n} Y_i}_{n_2} = \frac{n_1 - n_2}{n} = \frac{1}{18} \approx 0.056, \end{aligned} \quad (162)$$

which is exactly the same as the plug-in estimates derived in Problem 4.4.

Problems of Chapter 8

8.1 MLE for a Bernoulli-Like Model
The likelihood function is the joint probability of the data, viewed as a function of the model parameter θ,

$$\begin{aligned} \mathscr{L}_n(\theta) &= \prod_{i=1}^{n} p(X_i\,|\,\theta) = p(X_1\,|\,\theta)p(X_2\,|\,\theta)p(X_3\,|\,\theta)p(X_4\,|\,\theta)p(X_5\,|\,\theta) \\ &= \underbrace{p(1\,|\,\theta)}_{\theta}\underbrace{p(2\,|\,\theta)}_{1-\theta}\underbrace{p(1\,|\,\theta)}_{\theta}\underbrace{p(2\,|\,\theta)}\underbrace{p(1\,|\,\theta)}_{\theta} = \theta^3(1-\theta)^2. \end{aligned} \quad (163)$$

The log-likelihood[18] is then

$$\ell_n(\theta) = \log \mathscr{L}_n(\theta) = 3\log\theta + 2\log(1-\theta). \quad (164)$$

[18] It is not hard to maximize the likelihood directly.

Let's find the critical points of $\ell_n(\theta)$:

$$\ell_n'(\theta) = \frac{3}{\theta} - \frac{2}{1-\theta} = \frac{3-5\theta}{\theta(1-\theta)} = 0 \quad \Leftrightarrow \quad \theta = \frac{3}{5}. \quad (165)$$

Since the second derivative $\ell_n''(\theta)$ is negative for all $\theta \in (0,1)$,

$$\ell_n''(\theta) = -\frac{3}{\theta^2} - \frac{2}{(1-\theta)^2} < 0, \tag{166}$$

the log-likelihood is a concave function and $\theta = 3/5$ is its global maximizer. Therefore, the maximum likelihood estimate[19] is

[19] Coincides with the MOM estimate derived in Problem 7.3.

$$\hat{\theta}_{MLE} = \frac{3}{5}. \tag{167}$$

8.2 Poisson Distribution

(a) By definition, the likelihood function is

$$\mathscr{L}_n(\lambda) = \prod_{i=1}^{n} p(X_i|\lambda) = \prod_{i=1}^{n} e^{-\lambda}\frac{\lambda^{X_i}}{X_i!} = e^{-n\lambda}\frac{\lambda^{\sum_{i=1}^{n} X_i}}{\prod_{i=1}^{n} X_i!} \tag{168}$$

The log-likelihood is then

$$\ell_n(\lambda) = \log \mathscr{L}_n(\lambda) = -n\lambda + \log\lambda \sum_{i=1}^{n} X_i - \sum_{i=1}^{n} \log(X_i!) \tag{169}$$

Solving equation $\ell_n'(\lambda) = 0$ for λ gives a single solution (a candidate for being the maximum likelihood estimate):

$$\ell_n'(\lambda) = -n + \frac{\sum_{i=1}^{n} X_i}{\lambda} = 0 \quad \Leftrightarrow \quad \lambda = \frac{1}{n}\sum_{i=1}^{n} X_i = \overline{X}. \tag{170}$$

To show that $\lambda = \overline{X}$ is indeed the MLE, we need to check that $\ell''(\overline{X}) < 0$. The second derivative is

$$\ell''(\lambda) = -\frac{\sum_{i=1}^{n} X_i}{\lambda^2} \quad \text{and} \quad \ell''(\overline{X}) = -\frac{n^2}{\sum_{i=1}^{n} X_i}. \tag{171}$$

Since $X_i \in \{0,1,2,\ldots\}$ for all $i = 1,\ldots,n$, the sum $\sum_{i=1}^{n} X_i \geq 0$. If the sum is strictly positive, $\sum_{i=1}^{n} X_i > 0$, then the second derivative $\ell''(\overline{X}) < 0$ and, therefore, $\lambda = \overline{X}$ is the MLE. If $\sum_{i=1}^{n} X_i = 0$, then $X_1 = \ldots = X_n = 0$, $\overline{X} = 0$, and $\ell''(\overline{X})$ is not defined. In this "limiting" case, the likelihood function is[20]

[20] Recall that $0! = 1$.

$$X_1 = \ldots = X_n = 0 \quad \Rightarrow \quad \mathscr{L}_n(\lambda) = e^{-n\lambda}, \tag{172}$$

and it does not have a global maximizer for $\lambda > 0$. This situation is similar to that in Example 8.4. So, in the extreme case, where all $X_1 = \ldots = X_n = 0$, the MLE does not exist; otherwise, it is given by the sample mean,

$$\hat{\lambda}_{MLE} = \overline{X}. \tag{173}$$

(b) We can show directly, without referring to the general result of Section 8.6.1, that $\hat{\lambda}_{MLE}$ is a consistent estimate of λ, i.e. that $\hat{\lambda}_{MLE}$ converges in probability to the true value $\lambda_\star$ as the samples size $n \to \infty$. Indeed, using the law of large numbers[21],

$$\hat{\lambda}_{MLE} = \overline{X} = \frac{1}{n}\sum_{i=1}^{n} X_i \xrightarrow{\mathbb{P}} \mathbb{E}[X_i] = \lambda_\star, \qquad \text{as } n \to \infty. \tag{174}$$

[21] For a brief review of the LLN see Probability Refresher 2.12.

(c) The Fisher information is given by

$$\mathscr{I}(\lambda_\star) = -\mathbb{E}\left[\left.\frac{\partial^2 \log p(X\,|\,\lambda)}{\partial \lambda^2}\right|_{\lambda=\lambda_\star}\right], \quad \text{where } X \sim f(x\,|\,\theta_\star). \tag{175}$$

The logarithm of the Poissonian PMF is

$$\log p(x\,|\,\lambda) = -\lambda + x\log\lambda - \log x! \tag{176}$$

Its derivatives are

$$\frac{\partial \log p(x\,|\,\lambda)}{\partial \lambda} = -1 + \frac{x}{\lambda} \qquad \left.\frac{\partial^2 \log p(x\,|\,\lambda)}{\partial \lambda^2}\right|_{\lambda=\lambda_\star} = -\frac{x}{\lambda_\star^2}. \tag{177}$$

So, the Fisher information is

$$\mathscr{I}(\lambda_\star) = -\mathbb{E}\left[-\frac{X}{\lambda_\star^2}\right] = \frac{\mathbb{E}[X]}{\lambda_\star^2} = \frac{1}{\lambda_\star}. \tag{178}$$

(d) The variance of $\hat{\lambda}_{MLE}$ is

$$\mathbb{V}[\hat{\lambda}_{MLE}] = \mathbb{V}[\overline{X}] = \frac{1}{n^2}\sum_{i=1}^{n}\underbrace{\mathbb{V}[X_i]}_{\lambda_\star} = \frac{\lambda_\star}{n} = \frac{1}{n\frac{1}{\lambda_\star}} = \frac{1}{n\mathscr{I}(\lambda_\star)}. \tag{179}$$

So, for the Poisson model, (8.60) holds exactly for any samples size n.

(e) Using the central limit theorem[22], for large n we have

$$\hat{\lambda}_{MLE} = \overline{X} \dot{\sim} \mathscr{N}\left(\mathbb{E}[X_i], \frac{\mathbb{V}[X_i]}{n}\right) = \mathscr{N}\left(\lambda_\star, \frac{\lambda_\star}{n}\right) = \mathscr{N}\left(\lambda_\star, \frac{1}{n\mathscr{I}(\lambda_\star)}\right), \tag{180}$$

[22] For a brief review of the CLT see Probability Refresher 2.12.

which is in exact agreement with the general result (8.78) derived in Section 8.6.2.

(f) The MLE $\hat{\lambda}_{MLE}$ is unbiased for any n, not only asymptotically,

$$\mathbb{E}[\hat{\lambda}_{MLE}] = \mathbb{E}[\overline{X}] = \frac{1}{n}\sum_{i=1}^{n}\underbrace{\mathbb{E}[X_i]}_{\lambda_\star} = \lambda_\star. \tag{181}$$

(g) The MLE $\hat{\lambda}_{MLE}$ is efficient, since it is unbiased (see part (f)) and achieves the Cramér-Rao lower bound (see part (d)):

$$\mathbb{V}[\hat{\lambda}_{MLE}] = \frac{1}{n\mathscr{I}(\lambda_\star)}. \tag{182}$$

8.3 Uniform Distribution with Unknown Support (Revisited)

The solution of this problem is similar to Example 8.3, where model $\mathscr{U}[0,\theta]$ was consider and the MLE of its single parameter θ was derived. By definition, the likelihood function is

$$\mathscr{L}_n(\alpha,\beta) = \prod_{i=1}^{n} f(X_i\,|\,\alpha,\beta), \tag{183}$$

where f is the PDF of the uniform distribution on $[\alpha,\beta]$,

$$f(x\,|\,\alpha,\beta) = \begin{cases} \frac{1}{\beta-\alpha}, & \text{if } x \in [\alpha,\beta], \\ 0, & \text{if } x \notin [\alpha,\beta]. \end{cases} \tag{184}$$

In this formula, x is the argument of f and α and β are fixed parameters. Since in (183), $f(X_i\,|\,\alpha,\beta)$ is viewed as a function of α and β and X_i is fixed, it is useful to rewrite (184) in an equivalent form, where the value of $f(x\,|\,\alpha,\beta)$ depends explicitly on α and β, while x is viewed as a fixed point:

$$f(x\,|\,\alpha,\beta) = \begin{cases} \frac{1}{\beta-\alpha}, & \text{if } \alpha \leq x \text{ and } \beta \geq x, \\ 0, & \text{if } \alpha > x \text{ or } \beta < x. \end{cases} \tag{185}$$

By analogy with Example 8.3, let's try to imagine how the likelihood function looks[23]. First, the domain of $\mathscr{L}_n(\alpha,\beta)$, i.e. the set where $\mathscr{L}_n(\alpha,\beta)$ is defined, is the half-plane $\{(\alpha,\beta) : \beta > \alpha\}$. If α is larger than the smallest values in the sample, $\alpha > X_{(1)}$, then $f(X_{(1)}\,|\,\alpha,\beta) = 0$ and $\mathscr{L}_n(\alpha,\beta) = 0$, since $f(X_{(1)}\,|\,\alpha,\beta)$ is exactly one of the factors $f(X_1\,|\,\alpha,\beta),\ldots,f(X_n\,|\,\alpha,\beta)$ in (183). Similarly, if β is smaller than the largest value in the sample, $\beta < X_{(n)}$, then $f(X_{(n)}\,|\,\alpha,\beta) = 0$ and again $\mathscr{L}_n(\alpha,\beta) = 0$. The likelihood is not zero if and only if both $\alpha \leq X_{(1)}$ and $\beta \geq X_{(n)}$. In this case, all factors $f(X_1\,|\,\alpha,\beta),\ldots,f(X_n\,|\,\alpha,\beta)$ equal $1/(\beta-\alpha)$ and the likelihood equals $1/(\beta-\alpha)^n$. Thus, for $\beta > \alpha$,

[23] Try to sketch it on the margins as you go through this paragraph.

$$\mathscr{L}_n(\alpha,\beta) = \begin{cases} \frac{1}{(\beta-\alpha)^n}, & \text{if } \alpha \leq X_{(1)} \text{ and } \beta \geq X_{(n)}, \\ 0, & \text{otherwise}. \end{cases} \tag{186}$$

The likelihood function is schematically shown in Fig 11. It is a discontinuous function and it achieves its maximum value when $\alpha \leq X_{(1)}$, $\beta \geq X_{(n)}$, and $\beta - \alpha$ is as small as possible, which happens when

$$\hat{\alpha}_{MLE} = X_{(1)} \quad \text{and} \quad \hat{\beta}_{MLE} = X_{(n)}. \tag{187}$$

8.4 MLE vs Plug-In Estimate

(a) Using the equivariance property of MLEs (see Section 8.6.5) and the result (187) derived in Problem 8.3, we have

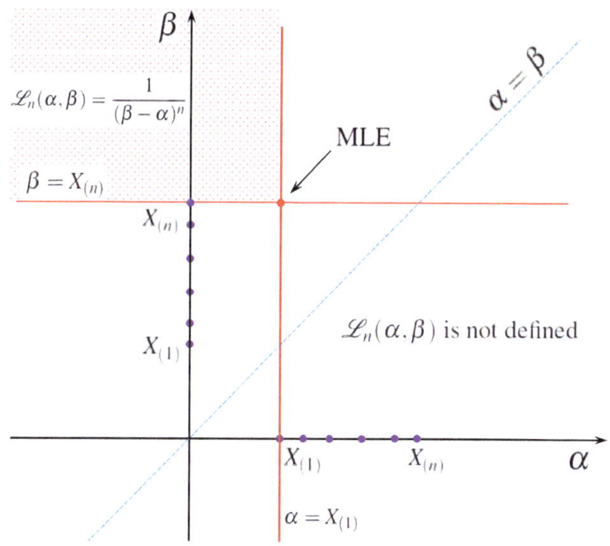

Fig. 11 Schematic representation of the likelihood function: $\mathscr{L}_n(\alpha,\beta)$ is not defined on and below the line $\alpha = \beta$, it equals zero everywhere above the line $\alpha = \beta$ except for the shaded region, and it equals $1/(\beta-\alpha)^n$ in the shaded region. The global maximum is achieved at $\hat{\alpha}_{MLE} = X_{(1)}$ and $\hat{\beta}_{MLE} = X_{(n)}$, represented by a red dot.

$$\mu = \frac{\alpha+\beta}{2} \quad \Rightarrow \quad \hat{\mu}_{MLE} = \frac{\hat{\alpha}_{MLE}+\hat{\beta}_{MLE}}{2} = \frac{X_{(1)}+X_{(n)}}{2}. \tag{188}$$

(b) Using the bias-variance decomposition, the mean squared error of the plug-in estimate $\hat{\mu}_n = \overline{X}$ is

$$\text{MSE}[\hat{\mu}_n] = \mathbb{B}[\hat{\mu}_n]^2 + \mathbb{V}[\hat{\mu}_n]. \tag{189}$$

Since the sample mean is an unbiased estimate of the mean, the bias is equal to zero,

$$\mathbb{E}[\hat{\mu}_n] = \mathbb{E}[\overline{X}] = \frac{1}{n}\sum_{i=1}^{n} \underbrace{\mathbb{E}[X_i]}_{\mu} = \mu \quad \Rightarrow \quad \mathbb{B}[\hat{\mu}_n] = 0. \tag{190}$$

The variance is

$$\mathbb{V}[\hat{\mu}_n] = \mathbb{V}[\overline{X}] = \frac{1}{n^2}\sum_{i=1}^{n} \underbrace{\mathbb{V}[X_i]}_{\frac{(\beta-\alpha)^2}{12}} = \frac{(\beta-\alpha)^2}{12n}. \tag{191}$$

Combining both terms and letting $n = 10$, $\alpha = \alpha_\star = 1$, and $\beta = \beta_\star = 3$, we have

$$\text{MSE}[\hat{\mu}_n] = \frac{(\beta_\star-\alpha_\star)^2}{12n} = \frac{1}{30} \approx 0.033. \tag{192}$$

The mean squared error of the MLE in (188) can be estimated using the law of large numbers as follows:

$$\begin{aligned}\text{MSE}[\hat{\mu}_{MLE}] = \mathbb{E}[(\hat{\mu}_{MLE} - \mu_\star)^2] &\approx \frac{1}{m}\sum_{i=1}^{n}(\hat{\mu}_{MLE}^{(i)} - \mu_\star)^2 \\ &= \frac{1}{m}\sum_{i=1}^{n}\left(\frac{X_{(1)}^{(i)} + X_{(n)}^{(i)}}{2} - \mu_\star\right)^2,\end{aligned} \tag{193}$$

where $\mu_\star = (\alpha_\star + \beta_\star)/2 = 2$ is the true value and $\hat{\mu}_{MLE}^{(i)} = (X_{(1)}^{(i)} + X_{(n)}^{(i)})/2$ is the value of the MLE computed based on the sample $X_1^{(i)}, \ldots, X_n^{(i)}$ generated from $\mathscr{U}[\alpha_\star, \beta_\star]$. The estimated value is

$$\text{MSE}[\hat{\mu}_{MLE}] \approx 0.015. \tag{194}$$

The MSE of the MLE $\hat{\mu}_{MLE}$ is approximately twice as small as the MSE of the plug-in estimate $\hat{\mu}_{MLE}$. As expected, the MLE is a more accurate estimate, since the parametric model (8.147) is correct in this case[24]. It is worth mentioning that the method of moments estimate of μ is

[24] See the discussion of the advantages of parametric models in Section 3.1.1.

$$\hat{\mu}_{MOM} = \frac{\hat{\alpha}_{MOM} + \hat{\beta}_{MOM}}{2} = \overline{X}, \tag{195}$$

where $\hat{\alpha}_{MOM}$ and $\hat{\beta}_{MOM}$ were derived in Problem 7.2 and are given in (157). So, the MOM and plug-in estimates coincide and are less accurate than the MLE estimate.

8.5 MLE and Confidence Interval

(a) Finding ψ in terms of μ is straightforward:

$$\begin{aligned}\psi &= \mathbb{E}[Y_i] = \mathbb{P}(Y_i = 1) = \mathbb{P}(X_i > 0) = 1 - \mathbb{P}(X_i \leq 0) \\ &= 1 - \mathbb{P}(\underbrace{X_i - \mu}_{\sim\mathscr{N}(0,1)} \leq -\mu) = 1 - \Phi(-\mu).\end{aligned} \tag{196}$$

Using the equivariance of MLEs and that $\hat{\mu}_{MLE} = \overline{X}$ (Example 8.2), we have

$$\hat{\psi}_{MLE} = 1 - \Phi(-\hat{\mu}_{MLE}) = 1 - \Phi(-\overline{X}). \tag{197}$$

(b) We can construct a confidence interval for $\psi_\star$ by first constructing a confidence interval for $\mu_\star$ and then transforming it to the interval for $\psi_\star$ using the derived functional relationship (196) between ψ and μ. The MLE based approximate $100(1-\alpha)\%$ confidence interval for $\mu_\star$, derived in Example 8.8, is

$$I_\alpha = \left(\overline{X} + z_{\frac{\alpha}{2}}\frac{\sigma}{\sqrt{n}},\ \overline{X} - z_{\frac{\alpha}{2}}\frac{\sigma}{\sqrt{n}}\right). \tag{198}$$

Since $\sigma^2 = 1$ and $\psi_\star = 1 - \Phi(-\mu_\star)$ or, equivalently, $\mu_\star = -\Phi^{-1}(1 - \psi_\star)$,

$$
\begin{aligned}
1-\alpha &\approx \mathbb{P}\left(\overline{X}+\frac{z_{\frac{\alpha}{2}}}{\sqrt{n}} < \mu_\star < \overline{X}-\frac{z_{\frac{\alpha}{2}}}{\sqrt{n}}\right) \\
&= \mathbb{P}\left(\overline{X}+\frac{z_{\frac{\alpha}{2}}}{\sqrt{n}} < -\Phi^{-1}(1-\psi_\star) < \overline{X}-\frac{z_{\frac{\alpha}{2}}}{\sqrt{n}}\right) \qquad (199) \\
&= \mathbb{P}\left(1-\Phi\left(-\overline{X}-\frac{z_{\frac{\alpha}{2}}}{\sqrt{n}}\right) < \psi_\star < 1-\Phi\left(-\overline{X}+\frac{z_{\frac{\alpha}{2}}}{\sqrt{n}}\right)\right),
\end{aligned}
$$

which means that

$$
I_\alpha = \left(1-\Phi\left(-\overline{X}-\frac{z_{\frac{\alpha}{2}}}{\sqrt{n}}\right),\ 1-\Phi\left(-\overline{X}+\frac{z_{\frac{\alpha}{2}}}{\sqrt{n}}\right)\right) \qquad (200)
$$

is an approximate $100(1-\alpha)\%$ confidence interval for $\psi_\star$.

8.6 Alternative Formula for the Fisher Information

To prove that (8.151) and (8.152) are equivalent, let's derive more explicit expressions for both expectations. The first expectation is

$$
\begin{aligned}
&\mathbb{E}\left[\left(\left.\frac{\partial \log f(X\,|\,\theta)}{\partial \theta}\right|_{\theta=\theta_\star}\right)^2\right] = \int f(x\,|\,\theta_\star)\left(\left.\frac{\partial \log f(x\,|\,\theta)}{\partial \theta}\right|_{\theta=\theta_\star}\right)^2 dx \\
&= \int f(x\,|\,\theta_\star)\left(\frac{\left.\frac{\partial f(x\,|\,\theta)}{\partial \theta}\right|_{\theta=\theta_\star}}{f(x\,|\,\theta_\star)}\right)^2 dx = \int \frac{1}{f(x\,|\,\theta_\star)}\left(\left.\frac{\partial f(x\,|\,\theta)}{\partial \theta}\right|_{\theta=\theta_\star}\right)^2 dx. \qquad (201)
\end{aligned}
$$

The second expectation is

$$
\begin{aligned}
&-\mathbb{E}\left[\left.\frac{\partial^2 \log f(X\,|\,\theta)}{\partial \theta^2}\right|_{\theta=\theta_\star}\right] = -\int f(x\,|\,\theta_\star)\left.\frac{\partial^2 \log f(x\,|\,\theta)}{\partial \theta^2}\right|_{\theta=\theta_\star} dx \\
&= -\int f(x\,|\,\theta_\star)\left.\frac{\partial}{\partial \theta}\right|_{\theta=\theta_\star}\frac{\frac{\partial f(x\,|\,\theta)}{\partial \theta}}{f(x\,|\,\theta)} dx \\
&= -\int f(x\,|\,\theta_\star)\frac{\left.\frac{\partial^2 f(x\,|\,\theta)}{\partial^2 \theta}\right|_{\theta=\theta_\star} f(x\,|\,\theta_\star) - \left(\left.\frac{\partial f(x\,|\,\theta)}{\partial \theta}\right|_{\theta=\theta_\star}\right)^2}{f(x\,|\,\theta_\star)^2} dx \\
&= -\int \left.\frac{\partial^2 f(x\,|\,\theta)}{\partial^2 \theta}\right|_{\theta=\theta_\star} dx + \int \frac{1}{f(x\,|\,\theta_\star)}\left(\left.\frac{\partial f(x\,|\,\theta)}{\partial \theta}\right|_{\theta=\theta_\star}\right)^2 dx \qquad (202) \\
&= -\left.\frac{\partial^2}{\partial \theta^2}\right|_{\theta=\theta_\star}\underbrace{\int f(x\,|\,\theta)dx}_{1} + \int \frac{1}{f(x\,|\,\theta_\star)}\left(\left.\frac{\partial f(x\,|\,\theta)}{\partial \theta}\right|_{\theta=\theta_\star}\right)^2 dx \\
&= \int \frac{1}{f(x\,|\,\theta_\star)}\left(\left.\frac{\partial f(x\,|\,\theta)}{\partial \theta}\right|_{\theta=\theta_\star}\right)^2 dx.
\end{aligned}
$$

So, it turns out that both definitions of the Fisher information lead to the same integral representations and, therefore, they are equivalent.

Problems of Chapter 9

9.1 ABC Example: Test Sizes and Error Probabilities

To find the sizes and error probabilities, we need to find the power function $\beta_{n_{\min}}(p)$ of the test with the rejection region $\mathscr{R}_{n_{\min}}$. By analogy with (9.20), we have

$$\begin{aligned}\beta_{n_{\min}}(p) &= \mathbb{P}(K \in \mathscr{R}_{n_{\min}} \,|\, p) = \mathbb{P}(K \in \{n_{\min}, \dots, n\} \,|\, p) \\ &= \sum_{k=n_{\min}}^{n} \mathbb{P}(K = k \,|\, p) = \sum_{k=n_{\min}}^{n} \binom{n}{k} p^k (1-p)^{n-k}.\end{aligned} \tag{203}$$

Using Definitions 9.1 and 9.2, the sizes and error probabilities are

$$\begin{aligned}\alpha_{n_{\min}} &= \mathbb{P}_{n_{\min}}(\text{Type I error}) = \beta_{n_{\min}}(p_0) = \sum_{k=n_{\min}}^{n} \binom{n}{k} p_0^k (1-p_0)^{n-k}, \\ \mathbb{P}_{n_{\min}}(\text{Type II error}) &= 1 - \beta_{n_{\min}}(p_1) = 1 - \sum_{k=n_{\min}}^{n} \binom{n}{k} p_1^k (1-p_1)^{n-k},\end{aligned} \tag{204}$$

where $p_0 = 0.5$ and $p_1 = 0.7$. Figure 12 shows how the error probabilities depend on parameter $n_{\min}$ that controls the size: both the test size $\alpha_{n_{\min}}$ and the geometric size of the rejection region $\mathscr{R}_{n_{\min}}$. Notice that $n_{\min} = 7$ corresponds to the special case (9.21) considered in Example 9.4.

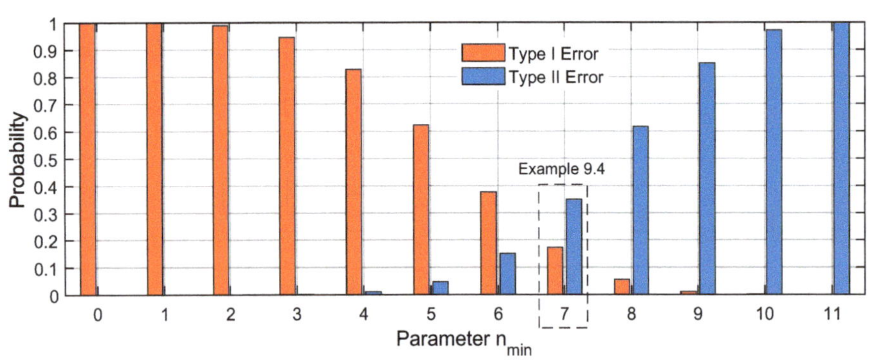

Fig. 12 Probabilities of the Type I and Type II errors as functions of $n_{\min}$. As expected, if $n_{\min}$ increases (the rejection region shrinks), then the probability if Type I error decreases and the probability of Type II error increases. Specifying the size of the test is equivalent to finding an acceptable trade-off between the two errors.

9.2 Probabilistic Interpretation of the P-value

The probabilistic interpretation of the p-value is given by (9.72), which, in our case, has the following form:

$$\mathfrak{p}(x) = \sup_{\mu \leq 0} \mathbb{P}(s(X) \geq s(x) \,|\, \mu) = \sup_{\mu \leq 0} \mathbb{P}(\overline{X} \geq \overline{x} \,|\, \mu). \tag{205}$$

According to the assumed stochastic model, $X_1, \ldots, X_n \overset{\text{iid}}{\sim} \mathcal{N}(\mu, \sigma^2)$. Therefore, the sample mean $\overline{X}$ is distributed according to the normal distribution with mean μ and variance σ^2/n,

$$\overline{X} \sim \mathcal{N}\left(\mu, \frac{\sigma^2}{n}\right) \implies \frac{\overline{X} - \mu}{\sigma/\sqrt{n}} \sim \mathcal{N}(0,1). \tag{206}$$

This allows to compute the probability that appears in the expression for the p-value,

$$\mathbb{P}(\overline{X} \geq \overline{x} \,|\, \mu) = \mathbb{P}\left(\frac{\overline{X} - \mu}{\sigma/\sqrt{n}} \geq \frac{\overline{x} - \mu}{\sigma/\sqrt{n}} \,\middle|\, \mu\right) = 1 - \Phi\left(\frac{\overline{x} - \mu}{\sigma/\sqrt{n}}\right). \tag{207}$$

Finally, since the standard normal CDF Φ is a monotonically increasing function,

$$\begin{aligned} \mathfrak{p}(x) &= \sup_{\mu \leq 0} \left(1 - \Phi\left(\frac{\overline{x} - \mu}{\sigma/\sqrt{n}}\right)\right) = 1 - \inf_{\mu \leq 0} \Phi\left(\frac{\overline{x} - \mu}{\sigma/\sqrt{n}}\right) \\ &= 1 - \Phi\left(\frac{\overline{x} - \mu}{\sigma/\sqrt{n}}\right)\bigg|_{\mu = 0} = 1 - \Phi\left(\frac{\sqrt{n}}{\sigma}\overline{x}\right), \end{aligned} \tag{208}$$

which, as expected, coincides with (9.83).

9.3 Testing the Uniform Distribution

(a) By Definition 9.1, the power function of a test with rejection region (9.86) is

$$\beta(\theta) = \mathbb{P}(X \in \mathcal{R} \,|\, \theta) = \mathbb{P}(X_{(n)} \geq c \,|\, \theta). \tag{209}$$

In Problem 3.1(b), we derived the distribution of the nth order statistic $X_{(n)}$:

$$F_{X_{(n)}}(x) = \mathbb{P}(X_{(n)} \leq x) = \left(\frac{x}{\theta}\right)^n. \tag{210}$$

Therefore, the power function is

$$\beta(\theta) = 1 - \mathbb{P}(X_{(n)} \leq c \mid \theta) = 1 - \left(\frac{c}{\theta}\right)^n. \tag{211}$$

(b) By Definition 9.2, the size of the test with power function $\beta(\theta)$ is

$$\alpha = \sup_{\theta \in \Theta_0} \beta(\theta). \tag{212}$$

In our case, $\Theta_0 = \{\theta_0\}$ and, therefore, the size is simply the value of the power function at $\theta = \theta_0$,

$$\alpha = \beta(\theta_0) = 1 - \left(\frac{c}{\theta_0}\right)^n. \tag{213}$$

To construct a test of size α, we need to solve the above equation for c, which gives

$$c = \theta_0(1-\alpha)^{1/n}. \tag{214}$$

So, the size α test based on the rejection region (9.86) has the following form:

$$\text{Reject } H_0 : \theta = \theta_0 \quad \Longleftrightarrow \quad X_{(n)} \geq \theta_0(1-\alpha)^{1/n}. \tag{215}$$

Notice that if $X_{(n)} > \theta_0$, then $X_{(n)} > \theta_0(1-\alpha)^{1/n}$ and the null hypothesis is rejected. This behavior is expected since $X_{(n)} > \theta_0$ and H_0 are logically incompatible.

(c) By Definition 9.3, the p-value of the actually observed data $X = x$ is the smallest size $\alpha_{\min}$ at which the test (215) rejects the null hypothesis. Therefore, to find the p-value of x, we need to solve for $\alpha_{\min}$ the following equation, so that x lies on the boundary of the rejection region[25],

[25] See (9.57).

$$x_{(n)} = \theta_0(1-\alpha_{\min})^{1/n}, \tag{216}$$

which gives[26]

[26] What if $x_{(n)} > \theta_0$?

$$\alpha_{\min} = \mathfrak{p}(x) = 1 - \left(\frac{x_{(n)}}{\theta_0}\right)^n. \tag{217}$$

Alternatively, we can find the p-value using its probabilistic interpretation (9.74) and the expression (210) for the CDF of $X_{(n)}$:

$$\mathfrak{p}(x) = \mathbb{P}(s(X) \geq s(x) \,|\, H_0) = \mathbb{P}(X_{(n)} \geq x_{(n)} \,|\, \theta = \theta_0) = 1 - \left(\frac{x_{(n)}}{\theta_0}\right)^n. \tag{218}$$

If $x_{(n)} = 0.48$, $n = 20$, and $\theta_0 = 1/2$, then $\mathfrak{p} = 0.56$, and the data does not provide any evidence against the null hypothesis.

9.4 Testing the Normal Distribution

(a) To construct a test of size α, we need to find the critical value c such that

$$\alpha = \sup_{\mu \in \Theta_0} \beta(\mu) = \sup_{\mu = \mu_0} \beta(\mu) = \beta(\mu_0). \tag{219}$$

The power function of the test with the rejection region (9.89) is

$$\begin{aligned}\beta(\mu) &= \mathbb{P}(X \in \mathscr{R} \,|\, \mu) = \mathbb{P}(\overline{X} \geq c \,|\, \mu) = \mathbb{P}\left(\frac{\overline{X}-\mu}{\sigma/\sqrt{n}} \geq \frac{c-\mu}{\sigma/\sqrt{n}} \,\middle|\, \mu\right) \\ &= 1 - \Phi\left(\frac{c-\mu}{\sigma/\sqrt{n}}\right), \qquad \mu \in \{\mu_0, \mu_1\}.\end{aligned} \tag{220}$$

So, (219) becomes

$$\alpha = 1 - \Phi\left(\frac{c-\mu_0}{\sigma/\sqrt{n}}\right) \implies c = \mu_0 + \frac{\sigma}{\sqrt{n}}\Phi^{-1}(1-\alpha). \quad (221)$$

So, the size α test based on the rejection region (9.89) has the following form:

$$\text{Reject } H_0 : \mu = \mu_0 \iff \overline{X} \geq \mu_0 + \frac{\sigma}{\sqrt{n}}\Phi^{-1}(1-\alpha). \quad (222)$$

(b) The power of this test under H_1 is the value of the power function (220) with c given by (221) computed at $\mu = \mu_1$,

$$\begin{aligned}\beta(\mu_1) &= 1 - \Phi\left(\frac{\mu_0 + \frac{\sigma}{\sqrt{n}}\Phi^{-1}(1-\alpha) - \mu_1}{\sigma/\sqrt{n}}\right)\\ &= 1 - \Phi\left(\Phi^{-1}(1-\alpha) - \frac{\sqrt{n}}{\sigma}(\mu_1 - \mu_0)\right).\end{aligned} \quad (223)$$

Notice that if $\mu_1 \approx \mu_0$, then the power $\beta(\mu_1) \approx \alpha$. This means that the probability of Type II error (failing to reject $H_0 : \mu = \mu_0$ when $H_1 : \mu = \mu_1$ is true) $\mathbb{P}(\text{Type II error}) \approx 1 - \beta(\mu_1) = 1 - \alpha$, which is large if $\alpha = \mathbb{P}(\text{Type I error})$ is small. This is somewhat expected, since if $\mu_1 \approx \mu_0$, then it is difficult to distinguish between them and one of the two error probabilities must be large.

On the other hand, if $\mu_1 \to \infty$, then the power $\beta(\mu_1) \to 1$, and, therefore, $\mathbb{P}(\text{Type II error}) \to 0$. The further away μ_1 is from μ_0, the less likely it is to make a type II error[27].

[27] The general form of the rejection region (9.89) was motivated by the fact that $\mu_1 > \mu_0$. What happens with the probability of type II error if, contrary to $\mu_1 > \mu_0$, $\mu_1 \to -\infty$?

(c) The power function of the size α test (222) is the power function (220) with the critical value c given by (221):

$$\begin{aligned}\beta(\mu) &= 1 - \Phi\left(\frac{c-\mu}{\sigma/\sqrt{n}}\right) = 1 - \Phi\left(\frac{\mu_0 + \frac{\sigma}{\sqrt{n}}\Phi^{-1}(1-\alpha) - \mu}{\sigma/\sqrt{n}}\right)\\ &= 1 - \Phi\left(\Phi^{-1}(1-\alpha) - \frac{\sqrt{n}}{\sigma}(\mu - \mu_0)\right), \quad \mu \in \{\mu_0, \mu_1\}.\end{aligned} \quad (224)$$

If $\mu = \mu_0$, then the value of the power does not depend on n, $\beta(\mu_0) = \alpha$. This is simply because the test was constructed to satisfy (219), so that the probability of Type I error is exactly α. If $\mu = \mu_1$ and $n \to \infty$, then, as we have already discussed in part (b), $\beta(\mu_1) \to 1$. So, as $n \to \infty$, the power function converges to

$$\beta(\mu) \to \begin{cases}\alpha, & \text{if } \mu = \mu_0,\\ 1, & \text{if } \mu = \mu_1.\end{cases} \quad (225)$$

(d) To find the p-value of the actually observed data $X = x$, we need to find $\alpha_{\min}$ such that x lies on the boundary of the rejection region (222). That is, we need to solve for $\alpha_{\min}$ the following equation:

$$\bar{x} = \mu_0 + \frac{\sigma}{\sqrt{n}}\Phi^{-1}(1-\alpha_{\min}) \implies \mathfrak{p}(x) = \alpha_{\min} = 1 - \Phi\left(\frac{\bar{x}-\mu_0}{\sigma/\sqrt{n}}\right). \quad (226)$$

Or, equivalently, we can use the probabilistic interpretation of the p-value (9.74):

$$\begin{aligned}\mathfrak{p}(x) &= \mathbb{P}(s(X) \geq s(x)\,|\,H_0) = \mathbb{P}(\overline{X} \geq \bar{x}\,|\,\mu = \mu_0) \\ &= \mathbb{P}\left(\frac{\overline{X}-\mu_0}{\sigma/\sqrt{n}} \geq \frac{\bar{x}-\mu_0}{\sigma/\sqrt{n}}\,\middle|\,\mu = \mu_0\right) = 1 - \Phi\left(\frac{\bar{x}-\mu_0}{\sigma/\sqrt{n}}\right).\end{aligned} \quad (227)$$

Figure 13 shows the dependence of the p-value on the observed sample mean $\bar{x}$ for $\mu_0 = -1$, $\mu_1 = 1$, $\sigma = 10$, and $n = 1000$. Notice that the p-value does not depend on the exact value of μ_1 and is the same for all $\mu_1 > \mu_0$.

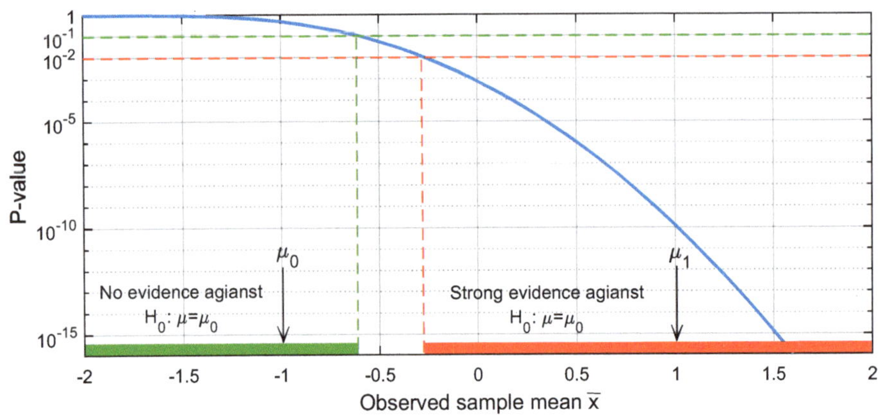

Fig. 13 The p-value (227) as a function of the sample mean $\bar{x}$ for $\mu_0 = -1$, $\sigma = 10$, $n = 10^3$.

Problems of Chapter 10

10.1 Wald Meets Poisson

(a) The Wald test for a parameter is based on an estimate of the parameter, which is unbiased and normally distributed (at least approximately)[28]. Since maximum likelihood estimates are generally asymptotically normal and asymptotically unbiased, let's use the MLE of λ for constructing the test. Recall that in Problem 8.2, we found that

$$\hat{\lambda}_{MLE} = \overline{X} \quad (228)$$

and that this estimate is indeed unbiased and asymptotically normally distributed.

As we derived in Section 10.1,

[28] Recall the key assumption (10.3).

$$\text{The size } \alpha \text{ Wald test rejects } H_0 : \lambda = \lambda_0 \quad \Longleftrightarrow \quad \left| \frac{\hat{\lambda}_{MLE} - \lambda_0}{\widehat{\text{se}}[\hat{\lambda}_{MLE}]} \right| \geq z_{1-\frac{\alpha}{2}}, \tag{229}$$

where $\widehat{\text{se}}[\hat{\lambda}_{MLE}]$ is an estimate of the standard error of $\hat{\lambda}_{MLE}$. Using (8.60),

$$\text{se}[\hat{\lambda}_{MLE}] \approx \frac{1}{\sqrt{n\mathscr{I}(\lambda_\star)}}, \tag{230}$$

where $\lambda_\star$ is the true value of λ and $\mathscr{I}(\lambda_\star)$ is the Fisher information. In Problem 8.2(c), we derived that $\mathscr{I}(\lambda_\star) = 1/\lambda_\star$. Therefore, the standard error is

$$\text{se}[\hat{\lambda}_{MLE}] \approx \sqrt{\frac{\lambda_\star}{n}}. \tag{231}$$

Finally, in Problem 8.2(b), we proved that $\hat{\lambda}_{MLE} = \overline{X}$ is a consistent estimate, $\hat{\lambda}_{MLE} \xrightarrow{\mathbb{P}} \lambda_\star$, i.e. it converges to the true value as the sample size increases. So, it is natural to estimate $\lambda_\star \approx \hat{\lambda}_{MLE}$. This yields the following estimate of the standard error:

$$\widehat{\text{se}}[\hat{\lambda}_{MLE}] \approx \sqrt{\frac{\overline{X}}{n}}. \tag{232}$$

Therefore, the Wald test of approximate size α rejects $H_0 : \lambda = \lambda_0$ if and only if

$$\left| \frac{\sqrt{n}(\overline{X} - \lambda_0)}{\sqrt{\overline{X}}} \right| \geq z_{1-\frac{\alpha}{2}}. \tag{233}$$

(b) Figure 14 shows the evolution of the type I error rate as a function of the number of independent experiments $m = 10, \ldots, 10^5$. As expected, when the number

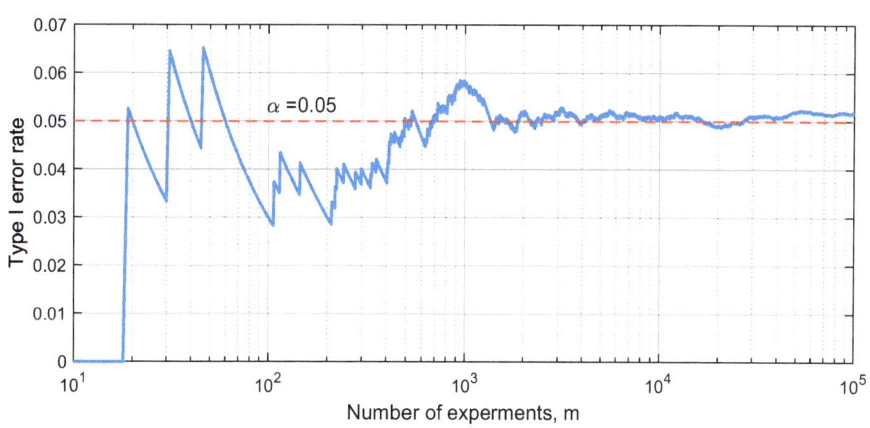

Fig. 14 The type I error rate as a function of the number of independent experiments m.

of experiments is large, the type I error rate is close to the size of the test $\alpha = 0.05$, which is the probability of the type I error. If $m = 10^5$, then the error rate is approximately 0.0517.

(c) One way to find the p-value $\mathfrak{p}(x)$ is to recall that it is the specific value of α for which the observed data $x = (x_1, \ldots, x_n)$ lies on the boundary of the rejection region defined by inequality (233). So, to find $\mathfrak{p}(x)$, we need to solve the following equation:

$$\left| \frac{\sqrt{n}(\bar{x} - \lambda_0)}{\sqrt{\bar{x}}} \right| = z_{1-\frac{\mathfrak{p}(x)}{2}}, \tag{234}$$

which yields

$$\mathfrak{p}(x) = 2\left(1 - \Phi\left(\left|\frac{\sqrt{n}(\bar{x} - \lambda_0)}{\sqrt{\bar{x}}}\right|\right)\right) = 2\Phi\left(-\left|\frac{\sqrt{n}(\bar{x} - \lambda_0)}{\sqrt{\bar{x}}}\right|\right). \tag{235}$$

Here we used the property $1 - \Phi(a) = \Phi(-a)$, which is induced by the symmetry of the standard normal distribution. Compare the derived expression for $\mathfrak{p}(x)$ with the general formula for the p-value based on the Wald test (10.20).

10.2 Wald Compares Drugs

(a) Let $\Delta p = p_A - p_B$ be the difference between the drug success probabilities. Then the testing problem $p_A = p_B$ versus $p_A \neq p_B$ can be equivalently rewritten as

$$H_0 : \Delta p = 0 \quad \text{vs} \quad H_1 : \Delta p \neq 0. \tag{236}$$

To construct the Wald test, first of all, we need an estimate of Δp, which is unbiased and normally distributed (at least approximately). Let's estimate p_A and p_B by their maximum likelihood estimates.

Both N_A and M_B can be written as a sum of IID Bernoulli random variables,

$$N_A = \sum_{i=1}^{n} X_i \quad \text{and} \quad M_B = \sum_{j=1}^{m} Y_j, \tag{237}$$

where $X_i \sim \mathscr{B}(p_A)$ equals 1 if patient i was cured after taking drug A and equals 0 otherwise, and, similarly, $Y_j \sim \mathscr{B}(p_B)$ equals 1 if patient j was cured after taking drug B and equals 0 otherwise. As we established in Example 8.1, the MLE of the Bernoulli parameter is given by the sample mean. Therefore, the MLEs of parameters p_A and p_B are

$$p_A \approx \hat{p}_A = \frac{N_A}{n} \quad \text{and} \quad p_B \approx \hat{p}_B = \frac{M_B}{m}. \tag{238}$$

Thanks to the CLT, both these estimates are asymptotically normal as $n, m \to \infty$.

Let's estimate Δp by the difference of the estimates:

$$\widehat{\Delta p} = \hat{p}_A - \hat{p}_B = \frac{N_A}{n} - \frac{M_B}{m}. \tag{239}$$

This estimate is asymptotically normal (as a difference of asymptotically normal estimates) and, therefore, is approximately normally distributed if n and m are sufficiently large. Moreover, it is unbiased for any n and m:

$$\mathbb{E}[\widehat{\Delta p}] = \frac{\mathbb{E}[N_A]}{n} - \frac{\mathbb{E}[M_B]}{m} = \frac{\sum_{i=1}^{n} \overbrace{\mathbb{E}[X_i]}^{p_A}}{n} - \frac{\sum_{j=1}^{m} \overbrace{\mathbb{E}[Y_j]}^{p_B}}{m} = p_A - p_B = \Delta p. \quad (240)$$

As we derived in Section 10.1,

$$\text{The size } \alpha \text{ Wald test rejects } H_0 : \Delta p = 0 \quad \Longleftrightarrow \quad \left| \frac{\widehat{\Delta p}}{\widehat{\mathrm{se}}[\widehat{\Delta p}]} \right| \geq z_{1-\frac{\alpha}{2}}, \quad (241)$$

where $\widehat{\mathrm{se}}[\widehat{\Delta p}]$ is an estimate of the standard error of $\widehat{\Delta p}$. The variance of $\widehat{\Delta p}$ is

$$\begin{aligned} \mathbb{V}[\widehat{\Delta p}] &= \mathbb{V}\left[\frac{N_A}{n} - \frac{M_B}{m}\right] = \frac{\mathbb{V}[N_A]}{n^2} + \frac{\mathbb{V}[M_B]}{m^2} = \frac{\sum_{i=1}^{n} \mathbb{V}[X_i]}{n^2} + \frac{\sum_{j=1}^{m} \mathbb{V}[Y_j]}{m^2} \\ &= \frac{\sum_{i=1}^{n} \left(\overbrace{\mathbb{E}[X_i^2]}^{p_A} - \overbrace{\mathbb{E}[X_i]^2}^{p_A^2}\right)}{n^2} + \frac{\sum_{j=1}^{m} \left(\overbrace{\mathbb{E}[Y_j^2]}^{p_B} - \overbrace{\mathbb{E}[Y_j]^2}^{p_B}\right)}{m^2} \\ &= \frac{p_A(1-p_A)}{n} + \frac{p_B(1-p_B)}{m}. \end{aligned} \quad (242)$$

We can estimate the standard error by replacing p_A and p_B with their estimates,

$$\widehat{\mathrm{se}}[\widehat{\Delta p}] = \sqrt{\frac{\hat{p}_A(1-\hat{p}_A)}{n} + \frac{\hat{p}_B(1-\hat{p}_B)}{m}} = \sqrt{\frac{N_A(n-N_A)}{n^3} + \frac{M_B(m-M_B)}{m^3}}. \quad (243)$$

Therefore, the Wald test of approximate size α rejects $H_0 : \Delta p = 0$ or, equivalently, $H_0 : p_A = p_B$ if and only if

$$\left| \frac{\frac{N_A}{n} - \frac{M_B}{m}}{\sqrt{\frac{N_A(n-N_A)}{n^3} + \frac{M_B(m-M_B)}{m^3}}} \right| \geq z_{1-\frac{\alpha}{2}}. \quad (244)$$

(b) Given the form of the size α test (244), the most straightforward way to find the p-value is, as in Problem 10.1(c), to find the specific value of α for which the observed data lies on the boundary of the rejection region. This leads to

$$\mathfrak{p}(n_A, n_B) = 2\Phi\left(-\left| \frac{\frac{n_A}{n} - \frac{m_B}{m}}{\sqrt{\frac{n_A(n-n_A)}{n^3} + \frac{m_B(m-m_B)}{m^3}}} \right|\right) \approx 0.162. \quad (245)$$

According to the conventional scheme for the interpretation of p-values[29], the ob-

[29] See Table 9.1.

served data ($n_A = 91, m_B = 191$) does not provide any evidence against the null hypothesis that the drugs have the same therapeutic efficacy.

10.3 Understanding the Permutation Test

(a) The permutation distribution is a discrete distribution $\mathbb{P}_\pi$ that puts probability mass $1/N!$ on each of the values $s_1, \ldots, s_{N!}$ of the test statistic s computed on all $N! = 4! = 24$ permutations of the data $Z = (X_1, X_2, Y_1, Y_2) = (2, 5, 1, 7)$. Table 4 lists all permutations and the corresponding values of s. The test statistic has three

Permutations	Values of s	Permutations	Values of s
$(2,5,1,7)$	$s_1 = 0.5 = s_{\text{obs}}$	$(1,2,5,7)$	$s_{13} = 4.5$
$(2,5,7,1)$	$s_2 = 0.5$	$(1,2,7,5)$	$s_{14} = 4.5$
$(2,1,5,7)$	$s_3 = 4.5$	$(1,5,2,7)$	$s_{15} = 1.5$
$(2,1,7,5)$	$s_4 = 4.5$	$(1,5,7,2)$	$s_{16} = 1.5$
$(2,7,5,1)$	$s_5 = 1.5$	$(1,7,2,5)$	$s_{17} = 0.5$
$(2,7,1,5)$	$s_6 = 1.5$	$(1,7,5,2)$	$s_{18} = 0.5$
$(5,2,1,7)$	$s_7 = 0.5$	$(7,2,5,1)$	$s_{19} = 1.5$
$(5,2,7,1)$	$s_8 = 0.5$	$(7,2,1,5)$	$s_{20} = 1.5$
$(5,1,2,7)$	$s_9 = 1.5$	$(7,5,2,1)$	$s_{21} = 4.5$
$(5,1,7,2)$	$s_{10} = 1.5$	$(7,5,1,2)$	$s_{22} = 4.5$
$(5,7,2,1)$	$s_{11} = 4.5$	$(7,1,2,5)$	$s_{23} = 0.5$
$(5,7,1,2)$	$s_{12} = 4.5$	$(7,1,5,2)$	$s_{24} = 0.5$

Table 4 Permutations of the data and the corresponding values of the test statistic $s = |\overline{X} - \overline{Y}|$.

distinct values: 0.5, 1.5, and 4.5, each occurring eight times. The permutation distribution is, therefore, the discrete uniform distribution on the set $\{0.5, 1.5, 4.5\}$. In other words, $\mathbb{P}_\pi(s = 0.5) = \mathbb{P}_\pi(s = 1.5) = \mathbb{P}_\pi(s = 4.5) = 1/3$.

(b) By definition, the p-value based on the permutation test is the proportion of $s_1, \ldots, s_{N!}$ that are larger than or equal to the observed value s_{obs},

$$\mathfrak{p} = \mathbb{P}_\pi(s \geq s_{\text{obs}}) = \frac{1}{N!} \sum_{i=1}^{N!} \mathbb{1}_{[s_{\text{obs}}, \infty)}(s_i). \tag{246}$$

In our case, $s_{\text{obs}} = 0.5$ is the smallest out of all permutation-based values of s. Therefore, the p-value $\mathfrak{p} = 1$.

10.4 Mark Twain and the Quintus Curtius Snodgrass Letters

(a) The p-value based on the two-sample t-test is given by[30]

[30] See equation (10.55).

$$\mathfrak{p}(T, S) = 2F_T\left(\frac{-|\overline{T} - \overline{S}|}{\sqrt{\frac{\hat{s}^2_{T,n}}{n} + \frac{\hat{s}^2_{S,m}}{m}}} \middle| d \right), \tag{247}$$

where $n = 8$ and $m = 10$ are the sample sizes (numbers of Twain's essays and Snodgrass's letters), $\hat{s}^2_{T,n}$ and $\hat{s}^2_{S,m}$ are the sample variances of T and S, and F_T is the CDF of the t-distribution with d degrees of freedom, which is given by the Welch-Satterthwaite equation (10.53). After plugging in all the numbers into (247), we get

$$\mathfrak{p}(T,S) = 0.0032. \tag{248}$$

(b) The p-value based on the two-sample Kolmogorov-Smirnov test is given by[31]

[31] See equation (10.93).

$$\mathfrak{p}(T,S) = 1 - F_K\left(\sqrt{\frac{nm}{n+m}}D_{\text{obs}}\right), \tag{249}$$

where F_K is the CDF of the Kolmogorov distribution (10.69) and $D_{\text{obs}} = 0.75$ is the observed value of statistic (10.86). In other words, $D_{\text{obs}} = 0.75$ is the largest absolute difference between the empirical CDFs constructed from samples S and T. Computing the p-value yields

$$\mathfrak{p}(T,S) = 0.0058. \tag{250}$$

(c) The total sample size $N = n + m = 18$ is too large for summing over all $N!$ permutations in (10.99)[32]. Instead, we can estimate the p-value by sampling from the permutation distribution and using the law of large numbers (10.100). Based on the sample of size $K = 10^6$, the p-values corresponding to statistics

[32] $18! \approx 6.4 \times 10^{15}$.

$$s_1(T,S) = |\overline{T} - \overline{S}| \quad \text{and} \quad s_2(T,S) = \sup_{z\in\mathbb{R}} \left|\hat{F}_{T,n}(z) - \hat{F}_{S,m}(z)\right| \tag{251}$$

are

$$\mathfrak{p}_1(T,S) = 0.0008 \quad \text{and} \quad \mathfrak{p}_2(T,S) = 0.0055. \tag{252}$$

All tests that we used in this problem lead to extremely small p-values and strong evidence against the null hypothesis that samples T and S were generated by the same "text-generating mechanism." This suggests, in line with the conclusion obtained in [47], that Mark Twain almost certainly did not write the Quintus Curtius Snodgrass letters.

10.5 Neyman-Pearson Lemma in Action

Since both the null and alternative hypotheses are simple, the Neyman-Pearson Lemma asserts that the likelihood ratio test (LRT) is the most powerful test at any level α. The LRT-based p-value $\mathfrak{p}(x)$ of the actually observed data x is

$$\mathfrak{p}(x) = \mathbb{P}(\Lambda(X) \geq \Lambda(x) \,|\, H_0) = \mathbb{P}(\Lambda(X) \geq \Lambda(x) \,|\, \theta = \theta_0), \tag{253}$$

where $\Lambda(X)$ is the likelihood ratio statistic,

$$\Lambda(X) = \frac{\mathscr{L}(\theta_1|X)}{\mathscr{L}(\theta_0|X)}, \tag{254}$$

where the likelihood function $\mathscr{L}(\theta\,|\,X)$ is given by

$$\mathscr{L}(\theta\,|\,X) = \prod_{i=1}^{n} f(X_i\,|\,\theta) = \prod_{i=1}^{n}(1+\theta)X_i^{\theta} = (1+\theta)^n \left(\prod_{i=1}^{n} X_i\right)^{\theta}. \tag{255}$$

Plugging this expression back into the formula for the likelihood ratio statistic, we have:

$$\Lambda(X) = \left(\frac{1+\theta_1}{1+\theta_0}\right)^n \left(\prod_{i=1}^{n} X_i\right)^{\theta_1-\theta_0}. \tag{256}$$

Since the factor $(1+\theta_1)^n/(1+\theta_0)^n > 0$ and $\theta_1 - \theta_0 > 0$, the p-value is

$$\mathfrak{p}(x) = \mathbb{P}\left(\prod_{i=1}^{n} X_i \geq \prod_{i=1}^{n} x_i \,\middle|\, \theta = \theta_0\right), \tag{257}$$

or, equivalently (expanding what exactly "given that $\theta = \theta_0$" means),

$$\mathfrak{p}(x) = \mathbb{P}\left(\prod_{i=1}^{n} X_i \geq \prod_{i=1}^{n} x_i\right), \quad \text{where} \quad X_1 \ldots, X_n \overset{\text{iid}}{\sim} f(x\,|\,\theta_0) = (1+\theta_0)x^{\theta_0}. \tag{258}$$

We can try to find the exact p-value by finding the exact distribution of the product $\prod_{i=1}^{n} X_i$ using the so-called *product distribution formula*. It can be shown[33] that if $X_1 \sim f_1(x)$ and $X_2 \sim f_2(x)$ are two independent continuous random variables, then the PDF of their *product* is

[33] See, for instance, [48].

$$f_{X_1X_2}(x) = \int_{-\infty}^{\infty} f_1(t) f_2\left(\frac{x}{t}\right)\frac{1}{|t|}dt. \tag{259}$$

In our case, X_1 and X_2 are identically distributed on the unit interval $(0,1)$ with PDF $f(x\,|\,\theta_0) = (1+\theta_0)x^{\theta_0}$, where $x \in (0,1)$. Therefore, the product X_1X_2 is also distributed on $(0,1)$ and its PDF is

$$\begin{aligned} f_{X_1X_2}(x\,|\,\theta_0) &= \int_{-\infty}^{\infty} \underbrace{f(t\,|\,\theta_0)}_{\neq 0 \iff t\in[0,1]} \underbrace{f\left(\frac{x}{t}\,\middle|\,\theta_0\right)}_{\neq 0 \iff x/t\in[0,1]} \frac{1}{|t|}dt \\ &= \int_{x}^{1} (1+\theta_0)t^{\theta_0}(1+\theta_0)\left(\frac{x}{t}\right)^{\theta_0}\frac{1}{t}dt \\ &= (1+\theta_0)^2 x^{\theta_0}\int_{x}^{1}\frac{1}{t}dt = -(1+\theta_0)^2 x^{\theta_0}\log x. \end{aligned} \tag{260}$$

Similarly, again using the product distribution formula, we can find the PDF of product $X_1X_2X_3 = X_1X_2 \times X_3$:

$$\begin{aligned} f_{X_1X_2X_3}(x\,|\,\theta_0) &= \int_{-\infty}^{\infty} f_{X_1X_2}(t\,|\,\theta_0) f_{X3}\left(\frac{x}{t}\,\Big|\,\theta_0\right)\frac{1}{|t|}dt \\ &= -\int_x^1 (1+\theta_0)^2 t^{\theta_0}\log t \times (1+\theta_0)\left(\frac{x}{t}\right)^{\theta_0}\frac{1}{t}dt \\ &= -(1+\theta_0)^3 x^{\theta_0}\underbrace{\int_x^1 \frac{\log t}{t}dt}_{-\frac{1}{2}(\log x)^2} = \frac{1}{2}(1+\theta_0)^3 x^{\theta_0}(\log x)^2. \end{aligned} \tag{261}$$

It can be shown by induction that the PDF of the full product $\prod_{i=1}^n X_i$ is given by

$$f_{\prod_{i=1}^n X_i}(x\,|\,\theta_0) = \frac{(-1)^{n-1}(1+\theta_0)^n}{(n-1)!}x^{\theta_0}(\log x)^{n-1}, \quad x \in (0,1). \tag{262}$$

This result allows us to express the p-value $\mathfrak{p}(x)$ explicitly in term of an integral:

$$\mathfrak{p}(x) = \mathbb{P}\left(\prod_{i=1}^n X_i \geq \prod_{i=1}^n x_i\right) = \frac{(-1)^{n-1}(1+\theta_0)^n}{(n-1)!}\int_{\prod_{i=1}^n x_i}^1 x^{\theta_0}(\log x)^{n-1}dx, \tag{263}$$

which, however, is not trivial to compute in closed-form in terms of elementary functions[34].

[34] Possible, as we will see below, but not trivial.

Instead of aiming at computing the p-value exactly, we can try to estimate its value using the central limit theorem:

$$\mathfrak{p}(x) = \mathbb{P}\left(\log\prod_{i=1}^n X_i \geq \log\prod_{i=1}^n x_i\right) = \mathbb{P}\left(\sum_{i=1}^n \log X_i \geq \sum_{i=1}^n \log x_i\right). \tag{264}$$

Since $X_1,\ldots,X_n$ are IID, so are $\log X_1,\ldots,\log X_n$. Their mean and variance are

$$\begin{aligned} \mu &= \mathbb{E}[\log X_i] = \int_0^1 \log x \times (1+\theta_0)x^{\theta_0}dx = -\frac{1}{1+\theta_0}, \\ \sigma^2 &= \mathbb{E}[(\log X_i)^2] - \mathbb{E}[\log X_i]^2 = \int_0^1 (\log x)^2 \times (1+\theta_0)x^{\theta_0}dx - \frac{1}{(1+\theta_0)^2} \\ &= \frac{2}{(1+\theta_0)^2} - \frac{1}{(1+\theta_0)^2} = \frac{1}{(1+\theta_0)^2}. \end{aligned} \tag{265}$$

If the sample size n is sufficiently large, then, thanks to the CLT, the distribution of $\sum_{i=1}^n \log X_i$ is approximately normal. Namely,

$$\frac{\frac{1}{n}\sum_{i=1}^n \log X_i - \mu}{\sigma/\sqrt{n}} \stackrel{\sim}{\sim} \mathcal{N}(0,1). \tag{266}$$

Therefore, the p-value is

$$\begin{aligned}\mathfrak{p}(x) &= \mathbb{P}\left(\sum_{i=1}^{n}\log X_i \geq \sum_{i=1}^{n}\log x_i\right)\\ &= \mathbb{P}\left(\frac{\frac{1}{n}\sum_{i=1}^{n}\log X_i-\mu}{\sigma/\sqrt{n}} \geq \frac{\frac{1}{n}\sum_{i=1}^{n}\log x_i-\mu}{\sigma/\sqrt{n}}\right)\\ &\approx 1-\Phi\left(\frac{\frac{1}{n}\sum_{i=1}^{n}\log x_i-\mu}{\sigma/\sqrt{n}}\right)\\ &= 1-\Phi\left(\frac{1+\theta_0}{\sqrt{n}}\sum_{i=1}^{n}\log x_i+\sqrt{n}\right).\end{aligned} \tag{267}$$

It is interesting to investigate how accurately the estimated value of the p-value (267) approximates the exact p-value (263). Upon careful consideration, it turns out that it is possible to compute the integral in (263) in closed form. Namely,

$$\int_a^1 x^{\theta_0}(\log x)^{n-1}dx = \frac{(-1)^n}{(1+\theta_0)^n}\left(\Gamma(n,-(1+\theta_0)\log a)-\Gamma(n,0)\right), \tag{268}$$

where $\Gamma(n,x)$ is the (upper) incomplete gamma function, which is given by (for integer $n>0$)

$$\Gamma(n,x) = (n-1)!\, e^{-x}\sum_{k=0}^{n-1}\frac{x^k}{k!}. \tag{269}$$

Substituting these expressions into (263) and simplifying, we obtain that the exact p-value is[35]

[35] "Well, here we are!" said Merry. "And very cheerless and uninviting it looks!" – J.R.R. Tolkien, *The Lord of the Rings*.

$$\mathfrak{p}(x) = 1-\exp\left((1+\theta_0)\sum_{i=1}^{n}\log x_i\right)\sum_{k=0}^{n-1}\frac{(-(1+\theta_0)\sum_{i=1}^{n}\log x_i)^k}{k!}. \tag{270}$$

Notice that both the exact (270) and approximate (267) p-values $\mathfrak{p}(x)$ depend on $x=(x_1,\dots,x_n)$ through $s=\sum_{i=1}^{n}\log x_i$. Figure 15 shows the exact and approximate p-values as functions of s for $\theta_0=5$ and different values of the sample size n. This figure suggests that the CLT-based estimate (267) approximates the exact p-value (270) rather accurately even when the sample size n is small.

10.6 Generalized Likelihood Ratio Test

The first step in constructing a GLRT is computing the likelihood function. For the normal model, it is given by[36]

[36] For the derivation, see (10.117).

$$\mathcal{L}(\mu|X) = \left(\frac{1}{\sqrt{2\pi}\sigma}\right)^n \exp\left(-\frac{\sum_{i=1}^{n}(X_i-\mu)^2}{2\sigma^2}\right). \tag{271}$$

The likelihood ratio statistic is

$$\Lambda = \frac{\sup_{\mu\in\mathbb{R}}\mathcal{L}(\mu|X)}{\sup_{\mu\leq 0}\mathcal{L}(\mu|X)} = \frac{\mathcal{L}(\hat{\mu}_{MLE}|X)}{\sup_{\mu\leq 0}\mathcal{L}(\mu|X)} = \frac{\mathcal{L}(\overline{X}|X)}{\sup_{\mu\leq 0}\mathcal{L}(\mu|X)}, \tag{272}$$

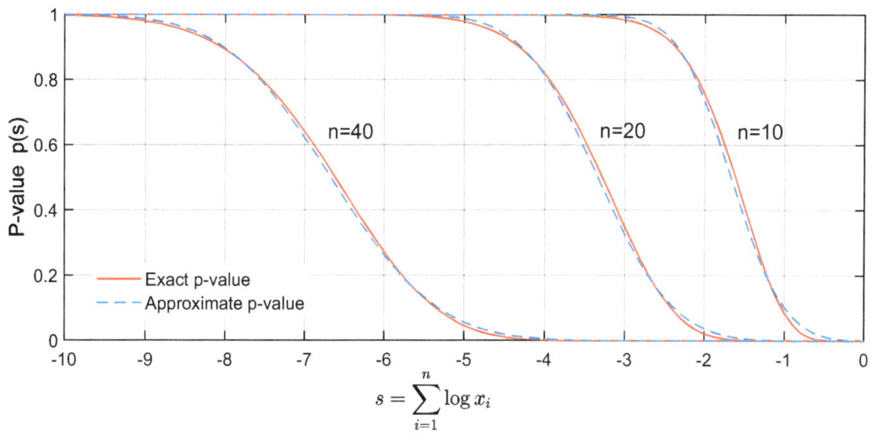

Fig. 15 The exact and approximate p-values for $\theta_0 = 5$ and sample sizes $n = 10, 20$, and 40.

where $\hat{\mu}_{MLE} = \overline{X}$ is the MLE of the mean[37]. To find the denominator of Λ, we need to solve the following constrained optimization problem:

[37] Recall Example 8.2.

$$\mathscr{L}(\mu|X) \to \max, \quad \text{subject to } \mu \leq 0, \tag{273}$$

or, equivalently,

$$\ell(\mu|X) = \log \mathscr{L}(\mu|X) = C - \frac{1}{2\sigma^2}\sum_{i=1}^{n}(X_i - \mu)^2 \to \max, \quad \text{subject to } \mu \leq 0, \tag{274}$$

where $C = -\frac{n}{2}\log(2\pi\sigma)$ is a constant that does not affect maximization with respect to μ. The graph of the log-likelihood function $\ell(\mu|X)$ for $\mu \in \mathbb{R}$ is a concave parabola with the global maximum at the critical point $\hat{\mu}_{MLE} = \overline{X}$. Therefore, under the constraint $\mu \leq 0$, the (log-) likelihood achieves its global maximum at $\mu = \overline{X}$ if $\overline{X} \leq 0$ or at $\mu = 0$ if $\overline{X} > 0$. So, the denominator of Λ is

$$\sup_{\mu \leq 0} \mathscr{L}(\mu|X) = \begin{cases} \mathscr{L}(\overline{X}|X), & \text{if } \overline{X} \leq 0, \\ \mathscr{L}(0|X), & \text{if } \overline{X} > 0. \end{cases} \tag{275}$$

Therefore, if $\overline{X} \leq 0$, then the likelihood ratio statistic $\Lambda = 1$; and if $\overline{X} > 0$, then

$$\begin{aligned} \Lambda = \frac{\mathscr{L}(\overline{X}|X)}{\mathscr{L}(0|X)} &= \frac{\exp\left(-\frac{\sum_{i=1}^{n}(X_i - \overline{X})^2}{2\sigma^2}\right)}{\exp\left(-\frac{\sum_{i=1}^{n}X_i^2}{2\sigma^2}\right)} = \exp\left(\frac{\sum_{i=1}^{n}X_i^2 - \sum_{i=1}^{n}(X_i - \overline{X})^2}{2\sigma^2}\right) \\ &= \exp\left(\frac{2\overline{X}\sum_{i=1}^{n}X_i - \sum_{i=1}^{n}\overline{X}^2}{2\sigma^2}\right) = \exp\left(\frac{n\overline{X}^2}{2\sigma^2}\right). \end{aligned} \tag{276}$$

So, the likelihood ratio statistic

$$\Lambda = \begin{cases} 1, & \text{if } \overline{X} \leq 0, \\ \exp\left(\frac{n\overline{X}^2}{2\sigma^2}\right), & \text{if } \overline{X} > 0. \end{cases} \tag{277}$$

The rejection region of the GLRT has the following general form:

$$\mathscr{R} = \{X \in \mathbb{R}^n \mid \Lambda \geq c\}. \tag{278}$$

Since the LRT statistic $\Lambda \geq 1$, we can always assume (for any GLRT) that the critical value $c > 1$. Otherwise, if $c \leq 1$, the rejection region coincides with the entire sample space, $\mathscr{R} = \mathbb{R}^n$, and the test always rejects H_0, regardless of the data, which makes the test absurd. For $c > 1$, the rejection region of the GLRT is

$$\begin{aligned} \mathscr{R} = \{X \in \mathbb{R}^n \mid \Lambda \geq c\} &= \left\{X \in \mathbb{R}^n \;\middle|\; \overline{X} > 0 \text{ and } \exp\left(\frac{n\overline{X}^2}{2\sigma^2}\right) \geq c\right\} \\ &= \left\{X \in \mathbb{R}^n \;\middle|\; \overline{X} \geq \sqrt{\frac{2\sigma^2 \log c}{n}} = c'\right\} = \{X \in \mathbb{R}^n \mid \overline{X} \geq c'\}. \end{aligned} \tag{279}$$

Thus, the rejection region of the GLRT has exactly the same form as the rejection region of the ad hoc test: both tests reject the null hypothesis precisely when the sample mean $\overline{X}$ is larger than or equal to some critical value, which is determined by specifying the size of the test. Consequently, the size α GLRT is also given by (10.207), and the ad hoc test constructed on the intuitive grounds turns out to be the GLRT.

10.7 Normal GLRT and Unknown Variance

The likelihood function is exactly the same as in the previous problem, except that now it is a function of two variables, both μ and σ^2,

$$\mathscr{L}(\mu, \sigma^2 | X) = \left(\frac{1}{2\pi\sigma^2}\right)^{\frac{n}{2}} \exp\left(-\frac{\sum_{i=1}^n (X_i - \mu)^2}{2\sigma^2}\right). \tag{280}$$

The likelihood ratio statistic is

$$\Lambda = \frac{\sup_{\mu \in \mathbb{R}, \sigma^2 \geq 0} \mathscr{L}(\mu, \sigma^2 | X)}{\sup_{\sigma^2 \geq 0} \mathscr{L}(\mu_0, \sigma^2 | X)} = \frac{\mathscr{L}(\hat{\mu}_{MLE}, \hat{\sigma}^2_{MLE} | X)}{\sup_{\sigma^2 \geq 0} \mathscr{L}(\mu_0, \sigma^2 | X)}, \tag{281}$$

where $\hat{\mu}_{MLE}$ and $\hat{\sigma}^2_{MLE}$ are the MLEs of the mean and variance[38],

[38] Recall Example 8.2.

$$\hat{\mu}_{MLE} = \overline{X} \quad \text{and} \quad \hat{\sigma}^2_{MLE} = \frac{1}{n} \sum_{i=1}^n (X_i - \overline{X})^2. \tag{282}$$

To find the denominator of Λ, we need to solve the following optimization problem:

$$\mathscr{L}(\mu_0, \theta | X) \to \max, \quad \text{subject to } \theta = \sigma^2 \geq 0, \tag{283}$$

or, equivalently,

$$\ell(\mu_0, \theta \,|\, X) = \log \mathscr{L}(\mu_0, \theta \,|\, X)$$
$$= -\frac{n}{2}\log(2\pi\theta) - \frac{1}{2\theta}\sum_{i=1}^{n}(X_i - \mu_0)^2 \to \max, \quad \text{subject to } \theta \geq 0. \tag{284}$$

Here, since μ is set to μ_0, the log-likelihood is a function of a single variable $\theta = \sigma^2$, and the goal is to maximize it with respect to θ. We can do this by using the standard Calculus procedure: calculate the derivative of ℓ, set it to zero, find the corresponding critical point $\hat{\theta}$, and check that the critical point is the global maximizer of ℓ. The derivative is

$$\frac{d\ell}{d\theta} = -\frac{n}{2\theta} + \frac{1}{2\theta^2}\sum_{i=1}^{n}(X_i - \mu_0)^2. \tag{285}$$

A unique solution of $d\ell/d\theta = 0$ (the critical point of ℓ) is given by

$$\hat{\theta} = \hat{\sigma}^2 = \frac{1}{n}\sum_{i=1}^{n}(X_i - \mu_0)^2. \tag{286}$$

Since the second derivative at $\hat{\theta}$ is negative,

$$\left.\frac{d^2\ell}{d\theta^2}\right|_{\theta=\hat{\theta}} = \left.\left(\frac{n}{2\theta^2} - \frac{1}{\theta^3}\sum_{i=1}^{n}(X_i - \mu_0)^2\right)\right|_{\theta=\hat{\theta}} = \frac{n}{2\hat{\theta}^2} - \frac{1}{\hat{\theta}^3}n\hat{\theta} = -\frac{n}{2\hat{\theta}^2} < 0, \tag{287}$$

the critical point $\hat{\theta}$ is the global maximizer of $\ell(\mu_0, \theta \,|\, X)$ and, therefore, of $\mathscr{L}(\mu_0, \theta \,|\, X)$.

Substituting (282) and (286) into (281), we can find the likelihood ratio statistic in terms of the data X and the null value μ_0:

$$\Lambda = \frac{\mathscr{L}(\hat{\mu}_{MLE}, \hat{\sigma}^2_{MLE} \,|\, X)}{\mathscr{L}(\mu_0, \hat{\sigma}^2 \,|\, X)} = \frac{\left(\frac{1}{2\pi\hat{\sigma}^2_{MLE}}\right)^{\frac{n}{2}} \exp\left(-\frac{\sum_{i=1}^{n}(X_i - \overline{X})^2}{2\hat{\sigma}^2_{MLE}}\right)}{\left(\frac{1}{2\pi\hat{\sigma}^2}\right)^{\frac{n}{2}} \exp\left(-\frac{\sum_{i=1}^{n}(X_i - \mu_0)^2}{2\hat{\sigma}^2}\right)}$$
$$= \left(\frac{\hat{\sigma}^2}{\hat{\sigma}^2_{MLE}}\right)^{\frac{n}{2}} \exp\left(-\frac{n}{2} + \frac{n}{2}\right) = \left(\frac{\sum_{i=1}^{n}(X_i - \mu_0)^2}{\sum_{i=1}^{n}(X_i - \overline{X})^2}\right)^{\frac{n}{2}}. \tag{288}$$

For constructing the GLRT of a given size and for computing the p-value, we need to know the null distribution of Λ, i.e. the distribution of Λ under the assumption that $H_0 : \mu = \mu_0$ is true. The expression for Λ, however, looks rather complicated and it is not immediately clear how to find its distribution. Do we know a random variable constructed from X_i, $\overline{X}$, and μ_0 — the main ingredients of Λ — with a known probability distribution[39]? Recall[40] that under H_0 the random variable

$$T = \frac{\overline{X} - \mu_0}{\hat{s}_n/\sqrt{n}} = \sqrt{n(n-1)}\frac{\overline{X} - \mu_0}{\sqrt{\sum_{i=1}^{n}(X_i - \overline{X})^2}} \sim \mathscr{T}(n-1) \tag{289}$$

[39] "Tea! Bless ordinary everyday afternoon tea!" – Agatha Christie.

[40] See (10.39).

follows the t-distribution with $(n-1)$ degrees of freedom. Let's try to express the likelihood ratio statistic Λ in terms of T. First, let's massage the numerator[41]:

[41] "Boil em, mash em, stick em in a stew." – Samwise Gamgee, *The Lord of the Rings*.

$$\begin{aligned}\sum_{i=1}^{n}(X_i-\mu_0)^2 &= \sum_{i=1}^{n}(X_i-\overline{X}+\overline{X}-\mu_0)^2 \\ &= \sum_{i=1}^{n}(X_i-\overline{X})^2+2(\overline{X}-\mu_0)\underbrace{\sum_{i=1}^{n}(X_i-\overline{X})}_{0}+\sum_{i=1}^{n}(\overline{X}-\mu_0)^2 \\ &= \sum_{i=1}^{n}(X_i-\overline{X})^2+n(\overline{X}-\mu_0)^2.\end{aligned} \tag{290}$$

Therefore,

$$\begin{aligned}\Lambda^{\frac{2}{n}} &= \frac{\sum_{i=1}^{n}(X_i-\mu_0)^2}{\sum_{i=1}^{n}(X_i-\overline{X})^2} = \frac{\sum_{i=1}^{n}(X_i-\overline{X})^2+n(\overline{X}-\mu_0)^2}{\sum_{i=1}^{n}(X_i-\overline{X})^2} \\ &= 1+n\frac{(\overline{X}-\mu_0)^2}{\sum_{i=1}^{n}(X_i-\overline{X})^2} = 1+\frac{1}{n-1}T^2.\end{aligned} \tag{291}$$

This relationship allows to rewrite the GLRT rejection region in terms of T, whose probability distribution is known. Namely,

$$\mathscr{R} = \{X\in\mathbb{R}^n \mid \Lambda\geq c\} = \left\{X\in\mathbb{R}^n \;\middle|\; |T|\geq\sqrt{(n-1)(c^{\frac{2}{n}}-1)}=c'\right\}. \tag{292}$$

This rejection region has exactly the same form as the rejection region of the Student's t-test (10.41): both tests reject the hull hypothesis precisely when $G=|T|$ is larger than or equal to some critical value, which defines the size of the test. This means that, in this problem, the GLRT test coincides with the t-test, and[42]

[42] See (10.42).

$$\text{The size } \alpha \text{ GLRT rejects } H_0 \quad\Longleftrightarrow\quad G=\left|\frac{\overline{X}-\mu_0}{\hat{s}_n/\sqrt{n}}\right|\geq t_{n-1,1-\frac{\alpha}{2}}, \tag{293}$$

where $t_{n,q}$ denotes the qth quantile of the t-distribution with n degrees of freedom, and[43] the corresponding p-value of the observed data $x=(x_1,\ldots,x_n)$ is

[43] See (10.47).

$$\mathfrak{p}(x) = 2F_T\left(-\left|\frac{\overline{x}-\mu_0}{\hat{s}_n/\sqrt{n}}\right| \;\middle|\; n-1\right), \tag{294}$$

where $F_T(x|n)$ is the CDF of the t-distribution with n degrees of freedom.

Notice the elegant similarity between the known variance case, (10.210) and (10.211), and the unknown variance case, (293) and (294). The second case is obtained from the first by replacing the normal distribution and the known variance σ^2 with the t-distribution and the sample variance $\hat{s}_n^2$ respectively.

Problems of Chapter 11

11.1 Regression Through the Origin

(a) The method of ordinary least squares finds the estimate $\hat{\beta}$ of the regression parameter β by minimizing the residual sum of squares, which for regression through the origin has the following form:

$$\begin{aligned} \mathrm{RSS}(\hat{\beta}) &= \sum_{i=1}^{n} \hat{E}_i^2 = \sum_{i=1}^{n} (Y_i - \hat{Y}_i)^2 = \sum_{i=1}^{n} (Y_i - \hat{\beta} X_i)^2 \\ &= \sum_{i=1}^{n} Y_i^2 - 2\hat{\beta} \sum_{i=1}^{n} X_i Y_i + \hat{\beta}^2 \sum_{i=1}^{n} X_i^2. \end{aligned} \tag{295}$$

The RSS is a quadratic function of $\hat{\beta}$, and it is minimized at

$$\hat{\beta} = \frac{\sum_{i=1}^{n} X_i Y_i}{\sum_{i=1}^{n} X_i^2}. \tag{296}$$

(b) Since prediction errors have zero mean, $\mathbb{E}[E_i] = 0$, the expected value of responses $\mathbb{E}[Y_i] = \mathbb{E}[\beta X_i + E_i] = \beta X_i$. Therefore, the expected value of the OLS estimate is[44]

$$\mathbb{E}[\hat{\beta}] = \mathbb{E}\left[\frac{\sum_{i=1}^{n} X_i Y_i}{\sum_{i=1}^{n} X_i^2}\right] = \frac{\sum_{i=1}^{n} X_i \mathbb{E}[Y_i]}{\sum_{i=1}^{n} X_i^2} = \frac{\beta \sum_{i=1}^{n} X_i^2}{\sum_{i=1}^{n} X_i^2} = \beta. \tag{297}$$

So, as in the case of the SLR model, the OLS estimate $\hat{\beta}$ is unbiased.

(c) Since prediction errors are independent and homoscedastic, $\mathbb{V}[E_i] = \sigma^2$, the responses are also independent and have constant variance $\mathbb{V}[Y_i] = \sigma^2$. Therefore, the variance of the OLS estimate is

$$\mathbb{V}[\hat{\beta}] = \mathbb{V}\left[\frac{\sum_{i=1}^{n} X_i Y_i}{\sum_{i=1}^{n} X_i^2}\right] = \frac{\sum_{i=1}^{n} X_i^2 \mathbb{V}[Y_i]}{(\sum_{i=1}^{n} X_i^2)^2} = \frac{\sigma^2}{\sum_{i=1}^{n} X_i^2}. \tag{298}$$

The standard error of $\hat{\beta}$ is, thus,

$$\mathrm{se}[\hat{\beta}] = \frac{\sigma}{\sqrt{\sum_{i=1}^{n} X_i^2}}. \tag{299}$$

(d) Recall that if the mean squared error of an estimate converges to zero as the sample size increases, then the estimate is consistent[45]. Using the bias-variance decomposition, the mean squared error of $\hat{\beta}$ is

$$\mathrm{MSE}[\hat{\beta}] = \mathbb{B}[\hat{\beta}]^2 + \mathrm{se}[\hat{\beta}]^2 = \frac{\sigma^2}{\sum_{i=1}^{n} X_i^2}. \tag{300}$$

[44] Recall that regression analysis is performed conditional on inputs. In other words, $X_1, \ldots, X_n$ are considered as deterministic quantities, as constants.

[45] The proof of this statement is provided in Chapter 4, after Definition 4.1.

Therefore, if the sum of squares $\sum_{i=1}^{n} X_i^2$ blows up to infinity[46] as the sample size increases, then $\mathrm{MSE}[\hat{\beta}]$ converges to zero, which guarantees the consistency of $\hat{\beta}$. So, a condition sufficient for consistency is

[46] "The only way to comprehend what mathematicians mean by Infinity is to contemplate the extent of human stupidity." – Voltaire.

$$\sum_{i=1}^{n} X_i^2 \to \infty \qquad \text{as } n \to \infty. \tag{301}$$

11.2 Controlling Inputs

The overall accuracy of an estimate is usually quantified by the mean squared error. Since the OLS estimate $\hat{\beta}_1$ is unbiased, its MSE coincides with its variance, which is given by

$$\mathbb{V}[\hat{\beta}_1] = \frac{\sigma^2}{S_{XX}}. \tag{302}$$

Therefore, to maximize the accuracy of $\hat{\beta}_1$, we need to minimize its variance or, equivalently, to maximize the sum of squares

$$S_{XX} = \sum_{i=1}^{n} (X_i - \overline{X})^2. \tag{303}$$

Expanding the square, we have:

$$S_{XX} = \sum_{i=1}^{n} X_i^2 - 2\overline{X} \underbrace{\sum_{i=1}^{n} X_i}_{n\overline{X}} + n(\overline{X})^2 = \sum_{i=1}^{n} X_i^2 - n(\overline{X})^2. \tag{304}$$

Since all $X_i \in [-1,1]$, the sum $\sum_{i=1}^{n} X_i^2 \leq n$. Besides, $(\overline{X})^2 \geq 0$. Therefore,

$$S_{XX} \leq n. \tag{305}$$

The maximum value $S_{XX} = n$ is achieved if we set one half of inputs to -1 and the other half to 1. Then $\sum_{i=1}^{n} X_i^2 = n$, $\overline{X} = 0$, and $S_{XX} = n$. Thus, $X_1 = \ldots = X_{50} = -1$ and $X_{51} = \ldots = X_{100} = 1$ provide the optimal choice of input values for maximizing the accuracy of the OLS estimate of the slope of the regression line.

11.3 Testing the Relationship Between Regression Coefficients

Let's introduce a new model parameter $\theta = \beta_1 - \psi\beta_0$. Then the testing problem (11.141) can be equivalently rewritten in terms of θ as follows:

$$H_0 : \theta = 0 \qquad \text{vs} \qquad H_1 : \theta \neq 0. \tag{306}$$

Since the null hypothesis is simple and it is straightforward to obtain a point estimate of θ, we can aim at constructing the Wald test. The point estimate of θ is

$$\hat{\theta} = \hat{\beta}_1 - \psi\hat{\beta}_0, \tag{307}$$

where $\hat{\beta}_0$ and $\hat{\beta}_1$ are the OLS estimates. Since the prediction errors $E_1,\ldots,E_n$ are IID normally distributed, the responses $Y_1,\ldots,Y_n$ are also normally distributed (independently, but not identically). Since both $\hat{\beta}_0$ and $\hat{\beta}_1$ are linear estimates[47], so is $\hat{\theta}$, and, therefore, it is also normally distributed,

[47] This means that $\hat{\beta}_0$ and $\hat{\beta}_1$ are linear combinations of $Y_1,\ldots,Y_n$.

$$\hat{\theta} \sim \mathscr{N}. \tag{308}$$

We can find the mean and the variance of $\hat{\theta}$ (and, therefore, its full distribution) using the properties of $\hat{\beta}_0$ and $\hat{\beta}_1$. Since the OLS estimates are unbiased, so is $\hat{\theta}$:

$$\mathbb{E}[\hat{\theta}] = \mathbb{E}[\hat{\beta}_1] - \psi\mathbb{E}[\hat{\beta}_0] = \beta_1 - \psi\beta_0 = \theta. \tag{309}$$

The variance of $\hat{\theta}$ is[48]

[48] In what follows, we use the fact that $\mathrm{Cov}(\overline{Y},\hat{\beta}_1) = 0$ established in (11.104).

$$\begin{aligned}
\mathbb{V}[\hat{\theta}] &= \mathbb{V}\left[\hat{\beta}_1 - \psi\hat{\beta}_0\right] = \mathbb{V}\left[\hat{\beta}_1\right] + \psi^2\mathbb{V}\left[\hat{\beta}_0\right] - 2\psi\mathrm{Cov}\left(\hat{\beta}_0,\hat{\beta}_1\right) \\
&= \frac{\sigma^2}{S_{XX}} + \psi^2\sigma^2\left(\frac{1}{n} + \frac{\overline{X}^2}{S_{XX}}\right) - 2\psi\mathrm{Cov}\left(\overline{Y} - \hat{\beta}_1\overline{X},\hat{\beta}_1\right) \\
&= \sigma^2\left(\frac{\psi^2}{n} + \frac{1+\psi^2\overline{X}^2}{S_{XX}}\right) + 2\psi\overline{X}\mathbb{V}\left[\hat{\beta}_1\right] \\
&= \sigma^2\left(\frac{\psi^2}{n} + \frac{1+\psi^2\overline{X}^2}{S_{XX}}\right) + 2\psi\overline{X}\frac{\sigma^2}{S_{XX}} = \sigma^2\left(\frac{\psi^2}{n} + \frac{(1+\psi\overline{X})^2}{S_{XX}}\right).
\end{aligned} \tag{310}$$

Therefore, after standardizing $\hat{\theta} \sim \mathscr{N}$, we have

$$\frac{\hat{\theta} - \theta}{\sigma\sqrt{\frac{\psi^2}{n} + \frac{(1+\psi\overline{X})^2}{S_{XX}}}} \sim \mathscr{N}(0,1). \tag{311}$$

Notice that if $\psi = 0$, then $\theta = \beta_1$, and (311) reduces to (11.76). Replacing the unknown standard deviation σ with its estimate $\hat{\sigma}_u = \sqrt{\frac{1}{n-2}\sum_{i=1}^n \hat{E}_i^2}$ yields

$$\frac{\hat{\theta} - \theta}{\hat{\sigma}_u\sqrt{\frac{\psi^2}{n} + \frac{(1+\psi\overline{X})^2}{S_{XX}}}} \mathrel{\dot\sim} \mathscr{N}(0,1). \tag{312}$$

Therefore, under the null hypothesis $H_0 : \theta = 0$, we have

$$\frac{\hat{\theta}}{\hat{\sigma}_u\sqrt{\frac{\psi^2}{n} + \frac{(1+\psi\overline{X})^2}{S_{XX}}}} = \frac{\hat{\beta}_1 - \psi\hat{\beta}_0}{\hat{\sigma}_u\sqrt{\frac{\psi^2}{n} + \frac{(1+\psi\overline{X})^2}{S_{XX}}}} \mathrel{\dot\sim} \mathscr{N}(0,1), \tag{313}$$

and the approximate size α Wald test rejects $H_0 : \beta_1 = \psi\theta_0$ if and only if

$$\left| \frac{\hat{\beta}_1 - \psi \hat{\beta}_0}{\hat{\sigma}_u \sqrt{\frac{\psi^2}{n} + \frac{(1+\psi \overline{X})^2}{S_{XX}}}} \right| \geq z_{1-\frac{\alpha}{2}}. \tag{314}$$

11.4 Effect of Ignoring the Quadratic Term

(a) The ith residual is

$$\hat{E}_i = Y_i - \hat{Y}_i = a + bX_i + cX_i^2 + E_i - \hat{\beta}_0 - \hat{\beta}_1 X_i. \tag{315}$$

Its expected value is

$$\mathbb{E}[\hat{E}_i] = a + bX_i + cX_i^2 - \mathbb{E}[\hat{\beta}_0] - X_i \mathbb{E}[\hat{\beta}_1]. \tag{316}$$

Note that it is not true that $\mathbb{E}[\hat{\beta}_0] = a$ and $\mathbb{E}[\hat{\beta}_1] = b$, as one might have been expecting. These equations hold if $Y_i = a + bX_i + E_i$, but in our case $Y_i = a + bX_i + cX_i^2 + E_i$. So, we need to derived the expressions for the expected values of the OLS estimate "from scratch." For the slope (11.49), we have:

$$\begin{aligned}
\mathbb{E}[\hat{\beta}_1] &= \mathbb{E}\left[\sum_{i=1}^{n} \frac{X_i - \overline{X}}{S_{XX}} Y_i\right] = \sum_{i=1}^{n} \frac{X_i - \overline{X}}{S_{XX}} \mathbb{E}[Y_i] = \sum_{i=1}^{n} \frac{X_i - \overline{X}}{S_{XX}} (a + bX_i + cX_i^2) \\
&= a \underbrace{\sum_{i=1}^{n} \frac{X_i - \overline{X}}{S_{XX}}}_{0} + b \underbrace{\sum_{i=1}^{n} \frac{(X_i - \overline{X})X_i}{S_{XX}}}_{1} + c \underbrace{\sum_{i=1}^{n} \frac{(X_i - \overline{X})X_i^2}{S_{XX}}}_{\xi(X)} = b + c\xi(X).
\end{aligned} \tag{317}$$

For the intercept (11.50), we have:

$$\begin{aligned}
\mathbb{E}[\hat{\beta}_0] &= \mathbb{E}\left[\sum_{i=1}^{n} \left(\frac{1}{n} - \frac{\overline{X}(X_i - \overline{X})}{S_{XX}}\right) Y_i\right] = \sum_{i=1}^{n} \left(\frac{1}{n} - \frac{\overline{X}(X_i - \overline{X})}{S_{XX}}\right) \mathbb{E}[Y_i] \\
&= \sum_{i=1}^{n} \left(\frac{1}{n} - \frac{\overline{X}(X_i - \overline{X})}{S_{XX}}\right) (a + bX_i + cX_i^2) \\
&= a\left(1 - \frac{\overline{X}}{S_{XX}} \underbrace{\sum_{i=1}^{n} (X_i - \overline{X})}_{0}\right) + b\left(\overline{X} - \frac{\overline{X}}{S_{XX}} \underbrace{\sum_{i=1}^{n} (X_i - \overline{X})X_i}_{S_{XX}}\right) \\
&+ c\left(\underbrace{\frac{1}{n}\sum_{i=1}^{n} X_i^2}_{\overline{X^2}} - \overline{X} \underbrace{\sum_{i=1}^{n} \frac{(X_i - \overline{X})X_i^2}{S_{XX}}}_{\xi(X)}\right) = a + c\left(\overline{X^2} - \overline{X}\xi(X)\right).
\end{aligned} \tag{318}$$

Note that if $c = 0$, then, as expected, $\mathbb{E}[\hat{\beta}_0] = a$ and $\mathbb{E}[\hat{\beta}_1] = b$.

Finally, the expected value of the ith residual is

$$\begin{aligned}\mathbb{E}[\hat{E}_i] &= a + bX_i + cX_i^2 - \left(a + c\left(\overline{X^2} - \overline{X}\xi(X)\right)\right) - X_i\left(b + c\xi(X)\right)\\ &= c\left(X_i^2 - \overline{X^2} - \xi(X)(X_i - \overline{X})\right).\end{aligned} \tag{319}$$

Again, if $c = 0$, then $\mathbb{E}[\hat{E}_i] = 0$, but in general this is not the case.

(b) The expected value of the average of all n residuals is

$$\begin{aligned}\mathbb{E}\left[\frac{1}{n}\sum_{i=1}^{n}\hat{E}_i\right] &= \frac{1}{n}\sum_{i=1}^{n}\mathbb{E}[\hat{E}_i] = \frac{1}{n}\sum_{i=1}^{n}c\left(X_i^2 - \overline{X^2} - \xi(X)(X_i - \overline{X})\right)\\ &= c\left(\underbrace{\frac{1}{n}\sum_{i=1}^{n}X_i^2}_{\overline{X^2}} - \overline{X^2} - \frac{\xi(X)}{n}\underbrace{\sum_{i=1}^{n}(X_i - \overline{X})}_{0}\right) = 0.\end{aligned} \tag{320}$$

This means that, despite the fact that the SLR model is not an adequate model for the data $(X_1, Y_1), \ldots, (X_n, Y_n)$ generated by the model $Y_i = a + bX_i + cX_i^2 + E_i$, the expected average discrepancy between Y_i and the OLS fitted value $\hat{Y}_i$ is zero.

11.5 The SLR Model in Action

(a) The OLS estimates (11.34) of the regression coefficients are $\hat{\beta}_0 = 1.078$ and $\hat{\beta}_1 = 0.088$. Both are close to their true values $\beta_0 = 1$ and $\beta_1 = 0.1$. Figure 16 shows the scatterplot of the data together with the fitted regression line.

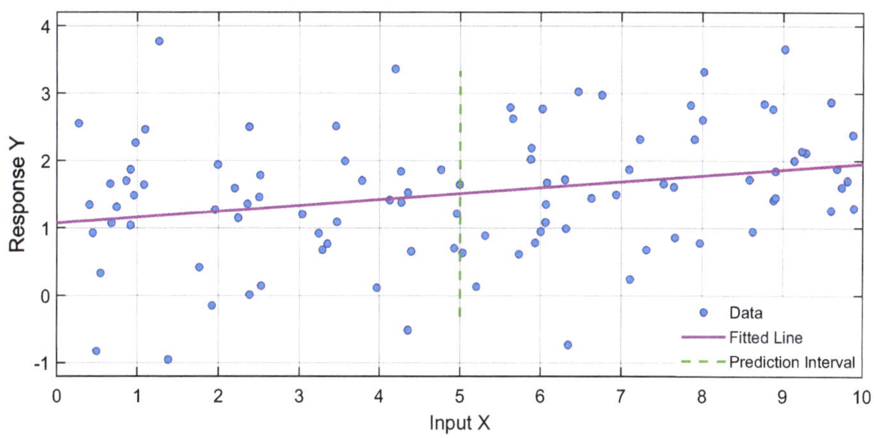

Fig. 16 The data, the OLS fitted line, and the prediction interval for the response to input $X = 5$.

(b) Based on the scatterplot it is indeed hard to decide whether Y depends on X. This is because the true value $\beta_1 = 0.1$ is relatively small compared to the variance $\sigma^2 = 1$, which results into significant scatter of responses. To formally test the significance of linear regression, we can use the Wald test or the t-test. The corresponding p-values (11.84) are

$$\mathfrak{p}_1^W = 0.005 \quad \text{and} \quad \mathfrak{p}_1^T = 0.006. \tag{321}$$

As expected, the p-values are approximately the same and indicate very strong evidence against the null hypothesis $H_0 : \beta_1 = 0$.

(c) The $100(1-\alpha)\%$ confidence intervals for β_1 and β_0, which are based on the normal distribution, are given by (11.95) and (11.97), respectively. The confidence intervals based on the t-distribution are obtained by replacing the normal quantile $z_{\frac{\alpha}{2}}$ with the t-quantile $t_{n-2,\frac{\alpha}{2}}$. The realizations of the 95% confidence intervals computed from the data are given in Table 5. Since $\mathscr{T}(n-2) \approx \mathscr{N}(0,1)$, the confidence

	95% CI for β_1	95% CI for β_0
Based on $\mathscr{N}$	$(0.0265, 0.1490)$	$(0.7199, 1.4354)$
Based on $\mathscr{T}$	$(0.0257, 0.1498)$	$(0.7155, 1.4398)$

Table 5 Confidence intervals for regression coefficients β_1 and β_0 computed from the data.

intervals based on the normal and t-distributions are very similar. All intervals contain the corresponding true values of parameters $\beta_1 = 0.1$ and $\beta_0 = 1$.

(d) The $100(1-\alpha)\%$ prediction interval for the response Y to input X is given by (11.114). The realization of the 95% prediction interval for $X = 5$ computed from the data is shown in Fig. 16 and is

$$I = (-0.307, 3.340). \tag{322}$$

The size of the interval matches the size of the scatter of responses well.

Problems of Chapter 12

12.1 Coherence of the OLS Estimates for SLR and MLR Models

If $p = 1$ and all inputs $X_1, \ldots, X_n$ are scalar, then the design matrix

$$\mathbb{X} = \begin{bmatrix} 1 & X_1 \\ \vdots & \vdots \\ 1 & X_n \end{bmatrix},$$

$$\mathbb{X}^T\mathbb{X} = \begin{bmatrix} 1 & \ldots & 1 \\ X_1 & \ldots & X_n \end{bmatrix} \begin{bmatrix} 1 & X_1 \\ \vdots & \vdots \\ 1 & X_n \end{bmatrix} = \begin{bmatrix} n & \sum_{i=1}^n X_i \\ \sum_{i=1}^n X_i & \sum_{i=1}^n X_i^2 \end{bmatrix} = n \begin{bmatrix} 1 & \overline{X} \\ \overline{X} & \overline{X^2} \end{bmatrix}, \tag{323}$$

$$\mathbb{X}^T Y = \begin{bmatrix} 1 & \ldots & 1 \\ X_1 & \ldots & X_n \end{bmatrix} \begin{bmatrix} Y_1 \\ \vdots \\ Y_n \end{bmatrix} = \begin{bmatrix} \sum_{i=1}^n Y_i \\ \sum_{i=1}^n X_i Y_i \end{bmatrix} = n \begin{bmatrix} \overline{Y} \\ \overline{XY} \end{bmatrix}.$$

Therefore, the normal equation $\mathbb{X}^T\mathbb{X}\hat{\beta} = \mathbb{X}^T Y$ becomes

$$\begin{bmatrix} 1 & \overline{X} \\ \overline{X} & \overline{X^2} \end{bmatrix} \begin{bmatrix} \hat{\beta}_0 \\ \hat{\beta}_1 \end{bmatrix} = \begin{bmatrix} \overline{Y} \\ \overline{XY} \end{bmatrix} \tag{324}$$

From the first equation, we have

$$\hat{\beta}_0 = \overline{Y} - \hat{\beta}_1 \overline{X}, \tag{325}$$

which is the same relationship as in the case of the SLR model. Substituting this into the second equation yields

$$\overline{X}(\overline{Y} - \hat{\beta}_1 \overline{X}) + \hat{\beta}_1 \overline{X^2} = \overline{XY}. \tag{326}$$

Solving this for $\hat{\beta}_1$, we obtain:

$$\hat{\beta}_1 = \frac{\overline{XY} - \overline{X} \cdot \overline{Y}}{\overline{X^2} - \overline{X}^2}. \tag{327}$$

To show that this is equivalent to $\hat{\beta}_1 = S_{XY}/S_{XX}$, let's express the sum of cross-products S_{XY} and the sum of squares S_{XX} in terms of the averages $\overline{XY}$, $\overline{X}$, $\overline{Y}$, and $\overline{X^2}$. Here we go:

$$\begin{aligned} S_{XY} &= \sum_{i=1}^{n}(X_i - \overline{X})(Y_i - \overline{Y}) = \sum_{i=1}^{n} X_i Y_i - \overline{X}\sum_{i=1}^{n} Y_i - \overline{Y}\sum_{i=1}^{n} X_i + n\overline{X} \cdot \overline{Y} \\ &= n\overline{XY} - n\overline{X} \cdot \overline{Y} - n\overline{X} \cdot \overline{Y} + n\overline{X} \cdot \overline{Y} = n(\overline{XY} - \overline{X} \cdot \overline{Y}), \\ S_{XX} &= \sum_{i=1}^{n}(X_i - \overline{X})^2 = \sum_{i=1}^{n} X_i^2 - 2\overline{X}\sum_{i=1}^{n} X_i + n\overline{X}^2 \\ &= n\overline{X^2} - 2n\overline{X}^2 + n\overline{X}^2 = n(\overline{X^2} - \overline{X}^2). \end{aligned} \tag{328}$$

Finally, we have

$$\hat{\beta}_1 = \frac{\overline{XY} - \overline{X} \cdot \overline{Y}}{\overline{X^2} - \overline{X}^2} = \frac{n(\overline{XY} - \overline{X} \cdot \overline{Y})}{n(\overline{X^2} - \overline{X}^2)} = \frac{S_{XY}}{S_{XX}}. \tag{329}$$

12.2 Equivariance of OLS

(a) We will solve parts (a) and (b) of this problem using different approaches. For this part, let's use the optimization definition of the OLS estimate:

$$\hat{\beta} = \arg \min_{\beta \in \mathbb{R}^{p+1}} \|\mathbb{X}\beta - Y\|. \tag{330}$$

Since the data transformation considered in this part of the problem is a rescaling of the jth column of $\mathbb{X}$, it is logical to rewrite the above formula directly in terms of the columns of $\mathbb{X}$. Since the product $\mathbb{X}\beta$ is simply the linear combination of columns of

$\mathbb{X}$ with coefficients $\beta_0, \ldots, \beta_p$, we have:

$$\hat{\beta} = \arg\min_{\beta \in \mathbb{R}^{p+1}} \underbrace{\left\| \sum_{i=1}^{n} \beta_i X^{(i)} - Y \right\|}_{\mathrm{RSS}(\beta_0,\ldots,\beta_p)} = \arg\min_{\beta \in \mathbb{R}^{p+1}} \mathrm{RSS}(\beta_0, \ldots, \beta_p). \tag{331}$$

Let $\hat{\hat{\beta}}$ be the new OLS estimate after the rescaling $X^{(j)} \to \tilde{X}^{(j)} = \alpha X^{(j)}$,

$$\begin{aligned} \hat{\hat{\beta}} &= \arg\min_{\beta \in \mathbb{R}^{p+1}} \underbrace{\left\| \sum_{i \neq j} \beta_i X^{(i)} + \alpha \beta_j X^{(j)} - Y \right\|}_{\mathrm{RSS}(\beta_0,\ldots,\beta_{j-1},\alpha\beta_j,\beta_{j+1},\ldots,\beta_p)} \\ &= \arg\min_{\beta \in \mathbb{R}^{p+1}} \mathrm{RSS}(\beta_0, \ldots, \beta_{j-1}, \alpha\beta_j, \beta_{j+1}, \ldots, \beta_p). \end{aligned} \tag{332}$$

Since the smallest possible value RSS attains at $\hat{\beta}$, the solution $\hat{\hat{\beta}}$ of the above optimization problem must be such that

$$(\hat{\hat{\beta}}_0, \ldots, \hat{\hat{\beta}}_{j-1}, \alpha\hat{\hat{\beta}}_j, \hat{\hat{\beta}}_{j+1}, \ldots, \hat{\hat{\beta}}_p) = (\hat{\beta}_0, \ldots, \hat{\beta}_{j-1}, \hat{\beta}_j, \hat{\beta}_{j+1}, \ldots, \hat{\beta}_p) \tag{333}$$

This means that $\hat{\hat{\beta}}_i = \hat{\beta}_i$ for all $i \neq j$, and $\hat{\hat{\beta}}_j = \alpha^{-1}\hat{\beta}_j$, as claimed.

(b) For this part, let's use linear algebra for deriving the relationship between the original $\hat{\beta}$ and transformed $\hat{\hat{\beta}}$ OLS estimates. Since $\operatorname{Ker}\mathbb{X} = \{0\}$,

$$\hat{\beta} = (\mathbb{X}^T\mathbb{X})^{-1}\mathbb{X}^T Y. \tag{334}$$

Since T is nonsingular, $\operatorname{Ker}\tilde{\mathbb{X}} = \operatorname{Ker}(\mathbb{X}T) = \{0\}$. Indeed,

$$v \in \operatorname{Ker}\tilde{\mathbb{X}} \iff \mathbb{X}Tv = 0 \iff Tv \in \operatorname{Ker}\mathbb{X} \iff Tv = 0 \iff v = 0. \tag{335}$$

Therefore, $\hat{\hat{\beta}}$ is given by the same formula as $\hat{\beta}$, except that $\mathbb{X}$ is replaced with $\tilde{\mathbb{X}}$[49]:

$$\begin{aligned} \hat{\hat{\beta}} &= (\tilde{\mathbb{X}}^T\tilde{\mathbb{X}})^{-1}\tilde{\mathbb{X}}^T Y = \left((\mathbb{X}T)^T(\mathbb{X}T)\right)^{-1}(\mathbb{X}T)^T Y \\ &= \left(T^T(\mathbb{X}^T\mathbb{X})T\right)^{-1} T^T\mathbb{X}^T Y = T^{-1}(\mathbb{X}^T\mathbb{X})^{-1}(T^T)^{-1}T^T\mathbb{X}^T Y \\ &= T^{-1}(\mathbb{X}^T\mathbb{X})^{-1}\mathbb{X}^T Y = T^{-1}\hat{\beta}. \end{aligned} \tag{336}$$

So, if the design matrix is multiplied by T from the right, then the OLS vector multiplied by T^{-1} from the left. This proof is of course applicable to part (a) as well. In the special case, where $T = D_\alpha^{(j)}$, the relationship $\hat{\hat{\beta}} = T^{-1}\hat{\beta}$ reduces to (333).

[49] A refreshing clash of notation is in order: in T^T, the T in the superscript denotes the transposition, and the other T denotes the transformation matrix. "Doublethink means the power of holding two contradictory beliefs in one's mind simultaneously, and accepting both of them." – George Orwell, *Nineteen Eighty-Four*.

12.3 Weighted Least Squares

Let's follow the process we used to derive the OLS estimate in Section 12.2 and see where this will lead us. To rewrite the weighted residual sum of squares in a matrix form, let W be the diagonal weight matrix:

$$W = \begin{bmatrix} w_1 & 0 & \dots & 0 \\ 0 & w_2 & \dots & 0 \\ \vdots & \vdots & \ddots & \vdots \\ 0 & 0 & \dots & w_n \end{bmatrix} \tag{337}$$

Then the wRSS is

$$\begin{aligned} \text{wRSS}(\beta) &= \sum_{i=1}^{n} w_i(X_i^T\beta - Y_i)^2 = (\mathbb{X}\beta - Y)^T W(\mathbb{X}\beta - Y) \\ &= \beta^T\mathbb{X}^T W\mathbb{X}\beta - \beta^T\mathbb{X}^T WY - Y^T W\mathbb{X}\beta + Y^T WY \\ &= \beta^T\mathbb{X}^T W\mathbb{X}\beta - 2\beta^T\mathbb{X}^T WY + Y^T WY. \end{aligned} \tag{338}$$

In the last equation, we used the fact that $\beta^T\mathbb{X}^T WY = Y^T W\mathbb{X}\beta$. If it is not clear why this is true, see how (12.16) was proved.

The gradient of wRSS with respect to β is

$$\nabla_\beta \text{wRSS}(\beta) = \nabla_\beta(\beta^T\mathbb{X}^T W\mathbb{X}\beta) - 2\nabla_\beta(\beta^T\mathbb{X}^T WY). \tag{339}$$

Using the properties of the gradient established in (12.19) and (12.22), we have:

$$\nabla_\beta(\beta^T\mathbb{X}^T W\mathbb{X}\beta) = 2\mathbb{X}^T W\mathbb{X}\beta \quad \text{and} \quad \nabla_\beta(\beta^T\mathbb{X}^T WY) = \mathbb{X}^T WY. \tag{340}$$

Therefore, the gradient is

$$\nabla_\beta \text{wRSS}(\beta) = 2\mathbb{X}^T W\mathbb{X}\beta - 2\mathbb{X}^T WY. \tag{341}$$

Setting the gradient of the wRSS to zero, we obtain that the WLS estimate $\hat{\beta}_W$ of the vector of regression coefficients is a solution of the following vector equation, which is the "weighted" analog of the normal equation $\mathbb{X}^T\mathbb{X}\beta = \mathbb{X}^T Y$,

$$\mathbb{X}^T W\mathbb{X}\beta = \mathbb{X}^T WY. \tag{342}$$

Note that if all observations are equally important and, thus, all weights are the same, $w_1 = \dots = w_n = w$, then this equation reduces to the standard normal equation.

To find an explicit expression for $\hat{\beta}_W$, it remains to show that the matrix $\mathbb{X}^T W\mathbb{X}$ is nonsingular, given that $\text{Ker}\,\mathbb{X} = \{0\}$. Let's show that its kernel is also trivial, $\text{Ker}(\mathbb{X}^T W\mathbb{X}) = \{0\}$ (this implies nonsingularity). Suppose that $v \in \mathbb{R}^{p+1}$ is in the kernel $\text{Ker}(\mathbb{X}^T W\mathbb{X})$, then[50]

[50] Compare what follows with (12.168).

$$\begin{aligned} v \in \text{Ker}(\mathbb{X}^T W\mathbb{X}) &\implies \mathbb{X}^T W\mathbb{X}v = 0 \implies v^T\mathbb{X}^T W\mathbb{X}v = 0 \\ &\implies (\mathbb{X}v)^T W \underbrace{\mathbb{X}v}_{u} = 0 \implies u^T Wu = 0 \end{aligned} \tag{343}$$

Since the weight matrix W is diagonal with positive diagonal elements $w_i > 0$,

$$u^T W u = \sum_{i=1}^{n} w_i u_i^2 = 0 \implies u = 0. \tag{344}$$

The final sequence of simple logical implications is

$$u = \mathbb{X} v = 0 \implies v \in \operatorname{Ker} \mathbb{X} \implies v = 0. \tag{345}$$

So, the matrix $\mathbb{X}^T W \mathbb{X}$ is nonsingular and, therefore, the WLS estimate is

$$\hat{\beta}_W = (\mathbb{X}^T W \mathbb{X})^{-1} \mathbb{X}^T W Y. \tag{346}$$

If all the weights are identical $w_1 = \ldots = w_n = w$, then the weight matrix $W = w\mathbb{I}_n$, and the WLS estimate, as expected, reduces to the OLS estimate:

$$\hat{\beta}_W = (\mathbb{X}^T w\mathbb{I}_n \mathbb{X})^{-1} \mathbb{X}^T w\mathbb{I}_n Y = (\mathbb{X}^T \mathbb{X})^{-1} \mathbb{X}^T Y = \hat{\beta}. \tag{347}$$

12.4 Testing for Significance and Confidence Intervals

Recall that the Wald test p-value $\mathfrak{p}_j^W$ and the confidence interval I_α^j are given by

$$\mathfrak{p}_j^W = 2\Phi\left(-\left|\frac{\hat{\beta}_j}{\hat{\sigma}_u \sqrt{\xi_j}}\right|\right) \quad \text{and} \quad I_\alpha^j = \left(\hat{\beta}_j + z_{\frac{\alpha}{2}} \hat{\sigma}_u \sqrt{\xi_j},\, \hat{\beta}_j - z_{\frac{\alpha}{2}} \hat{\sigma}_u \sqrt{\xi_j}\right). \tag{348}$$

Therefore (keep in mind that $z_{\frac{\alpha}{2}} < 0$),

$$\begin{aligned} \mathfrak{p}_j^W > \alpha &\iff 2\Phi\left(-\left|\frac{\hat{\beta}_j}{\hat{\sigma}_u \sqrt{\xi_j}}\right|\right) > \alpha \\ &\iff -\left|\frac{\hat{\beta}_j}{\hat{\sigma}_u \sqrt{\xi_j}}\right| > \Phi^{-1}\left(\frac{\alpha}{2}\right) = z_{\frac{\alpha}{2}} \iff |\hat{\beta}_j| < -z_{\frac{\alpha}{2}} \hat{\sigma}_u \sqrt{\xi_j}. \end{aligned} \tag{349}$$

But the last inequality means precisely that regardless of the sign of $\hat{\beta}_j$, zero is inside the interval I_α^j:

$$\hat{\beta}_j + z_{\frac{\alpha}{2}} \hat{\sigma}_u \sqrt{\xi_j} < 0 < \hat{\beta}_j - z_{\frac{\alpha}{2}} \hat{\sigma}_u \sqrt{\xi_j}. \tag{350}$$

12.5 The MLR Model in Action

(a) Figure 17 shows the scatterplots of the response versus the four predictors. Based on the figure, the response Y appears to be positively correlated with $X^{(1)}$ and $X^{(2)}$. However, its correlation with $X^{(3)}$ and $X^{(4)}$ is questionable.

(b) The OLS estimate $\hat{\beta}$ is a random vector, and its realizations depend on the training data $(\mathbb{X}, Y)$, which is randomly generated by (12.231) and (12.232) with $\beta = (1,2,3,0,0)$ and $\sigma^2 = 1$. One particular realization of $\hat{\beta}$ is

$$\hat{\beta} = (\mathbb{X}^T \mathbb{X})^{-1} \mathbb{X}^T Y = (1.12, 1.92, 3.02, -0.16, -0.15)^T. \tag{351}$$

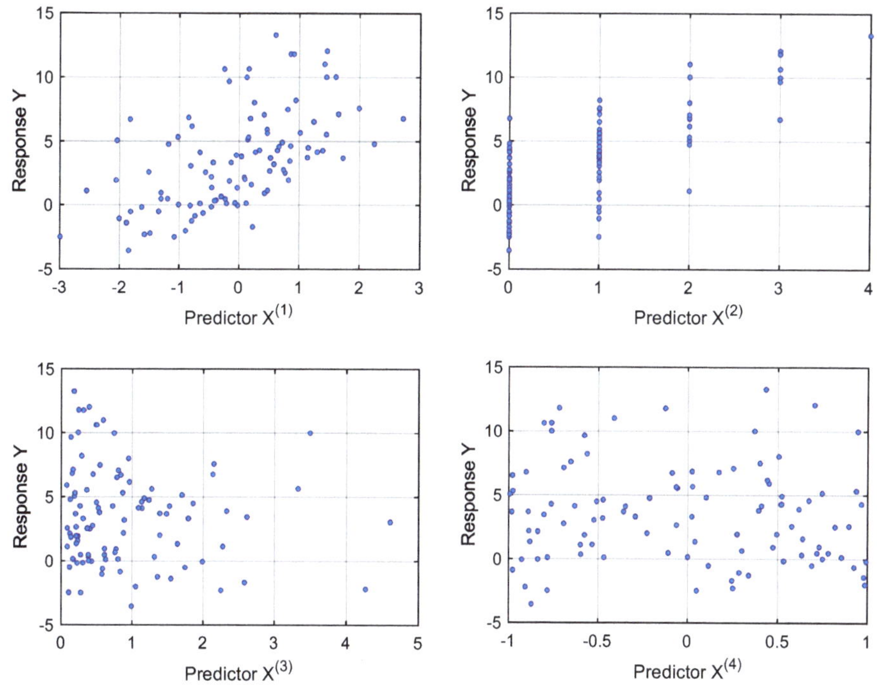

Fig. 17 Scatterplots of the training data: response Y versus predictors $X^{(1)}$, $X^{(2)}$, $X^{(3)}$, and $X^{(4)}$.

Your realization will, with probability 1, be different, but it should approximately equal the true value of β.

(c) Table 6 provides all the required statistics. The p-values associated with the

Predictor	OLS $\hat{\beta}_j$	z-score $\mathfrak{z}_j$	p-value p_j^W	p-value p_j^T	95% CI $I_{0.05}^j$
$X^{(1)}$	1.92	22.19	0	0	(1.75,2.09)
$X^{(2)}$	3.02	30.83	0	0	(2.83, 3.23)
$X^{(3)}$	-0.16	-1.52	0.130	0.133	(-0.38,0.05)
$X^{(4)}$	-0.15	-0.96	0.338	0.341	(-0.45,0.15)

Table 6 The OLS estimates, z-scores, p-values, and confidence intervals for the regression coefficients, computed from the randomly generated training data of size $n = 100$ via (12.231), (12.232).

first two predictors are not exactly zero, but they are so small that they can be viewed as zeros for all practical purposes. In other words, the data provides very strong evidence against the hypotheses that $X^{(1)}$ and $X^{(2)}$ are not relevant for predicting the response Y. On the contrary, the p-values corresponding to predictors $X^{(3)}$ and $X^{(4)}$ are relatively large, meaning that the training data provides no evidence against the hypotheses that they are not relevant (which is consistent with how the training data was generated). As in the previous part of the problem, the p-values that you obtain will be different, but the last two p-values are likely to be substantially larger than the first two p-values.

(d) The OLS estimate $\hat{\hat{\beta}}$ is given by

$$\hat{\hat{\beta}} = (\tilde{\mathbb{X}}^T\tilde{\mathbb{X}})^{-1}\tilde{\mathbb{X}}^T Y = (0.96, 1.93, 3.04)^T, \tag{352}$$

where $\tilde{\mathbb{X}}$ is the $n \times 3$ reduced design matrix obtained from the full design matrix $\mathbb{X}$ by removing the last two columns corresponding to predictors $X^{(3)}$ and $X^{(4)}$. Since $\|\hat{\beta} - \beta\| \approx 0.26$ and $\|\hat{\hat{\beta}} - \beta\| \approx 0.09$, the vector $\hat{\hat{\beta}}$ approximates the true vector of regression coefficients β more accurately than $\hat{\beta}$.

(e) The value of the F-statistic is

$$\mathfrak{f} = \frac{n-p-1}{p-\tilde{p}} \cdot \frac{\|\tilde{\mathbb{X}}\hat{\hat{\beta}} - Y\|^2 - \|\mathbb{X}\hat{\beta} - Y\|^2}{\|\mathbb{X}\hat{\beta} - Y\|^2} = 1.38, \tag{353}$$

where $p = 4$ and $\tilde{p} = 2$ are the numbers of predictors in the full and reduced models. The p-value from the F-test is

$$\mathfrak{p} = 1 - F_{p-\tilde{p},n-p-1}(\mathfrak{f}) = 1 - F_{2,95}(\mathfrak{f}) = 0.25. \tag{354}$$

Therefore, the training data provides no evidence against the null hypothesis: predictors $X^{(3)}$ and $X^{(4)}$ are not simultaneously significant even at the 20% level. This result is consistent with how the training data was generated.

(f) Let $\{(\dot{X}_1, \dot{Y}_1), \ldots, (\dot{X}_n, \dot{Y}_n)\}$ be the validation data set generated by first generating inputs via (12.231) and then generating the corresponding responses via (12.232) with $\beta = (1, 2, 3, 0, 0)^T$ and $\sigma^2 = 1$. The test error of the full regression model (fitted to the training data) on the validation set is

$$\text{Err}_f = \frac{1}{n}\sum_{i=1}^{n}(\dot{Y}_i - \dot{X}_i^T\hat{\beta})^2 = 1.44, \tag{355}$$

where $\dot{X}_i^T = (1, \dot{X}_i^{(1)}, \dot{X}_i^{(2)}, \dot{X}_i^{(3)}, \dot{X}_i^{(4)})$ and $\hat{\beta}$ is the OLS estimate (351). For the reduced model, the test error is

$$\text{Err}_r = \frac{1}{n}\sum_{i=1}^{n}(\dot{Y}_i - \tilde{\dot{X}}_i^T\hat{\hat{\beta}})^2 = 1.37, \tag{356}$$

where $\tilde{\dot{X}}_i^T = (1, \dot{X}_i^{(1)}, \dot{X}_i^{(2)})$ and $\hat{\hat{\beta}}$ is the OLS estimate (352).

Although the improvement in the prediction accuracy after dropping two irrelevant predictors does not appear substantial, the reduced model is still preferable since it is simpler than the full model and is easier to interpret.

Index

K. M. Zuev, *Fundamentals of Statistical Inference*,
International Series in Operations Research & Management Science 374,
https://doi.org/10.1007/978-3-032-03848-7

MIX
Papier aus verantwortungsvollen Quellen
Paper from responsible sources
FSC® C105338

If you have any concerns about our products,
you can contact us on
ProductSafety@springernature.com

In case Publisher is established outside the EU,
the EU authorized representative is:
Springer Nature Customer Service Center GmbH
Europaplatz 3, 69115 Heidelberg, Germany

Printed by Libri Plureos GmbH
in Hamburg, Germany